Volume 43

CRM
PROCEEDINGS &
LECTURE NOTES

Centre de Recherches Mathématiques
Université de Montréal

Additive Combinatorics

Andrew Granville
Melvyn B. Nathanson
József Solymosi
Editors

The Centre de Recherches Mathématiques (CRM) of the Université de Montréal was created in 1968 to promote research in pure and applied mathematics and related disciplines. Among its activities are special theme years, summer schools, workshops, postdoctoral programs, and publishing. The CRM is supported by the Université de Montréal, the Province of Québec (FCAR), and the Natural Sciences and Engineering Research Council of Canada. It is affiliated with the Institut des Sciences Mathématiques (ISM) of Montréal, whose constituent members are Concordia University, McGill University, the Université de Montréal, the Université du Québec à Montréal, and the École Polytechnique. The CRM may be reached on the Web at www.crm.umontreal.ca.

American Mathematical Society
Providence, Rhode Island USA

The production of this volume was supported in part by the Fonds pour la Formation de Chercheurs et l'Aide à la Recherche (Fonds FCAR) and the Natural Sciences and Engineering Research Council of Canada (NSERC).

2000 *Mathematics Subject Classification.* Primary 11-02; Secondary 0, 11P70, 28D05, 37A45.

Library of Congress Cataloging-in-Publication Data

CRM–Clay School on Additive Combinatorics (2006 : Université de Montréal)
  Additive combinatorics / Andrew Granville, Melvyn B. Nathanson, József Solymosi, editors.
    p. cm. — (CRM proceedings & lecture notes, ISSN 1065-8580 ; v. 43)
  Includes bibliographical references.
  ISBN 978-0-8218-4351-2 (alk. paper)
  1. Additive combinatorics—Congresses. 2. Combinatorial analysis—Congresses. I. Granville, Andrew. II. Nathanson, Melvyn B. (Melvyn Bernard), 1944– III. Solymosi, József, 1959– IV. Title. V. Series.

QA164.C75 2006
511′.5—dc22   2007060834

Copying and reprinting. Material in this book may be reproduced by any means for educational and scientific purposes without fee or permission with the exception of reproduction by services that collect fees for delivery of documents and provided that the customary acknowledgment of the source is given. This consent does not extend to other kinds of copying for general distribution, for advertising or promotional purposes, or for resale. Requests for permission for commercial use of material should be addressed to the Acquisitions Department, American Mathematical Society, 201 Charles Street, Providence, Rhode Island 02904-2294, USA. Requests can also be made by e-mail to reprint-permission@ams.org.

Excluded from these provisions is material in articles for which the author holds copyright. In such cases, requests for permission to use or reprint should be addressed directly to the author(s). (Copyright ownership is indicated in the notice in the lower right-hand corner of the first page of each article.)

© 2007 by the American Mathematical Society. All rights reserved.
The American Mathematical Society retains all rights
except those granted to the United States Government.
Copyright of individual articles may revert to the public domain 28 years
after publication. Contact the AMS for copyright status of individual articles.
Printed in the United States of America.
∞ The paper used in this book is acid-free and falls within the guidelines
established to ensure permanence and durability.
This volume was submitted to the American Mathematical Society
in camera ready form by the Centre de Recherches Mathématiques.
Visit the AMS home page at http://www.ams.org/
10 9 8 7 6 5 4 3 2 1   12 11 10 09 08 07

Contents

Preface	v
An Introduction to Additive Combinatorics *Andrew Granville*	1
Elementary Additive Combinatorics *József Solymosi*	29
Many Additive Quadruples *Antal Balog*	39
An Old New Proof of Roth's Theorem *Endre Szemerédi*	51
Bounds on Exponential Sums over Small Multiplicative Subgroups *Pär Kurlberg*	55
Montréal Notes on Quadratic Fourier Analysis *Ben Green*	69
Ergodic Methods in Additive Combinatorics *Bryna Kra*	103
The Ergodic and Combinatorial Approaches to Szemerédi's Theorem *Terence Tao*	145
Cardinality Questions About Sumsets *Imre Z. Ruzsa*	195
Open Problems in Additive Combinatorics *Ernest S. Croot III and Vsevolod F. Lev*	207
Some Problems Related to Sum-Product Theorems *Mei-Chu Chang*	235
Lattice Points on Circles, Squares in Arithmetic Progressions and Sumsets of Squares *Javier Cilleruelo and Andrew Granville*	241
Problems in Additive Number Theory. I *Melvyn B. Nathanson*	263
Double and Triple Sums Modulo a Prime *Katalin Gyarmati, Sergei Konyagin and Imre Z. Ruzsa*	271

Additive Properties of Product Sets in Fields of Prime Order
    A. A. Glibichuk and S. V. Konyagin   279

Many Sets Have more Sums than Differences
    Greg Martin and Kevin O'Bryant   287

Davenport's Constant for Groups of the Form $\mathbb{Z}_3 \oplus \mathbb{Z}_3 \oplus \mathbb{Z}_{3d}$
    Gautami Bhowmik and Jan-Christoph Schlage-Puchta   307

Some Combinatorial Group Invariants and Their Generalizations with Weights
    S. D. Adhikari, R. Balasubramanian, and P. Rath   327

Preface

Andrew Granville, Melvyn B. Nathanson, and József Solymosi

ABSTRACT. From March 30th to April 5th, 2006, a CRM–Clay School on *Additive Combinatorics* was held at the Université de Montréal, followed by a workshop from April 6th to April 12th, all as part of the 2005–2006 special year program, *Analysis in Number Theory*. The school was attended by roughly one hundred participants from sixteen countries around the world, the workshop by forty more people. The first part of this volume contains written versions of most of the lectures given at the school; the second half submitted contributions from the speakers at the workshop.

One of the most active areas in analysis today is the rapidly emerging new topic of "additive combinatorics." Building on Gowers' use of the Freĭman–Ruzsa theorem in harmonic analysis (in particular, his proof of Szemerédi's theorem), Green and Tao famously proved that there are arbitrarily long arithmetic progressions of primes, and Bourgain has given non-trivial estimates for hitherto untouchably short exponential sums. This new subject brings together ideas from harmonic analysis, ergodic theory, discrete geometry, combinatorics, graph theory, group theory, probability theory and number theory to prove some extraordinary results. The basis of the subject is not too difficult: it can be best described as the theory of adding together sets of numbers; in particular understanding the structure of the two original sets if their sum is small. Ideas from all of the above areas come in when providing proofs of key results like the Freĭman–Ruzsa theorem, and the Balog–Gowers–Szemerédi lemma.

Because the background is so broad, the school and conference attracted an eclectic mix of participants, bringing different skills and perspectives. It seems evident that this combustible mixture will continue to lead to exciting advances for some time to come.

The lectures at the school began with some elementary topics:

Andrew Granville: *The basics of additive combinatorics; The Freĭman–Ruzsa theorem; Uniform distribution and Roth's theorem.*
Jószef Solymosi: *Combinatorial discrete geometry and additive combinatorics; a proof of Roth's theorem.*
Antal Balog: *The Balog–Szemerédi–Gowers Theorem.*

We would like to thank the Centre de recherches mathématiques, NSERC (Canada), the National Science Foundation (USA), the Clay Mathematics Institute (USA), Dimatia (Czech Republic) and the Université de Montréal, for their generous and willing support of our school and workshop.

The main lecturers, who each gave a series of talks on a deep central theme, were

Ben Green: *Quadratic Fourier analysis.*
Bryna Kra: *Ergodic methods in combinatorial number theory.*
Terry Tao: *Combinatorial and ergodic techniques for proving Szemerédi-type theorems.*
Van Vu: *Structure of sumsets and applications.*

The school ended with lectures by several of the key figures in the early development of what we now call "additive combinatorics":

Imre Ruzsa: *Plünnecke and the others.*
Gregory Freiman: *Inverse additive number theory: Results and problems.*
Endre Szemerédi: *Another proof of Roth's theorem.*

The main lecture series of the school was extraordinarily successful. All of these lecturers made their talks highly accessible, which was reflected in high attendance throughout the meeting, and the buoyant atmosphere. We believe that the lecturers have brought that attitude to the write-ups of their contributions herein! We would particularly like to thank all of the lecturers for their superb talks, and the generosity with which they worked with the participants, helping them to understand this challenging material.

Many leading figures in additive combinatorics, ergodic theory, combinatorics, number theory and harmonic analysis arrived for the workshop, including Jean Bourgain, Tim Gowers, Mei-Chu Chang, Ron Graham, Trevor Wooley, Michael Lacey, Sanju Velani, Sergei Konyagin, Jaroslav Nesetril and many more. In addition, there were two beautiful lectures for the general public: in the first week, Terry Tao gave us further insights into *Long arithmetic progressions in the primes*; in the second week Manjul Bhargava gave a beguiling introduction to his work with Jonathan Hanke on *The representation of integers by quadratic forms*. There were a lot of announcements of exciting new work in the meeting on this very hot topic, most notably perhaps Green and Tao's announcement that they have a viable plan to extend their result on primes in arithmetic progressions to prove a "weak form" of the prime k-tuplets conjecture; and work by Helfgott, and then by Bourgain and Gambaud on the (non-abelian) group generation problem.

This was just the third workshop on additive combinatorics (the first was at the American Institute of Mathematics in September 2004, the second at the University of Bristol in September 2005). Over the next few years there will be several major programs in additive combinatorics and related areas, including during Fall 2007 at the Institute for Advanced Study in Princeton, during Spring 2008 at the Fields Institute in Toronto, and during Fall 2008 at the Mathematical Sciences Research Institute in Berkeley, California.

The meeting would not have been possible without the organizational skills of Louis Pelletier and his team from the Centre de recherches mathématiques, for which we are very grateful.

This book is split into three parts: the proceedings from the school, articles on open questions in the subject, and new research. We begin with basic articles, setting the scene, by the organizers, along with an explicit discussion of the Balog-Szemerédi-Gowers Theorem by Antal Balog, an old but unpublished proof of Roth's theorem by Szemerédi, and a discussion of Bourgain's bounds on exponential sums by Pär Kurlberg.

Andrew Granville: *An Introduction to Additive Combinatorics.*

Jószef Solymosi: *Elementary Additive Combinatorics.*

Antal Balog: *Many Additive Quadruples.*

Endre Szemerédi: *An Old New Proof of Roth's Theorem.*

Pär Kurlberg: *Bounds on Exponential Sums over Small Multiplicative Subgroups.*

Next we have three lecture series describing some of the most exciting ideas in the current development of the subject.

Ben Green: *Quadratic Fourier analysis.*

Bryna Kra: *Ergodic Methods in Combinatorial Number Theory.*

Terence Tao: *The Ergodic and Combinatorial Approaches to Szemerédi's Theorem.*

These are followed by several articles highlighting open questions in different aspects of additive combinatorics.

Imre Z. Ruzsa: *Cardinality Questions about Sumsets.*

Ernest S. Croot III and Seva Lev: *Open Problems in Additive Combinatorics.*

Mei-Chu Chang: *Some Problems Related to Sum-Product Theorems.*

Javier Cilleruelo and Andrew Granville: *Lattice Points on Circles, Squares in Arithmetic Progressions and Sumsets of Squares.*

Melvyn B. Nathanson: *Problems in Additive Number Theory.* I.

This is followed by several research articles corresponding to the proceedings of the workshop.

Katalin Gyarmati, Sergei Konyagin and Imre Z. Ruzsa: *Double and Triple Sums Modulo a Prime.*

A. A. Glibichuk and Sergei Konyagin: *Additive Properties of Product Sets in Fields of Prime Order.*

Greg Martin and Kevin O'Bryant: *Many Sets Have More Sums Than Differences.*

Gautami Bhowmik and Jan-Christoph Schlage-Puchta: *Davenport's Constant for Groups of the Form $\mathbb{Z}_3 \oplus \mathbb{Z}_3 \oplus \mathbb{Z}_{3d}$.*

S. D. Adhikari, R. Balasubramanian and P. Rath: *Some Combinatorial Group Invariants and Their Generalizations with Weights.*

An Introduction to Additive Combinatorics

Andrew Granville

ABSTRACT. This is a slightly expanded write-up of my three lectures at the Additive Combinatorics school. In the first lecture we introduce some of the basic material in Additive Combinatorics, and in the next two lectures we prove two of the key background results, the Freĭman–Ruzsa theorem and Roth's theorem for 3-term arithmetic progressions.

CONTENTS

1. Lecture 1. Introductory material   1
2. Lecture 2. The Freĭman–Ruzsa theorem   13
3. Lecture 3. Uniform distribution, Roth's theorem and beyond   18
References   26

1. Lecture 1. Introductory material

1.1. Basic definitions. Let A and B be subsets of G, an additive group. Typically we work with the integers $\mathbb{Z}$, or the integers mod N, that is $(\mathbb{Z}/N\mathbb{Z})$, though sometimes with other groups like $\mathbb{R}$ or $\mathbb{Z}^k$. The *sumset* of A and B is defined by

$$A + B := \{g \in G : \text{There exist } a \in A, b \in B \text{ such that } g = a + b\}.$$

Typically we write $A + B = \{a + b : a \in A, b \in B\}$ with the understanding that elements are not repeated in $A + B$. For example, $\{1, 2, 3\} + \{1, 3\} = \{2, 3, 4, 5, 6\}$.

The addition of sets, "+," is commutative if $(G, +)$ is commutative. It is also associative, and it is distributive over unions, that is, $A+(B\cup C) = (A+B)\cup(A+C)$.

2000 *Mathematics Subject Classification.* Primary 11B75; Secondary 05D05, 11H06, 11P70, 37A45.

This is the final form of the paper.

Other important definitions include

$$kA := A + A + \cdots + A;$$
$$b + A = \{b\} + A, \qquad \text{a } \textit{translate} \text{ of } A;$$
$$A - B = \{a - b : a \in A, b \in B\};$$
$$k \diamond A = \{ka : a \in A\}, \qquad \text{a } \textit{dilate} \text{ of } A;$$

and

$$A \diamond B = \{ab : a \in A, b \in B\}.$$

Having given all this notation we note that we will abuse it by writing $N\mathbb{Z}$ instead of $N \diamond \mathbb{Z}$, for the integers divisible by N.

Warm up exercises.

1.1 Show that if $A - A = \{0\}$ then $|A| = 1$.
1.2 Show that $k \diamond A \subseteq kA$; when are they equal?
1.3 Show that $|A| \leq |A + B| \leq |A||B|$.
1.4 When do we have $|A + B| = |A||B|$?

If $a \in A$ then $A+\{a\} \subset A+A$ and $A-\{a\} \subset A-A$, so that $|A+A|, |A-A| \geq |A|$. Also $|A + A|, |A - A| \leq |A \times A| \leq |A|^2$.

1.5 Improve these upper bounds for $|A + A|$ and for $|A - A|$.

One of our main objectives is to study the size and structure of sumsets in $\mathbb{Z}$. Above we have considered finite sets, but there is an interesting history of results on summing infinite sets: Define $A_{\geq m} := \{n \in A : n \geq m\}$. A set of integers A is a *basis of order h* if $h(A \cup \{0\}) \supseteq \mathbb{Z}_{\geq m}$. We now give several well-known examples. Let $\mathbb{P}$ be the set of primes.

Lagrange's theorem. $4\{n^2 : n \in \mathbb{Z}\} = \mathbb{Z}_{\geq 0}$

Goldbach's conjecture. $2\mathbb{P}_{\geq 3} = 2 \diamond \mathbb{Z}_{\geq 3}$ *or* $3(\mathbb{P} \cup \{0\}) = \mathbb{Z}_{\geq 2} \cup \{0\}$

Generalized twin prime conjecture. $\mathbb{P}_{\geq m} - \mathbb{P}_{\geq m} = 2 \diamond \mathbb{Z}$ *for all* m.

Therefore Lagrange's theorem states that the squares form a basis of order 4, and Goldbach's conjecture postulates that the primes form a basis of order 3.

1.2. If $A + B$ is small then A and B are...? Suppose that A and B are finite sets of integers, say A is $a_1 < a_2 < \cdots < a_r$, and B is $b_1 < b_2 < \cdots < b_s$. Then $A + B$ contains the $r + s - 1$ distinct elements

$$a_1 + b_1 < a_1 + b_2 < a_1 + b_3 < \cdots < a_1 + b_s < a_2 + b_s < \cdots < a_r + b_s,$$

so that

(1.1) $$|A + B| \geq |A| + |B| - 1.$$

Can we have equality in (1.1)? That is, what if $|A + B| = |A| + |B| - 1$? We will write down another list $r + s - 1$ distinct elements of $A + B$, namely

$$a_1 + b_1 < a_2 + b_1 < a_2 + b_2 < \cdots < a_2 + b_{s-1} < a_2 + b_s < \cdots < a_r + b_s.$$

If $|A + B| = r + s - 1$, then the terms in each list must be the same and so we have $a_1 + b_2 = a_2 + b_1$, and $a_1 + b_3 = a_2 + b_2$, etc., implying that $a_2 - a_1 = b_2 - b_1 =$

$b_3 - b_2 = \cdots$. In fact we can deduce that A and B are both arithmetic progressions with the same common difference; that is there exists a nonzero integer d such that
$$A = \{a + id : 0 \leq i \leq I - 1\} \quad \text{and} \quad B = \{b + jd : 0 \leq j \leq J - 1\}.$$
Thus A and B are highly structured. However if A is a large subset of $\{a + id : 0 \leq i \leq I - 1\}$ and B is a large subset of $\{b + jd : 0 \leq j \leq J - 1\}$, then we expect that $|A + B| = |A| + |B| + \Delta$ for some small Δ, yet A and B may not have much internal structure. The key thing is that they are both large subsets of arithmetic progressions with the same common difference.

Another interesting case is given by
$$\begin{aligned} A &= \{1, 2, \ldots, 10, 101, 102, \ldots, 110, 201, 202, \ldots, 210\} \\ &= 1 + \{0, 1, \ldots, 9\} + 100 \diamond \{0, 1, 2\}, \\ B &= 3 + \{0, 1, \ldots, 7\} + 100 \diamond \{0, 1, \ldots, 4\}, \end{aligned}$$

and
$$A + B = 4 + \{0, 1, 2, \ldots, 16\} + 100 \diamond \{0, 1, \ldots, 6\},$$
so that $|A| = 30$, $|B| = 40$ and $|A + B| = 119$. These are examples of a *generalized arithmetic progression*:
$$C := \{a_0 + a_1 n_1 + a_2 n_2 + \cdots + a_k n_k : 0 \leq n_j \leq N_j - 1 \text{ for } 1 \leq j \leq k\},$$
where $N_1, N_2, \ldots, N_k$ are integers ≥ 2. This generalized arithmetic progression is said to have *dimension* k and *volume* $N_1 N_2 \cdots N_k$; and is *proper* if its elements are distinct.

Most questions about the structure of A and B, when $A + B$ is small, are open! We study the structure of A when $A + A = 2A$ is small (i.e., the case $B = A$). For a generalized arithmetic progression C we have $|2C| < 2^k |C|$; and, indeed, if $A \subset C$ with $|A| \geq \delta |C|$ then
$$|2A| \leq |2C| < 2^k |C| \leq (2^k/\delta)|A|.$$

What about the converse? If $|2A|$ is a small multiple of $|A|$ then what possible A are there? A rather daring guess is that the only possible such A are large subsets of generalized arithmetic progressions; and indeed this is the Freĭman–Ruzsa theorem which we will prove in our next lecture.

Freĭman–Ruzsa theorem. *If $|2A|$ is "small" then A is a "large" subset of a generalized arithmetic progression.*

Precise quantifiers in an explicit version of this result are complicated and best left till we study it in more detail.

1.3. Densities. The Schnirelmann density of a set A of integers is given by
$$\sigma(A) := \inf_{n \geq 1} \frac{\#\{a \in A : 1 \leq a \leq n\}}{n},$$
so that $A(n) \geq n\sigma(A)$ for all $n \geq 1$. It is easy to see, by the pigeonhole principle that if $0 \in A \cap B$ and $\sigma(A) + \sigma(B) \geq 1$ then $A + B \supseteq \mathbb{Z}_{\geq 0}$. By counting the elements in $A + B$ of the form $a_i + b_j$ with $a_i \leq a_i + b_j < a_{i+1}$, Schnirelmann proved that if $1 \in A$ and $0 \in B$ then
$$\sigma(A + B) \geq \sigma(A) + \sigma(B) - \sigma(A)\sigma(B).$$

This is more usefully rewritten as $(1 - \sigma(A+B)) \leq (1 - \sigma(A))(1 - \sigma(B))$, since then we see that $(1 - \sigma(hA)) \leq (1 - \sigma(A))^h$. The last two results thus imply that if $1 \in A$ and $\sigma(A) > 0$ then A is a basis of order $2h$ where the integer h is chosen so that $(1 - \sigma(A))^h \leq \frac{1}{2}$. (Note that $\sigma(A) > 0$ implies that $1 \in A$; and that some condition like $1 \in A$ is necessary to avoid A, and hence hA, being a subset of the even integers.)

The lower density $\underline{d}(A)$ is defined by

$$\underline{d}(A) := \liminf_{n \to \infty} \frac{\#\{a \in A : 1 \leq a \leq n\}}{n},$$

We will prove that if $\underline{d}(A) > 0$ then for all $\epsilon \in (0, \underline{d}(A))$ there exists $r = r_\epsilon$ such that $\sigma(A_{(r)}) \geq \underline{d}(A) - \epsilon$, where $A_{(r)} = \{a - r : a \in A, a > r\}$. There exists an integer n_ϵ such that if $n \geq n_\epsilon$ then $\#\{a \in A : 1 \leq a \leq n\} \geq (\underline{d}(A) - \epsilon)n$. If there exists any $n \geq n_\epsilon$ with $\#\{a \in A : 1 \leq a \leq n\} < \underline{d}(A)n$ then there must be an $n \geq n_\epsilon$, say $n = m_\epsilon$, with $\rho_\epsilon := \#\{a \in A : 1 \leq a \leq n\}/n$ minimal. Hence if $n > m_\epsilon$ then $\#\{a \in A : m_\epsilon < a \leq n\} \geq \rho_\epsilon(n - m_\epsilon) \geq (\underline{d}(A) - \epsilon)(n - m_\epsilon)$. On the other hand if $\#\{a \in A : 1 \leq a \leq n\} \geq \underline{d}(A)n$ for all $n \geq n_\epsilon$ then either $\sigma(A) \geq \underline{d}(A)$, or there exists a maximal r_ϵ (which is necessarily $< n_\epsilon$) with $\#\{a \in A : 1 \leq a \leq r_\epsilon\} < \underline{d}(A)r_\epsilon$, and the result follows.

A standard, but careful sieve argument implies that at least $\frac{1}{4}$ of the even integers can be written as the sum of two primes; that is $\underline{d}(2\mathbb{P}_{\geq 3}) \geq 1/8$. Using the argument of the previous paragraph, and Schnirelmann's theorem, one can prove that the primes are a basis of order 11 (or less). It can also be shown that the k-th powers of integers form an additive basis.

For a finite set of integers S define *the cube $\bar{S}$* by

$$\bar{S} := \left\{ \sum_{s \in S} \epsilon_s s : \epsilon_s \in \{-1, 0, 1\} \text{ for all } s \in S \right\},$$

which is a generalized arithmetic progression of dimension $|S|$ and volume $3^{|S|}$.

Theorem. *If A is a set of integers with $\underline{d}(A) > 0$, then there exists a finite set of integers S such that $A - A + \bar{S} = \mathbb{Z}$.*

PROOF. If $A - A \neq \mathbb{Z}$ then there exists $m \notin A - A$, and so A and $m + A$ are disjoint. Let $A_1 = A \cup (m+A)$, so that $\underline{d}(A_1) = 2\underline{d}(A)$ and $A_1 - A_1 = A - A + \overline{\{m\}}$. If this is not $\mathbb{Z}$, define $A_2, A_3, \ldots$. Therefore $|S| \leq k$ where k is the largest integer for which $2^k \underline{d}(A) \leq 1$. □

Since $\underline{d}(2\mathbb{P}_{\geq 3}) \geq \frac{1}{8}$, we can deduce that there exists a set S_1 of no more than three integers for which

$$\mathbb{Z} = 2\mathbb{P}_{\geq 3} - 2\mathbb{P}_{\geq 3} + \bar{S_1}.$$

It is interesting to determine how small a set one needs to "complete" a given set in this manner. Thus above we added $\bar{S_1}$ to $2\mathbb{P}_{\geq 3} - 2\mathbb{P}_{\geq 3}$ to obtain $\mathbb{Z}$, though we believe that $\mathbb{P}_{\geq 3} - \mathbb{P}_{\geq 3} + \{0, 1\} = \mathbb{Z}$. For sums of squares we have $4\{n^2 : n \in \mathbb{Z}\} = \mathbb{Z}_{\geq 0}$; and one can show that $3\{n^2 : n \in \mathbb{Z}\} + \{0, 2\} = \mathbb{Z}_{\geq 0}$. A challenge is to find "thin" sets B and C for which $2\{n^2 : n \in \mathbb{Z}\} + B = \mathbb{Z}_{\geq 0}$, and for which $\mathbb{P} + C = \mathbb{Z}_{\geq 0}$.

1.4. The Dyson transformation. Many of the early papers in "additive number theory" were characterized by complicated, seemingly ad hoc, arguments. However, once Freeman Dyson introduced a simple map between pairs of sets, researchers found new, cleaner arguments in many of the essential questions: For $e \in A$ let $B_e := \{b \in B : b + e \notin A\}$, and define the *Dyson transformation* of A, B with respect to e to be

$$\delta_e(A) := A \cup (e + B) = A \cup (e + B_e), \quad \text{and} \quad \delta_e(B) := B \setminus B_e.$$

Notice that $B_e \subseteq B$ and $(e + B_e) \cap A = \varnothing$. There are several other observations to be made besides:

$$e + \delta_e(B) \subseteq A \subseteq \delta_e(A), \quad \text{and} \quad |\delta_e(A)| + |\delta_e(B)| = |A| + |B|;$$
$$A \cap (e + B) = e + \delta_e(B) = \delta_e(A) \cap (e + \delta_e(B)),$$

and

$$A \cup (e + B) = \delta_e(A) = \delta_e(A) \cup (e + \delta_e(B)),$$

as well as the nontrivial

$$\delta_e(A) + \delta_e(B) \subseteq A + B.$$

Using a sequence of Dyson transformations one can easily prove Mann's theorem, which is often more easily applied than Schnirelmann's theorem:

Mann's theorem. *If $0 \in A \cap B$ then*

$$\sigma(A + B) \geq \min\{1, \sigma(A) + \sigma(B)\}.$$

Note that this result does not extend directly to questions about lower density; that is, $\underline{d}(A + B) \geq \min\{1, \underline{d}(A) + \underline{d}(B)\}$ is not true in general: For example, if $A = B = \{n \equiv 0 \text{ or } 1 \pmod{m}\}$ then $A + B = \{n \equiv 0, 1 \text{ or } 2 \pmod{m}\}$. So, to understand set addition with respect to lower density, we certainly need to understand set addition mod N. Here the key result is

Cauchy–Davenport theorem. *If A and B are nonempty subsets of $\mathbb{Z}/N\mathbb{Z}$ where $0 \in B$, and $(b, N) = 1$ for all $b \in B \setminus \{0\}$, then*

$$|A + B| \geq \min\{N, |A| + |B| - 1\}.$$

PROOF. By induction on $|B|$: If $|B| = 1$ then $B = \{0\}$ so $A + B = A$ which is okay. We may assume that $1 \leq |A| \leq N - 1$. Now $A + B \neq A$ else for each $b \in B$, for all $a \in A$ there exists $a' \in A$ such that $a + b \equiv a' \pmod{N}$. Running through all $a \in A$ we obtain all $a' \in A$, and so taking the sum over all $a \in A$ we get $|A|b \equiv 0 \pmod{N}$. By selecting nonzero $b \in B$ we have $(b, N) = 1$, and so N divides $|A|$, which is impossible.

So take $e \in A$ for which $e + b \notin A$. By the induction hypothesis the result holds for the pair $\delta_e(A), \delta_e(B)$ (which are nonempty since $A \subseteq \delta_e(A)$ and $0 \in \delta_e(B)$), so that

$$|A+B| \geq |\delta_e(A)+\delta_e(B)| \geq \min\{N, |\delta_e(A)|+|\delta_e(B)|-1\} = \min\{N, |A|+|B|-1\}. \quad \square$$

Corollary. *If $A, B \subseteq \mathbb{Z}/p\mathbb{Z}$ with p prime then $|A + B| \geq \min\{p, |A| + |B| - 1\}$.*

There are just three cases in which we get equality (that is, $|A + B| = |A| + |B| - 1$) when $A + B$ is a proper subset of $\mathbb{Z}/p\mathbb{Z}$:

- Either A or B has just one element (that is, $|A| = 1$ or $|B| = 1$); or

- A and B are segments of arithmetic progressions with the same common difference (that is, $A = a + d \diamond \{0, 1, \ldots, r-1\}$, and $B = b + d \diamond \{0, 1, \ldots, s-1\}$ for some $r + s \le p$); or
- A and B are selected maximally so that $d \notin A + B$ (that is, $A \cup (d - B)$ is a partition of $\mathbb{Z}/p\mathbb{Z}$ for some integer d).

1.5. Simple inequalities for sizes of sumsets. The Freĭman–Ruzsa theorem tells us that if $|A + A| < C|A|$ then A is a large subset of a d-dimensional generalized arithmetic progression, G, for some d that can be bounded as a function of C. This implies that $A - A$ is a large subset of $G - G$, a generalized arithmetic progression that is at most twice as large (in each direction) as G, and so $|A - A| \le 2^d |G| \le 2^d C'|A|$ for some constant C' which depends only on C. Similarly $kA - lA$ is a large subset of $kG - lG$, also a d-dimensional generalized arithmetic progression, and so $|kA - lA| \le (k+l)^d C'|A|$.

In this section we derive consequences of this type directly, without using the relatively deep Freĭman–Ruzsa theorem; that is, our objective is to prove that if $|A + A| < C|A|$ then $|kA - lA| \le C_{k,l}|A|$ for some constant $C_{k,l}$ which depends only on C, k, l. We will see that there are several easy approaches to this problem. When we prove the Freĭman–Ruzsa theorem during the next lecture, we will use such inequalities in our proof. We start with the most basic question of this type:

1.5.1. *The relationship between $A + A$ and $A - A$.* We will prove a little later that

$$(1.2) \qquad \frac{1}{2} \le \log\left(\frac{|A+A|}{|A|}\right) \Big/ \log\left(\frac{|A-A|}{|A|}\right) \le 3;$$

we are interested in determining the strongest possible form of each of these inequalities. We give two examples

- For $A = \{0, 1, 3\}$ we have $A + A = \{0, 1, 2, 3, 4, 6\}$ and $A - A = \{-3, -2, -1, 0, 1, 2, 3\}$, so that $|A + A| = 6 < |A - A| = 7$.
- For $A = \{0, 2, 3, 4, 7, 11, 12, 14\}$ we have $A + A = [0, 28] \cap \mathbb{Z} \setminus \{1, 20, 27\}$ and $A - A = ([-14, 14] \cap \mathbb{Z}) \setminus \{-13, -6, 6, 13\}$, so that $|A - A| = 25 < |A + A| = 26$.

These isolated examples can be made into arbitrarily large examples by using the Cartesian product: The idea simply is to take $B = A^{(k)} = A \times \cdots \times A$, so in the first case $|B + B| = 6^k < |B - B| = 7^k$. One might object that B is not a subset of the integers but in fact the bijection $B \leftrightarrow C$ defined by $(a_0, \ldots, a_{k-1}) \leftrightarrow a_0 + a_1 7 + \cdots + a_{k-1} 7^{k-1}$ is also a bijection, when correctly interpreted, between the sets $B + B$ and $C + C$, and between $B - B$ and $C - C$. This map is called a Freĭman 2-isomorphism (the "2" since it remains a bijection when we add two elements of our set); we will discuss this in detail in our next lecture. We thus conclude from our examples that the constant "$\frac{1}{2}$" in (1.2) may not be increased above $\log(6/3)/\log(7/3) = .81806\ldots$; and that the constant "3" in (1.2) may not be decreased below $\log(26/8)/\log(25/8) = 1.03442\ldots$,

The lower bound $\frac{1}{2}$ in (1.2) cannot be increased at all: For given positive integer d let $A = A_d(T) := \{(x_1, \ldots, x_d) \in \mathbb{Z}_{\ge 0} : x_1 + \cdots + x_d \le T\}$ so that

$$|A| = \binom{T+d}{d} = \frac{T^d}{d!} + O_d(T^{d-1}), \quad |2A| = \binom{2T+d}{d} = \frac{(2T)^d}{d!} + O_d(T^{d-1}),$$

and, by counting the number of possibilities for the positive, negative and zero coordinates in $A - A$ separately,

$$|A - A| = \sum_{a+b+c=d} \binom{d}{a,b,c} \binom{T}{a} \binom{T}{b}$$

$$= \frac{T^d}{d!} \sum_{a=0}^{d} \binom{d}{a}^2 + O_d(T^{d-1}) = \binom{2d}{d} \frac{T^d}{d!} + O_d(T^{d-1}).$$

Therefore we have shown that there are examples of $|2A| = \alpha|A|$, with α arbitrarily large, for which $|A - A| \gg (\alpha^2/\sqrt{\log \alpha})|A|$.

1.5.2. *Some first bounds.* We begin by establishing that for any finite sets A, B, C inside an additive group G (whether commutative or not) we have

(1.3) $$|A - C||B| \leq |A - B||B - C|,$$

by showing that there is an injection $\phi\colon (A - C) \times B \to (A - B) \times (B - C)$: For each $\lambda \in A - C$ fix $a_\lambda \in A, c_\lambda \in C$ such that $a_\lambda - c_\lambda = \lambda$. Then define $\phi(\lambda, b) = (a_\lambda - b, b - c_\lambda)$. To see that this is an injection we show how to reconstruct λ and b given $a_\lambda - b$ and $b - c_\lambda$: First we have $\lambda = (a_\lambda - b) + (b - c_\lambda)$, so we obtain a_λ, c_λ, and thus b.

We now use (1.3) to obtain all sorts of useful inequalities:

- Taking $C = A$ gives $|A - A| \leq |A - B|^2/|B|$.
- Then taking $B = -A$ gives $|A - A|/|A| \leq (|A + A|/|A|)^2$, which is the lower bound in (1.2).
- Next taking $A = rA$, $B = -A$ with $C = -sA$ and then $C = sA$ implies:

$$|(r + s)A||-A| \leq |(r + 1)A||sA - A|,$$
$$|rA - sA||-A| \leq |(r + 1)A||(s + 1)A|.$$

With the choices $r = n - 2$, $s = 2$, and $r = 2$, $s = 1$, respectively, we obtain

$$\frac{|nA|}{|A|} \leq \frac{|(n-1)A|}{|A|} \frac{|2A - A|}{|A|} \leq \frac{|(n-1)A|}{|A|} \frac{|3A|}{|A|} \frac{|2A|}{|A|}.$$

We deduce that, for $n \geq 3$,

$$\frac{|nA|}{|A|} \leq \left(\frac{|3A|}{|A|}\right)^{n-2} \left(\frac{|2A|}{|A|}\right)^{n-3} \quad \text{for all } n \geq 3;$$

and then that

$$\frac{|rA - sA|}{|A|} \leq \left(\frac{|3A|}{|A|}\right)^{r+s-2} \left(\frac{|2A|}{|A|}\right)^{r+s-4} \quad \text{for all } r, s \geq 2.$$

This is almost what we asked for! We wanted bounds as a function of r, s and $|2A|/|A|$, and instead we have very easily obtained bounds in terms of these variables and $|3A|/|A|$. So the question becomes whether one can find an easy way to bound $|3A|/|A|$ in terms of $|2A|/|A|$? Certainly such bounds can be proved by straightforward combinatorial arguments, but we know of no proof that is quite so simple as that above. (Taking $r = 1$, $s = 2$ in the inequalities above, we see that we could replace $3A$ by $2A - A$ in these last few comments.) Relationships between these different quantities are explored in detail by Imre Ruzsa in his article in this volume [26].

1.5.3. *Representation numbers.* Denote the number of representations of n as a sum $a + b$, $a \in A$, $b \in B$ by

$$r_{A+B}(n) := \#\{(a,b) : a \in A, b \in B, n = a+b\},$$

and similarly $r_{kA+lB}(n)$, etc. There are several straightforward but useful identities: First, by counting all ordered pairs (a,b), $a \in A$, $b \in B$ we obtain

$$|A|\,|B| = \sum_x r_{A+B}(x) = \sum_y r_{A-B}(y).$$

The solutions to $a+b = a'+b'$ with $a, a' \in A$, $b, b' \in B$ are the same as the solutions to $a - b' = a' - b$, which are the same as the solutions to $a - a' = b' - b$, and so

$$E(A,B) := \sum_x r_{A+B}(x)^2 = \sum_y r_{A-B}(y)^2 = \sum_z r_{A-A}(z) r_{B-B}(z).$$

Therefore we obtain, by the Cauchy–Schwarz inequality, that

$$(|A|\,|B|)^2 = \left(\sum_x r_{A\pm B}(x)\right)^2 \leq |A \pm B| E(A,B).$$

Also note that

$$E(A,B) \leq \begin{cases} \max_x r_{A+B}(x) \sum_x r_{A+B}(x) = |A||B| \max_x r_{A+B}(x), \\ |A+B| \max_x r_{A+B}(x)^2. \end{cases}$$

Now we show that

$$r_{A+B}(x) \leq \frac{|A-B|^2}{|A+B|}$$

by exhibiting, for a given value of $x \in A+B$, an injection from $R_{A+B}(x) \times (A+B) \to (A-B) \times (A-B)$, where $R_{A+B}(x)$ is the set of representations of x as $a+b$, $a \in A$, $b \in B$. So fix a representation $a+b = x$, and for any $\lambda \in A+B$ fix $a_\lambda \in A$, $b_\lambda \in B$ such that $a_\lambda + b_\lambda = \lambda$. The map $(a,b,\{a_\lambda,b_\lambda\}) \to (a-b_\lambda, a_\lambda - b)$ is, indeed, an injection, because we can reconstruct our pre-image by noting that $\lambda = x + (a_\lambda - b) - (a - b_\lambda)$, from which we obtain a_λ and b_λ, then $a = (a-b_\lambda) + b_\lambda$ and $b = x - a$.

Combining the last three displayed equations we obtain

$$|A+B| \leq \frac{|A-B|^2}{\max_x r_{A+B}(x)} \leq \frac{|A-B|^2 |A||B|}{E(A,B)} \leq \frac{|A-B|^3}{|A||B|}.$$

Taking $B = A$ gives the upper bound in (1.2).

1.5.4. *Disjoint unions.* We start with an idea of Ruzsa that we shall see again.

Lemma 1.1. *There exists $X \subset B$ with $|X| \leq |A+B|/|A|$ such that $B \subset A - A + X$.*

PROOF. Choose $X \subset B$ to be as large as possible so that the sets $\{A + x : x \in X\}$ are disjoint. The union of these sets contains exactly $|A||X|$ elements, all in $A + B$, which implies that $|A| \cdot |X| \leq |A+B|$.

Now if $b \in B$ then $(A+b) \cap (A+x) \neq \varnothing$ for some $x \in X$, else X would not have been maximal, so $b \in A - A + x$, and we are done. $\square$

Take $B = A - 2A$ in Lemma 1.1 to get $2A - A \subset A - A + X$ where $X \subset 2A - A$ with $|X| \leq |2A - 2A|/|A|$ (replacing X by $-X$ for convenience). Add A to both sides to get
$$3A - A \subset 2A - A + X \subset A - A + 2X$$
and then, proceeding by induction, we obtain

(1.4) $\quad mA - nA \subset A - A + (m-1)X - (n-1)X \quad$ for all $m, n \geq 1$.

Now, since each $|rX| \leq |X|^r$, and as $|X| \leq |2A - 2A|/|A|$, we deduce that

(1.5) $\quad \dfrac{|mA - nA|}{|A|} \leq \dfrac{|A - A|}{|A|} \left(\dfrac{|2A - 2A|}{|A|} \right)^{m+n-2} \quad$ for all $m, n \geq 1$.

Another argument based on something similar to, but more complicated than, the above lemma (see of [31, Lemma 2.17, Proposition 2.18 and Corollary 2.19]), leads to the inequality
$$|2B - 2B| \leq |A + B|^4 |A - A|/|A|^4.$$
Taking $B = A$ in this formula, and then the first inequality in (1.2), we deduce from (1.5) that
$$\dfrac{|mA - nA|}{|A|} \leq \left(\dfrac{|2A|}{|A|} \right)^{6m+6n-10} \quad \text{for all } m, n \geq 1.$$
Finally, selecting $A = (n-1)A$, $C = -A$, $B = A - A$ in (1.3), and then substituting in (1.5) we obtain
$$\dfrac{|nA|}{|A|} \leq \dfrac{|A - A|}{|A|} \left(\dfrac{|2A - 2A|}{|A|} \right)^n \leq \left(\dfrac{|2A|}{|A|} \right)^{6n+2} \quad \text{for all } n \geq 1.$$
The strongest version of such an inequality that is known was first proved by Plünnecke [22], whose proof has been streamlined, over the years, by Ruzsa [24] and others (though it is still too complicated to give here):

Plünnecke – Ruzsa theorem. *For any $m, n \geq 0$ we have*
$$\dfrac{|mA - nA|}{|A|} \leq \left(\dfrac{|2A|}{|A|} \right)^{m+n}.$$

We may rephrase this as: If $|2A| \leq C|A|$ then $|mA - nA| \leq C^{m+n}|A|$.

This result can be given in the slightly stronger form: If $|A + B| \leq C|A|$ then $|mB - nB| < C^{m+n}|A|$ for all $m, n \geq 0$. Taking $B = A$ gives the above result. Taking $B = -A$ implies that the assumption $|A - A| \leq C|A|$ yields the same conclusion, and therefore we may replace the "≤ 3" by "≤ 2" in (1.2).

1.6. The Freĭman – Ruzsa theorem in groups, where the elements have bounded order. Take the union of (1.4) over all $m, n \geq 1$ to obtain $\langle A \rangle \subset A - A + \langle X \rangle$. However $X \subset 2A - A \subset \langle A \rangle$ and so
$$\langle A \rangle = A - A + \langle X \rangle.$$
Suppose that $|2A| \leq C|A|$. Then $|X| \leq |2A - 2A|/|A| \leq C^4$ by the Plünnecke–Ruzsa theorem (we can get $\leq C^6$ if we only use the results that are proved above). That is, the generalized arithmetic progression $\langle A \rangle$ belongs to a union of translates of the generalized arithmetic progression $\langle X \rangle$, which has (bounded) dimension $\leq$

C^4. If $A \subset G$, an abelian group in which the maximal order of any element is $\leq r$, then $|\langle X \rangle| \leq r^{|X|}$. Therefore

$$|\langle A \rangle| \leq |A - A| \, |\langle X \rangle| \leq C^2 |A| r^{|X|} \leq (C^2 r^{C^4}) |A|.$$

1.7. The Balog–Szemerédi (–Gowers) theorem. In many applications one does not have that $A + B$ is small, but rather that there is a large subset $G \subset \{(a,b) : a \in A, b \in B\}$ which contains $\gg |A| |B|$ elements, for which $S_G := \{a + b : (a,b) \in G\}$ is small. One then wishes to conclude something about the structure of large subsets of A and B. In the case that $|A| = |B|$ there is an important result of Balog and Szemerédi [2], strengthened by Gowers [13] (and subsequently by several others) with a much easier proof—Antal Balog's article in these proceedings [1] will discuss all this in detail. Here we simply state a version of this very flexible result, in order to get the flavour: Suppose that $|A| = |B| = n$ and that there exists $G \subset \{(a,b) : a \in A, b \in B\}$ containing $\geq \alpha n^2$ elements, for which $S_G := \{a + b : (a,b) \in G\} \leq n$. Then there exists $A' \subset A$, $B' \subset B$ with $|A'|, |B'| \geq (\alpha/16) n$ for which $|A' + B'| \leq (2^{23}/\alpha^5) \, n$, with

$$|G \cap \{(a', b') : a' \in A', b' \in B'\}| \geq (\alpha^2/128) n^2.$$

1.8. Discrete Fourier transforms. One of the most useful tools in additive combinatorics are Fourier transforms in $\mathbb{Z}/N\mathbb{Z}$: For a function $f \colon \mathbb{Z}/N\mathbb{Z} \to \mathbb{C}$ we define

$$\hat{f}(r) = \sum_{s=0}^{N-1} f(s) e\left(\frac{rs}{N}\right),$$

where $e(t) = \exp(2i\pi t)$. This has inverse

$$f(s) = \frac{1}{N} \sum_{r=0}^{N-1} \hat{f}(r) e\left(\frac{-rs}{N}\right).$$

One has

$$\sum_r \hat{f}(r) \bar{\hat{g}}(r) = N \sum_r f(r) \bar{g}(r).$$

Parseval's identity is the case $f = g$, namely $\sum_r |\hat{f}(r)|^2 = N \sum_r |f(r)|^2$.

We define the *convolution* of two functions to be

$$(f * g)(r) = \sum_{t-u=r} f(t) \overline{g(u)},$$

so that $\widehat{(f * g)} = \hat{f} \bar{\hat{g}}$, and

$$N \sum_r |(f * g)(r)|^2 = \sum_r |\hat{f}(r)|^2 |\hat{g}(r)|^2.$$

Taking $g = f$ we obtain

$$\sum_r |\hat{f}(r)|^4 = N \sum_{a+b=c+d} f(a) f(b) \overline{f(c) f(d)}.$$

Let A be a subset of $\mathbb{Z}/N\mathbb{Z}$, and then define $A(n)$ to be the *indicator* function of A; that is, $A(n) = 1$ if $n \in A$, and $A(n) = 0$ otherwise. Hence

$$\hat{A}(m) = \sum_{a \in A} e\left(\frac{am}{N}\right).$$

Noting that $(A * B)(n) = r_{A-B}(n)$ we deduce that

$$E(A, B) = \sum_n r_{A-B}(n)^2 = \sum_n |(A * B)(n)|^2 = \frac{1}{N} \sum_n |\hat{A}(n)|^2 |\hat{B}(n)|^2.$$

We also have

$$\hat{A}(m)\hat{B}(m) = \sum_n r_{A+B}(n) e\left(\frac{mn}{N}\right),$$

which can be inverted to give

$$r_{A+B}(n) = \frac{1}{N} \sum_m \hat{A}(m)\hat{B}(m) e\left(\frac{-mn}{N}\right);$$

a special case of which is

$$r_{kA-kA}(n) = \frac{1}{N} \sum_m |\hat{A}(m)|^{2k} e\left(\frac{-mn}{N}\right).$$

1.9. Sum-product formulas. I learnt to multiply by memorizing the multiplication tables; that is, we wrote down a table with the rows and columns indexed by the integers between 1 and N and the entries in the table were the row entry times the column entry.[1] Paul Erdős presumably learnt his multiplication tables rather more rapidly than the other students, and was left wondering: How many distinct integers are there in the N-by-N multiplication table? Note that if we take $A = \{1, 2, \ldots, N\}$, then we are asking how big is $A \diamond A$? Or, more specifically, since the numbers in the N-by-N multiplication table are all $\leq N^2$, what proportion of the integers up to N^2 actually appear in the table? That is,

Does $|A \diamond A|/N^2$ tend to a limit as $N \to \infty$?

Erdős showed that the answer is, yes, and that the limit is 0. His proof comes straight from "The Book."[2] Erdős's proof is based on the celebrated result of Hardy and Ramanujan that "almost all" positive integers $n \leq N$ have $\sim \log \log N$ (not necessarily distinct) prime factors (here "almost all" means for all but $o(N)$ values of $n \leq N$): Hardy and Ramanujan's result implies that "almost all" products ab with $a, b \leq N$ have $\sim 2 \log \log N$ prime factors, whereas "almost all" integers $\leq N^2$ have $\sim \log \log(N^2) \sim \log \log N$ prime factors! The result follows from comparing these two statements.

One can show that $|A \diamond A|$ is large whenever A is an arithmetic progression or, more generally, when A is a generalized arithmetic progression of not-too-large dimension. This led Erdős and Szemerédi to the conjecture that for any $\epsilon > 0$, there exists $c_\epsilon > 0$ such that

$$|A + A| + |A \diamond A| \geq c_\epsilon |A|^{2-\epsilon}.$$

Even more, Solymosi conjectured that if $|A| = |B| = |C|$ then

(1.6) $$|A + B| + |A \diamond C| \geq c_\epsilon |A|^{2-\epsilon};$$

and proved this for $\epsilon = 8/11$ [27]. We shall prove (1.6) for $\epsilon = \frac{3}{4}$. We begin by stating the

---

[1] In my primary school we took $n = 12$ which was the basic multiple needed for understanding U.K. currency at that time.

[2] Erdős claimed that the Supreme Being kept a book of all the best proofs, and only occasionally would allow any mortal to glimpse at "The Book."

Szemerédi–Trotter theorem. *We are given a set $\mathcal{C}$ of m curves in $\mathbb{R}^2$ such that*

- *Each pair of curves meet in $\leq b_1$ points;*
- *Any pair of points lie on $\leq b_2$ curves.*

For any given set $\mathcal{P}$ of n points, there are $\leq m + 4b_2 n + 4b_1 b_2^{1/3} (mn)^{2/3}$ pairs (π, γ) with point $\pi \in \mathcal{P}$ lying on curve $\gamma \in \mathcal{C}$.

Székely provided a gorgeous proof of this result, straight from *The Book*, via geometric and random graph theory. From this Elekes elegantly deduced that if $A, B, C \subset \mathbb{Z}$ then

(1.7) $$|A + B| + |A \diamond C| \geq \tfrac{2}{3}(|B||C|)^{1/4}(|A| - 1)^{3/4}.$$

PROOF. Let $\mathcal{P}$ be the set of points $(A + B) \times (A \diamond C)$; and $\mathcal{C}$ the set of lines $y = c(x - b)$ where $b \in B$ and $c \in C$. In this case we have $b_1 = b_2 = 1$ with

$$m = |B||C| \quad \text{and} \quad n = |A + B||A \diamond C|.$$

For fixed $b \in B$ and $c \in C$, all of the points $\{(a + b, ac) : a \in A\}$ in $\mathcal{P}$ lie on the line $y = c(x - b)$, so that

$$\#\{(\pi, \gamma) : \pi \in \mathcal{P} \text{ on } \gamma \in \mathcal{C}\} \geq |A|m.$$

Substituting this into the Szemerédi–Trotter theorem we obtain

$$(|A| - 1)m \leq 4n + 4(mn)^{2/3}.$$

If $m > 64n^{1/2}$ then $(|A| - 1)m \leq 4n + 4(mn)^{2/3} \leq (17/4)(mn)^{2/3}$ which yields $n^2 \geq (|A| - 1)^3 m/77$; and if $m \leq 64n^{1/2}$ then $(|A| - 1)m \leq 4n + 4(mn)^{2/3} \leq 68n$, which multiplied by the trivial $n \geq |A|^2$ yields the same. The result follows as $2/(77)^{1/4} > \tfrac{2}{3}$. □

Solymosi has proved (1.7) with several different counting arguments which do not involve the Szemerédi–Trotter theorem. Here we sketch one: Consider the set of distinct points $\{(a + b, ac) : a \in A, b \in B, c \in C\}$ in $\mathbb{R}^2$. We will suppose that we can partition $\mathbb{R}^2$ into a grid, with $|A|/3 + O(1)$ lines in each direction (that is lines of the form $x = r$ and of the form $y = s$), in which each box contains roughly equal numbers of points.[3] Now for each pair $b \in B$, $c \in C$ we will count the number of pairs of points $(b + a_i, ca_i)$, $(b + a_{i+1}, ca_{i+1})$ which belong to the same box where A is the set $a_1 < a_2 < \cdots < a_n$. Since $b + a_i < b + a_{i+1}$ and $|c|a_i < |c|a_{i+1}$ we see that the set of points $\{(b + a, ca) : a \in A\}$ can lie in no more than $2|A|/3 + O(1)$ boxes. But then the number of pairs of points $(b + a_i, ca_i)$, $(b + a_{i+1}, ca_{i+1})$ which belong to the same box is $\geq |A|/3 + O(1)$; so the total number of such pairs is $\gtrsim |A||B||C|/3$. Now for any two given points there is at most one triple b, c, i giving those two points else, taking the differences of x and y co-ordinates we have $a_{i+1} - a_i = a_{j+1} - a_j$ and $a_{i+1}/a_i = a_{j+1}/a_j$ which implies that $i = j$ and hence $b = b'$, $c = c'$. Therefore the total number of such pairs is no more than the total number of pairs in our boxes. There are $\sim (|A|/3)^2$ boxes with $\sim |A + B||A \diamond C|/(|A|/3)^2$ points in each box; and so with a total of $\sim |A + B|^2 |A \diamond C|^2/(2(|A|/3)^2)$ pairs of points. Combining these remarks we deduce that

$$|A + B|^2 |A \diamond C|^2 \gtrsim \frac{2}{27}|A|^3 |B||C|;$$

---

[3] This is not quite as easy as it sounds!

which implies the slight improvement $|A+B|+|A \diamond C| \gtrsim (|B||C|)^{1/4}|A|^{3/4}$ over (1.7).

Sum-product inequalities have also been proved over finite fields (by Bourgain, Katz, Tao [8], Konyagin, Chang, Glibichuk, ...): This was the basis for proving spectacularly strong bounds on exponential sums by Bourgain [5], Bourgain, Glibichuk and Konyagin [7], Bourgain and Chang [6] and others—see Kurlberg's article herein for a discussion of this proof [21]. These methods have been developed for nonabelian groups, in particular $\mathrm{SL}(2, \mathbb{Z}_p)$ by Helfgott [19], and then extended by Bourgain and Gamburd, Gowers,.... See Mei-Chu Chang's article herein for a discussion of these directions [10].

2. Lecture 2. The Freĭman–Ruzsa theorem

2.1. If $A+B$ is small what do A and B look like? We have already seen $|A+B| \geq |A|+|B|-1$ and that if $|A+B| = |A|+|B|-1$ then there exists an integer $d \geq 1$ such that

$$A = \{a+id : 0 \leq i \leq I-1\} \quad \text{and} \quad B = \{b+jd : 0 \leq j \leq J-1\}.$$

In other words A and B are segments of arithmetic progressions, both with the same common difference. Now if A' is a subset of A, and B' is a subset of B then $A'+B'$ is a large subset of $A+B$; so if A and B are segments of arithmetic progressions with the same common difference, then A' and B' can be chosen as large subsets with little particular structure and yet $A'+B'$ is relatively small. This construction generalizes to large subsets of generalized arithmetic progressions: A *generalized arithmetic progression* is a set of integers of the form

$$C := \{a_0 + a_1 n_1 + a_2 n_2 + \cdots + a_k n_k : 0 \leq n_j \leq N_j - 1 \text{ for } 1 \leq j \leq k\}$$

where $N_1, N_2, \ldots, N_k$ are given integers ≥ 2. We say that this generalized arithmetic progression has *dimension* k and *volume* $N_1 N_2 \cdots N_k$. It is called *proper* if the elements are distinct; that is if there are $N_1 N_2 \cdots N_k$ distinct elements in the generalized arithmetic progression. Our key observation is that $|2C| < 2^k |C|$ for a generalized arithmetic progression C of dimension k; so that if $A, B \subset C$ with $|A|, |B| \geq \delta |C|$ then

$$|A+B| \leq |2C| < 2^k |C| \leq (2^{k-1} \delta^{-1})(|A|+|B|).$$

What about the converse? If $|A+B|$ is a small multiple of $|A|+|B|$, what possibilities are there? Is it true that A and B are both large subsets of translates of the same low-dimensional generalized arithmetic progression? This question still remains open; we will restrict our attention to the case that $B = A$. In other words, if $|2A|$ is a small multiple of $|A|$ then is A necessarily a large subset of a low-dimensional generalized arithmetic progression? This question is answered by the wonderful

Freĭman–Ruzsa theorem. *If $|2A|$ is "small" then A is a "large" subset of a generalized arithmetic progression.*

This statement is a bit vague but, in essence, it is everything we asked for. The details are complicated and researchers have not yet found the best possible version so we leave all that until a little later.

This theorem was first announced by Freĭman and gained broad distribution in his book [11]. Just to dare guess at such a classification result is an extraordinary achievement, and all proofs to date require much ingenuity. The proof in Freĭman's book is deep, and is difficult to follow in places. Because of this, Freĭman's result did not quickly gain the prominence it deserves in combinatorial number theory. However in 1994, Ruzsa [25] came up with his own, much shorter and easier-to-follow proof, which caught many people's imagination. It is Ruzsa's paper that heralded the outpouring of research into this exciting area. It is for these reasons that I feel it is fair to give both Freĭman and Ruzsa credit for their extraordinary achievements by naming the theorem after them both.[4] The proof I give here is more-or-less that of Ruzsa, though incorporating some remarks from Ben Green's notes [15].

2.2. The geometry of numbers.

Given linearly independent vectors $\mathbf{x}_1, \mathbf{x}_2, \ldots, \mathbf{x}_k \in \mathbb{R}^k$, define a *lattice*

$$\Lambda := (\mathbf{x}_1 \diamond \mathbb{Z}) + (\mathbf{x}_2 \diamond \mathbb{Z}) + \cdots + (\mathbf{x}_k \diamond \mathbb{Z}).$$

We define $\det(\Lambda)$ to be the volume of

$$F := \{a_1 \mathbf{x}_1 + a_2 \mathbf{x}_2 + \cdots + a_k \mathbf{x}_k : 0 \leq a_i < 1 \text{ for all } i\},$$

the connection between Λ and F stemming from the fact that $F + \Lambda = \mathbb{R}^k$.

Blichfeldt's lemma. *If $L \subset \mathbb{R}^k$ is measurable with $\mathrm{vol}(L) > \det(\Lambda)$ then $L - L$ contains a nonzero point of Λ.*

Suppose that K is a centrally symmetric and convex subset of $\mathbb{R}^k$, so that

$$K = \tfrac{1}{2} \diamond K - \tfrac{1}{2} \diamond K.$$

Since $\mathrm{vol}(\tfrac{1}{2} \diamond K) = (1/2^k) \mathrm{vol}(K)$, Blichfeldt's lemma with $L = \tfrac{1}{2} \diamond K$ implies:

Minkowski I. *If $\mathrm{vol}(K) > 2^k \det(\Lambda)$ then K contains a nonzero point of Λ.*

Suppose we are given a lattice Λ in $\mathbb{R}^k$, as well as a closed, convex body $K \subset \mathbb{R}^k$. The *successive minima* $\lambda_1, \lambda_2, \ldots, \lambda_k$ of K with respect to Λ are the smallest values λ_j such that $\lambda_j \diamond K$ contains j linearly independent elements of Λ.

Minkowski II. *Suppose that K is a centrally symmetric, closed, convex subset of $\mathbb{R}^k$ and Λ a lattice of rank k. With the definitions as above, there exist linearly independent vectors $\mathbf{b}_1, \ldots, \mathbf{b}_k \in \Lambda$ where $\mathbf{b}_j$ lies on the boundary of $\lambda_j \diamond K$ for $j = 1, 2, \ldots, k$. Moreover, $\lambda_1 \cdots \lambda_k \mathrm{vol}(K) \leq 2^k \det(\Lambda)$.*

Let $\|t\|$ be the distance from t to the nearest integer (that is $\|t\| := \min_{m \in \mathbb{Z}} |t - m|$), and then $\|(x_1, \ldots, x_k)\| := \max_i \|x_i\|$. For given $r_1, \ldots, r_k$ let $\mathbf{r} := (r_1, \ldots, r_k)$. When $r_1, \ldots, r_k \in \mathbb{Z}/N\mathbb{Z} \setminus \{0\}$ and $\delta > 0$, we define the *Bohr neighbourhood* to be

$$B(r_1, \ldots, r_k; \delta) := \{s \in \mathbb{Z}/N\mathbb{Z} : \|\mathbf{r}s/N\| \leq \delta\};$$

in other words $s \in B(r_1, \ldots, r_k; \delta)$ if the least residue, in absolute value, of each $r_i s$ (mod N) belongs to the interval $[-\delta N, \delta N]$.

---

[4]Though some authors give credit only to Freĭman.

2.3. Structure of a Bohr set.

Theorem 2.1. *Suppose that N is prime with $r_1, \ldots, r_k \in \mathbb{Z}/N\mathbb{Z} \setminus \{0\}$, and that $0 < \delta < \frac{1}{2}$. Then $B(r_1, \ldots, r_k; \delta)$ contains a k-dimensional generalized arithmetic progression of volume $\geq (\delta/k)^k N$.*

PROOF. Let Λ be the lattice generated by $\mathbf{r}$ and $N\mathbb{Z}^k$ so that $\det(\Lambda) = N^{k-1}$. Let $K = \{(t_1, t_2, \ldots, t_k) : -1 \leq t_i \leq 1\}$, and then select $\lambda_1, \lambda_2, \ldots, \lambda_k$ and $\mathbf{b}_1, \ldots, \mathbf{b}_k \in \Lambda$ as in Minkowski II, so that we may write $\mathbf{b}_i = s_i \mathbf{r} + N\mathbb{Z}^k$ for some $s_i \pmod{N}$, for each i. Therefore
$$\|s_i r_j / N\| = \|(\mathbf{b}_i)_j / N\| \leq \|\mathbf{b}_i / N\| = \|\lambda_i / N\| \leq \lambda_i / N.$$
Let $P := \{\sum_{i=1}^k a_i s_i : |a_i| \leq \delta N / k\lambda_i\}$, which is a k-dimensional generalized arithmetic progression. If $s = \sum_{i=1}^k a_i s_i \in P$ then, for each j,
$$\left\|\frac{sr_j}{N}\right\| \leq \sum_{i=1}^k |a_i| \left\|\frac{s_i r_j}{N}\right\| \leq \sum_{i=1}^k \frac{\delta N}{k\lambda_i} \cdot \frac{\lambda_i}{N} = \delta,$$
so $s \in B(r_1, \ldots, r_k; \delta)$; that is $P \subset B(r_1, \ldots, r_k; \delta)$.

Using Minkowski II, and since there are at least t integers in the interval $[-t, t]$ for all $t \geq 0$, we have
$$\mathrm{Vol}(P) > \prod_{i=1}^k \frac{\delta N}{k\lambda_i} \geq \left(\frac{\delta N}{k}\right)^k \frac{\mathrm{vol}(K)}{2^k \det(\Lambda)} = \left(\frac{\delta}{k}\right)^k N,$$
as $\mathrm{vol}(K) = 2^k$ and $\det(\Lambda) = N^{k-1}$. □

Remark. Note that if $\delta < \frac{1}{2}$ then P is proper.

Bogolyubov's theorem. *Let $A \subset \mathbb{Z}/N\mathbb{Z}$ with $|A| = \alpha N$. Then $2A - 2A$ contains a generalized arithmetic progression of dimension $\leq \alpha^{-2}$ and volume $\geq (\alpha^2/4)^{\alpha^{-2}} N$.*

PROOF. Let R be the set of "large" Fourier coefficients, that is $R := \{r \pmod{N} : |\hat{A}(r)| \geq \alpha^{3/2} N\}$. By Parseval's identity we have
$$|A|N = \sum_r |\hat{A}(r)|^2 \geq \sum_{r \in R} |\hat{A}(r)|^2 \geq |R|\alpha^3 N^2$$
so that $|R| \leq \alpha^{-2}$. We also have
$$\sum_{r \notin R} |\hat{A}(r)|^4 \leq \max_{r \notin R} |\hat{A}(r)|^2 \sum_r |\hat{A}(r)|^2 < \alpha^3 N^2 \cdot |A|N = |A|^4 = |\hat{A}(0)|^4.$$
If $n \in B(R; \frac{1}{4})$ then $\|rn/N\| \leq \frac{1}{4}$, and hence $\cos(2\pi rn/N) \geq 0$ for all $r \in R$. Using that $\cos(2\pi rn/N) \geq -1$ for all $r \notin R$, that $|\hat{A}(-r)| = |\hat{A}(r)|$, and that $0 \in R$, we obtain
$$r_{2A-2A}(n) := \frac{1}{N} \sum_{r \pmod{N}} |\hat{A}(r)|^4 e\left(\frac{rn}{N}\right) = \frac{1}{N} \sum_{r \pmod{N}} |\hat{A}(r)|^4 \cos\left(2\pi \frac{rn}{N}\right)$$
$$\geq \frac{1}{N}\left(|\hat{A}(0)|^4 - \sum_{r \notin R} |\hat{A}(r)|^4\right) > 0.$$

Therefore $2A - 2A$ contains $B(R; \frac{1}{4})$, and hence contains the required arithmetic progression by Theorem 2.1. □

2.4. Freĭman homomorphisms. Suppose that A and B are both finite subsets of some ring like $\mathbb{Z}/s\mathbb{Z}$ or $\mathbb{Z}$ (perhaps different).

The map $\phi\colon A \to B$ is a *(Freĭman) k-homomorphism* if
$$\phi(x_1) + \cdots + \phi(x_k) = \phi(y_1) + \cdots + \phi(y_k)$$
whenever $x_i, y_i \in A$ satisfy
$$x_1 + x_2 + \cdots + x_k = y_1 + y_2 + \cdots + y_k.$$
ϕ is a *(Freĭman) k-isomorphism* if ϕ is invertible and ϕ and ϕ^{-1} are Freĭman k-homomorphisms. (Henceforth we drop the adjective "Freĭman.")

Examples. The reduction $\rho_p\colon \mathbb{Z} \to \mathbb{Z}/p\mathbb{Z}$ is a k-homomorphism for all k. If $A = \{a_1 < \cdots < a_n < a_1 + p/k\}$ then $\rho_p\big|_A$ is a k-isomorphism.

If $(q,p) = 1$ then $\mu_{q,p}\colon \mathbb{Z}/p\mathbb{Z} \to \mathbb{Z}/p\mathbb{Z}$, where $\mu_{q,p}(x) \equiv qx \pmod{p}$, is a k-isomorphism for all k.

The heart of our proof comes in the following remarkable lemma of Ruzsa which gives a Freĭman isomorphism between a large subset of our given set A, and some subset of the integers mod N. This allows us to work inside the integers mod N, where there are more convenient tools.

Ruzsa's lemma. *For any set of integers A and any prime $N > 2|kA - kA|$, there exists a subset A' of A, with $|A'| \geq |A|/k$, which is k-isomorphic to a subset of $\mathbb{Z}/N\mathbb{Z}$.*

PROOF. Select a prime $p > k(\max A - \min A)$, and any q with $(q,p) = 1$. Note that $\rho_p\big|_A$ is a k-isomorphism. By the pigeonhole principle, there exist $A' \subset A$ with $|A'| \geq |A|/k$ and
$$\mu_{q,p} \circ \rho_p\big|_{A'} \subset \left\{ x \in \mathbb{Z}/p\mathbb{Z} : \text{There exists } n \in \left[\frac{j-1}{k}p, \frac{j}{k}p\right) \text{ with } n \equiv x \pmod{p}\right\}$$
for some j. Therefore, taking ρ_p^{-1} here to give a residue in $[0, p)$, we have that
$$\Psi := \rho_p^{-1} \circ \mu_{q,p} \circ \rho_p\big|_{A'}$$
is a k-isomorphism such that
$$r := \sum_{i=1}^k \Psi(a_i) - \sum_{i=1}^k \Psi(a'_i) \in (-p, p),$$
with
$$r \equiv q\left(\sum_{i=1}^k a_i - \sum_{i=1}^k a'_i\right) \pmod{p}.$$
for all $a_1, a_2, \ldots, a_k, a'_1, a'_2, \ldots, a'_k \in A'$. (The reader should verify that this is indeed a k-isomorphism.)

Now define
$$\Phi^{(q)} := \rho_N \circ \Psi = \rho_N \circ \rho_p^{-1} \circ \mu_{q,p} \circ \rho_p,$$
which is a k-homomorphism; so the question becomes: *Is $\Phi^{(q)}\big|_{A'}$ a k-isomorphism?*

If not there exist integers $a_1, \ldots, a_k, a'_1, \ldots, a'_k \in A'$ for which $r \neq 0$ but $r \equiv 0 \pmod{N}$. In this case define $b := \sum_{i=1}^k a_i - \sum_{i=1}^k a'_i \in kA - kA$, so that $qb \equiv r \pmod{p}$ and $b \not\equiv 0 \pmod{p}$. Hence q is of the form $r/b \pmod{p}$ where $r \in (-p, p)$

with $r \neq 0$ and $N|r$, and where $b \in kA - kA$ with $b \not\equiv 0 \pmod{p}$: The number of such q is therefore

$$\leq \#\{r \in (-p,p) : r \neq 0, N|r\} \times \#\{b \in kA - kA : b \not\equiv 0 \pmod{p}\}$$
$$\leq \frac{2(p-1)}{N}|kA - kA| < p - 1.$$

Therefore there must exist values of q, $1 \leq q \leq p-1$ for which $\Phi^{(q)}|_{A'}$ is a k-isomorphism. □

2.5. The Freĭman–Ruzsa theorem. We recall the following result from our previous lecture:

Plünnecke–Ruzsa theorem. *If A and B are finite sets of integers for which $|A + B| \leq C|A|$ then $|kB - lB| \leq C^{k+l}|A|$.*

We are now ready to state and prove our main result:

Freĭman–Ruzsa theorem. *If $A \subset \mathbb{Z}/N\mathbb{Z}$ with $|A + A| \leq C|A|$ then A is contained in a generalized arithmetic progression of dimension $\leq d(C)$ and volume $\leq \nu(C)|A|$.*

Here $d(C)$ and $\nu(C)$ are constants which depend only on the constant C. Following the works of Bilu [3] and Chang [9] we know that we can take

$$d(C) = \lfloor C - 1 \rfloor, \quad \text{and} \quad \nu(C) = e^{O(C^2 (\log C)^3)}.$$

(Note that if $A = B + \{1, \ldots, n\}$ then $2A = 2B + \{2, \ldots, 2n\}$, so if B is a set of m very widely spaced integers then $|A| = mn$ and $|2A| = \binom{m+1}{2}(2n-1) \sim (m+1)|A|$ as $n \to \infty$. Now if A is contained in a generalized arithmetic progression of reasonable volume then it has dimension $\geq m = C - 1$, so we see that this bound on $d(C)$ cannot be much improved in general.)

PROOF. By the Plünnecke–Ruzsa theorem with $k = l = 8$ we have $|8A - 8A| \leq C^{16}|A|$. Therefore, by Ruzsa's lemma, there exists $A' \subset A$ with $|A'| \geq |A|/8$, which is 8-isomorphic to some $B \subset \mathbb{Z}/N\mathbb{Z}$ where N is prime with $2C^{16}|A| < N \leq 4C^{16}|A|$ (such a prime may be selected by Bertrand's postulate). So $|B| = \alpha N$ with $\alpha \geq 1/(32C^{16})$.

By Bogolyubov's theorem, $2B - 2B$ contains a generalized arithmetic progression of dimension $\leq \alpha^{-2}$ and volume $\gamma|A|$, with $1 \geq \gamma \geq (\alpha^2/8)^{\alpha^{-2}}$. Now B is 8-isomorphic to A' so $2B - 2B$ is 2-isomorphic to $2A' - 2A'$. Since any set which is 2-isomorphic to a d-dimensional generalized arithmetic progression is itself a d-dimensional generalized arithmetic progression, hence $2A' - 2A'$, and thus $2A - 2A$, contains a generalized arithmetic progression Q of dimension at most α^{-2} and volume $\gamma|A|$.

Let S be a maximal subset of A for which the sets $s + Q$, $s \in S$ are disjoint, so that $|S + Q| = |S||Q|$. Since S is maximal, if $a \in A$ there exists $s \in S$ and $q_1, q_2 \in Q$ such that $a + q_1 = s + q_2$, and therefore

$$A \subset S + Q - Q \subset Q - Q + \sum_{s \in S}\{0, s\},$$

a generalized arithmetic progression, of dimension $\leq |S| + \alpha^{-2}$ and volume $\leq 2^{|S|+\alpha^{-2}}\gamma|A|$. Now

$$S + Q \subset A + (2A - 2A) = 3A - 2A$$

so that
$$|S| = \frac{|S+Q|}{|Q|} \leq \frac{|3A-2A|}{\gamma|A|} \leq \frac{C^5}{\gamma}$$
by the Plünneke-Ruzsa theorem. Tracing through the above proof, we find that the result follows with volume $\nu(C) = 2^{d(C)}$ where $d(C) = C^{C^{48}}$. □

These bounds can be significantly improved by the following argument of Chang: The big bounds come as a consequence of the enormous size of S. We will improve this by replacing S and Q by S' and Q' where S' is significantly smaller than S, while Q' is a little bigger than Q: Let m be the smallest integer $\geq 2C$. Let $S_0 = S$ and $Q_0 = Q$. For any given $j \geq 0$, if $|S_j| \leq m$ then we stop the algorithm and let $r = j$. Otherwise we select any subset T_j of S_j of size m and let $Q_{j+1} = T_j + Q_j$. Now we select S_{j+1} to be a maximal subset of A for which the sets $s + Q_{j+1}$, $s \in S_{j+1}$ are disjoint, so that $|S_{j+1} + Q_{j+1}| = |S_{j+1}||Q_{j+1}|$. Note that this also implies that $|Q_{j+1}| = |T_j + Q_j| = |T_j||Q_j| = m|Q_j|$ for all j, so that $|Q_r| = m^r|Q|$. On the other hand $Q_{j+1} = T_j + Q_j \subset S_j + Q_j \subset A + Q_j$ for each j, so that $Q_r \subset rA + Q \subset (r+2)A - 2A$, which implies that $|Q_r| \leq C^{r+4}|A|$ by the Plünneke-Ruzsa theorem. Therefore $2^r \leq (m/C)^r \leq C^4|A|/|Q| = C^4/\gamma$ by the last two equations.

Now $A \subset S_r + Q_r - Q_r \subset S_r + \sum_{j=0}^{r-1}(T_j - T_j) + (Q - Q)$, which is a generalized arithmetic progression of dimension $\leq m(r+1) + \alpha^{-2}$, and volume $\leq 3^{m(r+1)+\alpha^{-2}}\gamma|A|$. Tracing through, we find that $r \ll C^{32} \log C$ so that we can take $d(C) = C^{33} \log C$ and $\nu(C) = C^{O(C^{33})}$.

This proof and the example that proceeded it suggest that there might be a result of the following kind (sometimes known as the *polynomial Freĭman-Ruzsa conjecture*): There exists a constant κ such that if $|2A| \leq C|A|$ then there exists a generalized arithmetic progression P of rank $\ll C^\kappa$ and volume $\ll C^\kappa|A|$ for which $|A \cap P| \gg C^{-\kappa}|A|$.

3. Lecture 3. Uniform distribution, Roth's theorem and beyond

3.1. Uniform distribution mod one.
We begin by discussing Hermann Weyl's famous criterion for recognizing uniform distribution mod one: Let $\{t\}$ be the fractional part of t, and $e(t) = e^{2i\pi t}$ so that $e(t) = e(\{t\})$. A sequence of real numbers $a_1, a_2, \ldots$ is *uniformly distributed mod one* if, for all $0 \leq \alpha < \beta \leq 1$ we have
$$\#\{n \leq N : \alpha < \{a_n\} \leq \beta\} \sim (\beta - \alpha)N \quad \text{as } N \to \infty.$$
To determine whether a sequence of real numbers is uniformly distributed we have the following extraordinary, and widely applicable, criterion:

Weyl's criterion. *A sequence of real numbers $a_1, a_2, \ldots$ is uniformly distributed mod one if and only if for every integer $b \neq 0$ we have*

(3.1) $$\left|\sum_{n \leq N} e(ba_n)\right| = o_b(N) \quad \text{as } N \to \infty.$$

In other words $\limsup_{N\to\infty}(1/N)|\sum_{n \leq N} e(ba_n)| = 0$.

Note that if $a_1, a_2, \ldots$ is uniformly distributed mod one then $ka_1, ka_2, \ldots$ is uniformly distributed mod one for all nonzero integers k.

An interesting example is where $a_n = f(n)$ for some polynomial $f(t) \in \mathbb{R}[t]$. It can be shown, using Weyl's criterion, that the sequence $a_1, a_2, \ldots$ is uniformly distributed mod one if and only if one or more of the coefficients of $f(t) - f(0)$ is irrational. Note that if all the coefficients of f are rational then there exists an integer $b > 0$ such that $b(f(t) - f(0)) \in \mathbb{Z}[t]$; but then each $e(bf(n)) = e(bf(0))$, and so (3.1) is not satisfied. If f is linear, that is $f = \gamma n + \delta$ with γ irrational then

$$\sum_{n \leq N} e(ba_n) = e(b\delta) \sum_{n \leq N} e(b\gamma n) = e(b(\gamma + \delta)) \cdot \frac{e(b\gamma N) - 1}{e(b\gamma) - 1},$$

the sum of a geometric progression, since $b\gamma$ is not an integer. Therefore

$$\left| \sum_{n \leq N} e(ba_n) \right| \leq \frac{2}{|e(b\gamma) - 1|} \asymp \frac{1}{\|b\gamma\|} \ll_b 1 = o_b(N)$$

as required, since $|e(t) - 1| \asymp \|t\|$, where $\|t\|$ denotes the distance from t to the nearest integer.

3.2. Uniform distribution mod N. For a given set, A, of residues mod N, define

$$\hat{A}(b) := \sum_{n \in A} e\left(\frac{bn}{N}\right).$$

Let $(t)_N$ denote the least nonnegative residue of $t \pmod{N}$ (so that $(t)_N/N = \{t/N\}$). The idea of uniform distribution mod N is surely something like: For all $0 \leq \alpha < \beta \leq 1$ and all $m \not\equiv 0 \pmod{N}$, we have

(3.2) $$\#\{a \in A : \alpha N < (ma)_N \leq \beta N\} \sim (\beta - \alpha)|A|.$$

One can only make sense of such a definition if $|A| \to \infty$ (since this is an asymptotic formula) but we are often interested in smaller sets A, indeed that are a subset of $\{1, 2, \ldots, N\}$; so we will work with something motivated by, but different from, (3.2). Let us see how far we can go in proving an analogy to Weyl's criterion.

For given subset A of the residues mod N define

$$\text{Error}(A) := \max_{\substack{0 \leq x < x+y \leq N \\ m \not\equiv 0 \pmod{N}}} \left| \frac{\#\{a \in A : x < (ma)_N \leq x+y\}}{|A|} - \frac{y}{N} \right|.$$

Theorem 3.1. *Suppose that N is prime. Fix $\delta > 0$.*
  (i) *If $\text{Error}(A) \leq \delta^2 |A|$ then $|\hat{A}(m)| \ll \delta |A|$ for any $m \not\equiv 0 \pmod{N}$.*
  (ii) *If $|\hat{A}(m)| \leq \delta^2 |A|$ for all $m \not\equiv 0 \pmod{N}$, and $|A| \geq N/e^{c/\delta}$ then $\text{Error}(A) \ll \delta |A|$, for some absolute constant $c > 0$.*

PROOF. For given integer $k \geq 1$, if $(ma)_N \in (x, x + N/k]$ then $e(ma/N) = e(x/N) + O(1/k)$. Therefore

$$\hat{A}(m) = \sum_{j=0}^{k-1} \sum_{\substack{a \in A \\ jN/k < (ma)_N \leq (j+1)N/k}} e(ma/N) = \sum_{j=0}^{k-1} \sum_{\substack{a \in A \\ jN/k < (ma)_N \leq (j+1)N/k}} (e(j/k) + O(1/k))$$

$$= \sum_{j=0}^{k-1} |A| \left(\frac{1}{k} + O(\text{Error}(A)) \right) e(j/k) + O\left(\frac{|A|}{k}\right) \ll |A|(k\,\text{Error}(A) + 1/k).$$

The result follows by taking $k \asymp 1/\delta$.

In the other direction we have, for integers x, y with $0 \leq x < x + y \leq N$

$$\sum_{\substack{a \in A \\ x < (ma)_N \leq x+y}} 1 = \sum_{j=1}^{y} \sum_{a \in A} \frac{1}{N} \sum_{r \pmod{N}} e\left(r\left(\frac{ma - x - j}{N}\right)\right)$$

$$= \frac{y}{N}|A| + \frac{1}{N} \sum_{\substack{r \pmod{N} \\ r \neq 0}} \hat{A}(rm) e\left(\frac{-rx}{N}\right) \sum_{j=1}^{y} e\left(\frac{-rj}{N}\right).$$

If r runs through the nonzero integers in $(-N/2, N/2]$ then

$$\left| e\left(\frac{-rx}{N}\right) \sum_{j=1}^{y} e\left(\frac{-rj}{N}\right) \right| \ll \frac{N}{|r|},$$

and so the second term above is, as $|\hat{A}(-rm)| = |\hat{A}(rm)|$,

$$\ll \sum_{r \neq 0} \frac{|\hat{A}(rm)|}{|r|} \ll \sum_{1 \leq r \leq R} \frac{|\hat{A}(rm)|}{r} + \sum_{R < r \leq N/2} \frac{|\hat{A}(rm)|}{r}$$

$$\leq (\log R + 1) \max_{s \neq 0} |\hat{A}(s)| + \left(\sum_{r \pmod{N}} |\hat{A}(rm)|^2 \right)^{1/2} \left(\sum_{r > R} \frac{1}{r^2} \right)^{1/2}$$

$$\ll (\log R) \delta^2 |A| + (|A|N/R)^{1/2} \ll \delta |A|$$

for $R \approx N/(\delta^2 |A|)$. □

To obtain an analogy to Weyl's criterion we think of an infinite sequence of pairs (A, N) with N prime and $N \to \infty$, where $|A| \gg N$. More precisely we have

Corollary. *For each prime N let A_N be a subset of the residues mod N with $|A_N| \gg N$. Then $\mathrm{Error}(A_N) = o(1)$ if and only if $|\hat{A}_N(m)| = o(N)$ for all $m \not\equiv 0 \pmod{N}$.*

One can therefore formulate an analogy to Weyl's criterion along the lines: The Fourier transforms of A are all small if and only if A and all of its dilates are uniformly distributed. (A *dilate* of A is the set $\{ma : a \in A\}$ for some $m \not\equiv 0 \pmod{N}$.) This idea is central to our recent understanding, in additive combinatorics, for proving that large sets contain 3-term arithmetic progressions; and finding appropriate analogies to this are essential to our understanding when considering k-term arithmetic progressions for $k \geq 3$. More on that later.

To give one example of how such a notion can be used, we ask whether a given set A of residues mod N contains a nontrivial 3-term arithmetic progression? In other words we wish to find solutions to $a + b = 2c$ with $a, b, c \in A$ where $a \neq b$.

Theorem 3.2. *If A is a subset of the residues $\pmod{N}$ where N is odd, for which $|\hat{A}(m)| < |A|^2/N - 1$ whenever $m \not\equiv 0 \pmod{N}$ then A contains nontrivial 3-term arithmetic progressions.*

PROOF. Since $(1/N) \sum_r e(rt/N) = 0$ unless t is divisible by N, whence it equals 1, we have that the number of 3-term arithmetic progressions in A is

$$\sum_{a,b,c \in A} \frac{1}{N} \sum_{r} e\left(\frac{r(a + b - 2c)}{N}\right) = \frac{1}{N} \sum_{r} \hat{A}(r)^2 \hat{A}(-2r).$$

The $r = 0$ term gives $|A|^3/N$. We regard the remaining terms as error terms, and bound them by their absolute values, giving a contribution (taking $m \equiv -2r \pmod{N}$)

$$\leq \frac{1}{N} \sum_r |\hat{A}(r)|^2 \cdot \max_{m \neq 0} |\hat{A}(m)| = |A| \max_{m \neq 0} |\hat{A}(m)|.$$

There are $|A|$ trivial 3-term arithmetic progressions (of the form a, a, a) so we have established that A has nontrivial 3-term arithmetic progressions when

$$|A|^3/N - |A| \max_{m \neq 0} |\hat{A}(m)| > |A|,$$

yielding the result. $\square$

Rather more generally we can ask for solutions to

(3.3) $$ia + jb + kc \equiv l \pmod{N}$$

where $(ijk, N) = 1$ with $a \in A$, $b \in B$, $c \in C$ and $A, B, C \subset \mathbb{Z}/N\mathbb{Z}$. We count the above set as

$$\sum_{\substack{a \in A, b \in B \\ c \in C}} \frac{1}{N} \sum_r e\left(\frac{r(ia+jb+kc-m)}{N}\right) = \frac{1}{N} \sum_r e\left(\frac{-rl}{N}\right) \hat{A}(ir) \hat{B}(jr) \hat{C}(kr).$$

The $r = 0$ term contributes $(1/N)\hat{A}(0)\hat{B}(0)\hat{C}(0) = |A|\,|B|\,|C|/N$. The total contribution of the other terms can be bounded above by

$$\frac{1}{N} \sum_{r \neq 0} |\hat{A}(ir)|\,|\hat{B}(jr)|\,|\hat{C}(kr)|$$

$$\leq \frac{1}{N} \max_{m \neq 0} |\hat{A}(m)| \sum_r |\hat{B}(jr)||\hat{C}(kr)|$$

$$\leq \frac{1}{N} \max_{m \neq 0} |\hat{A}(m)| \left(\sum_t |\hat{B}(t)|^2\right)^{1/2} \left(\sum_u |\hat{C}(u)|^2\right)^{1/2}$$

$$= \frac{1}{N} \max_{m \neq 0} |\hat{A}(m)|(N|B|N|C|)^{1/2} = (|B|\,|C|)^{1/2} \max_{m \neq 0} |\hat{A}(m)|$$

using the Cauchy–Schwarz inequality. Therefore there are $\geq |A|\,|B|\,|C|/2N$ solutions to (3.3) provided

(3.4) $$|\hat{A}(m)| \leq \frac{(|B|\,|C|)^{1/2}}{2N}|A| \quad \text{for every } m \not\equiv 0 \pmod{N}.$$

3.3. Roth's theorem. In 1953, Roth [23] proved that for any $\delta > 0$ if N is sufficiently large then any subset A of $\{1, \ldots, N\}$ with more than δN elements contains a nontrivial 3-term arithmetic progression. We shall prove Roth's theorem in this section.

In 1975 Szemerédi [28] generalized this to obtain nontrivial k-term arithmetic progressions. This was reproved by Furstenberg [12] in 1977, and there have been recent proofs by Gowers [13, 14], Tao, and many others. See the article herein by Tao [32] for an inspiring discussion of these proofs; and Kra's article [20] for developments of Furstenberg's ideas.

In Roth's proof, as we will see below, one can take $\delta \approx 1/\log \log N$. This was improved (but remained unpublished until this volume) by Szemerédi [30] to $\delta \approx 1/\exp(\sqrt{c \log \log N})$. In the late eighties, both Heath-Brown [18] and Szemerédi

[29] showed one can take $\delta \approx 1/(\log N)^c$ for some small $c > 0$. The best result known, due to Bourgain [4], is that one can take

$$\delta \approx \sqrt{\frac{\log \log N}{\log N}}.$$

To start our proof of Roth's theorem we note that the result is easy for $\delta > \frac{2}{3}$ since then A contains a subset of the form $\{a, a+1, a+2\}$. For smaller δ we shall either prove directly that it has 3-term arithmetic progressions by the methods of the previous section, or that there is a large arithmetic progression of length N_1 which contains $\delta_1 N_1$ terms of A with $\delta_1 > \delta(1 + c\delta)$ for some $c > 0$. This can be used to construct a subset A_1 of the first N_1 integers, of size $\delta_1 N_1$, which must have a 3-term arithmetic progression by an appropriate induction hypothesis (as δ_1 is significantly larger than δ), so that A also does.

Replace N by the smallest prime $\geq N$ which can be done with negligible change in our hypothesis. Let us assume that A is a subset of the integers up to N, containing at least δN elements, but which has no three term arithmetic progression. We will suppose that we have proved Roth's theorem for any constant $\delta' > \delta(1 + c\delta)$.

- If $\#\{a \in A : 0 < a < N/3\} \geq (1 + c\delta)|A|/3$ then $A_1 := \{a \in A : 0 < a \leq N/3\}$.
- If $\#\{a \in A : 2N/3 < a < N\} \geq (1 + c\delta)|A|/3$ then $A_1 := \{N - a : a \in A, 2N/3 < a < N\}$.

In these cases $N_1 = [N/3]$, and the result follows from our hypothesis. Otherwise we let $B := \{a \in A : N/3 < a < 2N/3\}$, so that $|B| > (1 - 2c\delta)|A|/3$. There are no solutions to $a + b \equiv 2d \pmod{N}$ with $a \in A$ and $b, d \in B \subset A$, all distinct. For if $b, d \in B$ then $0 < 2d - b < N$ and so $a + b = 2d$, hence $a = b = d$ by our assumption that A has no nontrivial 3-term arithmetic progressions.

This implies that there must exist $m \not\equiv 0 \pmod{N}$ such that $|\hat{A}(m)| > \delta(1 - 2c\delta)|A|/6$ else we have many nontrivial solutions to (3.3) (with $i = j = 1$, $k = -2, l = 0$) by (3.4). But then A is not uniformly distributed mod N; in particular, $\mathrm{Error}(A) \gg \delta^2 |A|$ by Theorem 3.1(i). In other words there is some dilate of A and some long interval which does not contain the expected number of elements of the dilate A; in fact it is out by a constant factor. However we need slightly more than that: We need an interval that has *too many* elements of A by a constant factor and so we make one more observation: Select an integer $l \gg 1/\delta$, and define

$$A_j := \left\{ a \in A : (ma)_N \in \left(\frac{jN}{l}, \frac{(j+1)N}{l} \right] \right\}$$

for $0 \leq j \leq l - 1$, so that if a is counted by A_j then $e(ma/N) = e(j/l) + O(1/l)$. Therefore

$$\hat{A}(m) = \sum_{j=0}^{l-1} \left(\# A_j - \frac{|A|}{l} \right) e\left(\frac{j}{l} \right) + O\left(\frac{|A|}{l} \right),$$

implying that

$$\sum_{j=0}^{l-1} \left| \# A_j - \frac{|A|}{l} \right| \geq \left| \sum_{j=0}^{l-1} \left(\# A_j - \frac{|A|}{l} \right) e\left(\frac{j}{l} \right) \right| \geq \hat{A}(m) - O\left(\frac{|A|}{l} \right) \gg \delta |A|.$$

Adding this to $\sum_j (\# A_j - |A|/l) = 0$, we find that there exists j for which

$$\left(\# A_j - \frac{|A|}{l}\right) \gg \delta \frac{|A|}{l}.$$

What we would like to do now is to define $A' := \{i : [jN/l] + i \in A_j\}$, a subset of $\{1, 2, \ldots, N'\}$ where $N' = [N/l]$, with $|A'| \geq (1 + c\delta)\delta N'$ and then assert that A' contains no nontrivial 3-term arithmetic progressions. To prove this last remark, we proceed by noting that if $u, v, w \in A'$ for which $u + w = 2v$ then there exist $a, b, c \in A$ such that $ma \equiv [jN/l] + u \pmod{N}$, $mb \equiv [jN/l] + v \pmod{N}$, $mc \equiv [jN/l] + w \pmod{N}$ so that $m(a + c - 2b) \equiv u + w - 2v \equiv 0 \pmod{N}$, and therefore $a + c \equiv 2b \pmod{N}$. *However* there is no guarantee that this implies that $a + c = 2b$ (as above), since there may be "wraparound" (that is, $a + c$ might equal $2b \pm N$ or $2b \pm 2N$ or...), and so we need to refine our construction to be able to make this final step.

The trick is to use the well-known result that if $RS = N$ where R and S are real numbers > 1 then there exist integers r and s, with $0 < r < R$ and $0 < s < S$, such that $\pm m \equiv s/r \pmod{N}$. (PROOF. There are more than N integers of the form $j + im$ with $0 \leq i < R$ and $0 \leq j < S$, so two must be congruent mod N. Thus their difference $s \pm rm \equiv 0 \pmod{N}$.) For convenience we will assume $m \equiv s/r \pmod{N}$ where $R = \sqrt{N/\delta^3}, S = \sqrt{N\delta^3}$, with $x = [jN/l]$ and $y = [N/l]$ and $l \asymp 1/\delta$, so that

$$\#\{a \in A : x < (ma)_N \leq x + y\} \geq (1 + c\delta)\delta y.$$

We begin by partitioning this set depending only on the value of $(ma)_N \pmod{s}$: For $1 \leq i \leq s$ let $\alpha_i = ((x+i)/m)_N$, and then define

$$A_i := \left\{a \in A : a \equiv \alpha_i + jr \pmod{N} \text{ and } 0 \leq j \leq \left[\frac{y-i}{s}\right]\right\}.$$

Note that $ma \equiv m(\alpha_i + jr) \equiv x + (i + js)$ so that $x < (ma)_N \leq x + y$ for $a \in A$. Therefore there exists some value of i for which $\# A_i \geq (1 + c\delta)\delta y/s$. Even within A_i we still have the possibility of the "wraparound problem"; so we deal with this by partitioning A_i:

Let $K = [(\alpha_i + ry/s)/N]$ so that $\alpha_i \leq \alpha_i + jr \leq \alpha_i + ry/s < (K+1)N$. For each $0 \leq k \leq K$ define

$$A_{i,k} := \{a \in A_i : kN < \alpha_i + jr \leq (k+1)N\}.$$

Let $\alpha_{i,0} = \alpha_i - r$, and let $\alpha_{i,k}$ be the largest integer $\leq kN$ which is $\equiv \alpha_i \pmod{r}$ for $1 \leq k \leq K$. Then $A_{i,k} = \{a \in A_i : a = \alpha_{i,k} + jr, 1 \leq j \leq J_k + O(1)\}$ where $J_0 = N/r - \alpha_i/r$, $J_k = N/r$ for $1 \leq k \leq K - 1$, and $J_K = y/s - KN/r + \alpha_i/r$. We let T be the set of indices k, $1 \leq k \leq K - 1$ together with $k = 0$ provided $J_0 > c\delta^2 y/4s$, and with $k = K$ provided $J_K > c\delta^2 y/4s$. Note that

$$\sum_{k \in T} \# A_{i,k} \geq \# A_i - c\delta^2 y/2s \geq (1 + c\delta/2)\delta y/s \geq (1 + c\delta/2)\delta \sum_{k \in T} J_k.$$

Thus there exists $k \in K$ such that $\# A_{i,k} \geq (1 + c\delta/2)\delta J_k$. Now define $N' = [J_k]$ and $A' = \{j : 1 \leq j \leq N', \alpha_{i,k} + jr - kN \in A\}$, a subset of $\{1, 2, \ldots, N'\}$, so that $\# A' = \# A_{i,k} \geq (1 + c\delta/2)\delta N'$. We claim that A' does not contain any nontrivial 3-term arithmetic progressions; else if $u + v = 2w$ with $u, v, w \in A'$ then $a = \alpha_{i,k} + ur - kN, b = \alpha_{i,k} + vr - kN, c = \alpha_{i,k} + wr - kN \in A$ and $a + b = 2c$, contradicting the fact that A does not contain any nontrivial 3-term arithmetic progressions. Note that $N' \geq \min\{N/r, c\delta^2 y/4s\} \gg \min\{N/R, \delta^2 N/lS\} \gg \sqrt{\delta^3 N}$.

We have obtained the induction hypothesis that we wanted. If we iterate we find that we increase the constant $\delta = 2^{-n}$ to $2\delta = 2^{-(n-1)}$ in $\asymp 1/\delta$ iterations by which time the size of our set is roughly 2^{-3n} times N to the power $(\frac{1}{2})^{2^{n+O(1)}}$. Thus when we get all the way up to $\delta = 1$ the size of our set is N to the power $(\frac{1}{2})^{2^{n+O(1)}}$. To ensure that this is not negligible we must have $2^{2^{n+O(1)}} = o(\log N)$; that is $2^n \ll \log \log N$ and so $\delta = 2^{-n} \gg 1/\log \log N$.

In the other direction we have

Behrend's theorem. *There exists a subset $A \subset \{1, \ldots, N\}$ with $\#A \geq N/\exp(c\sqrt{\log N})$, such that A has no nontrivial 3-term arithmetic progression.*

PROOF. Let $T := \{(x_0, \ldots, x_{n-1}) \in \mathbb{Z}^n : 0 \leq x_i < d\}$ and $T_k := \{\mathbf{x} \in T : |\mathbf{x}|^2 = k\}$. We have $|T| = d^n$, and $|\mathbf{x}|^2 < nd^2$ for every $\mathbf{x} \subset T$, so there exists a positive integer k for which T_k has $\geq d^{n-2}/n$ elements. Let

$$A := \{x_0 + x_1(2d) + \cdots + x_{n-1}(2d)^{n-1} : \mathbf{x} \in T_k\}.$$

If $a + b = 2c$ with $a, b, c \in A$ then $a_0 + b_0 \equiv 2c_0 \pmod{2d}$ and $-2d < a_0 + b_0 - 2c_0 < 2d$ so that $a_0 + b_0 = 2c_0$; similarly one proves that $a_1 + b_1 = 2c_1$, and indeed $a_i + b_i = 2c_i$ for each $i \geq 0$. But then $\mathbf{a} + \mathbf{b} = 2\mathbf{c}$, that is $\mathbf{a}, \mathbf{b}, \mathbf{c} \in T_k$ are collinear, which is impossible as T_k is a sphere! Therefore A contains no nontrivial 3-term arithmetic progressions.

The elements of A are all $\leq (d-1)(1 + 2d + \cdots + (2d)^{n-1}) < N := 2^{n-1}d^n$. The result follows by taking $n \approx \sqrt{\log N}$ and $d = [(2N)^{1/N}/2]$. □

For each integer $N \geq 1$, define $R(N)$ to be the size of the largest subset A of $\{1, \ldots, N\}$ which does not contain any nontrivial 3-term arithmetic progressions. We know, after Behrend and Bourgain, that

$$N\sqrt{\frac{\log \log N}{\log N}} \gg R(N) \gg \frac{N}{\exp(c\sqrt{\log N})};$$

the question is whether $R(N)$ is really near to one of these bounds, or somewhere in-between. There does not seem to be any convincing heuristic to predict the truth; at the school we asked the lecturers to all venture a guess — it seemed that people's intuitions varied substantially! It would be most exciting if one could prove that $R(N) \leq (1-\epsilon)N/\log N$ for sufficiently large N since this would give an "automatic proof" that there are infinitely many three term arithmetic progressions of primes.

Finally we prove the following slight strengthening of Roth's theorem

Varnavides' theorem. *Fix $1 \geq \delta > 0$. There exist constants $C(\delta) > 0$ and $N(\delta)$ such that if $N \geq N(\delta)$ and $A \subset \{1, \ldots, N\}$ with $\#A \geq \delta N$, then A has at least $C(\delta)N^2$ 3-term arithmetic progressions.*

PROOF. By Roth's theorem we know that there exists an integer M such that any set of $\delta M/2$ integers from an arithmetic progression of length M contains a nontrivial three term arithmetic progression. We will apply this result to the subset of A lying in each arithmetic progression of length M taken from the integers in $\{1, \ldots, N\}$. Let $\mathcal{P}(b, d)$ be the arithmetic progression $b, b+d, \ldots, b+(M-1)d$, for $1 \leq b \leq N - (M-1)d$, and let $A(b, d)$ be the number of elements of A in $\mathcal{P}(b, d)$.

Since every element of A from the interval $((M-1)d, N-(M-1)d]$ is counted in exactly M of these arithmetic progressions, we deduce that $\sum_{1 \leq b \leq N-(M-1)d} A(b,d) \geq M(\delta N - 2(M-1)d)$. Since each $A(b,d) \leq M$, we can deduce that there are $\geq \delta N/2$ values of b for which $A(b,d) \geq \delta M/2$, provided $N \geq 6(M-1)d/\delta^2$. Now, each of these contains a nontrivial three term arithmetic progression, making for a total of $\geq \delta^3 N^2/12(M-1)$ nontrivial three term arithmetic progressions, when we consider all $d \leq \delta^2 N/6(M-1)$, though many of these may have been counted more than once. Now if $a, a+D, a+2D$ is counted in some $A(b,d)$ then d divides D and $2D/d \leq M-1$; and it is counted in $A(b,d)$ for no more than $M - 2D/d$ values of b. Writing $D/d = h$, we find that $a, a+D, a+2D$ has been counted no more than $\leq \sum_{1 \leq h \leq M/2}(M-2h) \leq (M/2)^2$ times. Therefore A contains $\geq \delta^3 N^2/3M^3$ distinct nontrivial three term arithmetic progressions. $\square$

By Bourgain's result we may take $M = (1/\delta)^{c/2\delta^2}$ for some constant $c > 0$, and therefore $C(\delta) \geq \delta^{c/\delta^2}$ in Varnavides' theorem. A small modification of the proof of Behrend's theorem implies that $C(\delta) \leq \delta^{c' \log(1/\delta)}$, for some constant $c' > 0$.

3.4. Large Fourier coefficients. We saw in the previous section that proving Roth's theorem is difficult only in the case that there are large Fourier coefficients, $\hat{A}(m)$, with $m \not\equiv 0 \pmod{N}$. It is worth noting a few other results which reflect consequences of having large Fourier coefficients:

An easy one to prove is that for any $\eta > 0$ there exists $\delta > 0$ such that

$$r_{A-A}(n) > (1-\eta)|A| \quad \text{if and only if} \quad \sum_{m:|(mn)_N| \leq \epsilon N} |\hat{A}(m)|^2 \geq (1-\delta) \sum_m |\hat{A}(m)|^2.$$

(See Lecture 1 for further discussion of $r_{A-A}(n)$, the number of representations of n as $a - a'$ with $a, a' \in A$.)

A manifestation of the *uncertainty principle* (which roughly states that a nontrivial function and its Fourier transforms cannot all be too small) is given by: If $A \subset \mathbb{Z}/N\mathbb{Z}$ has no elements in $(x-L, x+L)$ then there exists m, $0 < m < (N/L)^2$ such that $|\hat{A}(m)| \geq (L/2N)|A|$.

In many proofs it is important to know how often $|\hat{A}(m)|$ can be large? Let $R := \{r \pmod{N} : |\hat{A}(r)| > \rho|A|\}$. From Parseval's identity we see that

$$|A|N = \sum_m |\hat{A}(m)|^2 \geq \sum_{m \in R} \rho^2 |A|^2,$$

so that $|R| \leq \rho^{-2} N/|A|$. Note that if $r, s \in R$ then this says that the numbers $(ra)_N$, $a \in A$ and $(sa)_N$, $a \in A$ have a bias towards being close to certain values x and y respectively. In that case we might expect that the numbers $((r+s)a)_N$, $a \in A$ have a bias towards $(x+y)_N$ so that $r + s \in R$. Therefore we might expect that R has some lattice structure, an intuition that is verified by Chang's result [9] that R is contained in a cube of dimension $\leq 2\rho^{-2} \log(N/|A|)$ (cubes, that is sets of numbers $\{\sum_{s \in S} \epsilon_s s : \epsilon_s \in \{-1, 0, 1\}\}$ for given S, were discussed in the previous two lectures.)

3.5. Four term arithmetic progressions. One can prove (using the proof of Theorem 3.2) that if $A \gg N$ and $\hat{A}(m) = o(N)$ for all $m \not\equiv 0 \bmod N$ then A has $\sim |A|^3/N$ 3-term arithmetic progressions. This leads one to ask:

What about 4-term arithmetic progressions?

Does $\hat{A}(m) = o(N)$ imply that A has $\sim |A|^4/N^2$ 4-term arithmetic progressions (that is, the expected number)? As an example consider the set

$$A_\delta := \left\{ n \pmod{N} : \left\| \frac{n^2}{N} \right\| < \frac{\delta}{2} \right\}$$

for N prime. For $J = \delta N/2$ we have

$$\hat{A}_\delta(m) = \sum_{n \pmod{N}} e\left(\frac{mn}{N}\right) \sum_{-J<j<J} \frac{1}{N} \sum_{r \pmod{N}} e\left(r\frac{(j-n^2)}{N}\right)$$

so that

$$|\hat{A}_\delta(m)| \leq \frac{1}{N} \sum_{r \pmod{N}} \sum_{-J<j<J} e\left(\frac{rj}{N}\right) \sum_{n \pmod{N}} e\left(\frac{mn - rn^2}{N}\right).$$

Now $\sum_n e(mn/N) = 0$ if $m \neq 0$, and $= N$ if $m = 0$; and if $r \neq 0$ then $\sum_n e((mn - rn^2)/N)$ is a Gauss sum and so has absolute value $\sqrt{N}$. Moreover $|\sum_{-J \leq j \leq J} e(rj/N)| \ll N/|r|$ for $1 \leq |r| \leq N/2$. Inputting all this into the equation above we obtain $|\hat{A}_\delta(m)| \ll \sqrt{N} \log N$ for each $m \not\equiv 0 \pmod{N}$ and $\# A_\delta = |\hat{A}_\delta(0)| = \delta N + O(\sqrt{N} \log N)$. It follows from the proof of Theorem 3.2 that A_δ has $\sim \delta^3 N^2$ 3-term arithmetic progressions $a, a+d, a+2d$. Now

$$(a+3d)^2 = 3(a+2d)^2 - 3(a+d)^2 + a^2$$

so if $a, a+d, a+2d \in A_\delta$ then $\|(a+3d)^2/N\| < 7\delta/2$, and hence $a, a+d, a+2d, a+3d \in A_{7\delta}$. But this implies that $A_{7\delta}$ has $\geq \{1+o(1)\}\delta^3 N^2$ 4-term arithmetic progressions far more than the expected, $\sim (7\delta)^4 N^2$, once δ is sufficiently small.

Thus we have shown, from this example, that in order to prove that a set of residues of positive density has the expected number of 4-term arithmetic progressions it is insufficient to simply assume that all of the Fourier transforms are small. What else we need to assume is at the heart of the subject of additive combinatorics — see Ben Green's article in these proceedings [16].

Acknowledgments. Thanks to Jason Lucier and Imre Ruzsa for their careful reading of these notes.

References

1. A. Balog, *Many additive quadruples*, in this book.
2. A. Balog and E. Szemerédi, *A statistical theorem of set addition*, Combinatorica **14** (1994), no. 3, 263–268.
3. Yu. Bilu, *Structure of sets with small sumsets*, Astérisque **258** (1999), 77–108.
4. J. Bourgain, *On triples in arithmetic progression*, Geom. Funct. Anal. **9** (1999), no. 5, 968–984.
5. ———, *Mordell's exponential sum estimate revisited*, J. Amer. Math. Soc. **18** (2005), 477–499.
6. J. Bourgain and M.-C. Chang, *Exponential sum estimates over subgroups and almost subgroups of $\mathbb{Z}_Q^*$, where Q is composite with few prime factors*, Geom. Funct. Anal. **16** (2006), 327–366.
7. J. Bourgain, A. A. Glibichuk, and S. V. Konyagin, *Estimates for the number of sums and products and for exponential sums in fields of prime order*, J. London Math. Soc. (2) **73** (2006), 380–398.
8. J. Bourgain, N. Katz, and T. Tao, *A sum-product estimate in finite fields and their applications*, Geom. Funct. Anal. **14** (2004), no. 1, 27–57.
9. M.-C. Chang, *A polynomial bound in Freĭman's theorem*, Duke Math. J. **113** (2002), 399–419.
10. ———, *Some problems related to sum-product theorems*, in this book.

11. G. A. Freĭman, *Foundations of a structural theory of set addition*, Transl. Math. Monogr., vol. 37, Amer. Math. Soc., Providence, R.I., 1973.
12. H. Furstenberg, *Ergodic behavior of diagonal measures and a theorem of Szemerédi on arithmetic progressions*, J. Analyse Math. **31** (1977), 204–256.
13. W. T. Gowers, *A new proof of Szemerédi's theorem for arithmetic progressions of length four*, Geom. Funct. Anal. **8** (1998), no. 3, 529–551.
14. _____, *A new proof of Szemerédi's theorem*, Geom. Funct. Anal. **11** (2001), no. 3, 465–588.
15. Ben Green, *Structure theory of set addition*, available at http://www-math.mit.edu/~green/icmsnotes.pdf.
16. _____, *Quadratic Fourier analysis*, in this book.
17. H. Halberstam and Roth K., *Sequences*, Springer-Verlag, London, 1966.
18. D. R. Heath-Brown, *Integer sets containing no arithmetic progressions*, J. London Math. Soc. (2) **35** (1987), no. 3, 385–394.
19. H. Helfgott, *Growth and generation in* $SL_2(\mathbb{Z}/p\mathbb{Z})$, Ann. of Math. (2), to appear.
20. B. Kra, *Ergodic methods in combinatorial number theory*, in this book.
21. P. Kurlberg, *Bounds on exponential sums over small multiplicative subgroups*, in this book.
22. H. Plünnecke, *Eigenschaften und Abschätzungen von Wirkingsfunktionen*, BMwF-GMD-22, Gesellschaft für Mathematik und Datenverarbeitung, Bonn, 1969.
23. K. F. Roth, *On certain sets of integers*, J. London Math. Soc. **28** (1953), 104–109.
24. I. Z. Ruzsa, *An application of graph theory to additive number theory*, Sci. Ser. A Math. Sci. (N.S.) **3** (1989), 97–109.
25. _____, *Generalized arithmetical progressions and sumsets*, Acta Math. Acad. Sci. Hungar. **65** (1994), no. 4, 379–388.
26. _____, *Cardinality questions about sumsets*, in this volume.
27. J. Solymosi, *On the number of sums and products*, Bull. London Math. Soc. **37** (2005), no. 4, 491–494.
28. E. Szemerédi, *On sets of integers containing no k elements in arithmetic progression*, Acta Arith. **27** (1975), 199–245.
29. _____, *Integer sets containing no arithmetic progressions*, Acta Math. Acad. Sci. Hungar. **56** (1990), no. 1-2, 155–158.
30. E. Szemerédi, *An old new proof of Roth's theorem*, in this book.
31. T. C. Tao and V. H. Vu, *Additive combinatorics*, Cambridge Stud. Adv, Math, vol. 105, Cambridge Univ. Press, Cambridge, 2006.
32. T. Tao, *The ergodic and combinatorial approaches to Szemerédi's theorem*, in this book.
33. H. Weyl, *Über ein Problem aus dem Gebeit der diophantischen Approximationen*, Nachr. Ges. Wiss. Göttingen Math.-Phys. Kl. **1914**, 234–244.

DÉPARTMENT DE MATHÉMATIQUES ET DE STATISTIQUE, UNIVERSITÉ DE MONTRÉAL, C.P. 6128, SUCC. CENTRE-VILLE, MONTRÉAL, QC H3C 3J7, CANADA

E-mail address: andrew@dms.umontreal.ca

Elementary Additive Combinatorics

József Solymosi

1. Introduction

This note is the extended version of the introductory lectures I gave at the Montreal Workshop on Additive Combinatorics. The reader is invited to solve the exercises.

2. Affine cubes

Definition 1. We say that B_d is an affine cube of reals with dimension d, if there are real numbers $x_0, x_1, \ldots, x_d$, such that

$$B_d = \left\{ x_0 + \sum_{i \in I} x_i \ \Big| \ I \subset [1, 2, \ldots, d] \right\}.$$

Exercise 1. Prove that if $|B_d| < 2^d$ then B_d contains a 3-term arithmetic progression.

Affine cubes of integers were introduced by Hilbert [16], who proved that for any partition of the integers into finitely many classes, one class will always contain arbitrary large affine cubes of integers. His result was extended in various directions, the most famous of which are Schur's theorem [23] and van der Waerden's theorem [31]. We will discuss these but first we prove the following extension of Hilbert's theorem.

Theorem 1. *For every d there is a $\delta > 0$ and a threshold $n_0 = n_0(d)$ such that for any set of reals, A, if the sumset of A is small, that is $|A + A| \leq |A|^{1+\delta}$ and $|A| \geq n_0$, then A contains an affine cube with dimension d.*

We describe the proof with Figure 1.

Exercise 2. How small should $|A + A|$ be to guarantee a B_2 in A?

Exercise 3. Express δ and n_0 as functions of d.

2000 *Mathematics Subject Classification.* Primary 05D05, 05D10.

This research is partially supported by NSERC and OTKA grants and a Sloan Research Fellowship.

This is tye final form of the paper.

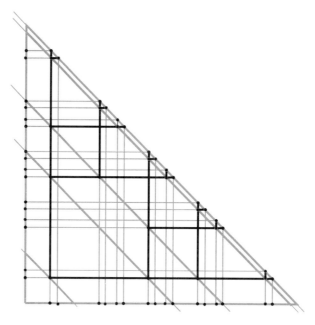

FIGURE 1. If the sumset is small, then the Cartesian product, $A \times A$, can be covered by a few lines only, of slope -1. Select one, which covers the most points of the Cartesian product. Consider the points lying on this line and the smaller Cartesian product defined by these points. Repeat the process, until the most popular line with slope -1 contains only one point of the last Cartesian product. Going back from this point along the nested Cartesian products we get a binary tree. It is easy to check that the projection of the tree onto the horizontal (or vertical) line determines an affine cube in A.

The bound one can get from our "visual" proof is not far from the best possible. For reference, we state the following bounds from [15]. For $d \geq 3$ there exists $n_0 \leq \left((2^d - 2)/\ln 2\right)^2$ so that for every $n \geq n_0$, if $A \subset [1, n]$ satisfies

(2.1) $$|A| \geq 2n^{1-1/2^{d-1}}$$

then A contains an affine cube with dimension d. On the other hand there exists a $B \subset [1, n]$ which contains no affine cube with dimension d, such that

(2.2) $$|B| \geq \tfrac{1}{8} n^{1-d/2^{d-1}}.$$

Exercise 4. Prove that there exists a $B \subset [1, n]$ which contains no affine cube with dimension d, such that

$$|B| \geq cn^{1-d/2^{d-1}},$$

for a universal $c > 0$.

While proving the existence of large affine cubes in dense subsets of integers is easy, showing the lack of affine cubes in certain sets seems to be a difficult task. For example, we state the following conjecture.

Conjecture 2. *There is a finite number, d, such that the set of perfect squares doesn't contain an affine cube with dimension greater than d.*

If true, then Conjecture 2 together with Theorem 1 would imply a conjecture of Imre Ruzsa.

Conjecture 3 (Ruzsa). *There is a positive real $\delta > 0$ such that for any finite subset of perfect squares, A, the sumset is large, $|A + A| \geq |A|^{1+\delta}$.*

Conjecture 2 is supported by the next observation. A large affine cube in the squares would rise to many rational points on the curve
$$y^2 = (x^2 + a)(x^2 + b)(x^2 + a + b).$$
However this curve has genus 2, and the Uniformity conjecture [4] (if true) gives a universal upper bound on the number of rational points, independent of a and b. For more details about these conjectures we refer to the paper of Javier Cilleruelo and Andrew Granville in this volume [5].

Definition 2. We say that Γ is an IP-set or Hindman set, if there is an infinite sequence of natural numbers $n_1, n_2, \ldots$ such that
$$\Gamma = \left\{ \sum_{i \in I} n_i \,\bigg|\, I \subset \mathbb{N}, |I| < \infty \right\}.$$

Sometimes we use the finite version, Γ_d (which is an affine cube with $x_0 = 0$):
$$\Gamma_d = \left\{ \sum_{i \in I} n_i \,\bigg|\, I \subset [1, \ldots, d] \right\}.$$

Theorem 4 (Hindman's theorem [17]). *If $\mathbb{N}$ is partitioned into finitely many classes, then one class contains an IP set.*

The finite version of Hindman's theorem, that one class contains an IP$_d$-set, Γ_d, for any d, was proved by Folkman [7] using elementary methods.

The simplest nontrivial case of Folkman's theorem is when $d = 2$.

Theorem 5 (Schur's theorem [23]). *For every natural number k, there is an n such that if $[1, n]$ is partitioned into k classes, then one class contains x, y, and $x + y$. (That is, in any k-coloring of the integers, there is a monochromatic solution to the equation $x + y = z$.)*

PROOF. Given a k-coloring (partitioning) of $[1, n]$, let us define a k-coloring of the edges of the complete graph on n vertices, K_n. The edge between two vertices v_i, v_j, $0 < i < j \leq n$, is colored by the color of $j - i$. A monochromatic solution of the equation $x + y = z$ is equivalent to a monochromatic triangle in K_n. All that we need to show is that for every natural number, k, there is an $n_0 = n_0(k)$ such that if $n \geq n_0$, then K_n contains a monochromatic triangle. We prove it by induction. If $k = 1$ and $n = 3$ then there is a monochromatic triangle. Let us suppose that for k there exists an $n_0(k)$. Set $n_0(k + 1) = (k + 1)n_0(k) + 1$. Choose an arbitrary vertex v in $K_{n_0(k+1)}$. Select the most popular color, s, among the edges incident to v. At least $n_0(k)$ vertices are connected by a color s edge to v. Among the $n_0(k)$ vertices any edge of color s would give a monochromatic triangle with v. On the other hand, if color s is missing from this subgraph then there is a monochromatic triangle by the induction hypothesis. □

The proof of Folkman's theorem is not much more complicated. Instead of proving it directly, we prove a variant which allows an easier induction argument.

Theorem 6. *For every d and k there is a number $n_0(k,d)$ such that the following is true. If $n \geq n_0$ and $[1,n]$ is k-colored, then one can find a IP_d set, $\Gamma_d \subset [1,n]$,*

$$\Gamma_d = \left\{ \sum_{i \in I} n_i \ \bigg| \ I \subset [1,\ldots,d] \right\},$$

with the property that the color of any element, $\sum_{i \in I} n_i$, depends only on the largest element of I.

For example, $n_1 + n_2 + n_3$, $n_1 + n_3$, $n_2 + n_3$, and n_3 all have the same color. An IP_d set with this coloring property is called weakly monochromatic in k colors. Every weakly monochromatic IP_d set in k colors contains a monochromatic $\mathrm{IP}_{\lceil d/k \rceil}$ set, so proving Theorem 6 gives a proof of Folkman's theorem. We will prove Theorem 6 in the next section, because we are going to use van der Waerden's theorem in the induction step.

Conjecture 7 (Hindman). *For every natural number k, there is an n such that if the set of the first n natural numbers, $[1,n]$, is partitioned into k classes, then one class contains x, y, xy and $x+y$.*

3. van der Waerden's theorem

In 1927 van der Waerden proved a conjecture, which is another extension of Hilbert's theorem.

Theorem 8 (van der Waerden's theorem). *If $\mathbb{N}$ is partitioned into finitely many classes, then one class contains arbitrary long arithmetic progressions.*

The theorem has simple elementary proofs, see, e.g., [15], however here we present a topological proof, to illustrate connections to topological dynamics.

3.1. A taste of topological dynamics. Let us remind the reader that a topological space is compact if each of its open covers has a finite subcover. A metric space is compact if and only if it is complete, i.e., every Cauchy sequence in it converges, and is totally bounded, i.e., for every radius $r > 0$ there exist a set of finitely many open balls of radius r covering the space. A compact dynamical system consists of a compact metric space X and a continuous function $f \colon X \to X$. The field of topological dynamics studies the properties of compact dynamical systems. The proof of Theorem 4, Hindman's theorem, is one of the early applications of topological dynamics to additive combinatorics. Here is another classical theorem, which provides an alternative way to prove van der Waerden's theorem.

Theorem 9 (Furstenberg and Weiss [9]). *X is a compact metric space, and T is a continuous map, $T \colon X \to X$. For any natural number k and $\epsilon > 0$ there exist a natural number n and $x \in X$ such that $\mathrm{dist}(x, T^{in}x) < \epsilon$ for any $1 \leq i \leq k$.*

In order to use the theorem to prove van der Waerden, we have to define X and T based on the partitions of integers. Let us consider the t-partitions of the integers.

$$\Omega = \{f : \mathbb{Z} \to [t]\},$$

where T is the "shift operator,"
$$Tf : Tf(n) = f(n+1).$$
The distance is defined by $\text{dist}(f,g) = 1/(k+1)$, where $k \geq 0$ is the smallest integer for which $f(k) \neq g(k)$.

Given a t-partition, $x \in \Omega$, of the integers, we want to show that a partition class contains k-term arithmetic progression; in other words that there exist integers a and b such that $x(a) = x(a+d) = \cdots = x(a + (k-1)d)$. We let $X = \text{cl}(x)$ be the topological closure of the set $\{T^u x \mid u \in \mathbb{Z}\}$.

Exercise 5. Prove that X is a compact metric space and that T is continuous.

It follows from Theorem 9, that for some $x' \in X$, there is an n such that $\text{dist}(x', T^{in}x') < 1$ for any $1 \leq i \leq k$. That is, $x'(0) = x'(n) = \cdots = x'(kn)$. Since $x' \in \text{cl}(x)$, therefore for any $\delta > 0$ there is an integer u such that
$$\text{dist}(x', T^u x) \leq \delta.$$
In particular, there is a $u \in \mathbb{Z}$ such that $\text{dist}(x', T^u x) \leq 1/(kn+1)$. This means that x' and $T^u x$ are identical on the interval $[-kn, kn]$. Thus $T^u x(0) = T^u x(n) = \cdots = T^u x(kn)$, or equivalently, $x(u) = x(u+n) = \cdots = x(u+kn)$. We conclude that the elements of the k-term arithmetic progression $u, u+n, \ldots, u+kn$ are in one partition class.

Exercise 6. Use van der Waerden's theorem to prove Theorem 9. (That is, show that the two theorems are equivalent.)

Definition 3. A subset of natural numbers, $X \subset \mathbb{N}$, is an AP-set if it contains arbitrarily long arithmetic progressions.

Reminder. A set U is an ultrafilter on $\mathbb{N}$ if
(a) The empty set is not an element of U.
(b) If $A, B \subset N$, $A \subset B$, and $A \in U$, then $B \in U$.
(c) If $A, B \in U$, then $A \bigcap B \in U$.
(d) For any $A \subset \mathbb{N}$, either A or $\bar{A}$ is an element of U.

Exercise 7. Show that there is an ultrafilter on $\mathbb{N}$ consisting only of AP-sets.

The interested reader can learn more about topological and ergodic methods in additive combinatorics from the article of Bryna Kra in this book [18].

3.2. Back to combinatorics. It is very difficult to find reasonable bounds for van der Waerden's theorem. Here we show that one needs at least $c \log \log p$ colors to color $\mathbb{Z}_p$ in order to avoid a 3-term arithmetic progression (mod p). The best known bounds follow from bounds on Roth's theorem. For the details check Andrew Granville's note in this book [14].

Theorem 10. *There is a $c > 0$ such that if the elements of $\mathbb{Z}_p$ are colored using no more than $c \log \log p$ colors, then there is a monochromatic 3-term arithmetic progression in $\mathbb{Z}_p$.*

PROOF. Suppose that we are given a coloring of $\mathbb{Z}_p$ by L colors. We now show that if $L \leq c \log \log p$ then there is a monochromatic 3-term arithmetic progression, AP3: Let $\mathcal{C}_0$ be the largest color-class, so that $|\mathcal{C}_0| \geq p/L$. There is a d, such that for at least $\binom{p/L}{2}/p$ pairs of elements $a, b \in \mathcal{C}_0$, we have $b - a = d$. Define T_1 as the set of the third elements of the 3-term arithmetic progressions defined by the pairs, $a, b \in \mathcal{C}_0$;

$$T_1 = \{c : c = b + d, a, b \in \mathcal{C}_0, b - a = d\}.$$

If $T_1 \cap \mathcal{C}_0 \neq \emptyset$, then there is a monochromatic AP3. Let us define T_{i+1} recursively. Find the most popular color, $\mathcal{C}_i$, in T_i. $|\mathcal{C}_i| \geq |T_i|/L$. There is a d, such that for at least $\binom{|T_i|/L}{2}/p$ pairs of elements $a, b \in \mathcal{C}_i$, we have $b - a = d$. Define T_{i+1} as the set of the third elements of the 3-term arithmetic progressions defined by the pairs, (a, b);

$$T_{i+1} = \{c : c = b + d, a, b \in \mathcal{C}_i, b - a = d\}.$$

If $\mathcal{C}_{i+1}$ has the same color as any earlier $\mathcal{C}_j$, then there is a monochromatic AP3. After $c \log \log p$ steps we have run out of colors but there are still elements to choose. □

Exercise 8. We are given an infinite sequence of natural numbers, $S = \{n_1, n_2, \ldots\}$, such that $1 \leq n_{i+1} - n_i \leq 1000$ for any index $i \geq 1$. Prove that S contains arbitrary long arithmetic progressions.

PROOF OF THEOREM 6. The case $d = 1$ is trivial for any k. Let us suppose that $n_0(k, d)$ exists for a $d \geq 1$ and a $k \geq 1$. We show that then $n_0(k, d+1)$ also exists. It follows from van der Waerden's theorem that one can set $n_0(k, d+1)$ so large that any coloring of $[1, n_0(k, d+1)]$ by k colors gives a monochromatic arithmetic progression of length $n_0(k, d)$. For a given coloring let $a + i\Delta$ denote the monochromatic arithmetic progression, for $0 \leq i \leq n_0(k, d)$. The arithmetic progression $i\Delta$ $(0 \leq i \leq n_0(k, d))$ contains a weakly monochromatic IP_d set by the induction hypothesis. Extend the generator set of this IP_d set with $n_{d+1} = a$. Then

$$\Gamma_{d+1} = \left\{ \sum_{i \in I} n_i \;\middle|\; I \subset [1, \ldots, d+1] \right\}$$

is a weakly monochromatic IP_{d+1} set. □

4. Density versions

There is a density version of van der Waerden's theorem, conjectured by Erdős and Turán [6]. This is a central theorem in additive combinatorics. It has several proofs, combinatorial, ergodic, and analytical proofs, but all proofs are quite involved. (The reader can find several distinct proofs in this book. In particular Terry Tao's article explains the connections between different proof techniques.) The theorem was proved by Endre Szemerédi [26]. His proof was elementary. Later Hillel Furstenberg [8] found a new proof using ergodic theory, and recently Tim Gowers [10] proved the theorem using Fourier analysis. Very recently a new combinatorial proof was discovered, we will see more details about this proof below.

Theorem 11 (Szemerédi [26]**).** *Let k be a positive integer and let $c > 0$. There exists a positive integer $n_0 = n_0(k, c)$ such that every subset S of $[1, n_0]$ of size at least cn_0 contains an arithmetic progression of length k.*

Even the $k = 3$ case is difficult. It was first proved by Klaus Roth [21]. The magnitude of $n_0(3, c)$ is not known. The best upper bound is due to Jean Bourgain [3]. If the density of S is at least $\sqrt{\log \log n_0 / \log n_0}$, then there is a 3-term arithmetic progression in S. The lower bound hasn't been improved since 1946.

A construction of Felix Behrend [2] shows that there is an S with density at least $e^{-c\sqrt{\log n_0}}$, for any n_0, which contains no 3-term arithmetic progression.

The next exercise is not easy. The solution can be found in [13, p. 46].

Exercise 9. Use Theorem 1 to prove that every subset of integers of positive density contains a 3-term arithmetic progression.

For a graph $G = (V, E)$ and two disjoint sets $V_1, V_2 \subset V$, we denote by $E(V_1, V_2)$ the set of edges with one endpoint in V_1 and one endpoint in V_2. The density $d(V_1, V_2)$ is given by
$$d(V_1, V_2) = \frac{|E(V_1, V_2)|}{|V_1||V_2|}.$$
We say that the graph induced by V_1, V_2 is ε-regular if for all $V_1^* \subset V_1$ and $V_2^* \subset V_2$ with $|V_1^*| \geq \varepsilon|V_1|$ and $|V_2^*| \geq \varepsilon|V_2|$, we have
$$|d(V_1^*, V_2^*) - d(V_1, V_2)| \leq \varepsilon.$$

Theorem 12 (Szemerédi's Regularity Lemma [27]). *For any $\varepsilon > 0$ there is a number, $t = t(\varepsilon)$, such that any graph's vertex set can be partitioned into t almost equal vertex classes such that with only εt^2 exemptions the bipartite graphs between the classes are ε-regular.*

The Regularity Lemma was a byproduct of Szemerédi's proof of Theorem 11. Regularity is used so that much of a given graph appears to be random-like when viewed in this way. A random graph on n vertices, where every edge is selected independently with probability δ, contains about $\delta^3 \binom{n}{3}$ triangles.

Exercise 10. We are given a tri-partite graph G with three equal sized vertex sets V_1, V_2, V_3. Suppose that all three subgraphs induced by the V_i, V_j pairs are ε-regular and $d(V_i, V_j) \geq \delta$, for a given $\varepsilon, \delta > 0$. Prove that there is a $c > 0$, which depends only on ε and δ, such that the number of triangles in G is at least $c|V_1|^3$.

We are given a k-uniform hypergraph H_k^n on an n-element vertex set $V(H_k^n)$. A *clique*, K_{k+1}, is a $k+1$-element subset of $V(H_k^n)$ such that any k-tuple of K_{k+1} is an edge of the hypergraph H_k^n. Two cliques are said to be edge-disjoint if they don't have a common edge. Any set of pairwise edge-disjoint cliques in H_k^n has cardinality at most $1/(k+1)\binom{n}{k}$ since every clique has $k+1$ edges. The most important corollary of the Hypergraph Regularity Lemma [12, 20] is that if a hypergraph H_k^n contains a large set, S, of pairwise edge-disjoint cliques, then it contains many cliques. In particular, it contains at least one clique which is not in S.

Theorem 13 (Removal lemma). *For any $c > 0$ real number and $k \geq 2$ integer, there is a $\delta > 0$ which depends on c and k only, such that the following is true. If H_k^n contains a set, S, of pairwise edge-disjoint cliques with cardinality $|S| \geq c\binom{n}{k}$, then H_k^n contains at least $\delta\binom{n}{k+1}$ cliques.*

This result is known as the "Removal lemma" because one can reformulate it in the following way. If H_k^n contains only a few, $o(n^{k+1})$, cliques then one can remove $o(n^k)$ edges to make H_k^n triangle free. (We say that "$f(n)$ is little oh $g(n)$," which we write $f(n) = o(g(n))$, if $\lim_{n \to \infty} f(n)/g(n) = 0$.)

Exercise 11 ([22]). Prove the Removal lemma for $k = 2$. (Use the Regularity Lemma and Exercise 10.)

When n is large enough, then $\delta\binom{n}{k+1}$ is larger than $1/(k+1)\binom{n}{k}$, so there is at least one clique in H_k^n which is not in S. It is not known how δ depends on c. Even in the simplest, $k = 2$, case the gap between the best known upper and lower bounds is huge. A typical application of the result goes like this: We want to prove that a given hypergraph contains two cliques sharing an edge. If we can show that there is a large set of pairwise edge-disjoint cliques then we are done by the Removal lemma since there will be so many cliques. To illustrate the method, we prove the following generalization of Roth's theorem.

Theorem 14 ([1,24]). *For every $c > 0$ there is an $n_0 = n_0(c)$ such that if a subset S of the integer grid $[1, n_0] \times [1, n_0]$, has positive density, that is $|S| \geq cn_0^2$, then S contains a right angled isosceles triangle: that is, three points, $(x, y), (x + d, y), (x, y + d) \in S$, for some $d \neq 0$.*

We describe the proof with Figure 2 and Figure 3.

Exercise 12. Show that Theorem 14 implies Roth's theorem.

Exercise 13 ([25]). Use the Removal lemma with $k = 3$ to prove that every dense subset of the integer grid contains a square. (Hint: Use planes and a four-partite, 3-uniform hypergraph instead of lines and the three-partite graph in the proof of Theorem 14.)

Exercise 14. Use the Removal lemma to prove Szemerédi's theorem about long arithmetic progressions in subsets of integers of positive density. (Theorem 11.)

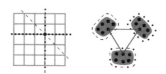

FIGURE 2. Take a tri-partite graph where the three vertex sets of the graph are the horizontal, vertical, and slope -1 lines which are incident to at least one point from S. In the graph two vertices are connected by an edge if the crossing point of the corresponding lines is a point of S. A triangle in the graph corresponds to three lines such that any two intersect in a point of S.

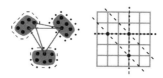

FIGURE 3. If there are two triangles sharing an edge, then at least one triangle is not degenerate, that is it is not defined by one single point of S, so we have an isosceles equilateral triangle in S. The degenerate triangles are edge disjoint, and there are cn_0^2 of them. By the Removal lemma, there are at least δn_0^3 triangles in S. If $n_0 > c/\delta$ then $\delta n_0^3 > cn_0^2$ and there is at least one nondegenerate triangle in S. In fact if $n_0 > 2c/\delta$ then there are at least $\delta n_0^3/2$ nondegenerate triangles in S

References

1. M. Ajtai and E. Szemerédi, *Sets of lattice points that form no squares*, Studia Sci. Math. Hungar. **9** (1974), 9–11.
2. F. A. Behrend, *On sets of integers which contain no three terms in arithmetic progression*, Proc. Nat. Acad. Sci. U.S.A. **32** (1946), 331–332.
3. J. Bourgain, *On triples in arithmetic progression*, Geom. Funct. Anal. **9** (1999), no. 5, 968–984.
4. L. Caporaso, J. Harris, and B. Mazur, *Uniformity of rational points*, J. Amer. Math. Soc. **10** (1997), no. 1, 1–35.
5. J. Cilleruelo and A. Granville, *Lattice points on circles, squares in arithmetic progressions, and sumsets of squares*, in this volume.
6. P. Erdős and P. Turán, *On some sequences of integers*, J. London Math. Soc. **11** (1936), 261–264.
7. J. Folkman, *Graphs with monochromatic complete subgraph in every edge coloring*, SIAM J. Appl. Math. **18** (1970), 19–24.
8. H. Furstenberg, *Ergodic behavior of diagonal measures and a theorem of Szemerédi on arithmetic progressions*, J. Analyse Math. **31** (1977), 204–256.
9. H. Furstenberg and B. Weiss, *Topological dynamics and combinatorial number theory*, J. Analyse Math. **34** (1978), 61–85.
10. W. T. Gowers, *A new proof of Szemerédi's theorem*, Geom. Funct. Anal. **11** (2001), no. 3, 465–588.
11. _____, *Quasirandomness, counting and regularity for 3-uniform hypergraphs*, Combin. Probab. Comput. **15** (2006), no. 1-2, 143–184.
12. _____, *Hypergraph regularity and the multidimensional Szemerédi theorem*, preprint.
13. R. Graham, B. Rothschild, and J. H. Spencer, *Ramsey theory*, 2nd ed., Wiley-Intersci. Ser. Discrete Math., John Wiley & Sons, New York, 1990.
14. A. Granville, *An introduction to additive combinatorics*, in this volume.
15. D. S. Gunderson and V. Rödl, *On extremal problems for affine cubes of integers*, Combin. Probab. Comput. **7** (1998), no. 1, 65–79.
16. D. Hilbert, *Über die Irreduzibilatät ganzer rationaler Functionen mit ganzzahligen Koeffizienten*, J. Reine Angew. Math. **110** (1892), 104–129.
17. N. Hindman, *Finite sums from sequences within cells of a partition of N*, J. Combin. Theory Ser. A **17** (1974), 1–11.
18. B. Kra, *Ergodic methods in combinatorial number theory*, in this volume.
19. F. P. Ramsey, *On a problem of formal logic*, Proc. London Math. Soc. **30** (1929), no. 2, 264–286.
20. V. Rödl, B. Nagle, J. Skokan, M. Schacht, and Y. Kohayakawa, *The hypergraph regularity method and its applications*, Proc. Nat. Acad. Sci. U.S.A. **102** (2005), no. 23, 8109–8113.
21. K. F. Roth, *On certain sets of integers*, J. London Math. Soc. **28** (1953), 104–109.
22. I. Ruzsa and E. Szemerédi, *Triple systems with no six points carrying three triangles*, Combinatorics, Vol. II (Keszthely, 1976) (A. Hajnal and V. T. Sós, eds.), Colloq. Math. Soc. János Bolyai, vol. 18, North-Holland, Amsterdam–New York, 1978, pp. 939–945.
23. I. Schur, *Über die Kongruenz $x^m + y^m = z^m \pmod{p}$*, Jber. Deutsch. Math.-Verein. **25** (1916), 114–116.
24. J. Solymosi, *Note on a generalization of Roth's theorem*, Discrete and Computational Geometry (B. Aronov, S. Basu, J. Pach, and M. Sharir, eds.), Algorithms Combin., vol. 25, Springer, Berlin, 2003, pp. 825–827.
25. _____, *A note on a question of Erdős and Graham*, Combin. Probab. Comput. **13** (2004), no. 2, 263–267.
26. E. Szemerédi, *On sets of integers containing no k elements in arithmetic progression*, Acta Arith. **27** (1975), 199–245.
27. _____, *Regular partitions of graphs*, Problémes combinatoires et théorie des graphes (Orsay, 1976), Colloq. Internat. CNRS, vol. 260, CNRS, Paris, 1978, pp. 399–401.
28. T. Tao, *Szemerédi's regularity lemma revisited*, Contrib. Discrete Math. **1** (2006), no. 1, 8–28.
29. _____, *A variant of the hypergraph removal lemma*, J. Combin. Theory Ser. A **113** (2006), no. 7, 1257–1280.
30. _____, *The ergodic and combinatorial approaches to Szemerédi's theorem*, in this book.

31. B. L. van der Waerden, *Beweis einer Baudetschen Vermutung*, Nieuw Arch. Wisk. **15** (1927), 212–216.

DEPARTMENT OF MATHEMATICS, UNIVERSITY OF BRITISH COLUMBIA, 1984 MATHEMATICS ROAD., VANCOUVER, BC V6T 1Z2, CANADA
  E-mail address: solymosi@math.ubc.ca

Many Additive Quadruples

Antal Balog

1. Introduction

Let A and B be finite subsets of an additive abelian group Z (for example, $\mathbb{Z}$). If there are "many" additive quadruples, that is many solutions to the equation $a_1 + b_1 = a_2 + b_2$, where $a_1, a_2 \in A$ and $b_1, b_2 \in B$, does this impose severe restrictions on the possible sets A and B? More precisely we suppose that $G \subset A \times B$ is "large", and $S = \{a + b : (a,b) \in G\}$ is "small." We may view G as the set of edges of a bipartite graph on the two (disjoint) vertex sets A and B (which should be considered to be disjoint even if they are not disjoint as subsets of Z). Then S is the set of sums of the endpoints of each edge of G.

In this article, we prove that if G is large and S is small then A and B must indeed have some "weak" structure. There are various results of this kind (the details vary according to the precise meaning of the words "many," "small," "large" and "weak"), all stating there must be "large" subsets $A' \subset A$ and $B' \subset B$ such that the sum set $A' + B' = \{a + b : a \in A', b \in B'\}$ is "small." One can now appeal to the structure theorem of Freĭman [10] and Ruzsa [25, 26] to describe A' and B' more precisely.

The first such result was proved by the author and E. Szemerédi in [1]. It quickly turned out that little variations are needed for various applications. The two most important variations are due to Laczkovich–Ruzsa [18] and Gowers [11, 12]. They all follow the pattern of the original proof but Gowers ingeniously replaced the use of difficult results on graph regularity by the relatively easy calculation of a certain average. This change made it possible to derive excellent quantitative connections between the different parameters. Bourgain [2], and Bourgain, Katz and Tao [3], introduced the use of such results in several new problems, such as the dimension of Kakeya sets and sum–product estimates. Some authors improved on the quantitative aspect of these results, see for example Chang [4], Green [13], Lev [19] (and probably many others). Recently Elekes and Ruzsa [9] provided a common gener-

---

2000 *Mathematics Subject Classification.* 11P70, 11P75.

This article is based on a lecture given at the CRM–Clay School on Additive Combinatorics, Montreal, on April 1, 2006. The author is grateful for the support of the CRM, the University of Montreal, and also a HNSFR grant. Special thanks go to Andrew Granville, who carefully read and revised the manuscript.

This is the final form of the paper.

alization of most of the variants, though it is not very competitive quantitatively. In this article we give a quantitatively strong version with a relatively straightforward proof, following the ideas of Gowers and Bourgain. We also provide many references, including papers that we do not explicitly refer to, but which contain characteristic applications of this type of result.

2. Statement of the main results

Theorem 1. *Let A and B be finite subsets of an additive abelian group, Z, and G be a subset of $A \times B$ with $S = \{a + b : (a, b) \in G\}$. If $|A|, |B|, |S| \leq N$ and $|G| \geq \alpha N^2$ then there exist $A' \subset A$ and $B' \subset B$ such that*

(1)
$$\text{(i)} \quad |A' + B'| \leq \frac{2^{23}}{\alpha^5} N, \qquad \text{(ii)} \quad |A'|, |B'| \geq \frac{\alpha}{16} N,$$
$$\text{(iii)} \quad |(A' \times B') \cap G| \geq \frac{\alpha^2}{128} N^2.$$

Theorem 2. *Let A and B be finite subsets of an additive abelian group, Z, and $|A|, |B| \leq N$. If there are $\geq 2\alpha N^3$ solutions to the equation $a_1 + b_1 = a_2 + b_2$, in $a_1, a_2 \in A$ and $b_1, b_2 \in B$, then there exist $A' \subset A$ and $B' \subset B$ such that*

$$\text{(i)} \quad |A' + B'| \leq \frac{2^{23}}{\alpha^5} N, \qquad \text{(ii)} \quad |A'|, |B'| \geq \frac{\alpha}{16} N.$$

Throughout we let
$$r_G(n) = \#\{n = a + b : (a, b) \in G\} \leq N$$
be the number of representations of n as the sum of endpoints of edges in G, and let
$$S_G(\alpha) = \{n : r_G(n) \geq \alpha N\}$$
be the set of "popular" sums in G.

DEDUCTION OF THEOREM 2 FROM THEOREM 1. Let $r(n) = r_{A \times B}(n)$. If $|S_{A \times B}(\alpha)| > N$ then let S be a subset of $S_{A \times B}(\alpha)$ of size N, so that $\sum_{n \in S} r(n) \geq N \cdot \alpha N = \alpha N^2$. Otherwise let $S = S_{A \times B}(\alpha)$. Since $\sum_n r(n) = |A||B| \leq N^2$, the hypotheses of Theorem 2 imply that

$$2\alpha N^3 \leq \sum_{a_1 + b_1 = a_2 + b_2} 1 = \sum_n r(n)^2 = \sum_{n \in S} r(n)^2 + \sum_{n \notin S} r(n)^2$$
$$\leq N \sum_{n \in S} r(n) + \alpha N \sum_{n \notin S} r(n) \leq N \sum_{n \in S} r(n) + \alpha N^3,$$

so that $\sum_{n \in S} r(n) \geq \alpha N^2$. Now let $G = \{(a, b) : a \in A, b \in B \text{ and } a + b \in S\}$ in either case, so that $|G| = \sum_{n \in S} r(n) \geq \alpha N^2$ and the hypotheses of Theorem 1 are satisfied, so that (i) and (ii) follow. $\square$

The proof of Theorem 1 rests on the repeated use of the following averaging argument:

Lemma 1. *Suppose that S has $\leq N$ elements, and that $f : S \to \mathbb{R}$ is a function with $0 \leq f(s) \leq M$ for all $s \in S$. If $\sum_{s \in S} f(s) \geq \alpha MN$ and $S' = \{s \in S : f(s) \geq (\alpha/2)M\}$ then*

(2)
$$\sum_{s \in S'} f(s) \geq \frac{\alpha}{2} MN, \quad \text{so that } |S'| \geq \frac{\alpha}{2} N.$$

PROOF. We have
$$\alpha MN \le \sum_{s \in S} f(s) = \sum_{s \in S'} f(s) + \sum_{s \notin S'} f(s) \le \sum_{s \in S'} f(s) + \frac{\alpha}{2} MN,$$
so that
$$\frac{\alpha}{2} MN \le \sum_{s \in S'} f(s) \le M|S'|. \qquad \square$$

3. Proof of Theorem 1

Our main technical result is as follows:

Proposition 1. *Let A and B be finite subsets of an additive abelian group, Z, and G be a subset of $A \times B$ with $S = \{a + b : (a, b) \in G\}$. If $|A|, |B|, |S| \le N$ and $|G| \ge \alpha N^2$ then there exist $A' \subset A$ and $B' \subset B$, satisfying (ii) and (iii) of Theorem 1, with the following property: If $a' \in A'$ and $b' \in B'$ then there are at least $\alpha^5 N^2 / 2^{20}$ pairs $a \in A, b \in B$ for which*
$$r_G(a + b'), r_G(a + b), r_G(a' + b) \ge \frac{\alpha}{2} N.$$

DEDUCTION OF THEOREM 1. We will prove the result when $|G| = \alpha N^2$. The result then follows whenever $|G| \ge \alpha N^2$ since all of the conclusions of Theorem 1 are appropriately monotone in α.

For each $c \in A' + B'$ select a unique $a' = a'(c) \in A'$ and $b' = b'(c) \in B'$ for which $c = a' + b'$. We obtain a lower bound on $|G|^3$ by counting the triples $(a_1, b_1), (a_2, b_2), (a_3, b_3) \in G$ as follows: For each $c \in A' + B'$ and each pair $a \in A$, $b \in B$ for which $r_G(a + b'), r_G(a + b), r_G(a' + b) \ge (\alpha/2) N$, count the triples $(a_1, b_1), (a_2, b_2), (a_3, b_3) \in G$ with $a_1 + b_1 = a + b'$, $a_2 + b_2 = a + b$ and $a_3 + b_3 = a' + b$. We claim that each such triple is counted just once in this way since, for any given triple, we have $(a_1 + b_1) - (a_2 + b_2) + (a_3 + b_3) = c$ and consequently $a' = a'(c)$, $b' = b'(c)$, $a = a_1 + b_1 - b'$ and $b = a_3 + b_3 - a'$ are all determined. Therefore
$$\alpha^3 N^6 = |G|^3 \ge \sum_{c \in A'+B'} \frac{\alpha^5 N^2}{2^{20}} r_G(a+b') r_G(a+b) r_G(a'+b) \ge |A' + B'| \frac{\alpha^8 N^5}{2^{23}}$$
by Proposition 1. This implies part (i) of Theorem 1. $\qquad \square$

PROOF OF PROPOSITION 1. Consider the bipartite graph with vertex sets A and B and edges $(a, b) \in G$. We will obtain a sequence of subgraphs $G \supseteq G_1 \supseteq G_2 \supseteq G_3 \supseteq G_4 \supseteq G_5$ by successively pruning vertices and/or edges to gain better properties for the graphs G_n (which have corresponding vertex sets A_n, B_n), and then we will take $A' = A_5$, $B' = B_5$. For $a \in A_j$ we define $N_j(a) = \{b \in B_j : (a, b) \in G_j\}$ the "shadow" of a in B_j, and, similarly, for $b \in B_j$ we define $N_j(b) = \{a \in A_j : (a, b) \in G_j\}$, for each j. We will follow the changes in the $N_j(a)$ and $N_j(b)$, as j increases, with some care.

Note that $r_G(n)$ is supported on S, so that $\sum_{n \in S} r_G(n) = |G| \ge \alpha N^2$.

Let $S_1 = S_G(\alpha/2)$ and $G_1 = \{(a, b) \in G : a + b \in S_1\}$, the set of edges with popular sums of vertices, with $A_1 = A$ and $B_1 = B$. Lemma 1 implies that
$$(3) \qquad |G_1| = \sum_{n \in S_1} r_G(n) \ge \frac{\alpha}{2} N^2.$$

Note that $|N_1(a)| \leq N$. By (3) we have $\sum_{a \in A} |N_1(a)| = |G_1| \geq (\alpha/2)N^2$. Let $A_2 = \{a \in A : |N_1(a)| \geq (\alpha/4)N\}$, the set of vertices in A_1 with big shadows, let $B_2 = B$ and $G_2 = \{(a,b) \in G_1 : a \in A_2\}$. Lemma 1 implies that

$$|G_2| = \sum_{a \in A_2} |N_1(a)| \geq \frac{\alpha}{4}N^2 \quad \text{with } |A_2| \geq \frac{\alpha}{4}N. \tag{4}$$

Note that $N_2(a) = N_1(a)$ for all $a \in A_2$.

We now come to the only delicate step in the proof. For a real number $\gamma > 0$, to be chosen later, let

$$Y := \{(a,a') \in A_2 \times A_2 : |N_2(a) \cap N_2(a')| \leq \gamma N\},$$

the set of pairs of elements of A_2 whose shadows have a small intersection. We expect Y to be small if γ is small enough. We will select b^* with a large shadow, that is with $|N_2(b^*)| \geq (\alpha/8)N$, such that few pairs of elements in $N_2(b^*)$ have shadows with a small intersection, that is we want

$$\left|(N_2(b^*) \times N_2(b^*)) \cap Y\right| \leq \frac{8\gamma}{\alpha} N |N_2(b^*)|. \tag{5}$$

We then let $A_3 = N_2(b^*)$, $G_3 = \{(a,b) \in G_1 : a \in A_3\}$ and $B_3 = B$, so that, since $N_3(a) = N_1(a)$, we have

$$|(A_3 \times A_3) \cap Y| \leq \frac{8\gamma}{\alpha} N|A_3|, \quad |A_3| \geq \frac{\alpha}{8}N, \quad |G_3| \geq \frac{\alpha}{4}N|A_3|. \tag{6}$$

We need to prove that such a b^* exists: First note that

$$\sum_{b \in B} \left|(N_2(b) \times N_2(b)) \cap Y\right| = \sum_{(a,a') \in Y} \sum_{b \in N_2(a) \cap N_2(a')} 1 \tag{7}$$

$$= \sum_{(a,a') \in Y} |N_2(a) \cap N_2(a')| \leq \gamma N |Y| \leq \gamma N^3;$$

that is, on average, there are not too many pairs whose shadows have small intersection. Equation (4) implies that $\sum_{b \in B} |N_2(b)| = |G_2| \geq (\alpha/4)N^2$, so for $B^* = \{b \in B : |N_2(b)| \geq (\alpha/8)N\}$, Lemma 1 implies that

$$\sum_{b \in B^*} |N_2(b)| \geq \frac{\alpha}{8}N^2. \tag{8}$$

Therefore there exists $b^* \in B^*$ satisfying (5), as desired, else by (7), then the contrapositive to (5), and then (6), we have

$$\gamma N^3 \leq \frac{8\gamma}{\alpha} N \sum_{b \in B^*} |N_2(b)| < \sum_{b \in B^*} \left|(N_2(b) \times N_2(b)) \cap Y\right| \leq \gamma N^3,$$

a contradiction.

One last bit of pruning is needed for the A vertices: Fix $\delta = 16\gamma/\alpha > 0$ and let

$$A_4 = \{a \in A_3 : \#\{a' \in A_3 : (a,a') \in Y\} \leq \delta N\},$$

the set of elements of A_3 whose shadow has a large intersection with the shadows of most of the other elements of A_3. Now

$$\frac{\delta}{2} N|A_3| \geq |(A_3 \times A_3) \cap Y| = \sum_{a \in A_3} \sum_{\substack{a' \in A_3 \\ (a,a') \in Y}} 1 \geq \sum_{a \in A_3 \setminus A_4} \sum_{\substack{a' \in A_3 \\ (a,a') \in Y}} 1 \geq \delta N |A_3 \setminus A_4|,$$

so that
$$|A_4| = |A_3| - |A_3 \setminus A_4| \geq |A_3| - \frac{|A_3|}{2} = \frac{|A_3|}{2}.$$
Therefore taking $G_4 = \{(a,b) \in G_1 : a \in A_4\}$ and $B_4 = B$ we obtain that for any $a \in A_4$ we have

(9) $\#\{a' \in A_4 : (a,a') \in Y\} \leq \frac{16\gamma}{\alpha}N, \quad |A_4| \geq \frac{\alpha}{16}N, \quad |G_4| \geq \frac{\alpha}{4}N|A_4|,$

since $N_4(a) = N_1(a)$.

Finally let $B_5 = \{b \in B : |N_4(b)| \geq (\alpha/8)|A_4|\}$ and $A_5 = A_4$, with $G_5 = \{(a,b) \in G_4 : b \in B_5\}$. Note that $|N_4(b)| \leq |A_4|$ for all $b \in B$, and $\sum_{b \in B} |N_4(b)| = |G_4| \geq (\alpha/4)N|A_4|$, so that by Lemma 1 we have

(10) $|G_5| = \sum_{b \in B_5} |N_4(b)| \geq \frac{\alpha}{8}N|A_4| \geq \frac{\alpha^2}{128}N^2 \quad \text{and} \quad |B_5| \geq \frac{\alpha}{16}N.$

Since $A' = A_5$, $B' = B_5$ and $(A' \times B') \cap G = G_5$ both (ii) and (iii) follow from the last two displayed equations.

Let $\gamma = \alpha^3/2^{12}$. Fix any pair of elements $a' \in A'$ and $b' \in B'$. By the choice of B', the shadow of b' is big, that is $|N_5(b')| \geq (\alpha/8)|A_5|$, while

$$\#\{a \in A_5 : (a,a') \in Y\} \leq \frac{16\gamma}{\alpha}N = \frac{\alpha^2}{2^8}N \leq \frac{\alpha}{16}|A_5| \leq \frac{1}{2}|N_5(b')|$$

using (9). Hence there are at least $\frac{1}{2}|N_5(b')|$ elements $a \in N_5(b')$ such that $(a,a') \notin Y$, that is $|N_4(a) \cap N_4(a')| > \gamma N = (\alpha^3/2^{12})N$ (using the fact that $N_4(a) = N_2(a)$). Therefore the number of pairs $a \in A$, $b \in B$ for which $(a',b), (a,b), (a,b')$ are all in G_4 is

$$\geq \frac{1}{2}|N_5(b')| \cdot \frac{\alpha^3}{2^{12}}N \geq \frac{\alpha}{2^4}|A_5| \cdot \frac{\alpha^3}{2^{12}}N \geq \frac{\alpha}{2^4} \cdot \frac{\alpha}{2^4}N \cdot \frac{\alpha^3}{2^{12}}N = \frac{\alpha^5}{2^{20}}N^2.$$

This implies the result since $G_4 \subset G$. $\square$

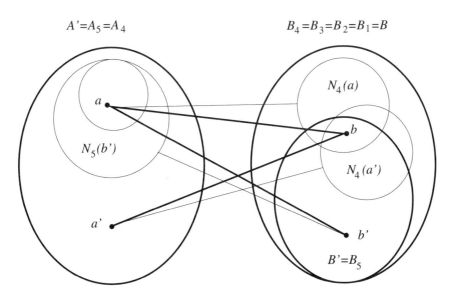

Figure 1

4. Variants

As we mentioned in the introduction, there are many useful variations of Theorem 1, differing in the precise nature of the conclusions. In Theorem 1 we take strong but partial information about *sums* (that is that $|S| \leq N$) and deduce strong, complete information about sums in certain large subsets (conclusion (i)), but we could ask for similarly strong, complete information about *differences* in certain large subsets, for example, asking that $A' - B'$ is small, or even that $A' - A'$ or $A' + A'$ is small. Such theorems can be proved as a consequence of Theorem 1 since Ruzsa [21, 22][1] has proved several results bounding the size of $A' - B'$, $A' + A'$ and $A' - A'$ in terms of $A' + B'$. However the exponents in the results then obtained are far larger than if we simply modify the proof of Theorem 1 to these new situations, which is what we will do in the rest of the article. We will state the theorems precisely but only sketch the proofs (we recommend that the interested reader recover the full proofs from our hints).

At this point there is a trade off, obtaining better exponents at the expense of simplicity, due to a more sophisticated averaging argument (due to Bourgain [2] and Chang [4]). For example, this leads to the quantity

$$\epsilon = \epsilon(\alpha) = \frac{3}{2\log(32/\alpha)}$$

appearing in the statements of Theorems 3, 5 and 6, which can be replaced by α if we instead use the familiar Lemma 1.

Theorem 3. *Let A and B be finite subsets of an additive abelian group, Z, and G be a subset of $A \times B$ with $S = \{a + b : (a, b) \in G\}$. If $|A|, |B|, |S| \leq N$ and $|G| \geq \alpha N^2$ then there are $A' \subset A$ and $B' \subset B$ such that*

(11)
$$\text{(i)} \quad |A' - B'| \leq \frac{2^{36}}{\epsilon^2 \alpha^7} N, \qquad \text{(ii)} \quad |A'| \geq \frac{\alpha}{16} N, \quad |B'| \geq \frac{\epsilon \alpha^2}{2^{13}} N$$

$$\text{(iii)} \quad |(A' \times B') \cap G| \geq \frac{\epsilon \alpha^4}{2^{20}} N^2.$$

Theorem 4. *Let A and B be finite subsets of an additive abelian group, Z, and G be a subset of $A \times B$ with $S = \{a + b : (a, b) \in G\}$. If $|A|, |B|, |S| \leq N$ and $|G| \geq \alpha N^2$ then there is an $A' \subset A$ such that*

(12)
$$\text{(i)} \quad |A' - A'| \leq \frac{2^{26}}{\alpha^5} N, \qquad \text{(ii)} \quad |A'| \geq \frac{\alpha}{16} N.$$

Theorem 5. *Let A and B be finite subsets of an additive abelian group, Z, and G be a subset of $A \times B$ with $S = \{a + b : (a, b) \in G\}$. If $|A|, |B|, |S| \leq N$ and $|G| \geq \alpha N^2$ then there is an $A' \subset A$ such that*

(13)
$$\text{(i)} \quad |A' + A'| \leq \frac{2^{37}}{\epsilon \alpha^7} N, \qquad \text{(ii)} \quad |A'| \geq \frac{\epsilon \alpha^3}{2^{15}} N.$$

The deduction of Theorem 1 from Proposition 1 rested on the identity

$$a' + b' = (a + b') - (a + b) + (a' + b),$$

and then counting solutions $(a_1, b_1), (a_2, b_2), (a_3, b_3) \in G$ to $a_1 + b_1 = a + b'$, $a_2 + b_2 = a + b$ and $a_3 + b_3 = a' + b$. To prove Theorem 4 we instead use the identity

$$a' - a'' = (a' + b) - (a + b) + (a + b') - (a'' + b')$$

---
[1] The relevant results are discussed in Section 1.5 of Granville's article in this volume.

and then count solutions $(a_1,b_1), (a_2,b_2), (a_3,b_3), (a_4,b_4) \in G$ to $a_1 + b_1 = a' + b$, $a_2 + b_2 = a + b$, $a_3 + b_3 = a + b'$ and $a_4 + b_4 = a'' + b'$. As in the proof of Theorem 1, we need $(a, b), (a', b), (a', b'), (a'', b')$ to be edges of G with popular sums of their endpoints. We follow the steps of the proof of Theorem 1 to build $G' = G_4$, the graph in which the shadow of any $a \in A_4$ has a large intersection with the shadow of most other elements of A_4. Note that (9) gives (ii). We now select $\gamma = \alpha^2/2^9$, so that for any given $a', a'' \in A'$, we have

$$\#\{a \in A_3 : (a,a') \in Y \text{ or } (a,a'') \in Y\} \leq \frac{32\gamma}{\alpha}N = \frac{\alpha}{2^4}N \leq \frac{|A_3|}{2}.$$

Therefore there are $\geq |A_3|/2$ elements $a \in A_3$ for which $(a, a'), (a, a'') \notin Y$; that is there are more than γN values of $b \in B$ for which $(a, b), (a', b) \in G_3$ and there are more than γN values of $b' \in B$ for which $(a, b'), (a'', b') \in G_3$. Imitating the deduction of Theorem 1 from Proposition 1, we have that

$$|G|^4 \geq |A' - A'| \cdot \frac{|A_3|}{2} \cdot (\gamma N)^2 \cdot \left(\frac{\alpha}{2}N\right)^4,$$

which implies (i).

To prove Theorems 3 and 5 we need a more substantial twist on the tale. For parity reasons one cannot express $a' - b'$ or $a' + a''$ as a sum of terms of the form $\pm(a+b)$; so we will consider an extra term, of the form $a-b$, as well, in the identities

(14) $\qquad a' - b' = (a' + b) - (a + b) + (a - b'),$

(15) $\qquad a' + a'' = (a' + b') - (a + b') + (a + b) + (a'' - b),$

All of the terms on the right-hand side of these equations, with "+" inside, represent edges of G with popular sums, and pairs with "−" inside represent popular differences. To this end we color the edges of our original graph, G_1, blue, and superimpose red edges on the same vertices connecting a with b if and only if $a - b$ is a popular difference (for convenience we let H be the graph containing only the red edges, $R_j(a)$ be the red shadow of $a \in A_j$, that is the set of $b \in B_j$ attached to a by a red edge, and similarly for $b \in B_j$ we let $R_j(b)$ be the set of $a \in A_j$ attached to b by a red edge).

To prove Theorem 3 we show that there are many paths of length three, starting with blue edges (a', b) and (a, b), and finishing with a red edge (a, b'). This will give us many solutions to the equation

$$a - b = (a_1 + b_1) - (a_2 + b_2) + (a_3 - b_3),$$

and then we proceed to deduce Theorem 3, just as we deduced Theorem 1 from Proposition 1.

To prove Theorem 5 we show that there are many paths of length four, starting with blue edges $(a', b), (a, b)$ and (a, b'), and finishing with a red edge (a'', b'). This will give us many solutions to the equation

$$a + a'' = (a_1 + b_1) - (a_2 + b_2) + (a_3 + b_3) + (a_4 - b_4),$$

and then we proceed to deduce Theorem 5, just as we deduced Theorem 1 from Proposition 1.

In these proofs we continue the construction from the proof of Proposition 1 where we left off, starting up at G_5 again. Remember that every element of B_5 has a big blue shadow; and that the shadow of every element of A_4 has a large intersection with the blue shadows of most of the elements of A_3. Therefore no matter how we select $B' \subset B_5$, there are many blue edges connecting A_4 and B' as long as B' is big enough.

We can now turn to the red edges of our graph. The frequency of popular differences and of popular sums are linked through the identity,

$$a_1 - b_1 = a_2 - b_2 \quad \text{if and only if} \quad a_1 + b_2 = a_2 + b_1.$$

Counting the total number of solutions to both equations, leads to the identity

$$\sum_{n \in A-B} d(n)^2 = \sum_{n \in A+B} r(n)^2,$$

where $r(n) := r_{A_4 \times B_5}(n) = \#\{n = a + b : (a,b) \in A_4 \times B_5\}$ and

$$d(n) := \#\{n = a - b : (a,b) \in A_4 \times B_5\}.$$

Now letting $r_0(n) := r_{(A_4 \times B_5) \cap G}(n)$ ($\leq r(n)$), which is only supported on S, we obtain, using the Cauchy–Schwarz inequality

$$\sum_{n \in A-B} d(n)^2 = \sum_{n \in A+B} r(n)^2 \geq \sum_{n \in S} r_0(n)^2 \geq \frac{\left(\sum_{n \in S} r_0(n)\right)^2}{|S|}.$$

Now $\sum_{n \in S} r_0(n) = |G_5| \geq (\alpha/8)N|A_4| \geq (\alpha^2/128)N^2$ by (10), so that $\sum_{n \in A-B} d(n)^2$ is large, and therefore there are many popular differences, that is values of $n \in A_4 - B_5$ for which $d(n)$ is large. The red edges, H, are therefore those $(a,b) \in A_4 \times B_5$ for which $d(a-b)$ is appropriately large. We can give a lower bound for the number of such popular differences using Lemma 1, but one can get a significantly better lower bound (involving ϵ) by using the second moment information obtained just above.

To complete the proof of Theorem 3 let B_6 be the set of elements $b \in B_5$ with a large red shadow, that is for which $|R_5(b)|$ is "big"; by Lemma 1 we know that B_6 is large. Now let $A' = A_6 = A_4$ and $B' = B_6$. If the value of γ is chosen appropriately then one can show that for any $a' \in A'$, $b' \in B'$, at least half of the elements $a \in R_6(b')$ are such that there are many elements b in $N_6(a) \cap N_6(a')$, and therefore we have a large number of appropriate solutions to (14), as desired. Note that (iii) follows from the fact that every element of B' has a big blue shadow in $G_5 \subset G$, so that there are many blue edges between A' and B'.

To complete the proof of Theorem 5 let A_6 be the set of elements $a \in A_5$ with a large red shadow (in B_5), that is for which $|R_5(a)|$ is "big"; by Lemma 1 we know that A_6 is large. Now let $A' = A_6$ and $B' = B_6 = B_5$. If the value of γ is chosen appropriately then one can show that for any $a', a'' \in A'$, for all $b \in R_6(a'')$, one has, for at least half of the elements $a \in N_6(b)$, that there are many elements b' in $N_6(a) \cap N_6(a')$.

In the proofs of all of these three theorems there are many details that are left as an exercise to the reader.

5. Another variant, $A = B$

Taking $A = B$ in Theorem 1 gives two large subsets $A' \subset A$ and $B' \subset A$ with small $A' + B'$, but for applications we need only one subset $A' \subset A$ with small $A' + A'$.

Although taking $A = B$ in Theorem 5 gives one such a result, there are many applications in which we need a large lower bound on the size of $(A' \times A') \cap G$ (as in Theorem 1(iii)), and nothing like this seems feasible from the proof of Theorem 5.

We will modify the above proof one more time to get such a result, this time needing paths on three blue edges followed by one red edge (though there are several variations possible, such as one blue edge followed by three red edges, which would lead to a similar result though with slightly different exponents on α and 2).

Theorem 6. *Let C be a finite subset of an additive abelian group, Z, and G be a subset of $C \times C$ with $S = \{a + b : (a,b) \in G\}$. If $|C|, |S| \leq N$ and $|G| \geq \alpha N^2$ then there is a $C' \subset C$ such that*

(16)
$$\text{(i)} \quad |C' + C'| \leq \frac{2^{37}}{\epsilon^2 \alpha^7} N, \qquad \text{(ii)} \quad |C'| \geq \frac{\epsilon \alpha^3}{2^{15}} N,$$
$$\text{(iii)} \quad |(C' \times C') \cap G| \geq \frac{\epsilon \alpha^4}{2^{22}} N^2.$$

This implies Theorem 1 (though, unsurprisingly, with weaker bounds): If A and B satisfy the hypotheses of Theorem 1, let $C = A \cup B$ in Theorem 6. Then the edges of G connect the A half of C to the B half of C, and the hypotheses of Theorem 6 are satisfied (with N and α modified accordingly). We can then take $A' = A \cap C'$ and $B' = B \cap C'$ so that Theorem 6(iii) implies (a weak version of) both Theorem 1(ii) and Theorem 1(iii), while trivially $A' + B' \subset C' + C'$, so that Theorem 6(i) implies (a weak version of) Theorem 1(i). Similarly Theorem 5 follows (with weaker bounds).

The proof of Theorem 6 uses many of the things we have done so far, and more! First we follow the steps of the proof of Theorem 1 with $A = B = C$, that is we have two copies of C as the two sides of our bipartite graph. In the proof of Proposition 1 we obtained graphs $G_4 \supseteq G_5$ in which any $a \in A_4$ and $b \in B_5$ are connected by many blue paths of length three, $A_4 + B_5$ is small, and $N_5(b)$ is large for all $b \in B_5$. Next, as in the proof of Theorem 5 we find a large $B_6 \subset B_5$ such that every $b \in B_6$ has a big red shadow, that is $R_4(b)$ is large. We let $B' = B_6$ and $A_6 = A_4$. Therefore, for any $b', b'' \in B'$ there are many blue paths of length three connecting b' with any $a' \in R_6(b'')$, thus giving rise to many solutions to the equation
$$b' + b'' = (a + b') - (a + b) + (a' + b) - (a' - b'').$$
Arguing as before, this implies that $B' + B'$ is small, and that there are many blue edges between A_6 and B'.

We have to repeat this argument but now representing $a' + a''$. We show that there is a large $A_7 \subset A_6$ such that every $a \in A_7$ has a big blue shadow, that is $N_6(a)$ is large. Moreover there is a large $A_8 \subset A_7$ such that every $a \in A_8$ has a big red shadow, that is $R_6(a)$ is large. We let $A' = A_8$, while $B_8 = B_7 = B_6 = B'$. Then, for any $a', a'' \in A'$ there are many blue paths of length three connecting a' with any $b' \in R_6(a'')$, thus giving rise to many solutions to the equation
$$a' + a'' = (a' + b) - (a + b) + (a + b') + (a'' - b').$$
This implies that $A' + A'$ is small, and that there are many blue edges between A' and B'. The result now follows taking $C' = A' \cup B'$. It is certainly a challenging exercise to complete the details in this proof.

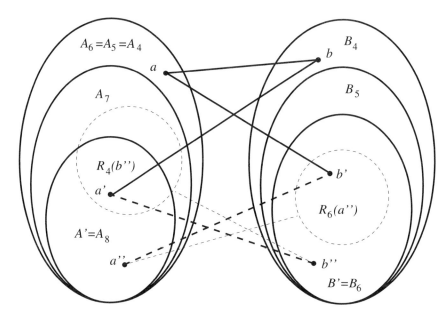

FIGURE 2

References

1. A. Balog and E. Szemerédi, *A statistical theorem of set addition*, Combinatorica **14** (1994), no. 3, 263–268.
2. J. Bourgain, *On the dimension of Kakeya sets and related maximal inequalities*, Geom. Funct. Anal. **9** (1999), no. 2, 256–282.
3. J. Bourgain, N. Katz, and T. Tao, *A sum-product estimate in finite fields and their applications*, Geom. Funct. Anal. **14** (2004), no. 1, 27–57.
4. M. C. Chang, *On problems of Erdős and Rudin*, J. Funct. Anal. **207** (2004), no. 2, 444–460.
5. G. Elekes, *On linear combinatorics. I: Concurrency — an algebraic approach*, Combinatorica **17** (1997), no. 4, 447–458.
6. _____, *On the number of distinct radii of circles determined by triples and on parameters of other curves*, Studia Sci. Math. Hungar. **40** (2003), no. 1-2, 195–203.
7. G. Elekes and Z. Király, *On the combinatorics of projective mappings*, J. Alg. Combin. **14** (2001), no. 3, 183–197.
8. G. Elekes and L. Rónyai, *A combinatorial problem on polynomials and rational functions*, J. Combin. Theory Ser. A **89** (2000), no. 1, 1–20.
9. G. Elekes and I. Z. Ruzsa, *The structure of sets with few sums along a graph*, J. Combin. Theory Ser. A **113** (2006), no. 7, 1476–1500.
10. G. A. Freĭman, *Foundations of a structural theory of set addition*, Transl. Math. Monogr., vol. 37, Amer. Math. Soc., Providence, R.I., 1973.
11. W. T. Gowers, *A new proof of Szemerédi's theorem for arithmetic progressions of length four*, Geom. Funct. Anal. **8** (1998), no. 3, 529–551.
12. _____, *A new proof of Szemerédi's theorem*, Geom. Funct. Anal. **11** (2001), no. 3, 465–588.
13. B. J. Green, *Notes on the Bourgain–Katz–Tao theorem*, available at http://www.dpmms.cam.ac.uk/~bjg23.
14. B. J. Green and T. C. Tao, *An inverse theorem for the Gowers U^3-norm, with applications*, available at arXiv:math.NT/0503014.
15. H. A. Helfgott, *Growth and generation in $SL_2(\mathbb{Z}/p\mathbb{Z})$*, available at arXiv:math.GR/0509024.
16. N. Katz and T. Tao, *Bounds on arithmetic projections, and applications to the Kakeya conjecture*, Math. Res. Lett. **6** (1999), 625–630.
17. _____, *A new bound on partial sum-sets and difference-sets, and applications to the Kakeya conjecture*, available at arXiv:math.CO/9906097.

18. M. Laczkovich and I. Z. Ruzsa, *The number of homothetic subsets*, The Mathematics of Paul Erdős (R. L. Graham and J. Nešetřil, eds.), Algorithms Combin., vol. 14, Springer, Berlin, 1997, pp. 294–302.
19. V. F. Lev, *The (Gowers–) Balog–Szemerédi theorem*, available at http://math2.haifa.ac.il/~seva/notes.html.
20. G. Mockenhaupt and T. Tao, *Restriction and Kakeya phenomena for finite fields*, Duke Math. J. **121** (2004), no. 1, 35–74.
21. I. Z. Ruzsa, *On the cardinality of $A + A$ and $A - A$*, Combinatorics, Vol. II (Keszthely, 1976) (A. Hajnal and V. T. Sós, eds.), Colloq. Math. Soc. János Bolyai, vol. 18, North-Holland, Amsterdam–New York, 1978, pp. 933–938.
22. _____, *An application of graph theory to additive number theory*, Sci. Ser. A Math. Sci. (N.S.) **3** (1989), 97–109.
23. _____, *Addentum to: An application of graph theory to additive number theory*, Sci. Ser. A Math. Sci. (N.S.) **4** (1990/91), 93–94.
24. _____, *Arithmetical progressions and the number of sums*, Period. Math. Hungar. **25** (1992), no. 1, 105–111.
25. _____, *Generalized arithmetical progressions and sumsets*, Acta Math. Acad. Sci. Hungar. **65** (1994), no. 4, 379–388.
26. _____, *An analog of Freĭman's theorem in groups*, Structure Theory of Set Addition (J.-M. Deshouillers, B. Landreau, and A. A. Yudin, eds.), Astérisque, vol. 258, Soc. Math. France, Paris, 1999, pp. 323–326.
27. J. Solymosi, *Arithmetic progressions in sets with small sumsets*, Combin. Probab. Comput. **15** (2006), no. 4, 597–603.
28. B. Sudakov, E. Szemerédi, and V. H. Vu, *On a question of Erdős and Moser*, Duke Math. J. **129** (2005), no. 1, 129–155.
29. T. C. Tao, *Product set estimates for noncommutative groups*, available at arXiv:math.CO/0601431.

ALFRÉD RÉNYI INSTITUTE OF MATHEMATICS, HUNGARIAN ACADEMY OF SCIENCES, POB 127, 1364 BUDAPEST, HUNGARY

E-mail address: balog@renyi.hu

An Old New Proof of Roth's Theorem

Endre Szemerédi

In 1953 Roth [3] proved that for any fixed $\delta > 0$, if N is sufficiently large and A is any subset of $\{1, 2, \ldots, N\}$ of size $\geq \delta N$ then A contains a non-trivial 3-term arithmetic progression. In the 1980s I came up with an alternate proof that is in some aspects a little simpler but which I did not publish. This school gives me another opportunity to present this approach.

We suppose that $A \subset \{1, 2, \ldots, N\}$ with $|A| = \delta N$ (where $|A| \geq 1000\sqrt{N}$), and that A does not contain a non-trivial 3-term arithmetic progression. As usual we define $e(t) = e^{2i\pi t}$ and
$$\hat{A}(\alpha) = \sum_{a \in A} e(a\alpha).$$
The number of solutions to $a + c = 2b$ with $a, b, c \in A$ is given by

$$(1) \qquad |A| = \sum_{a,b,c \in A} \int_0^1 e\big(\alpha(a + c - 2b)\big) \, d\alpha = \int_0^1 \hat{A}(\alpha)^2 \hat{A}(-2\alpha) \, d\alpha$$

(the $|A|$ comes from the solutions with $a = b = c$). We will partition $\mathbb{R}/\mathbb{Z}$ into the arcs $I_j := \big[(2j-1)/(2MN), (2j+1)/(2MN)\big)$ for $j = 0, 1, \ldots, NM - 1$ where M is the smallest integer $\geq 2\pi/\delta\eta$, with $\eta = 10^{-6}$. For real number t denote by $\|t\|$ the distance from t to the nearest integer. Note that $|e(t) - 1| = 2|\sin(\pi t)| = 2|\sin(\pi\|t\|)| \leq 2\pi\|t\|$. Hence if $\alpha \in I_j$, that is $\alpha = \frac{j}{MN} + \beta$ where $|\beta| \leq 1/2MN$, then

$$(2) \qquad |\hat{A}(j/MN) - \hat{A}(\alpha)| \leq \sum_{a \in A} |e(a\beta) - 1| \leq \sum_{a \in A} 2\pi\|a\beta\|$$
$$\leq |A| 2\pi N / 2MN \leq \eta \delta^2 N / 2.$$

Let J be the set of integers in $[0, MN)$ for which $|\hat{A}(j/MN)| \geq \eta\delta^2 N$; and then define the *major arc*, $\mathcal{M}$ to be the union of the I_j with $j \in J$. From (2) we deduce that

$$|\hat{A}(\alpha)| \geq \eta\delta^2 N/2 \quad \text{if } \alpha \in \mathcal{M}; \quad \text{and} \quad |\hat{A}(\alpha)| \leq 3\eta\delta^2 N/2 \quad \text{if } \alpha \notin \mathcal{M}.$$

2000 *Mathematics Subject Classification.* Primary 11B25; Secondary 11K38, 11K45.

This article was written by Antal Balog and Andrew Granville, based on the lecture given by the author at the school.

This is the final form of the paper.

©2007 American Mathematical Society

From the second of these inequalities we deduce that

(3) $$\left| \int_{\substack{0 \\ \alpha \notin \mathcal{M}}}^{1} \hat{A}(\alpha)^2 \hat{A}(-2\alpha) d\alpha \right| \leq \max_{\alpha \notin \mathcal{M}} |\hat{A}(\alpha)| \cdot \int_0^1 |\hat{A}(\alpha)| \cdot |\hat{A}(-2\alpha)| \, d\alpha$$
$$\leq \frac{3}{2} \eta \delta^2 N \left(\int_0^1 |\hat{A}(\alpha)|^2 \, d\alpha \int_0^1 |\hat{A}(-2\alpha)|^2 \, d\alpha \right)^{1/2}$$
$$= \frac{3}{2} \eta \delta^3 N^2$$

by Parseval's identity that $\int_0^1 |\hat{A}(\alpha)|^2 \, d\alpha = |A|$. From the first of the inequalities we have that

$$\delta N = |A| = \int_0^1 |\hat{A}(\alpha)|^2 \, d\alpha \geq \int_{\alpha \in \mathcal{M}} |\hat{A}(\alpha)|^2 \, d\alpha \geq |\mathcal{M}|(\eta \delta^2 N/2)^2,$$

so that $|\mathcal{M}| \leq 4/(\eta^2 \delta^3 N)$; and thus $k := |J| \leq 4M/\eta^2 \delta^3 \lesssim 8\pi/\delta^4 \eta^3$. (Here we use the notation "$\lesssim$" (and later "$\sim$") instead of "$\leq$" (and later "$=$," respectively), when there may be other terms that are negligible compared to the main term.)

We now claim that there exists a positive integer $q \leq Q$ for which

(4) $$\left\| \frac{qj}{MN} \right\| \leq Q^{-1/k} \quad \text{for each } j \in J.$$

To see this consider the vectors w_i in $(\mathbb{R}/\mathbb{Z})^k$ with coordinates indexed by $j \in J$, where the jth coordinate is ij/mn (mod 1). If we cut the space up into the Q k-dimensional minicubes given by cutting up each dimension into sides of length $Q^{-1/k}$, then at least two of the vectors from $w_0, w_1, \ldots, w_Q$ belong to the same minicube, by the pigeonhole principle. If these vectors are w_h and w_i with $0 \leq h < i \leq Q$ then let $q = i - h$ so that (4) holds as claimed.

Take $L = [N^{1/3k}/8M]$ and $Q = (8LM)^k$, so that $Q \leq N^{1/3}$. If $\alpha \in I_j$ with $j \in J$ then $\|q\alpha\| \leq \|qj/MN\| + \|q/2MN\| \leq Q^{-1/k} + Q/2MN$, and thus if ℓ is an integer for which $|\ell| \leq 4L$ then $\|\alpha q \ell\| \leq 4L(Q^{-1/k} + Q/2MN) \leq 1/M$, since $4LQ \leq Q^{1+1/k} \leq N^{2/3}$ as well. Therefore

$$\left| \int_0^1 \hat{A}(\alpha)^2 \hat{A}(-2\alpha) e(\alpha q \ell) \, d\alpha - \int_0^1 \hat{A}(\alpha)^2 \hat{A}(-2\alpha) \, d\alpha \right|$$
$$\leq 2\pi \int_{\alpha \in \mathcal{M}} |\hat{A}(\alpha)|^2 \cdot |\hat{A}(-2\alpha)| \cdot \|\alpha q \ell\| \, d\alpha + 2 \int_{\alpha \notin \mathcal{M}} |\hat{A}(\alpha)|^2 \cdot |\hat{A}(-2\alpha)| \, d\alpha$$
$$\leq 2\pi \delta^2 N^2 \max_{\alpha \in \mathcal{M}} \|\alpha q \ell\| + 3\eta \delta^3 N^2 \leq 4\eta \delta^3 N^2$$

by (3). We deduce that for any $|r|, |s|, |t| \leq L$ (taking $\ell = r + t - 2s$ above) we have

(5) $$\#\{a, b, c \in A : (a + rq) + (c + tq) = 2(b + sq)\} \leq 5\eta \delta^3 N^2,$$

using (1), since $\delta N \geq \sqrt{N/\eta}$ by assumption.

This suggests that for most 3-term arithmetic progressions of integers $u + w = 2v$ there cannot be many $a = u - rq, b = v - sq, c = w - tq \in A$, which seems implausible if A is reasonably distributed in segments of residue classes mod q. To show this define

$$\kappa(n) = \#\{r : |r| \leq L, n - rq \in A\}.$$

One expects that $\kappa(n)$ is roughly $\delta(2L+1)$ for most integers n. We will now prove that most integers belong to

$$B = \left\{ n : 1 \leq n \leq N, \kappa(n) > \frac{\delta}{8}(2L+1) \right\}$$

unless $\kappa(n)$ is surprisingly large for some n. Let $A(m) = 1$ if $m \in A$, and $= 0$ otherwise. Note that

$$\sum_{n=1}^{N} \kappa(n) = \sum_{n=1}^{N} \sum_{r=-L}^{L} A(n-rq) = \sum_{a \in A} \#\{r : |r| \leq L, \ 1 \leq a+rq \leq N\}$$

$$\geq (2L+1) \#\{a \in A : Lq < a < N - Lq\} \geq (2L+1)(\delta N - 2Lq).$$

Now assume that each $\kappa(n) \leq (9\delta/8)(2L+1)$ so that

$$\sum_{n=1}^{N} \kappa(n) \leq |B| \frac{9\delta}{8}(2L+1) + (N-|B|) \frac{\delta}{8}(2L+1).$$

We can combine the last two inequalities to obtain $|B| \geq 7N/8 + O(N^{2/3})$. On the other hand, by (5) we have, writing $a = u - rq$, $b = v - sq$, $c = w - tq$,

$$5\eta \delta^3 N^2 (2L+1)^3 \geq \sum_{|r|,|s|,|t| \leq L} \#\{a,b,c \in A : (a+rq) + (c+tq) = 2(b+sq)\}$$

$$= \sum_{u+w=2v} \kappa(u)\kappa(v)\kappa(w) \geq \sum_{\substack{u+w=2v \\ u,v,w \in B}} \kappa(u)\kappa(v)\kappa(w)$$

$$\geq \left(\frac{\delta}{8}(2L+1) \right)^3 \#\{u,v,w \in B : u+w = 2v\};$$

that is

(6) $$\#\{u,v,w \in B : u+w = 2v\} \leq 5 \cdot 8^3 \eta N^2 < N^2/300.$$

We can bound $\#\{u,v,w \in B : u+w = 2v\}$ from below by taking all $\sim N^2/4$ solutions to $u+w = 2v$ with $1 \leq u,v,w \leq N$, and then subtracting, for each $u \notin B$ the number of v for which $1 \leq 2v - u \leq N$ (that is $(N-|B|) \times N/2$) and similarly for w, and then subtracting, for each $v \notin B$ the number of $u, w \in B$ for which $u + w \in B$ (which is no more than $(N-|B|) \times |B|$). Thus

$$\#\{u,v,w \in B : u+w = 2v\} \gtrsim N^2/4 - (N^2 - |B|^2) \gtrsim N^2/64$$

as $|B| \gtrsim 7N/8$, which contradicts (6). Therefore the assumption is false, so that there exists n with $\kappa(n) > (9\delta/8)(2L+1)$.

We deduce that the set

$$A_0 := \{r + L + 1 : n - rq \in A\} \subset \{1, \ldots, 2L+1\}$$

has $\geq \frac{9}{8}\delta(2L+1)$ elements, but no 3-term arithmetic progression. Let $N_1 := [N^{\delta^4/10^{20}}]$, which is smaller than $2L+1$. Select the subinterval $[s+1, s+N]$ of $[1, 2L+1]$ containing the most elements of A_0, so that

$$A_1 := \{j : 1 \leq j \leq N \text{ and } s + j \in A_0\}$$

does not contain any non-trivial 3-term arithmetic progressions, and has $\gtrsim \frac{9}{8}\delta N_1$ elements. We have therefore proved the following:

If A is a subset of $\{1, 2, \ldots, N\}$, with δN elements, which does not contain a non-trivial 3-term arithmetic progression, then there exists a subset A_1 of $\{1, 2, \ldots, N_1\}$, with $\gtrsim \frac{9}{8}\delta N_1$ elements, which does not contain a non-trivial 3-term arithmetic progression.

Suppose that $\delta \geq \delta_g = (\frac{8}{9})^g$. If we iterate the above result j times then we have a subset $A_j \subset \{1, 2, \ldots, N_j\}$ containing $\delta_{g-j} N_j$ elements, no three of which form an arithmetic progression, where $N_j \sim N^{\eta_j}$ with $\eta_j := (\frac{8}{9})^{2((2g+1)j - j^2)}/10^{20j}$. Therefore A_g contains all the integers up to N_g and so must contain many three term arithmetic progressions, a contradiction, provided N_g is sufficiently large. This will be the case if $\eta_g \gg 1/\log N$ which follows provided $g < \left(\log \log N / (2 \log(\frac{9}{8}))\right)^{1/2} + O(1)$. Hence we may take any

$$\delta \gg 1/\exp\left(c\sqrt{\log \log N}\right)$$

where $c = \sqrt{\frac{1}{2} \log \frac{9}{8}}$. One can optimize our argument to slightly increase the value of c.

We have therefore proved the following result:

Theorem. *There exists a constant $c > 0$ such that if A is a subset of $\{1, 2, \ldots, N\}$ with N sufficiently large, where A contains at least*

$$N/\exp\left(c\sqrt{\log \log N}\right)$$

elements, then A contains a non-trivial three-term arithmetic progression.

Stronger results are proved in [1, 2, 4].

References

1. J. Bourgain, *On triples in arithmetic progression*, Geom. Funct. Anal. **9** (1999), no. 5, 968–984.
2. D. R. Heath-Brown, *Integer sets containing no arithmetic progressions*, J. London Math. Soc. (2) **35** (1987), no. 3, 385–394.
3. K. F. Roth, *On certain sets of integers*, J. London Math. Soc. **28** (1953), 104–109.
4. E. Szemerédi, *Integer sets containing no arithmetic progressions*, Acta Math. Acad. Sci. Hungar. **56** (1990), no. 1-2, 155–158.

ALFRÉD RÉNYI INSTITUTE OF MATHEMATICS, HUNGARIAN ACADEMY OF SCIENCES, POB 127, 1364 BUDAPEST, HUNGARY

E-mail address: szemered@nyuszik.rutgers.edu

Bounds on Exponential Sums over Small Multiplicative Subgroups

Pär Kurlberg

ABSTRACT. We show that there is significant cancellation in certain exponential sums over small multiplicative subgroups of finite fields, giving an exposition of the arguments by Bourgain and Chang [6].

1. Introduction

Let $\psi \colon \mathbb{F}_p \to \mathbb{C}$ be any non-trivial additive character in $\mathbb{F}_p$ (that is, $\psi(x) = \exp(2\pi i x \xi/p)$ for all $x \in \mathbb{F}_p$, for some $\xi \in \mathbb{F}_p^\times$), and let H be a subset of $\mathbb{F}_p$. We are interested in obtaining good upper bounds for

$$\left| \sum_{x \in H} \psi(x) \right|;$$

that is, bounds that are significantly smaller than $|H|$. A traditional analytic number theory approach when H is the multiplicative subgroup of $\mathbb{F}_p$ of index m is to "complete the sum": We have

$$\frac{1}{m} \sum_{\substack{\chi \,(\mathrm{mod}\, p) \\ \chi^m = \chi_0}} \chi(n) = \begin{cases} 1 & \text{if } n \in H, \\ 0 & \text{otherwise}; \end{cases}$$

where the sum runs through the Dirichlet characters (mod p) with order dividing m. Therefore

$$\sum_{x \in H} \psi(x) = \sum_{n \in \mathbb{F}_p} \psi(n) \frac{1}{m} \sum_{\chi : \chi^m = \chi_0} \chi(n) = \frac{1}{m} \sum_{\chi : \chi^m = \chi_0} \sum_{n \in \mathbb{F}_p} \psi(n) \chi(n).$$

The last sum, $\sum_{n \in \mathbb{F}_p} \psi(n)\chi(n)$, is a Gauss sum when $\chi \neq \chi_0$ and is known to have absolute value $\sqrt{p}$; and $\sum_{n \in \mathbb{F}_p} \psi(n)\chi_0(n) = -1$. We deduce that

$$\left| \sum_{x \in H} \psi(x) \right| < \sqrt{p}.$$

2000 *Mathematics Subject Classification.* Primary 11L07; Secondary 11T23.

The author was partially supported by grants from the Göran Gustafsson Foundation, the Royal Swedish Academy of Sciences, and the Swedish Research Council.

This is the final form of the paper.

©2007 American Mathematical Society

This is non-trivial when H has substantially more than $p^{1/2}$ elements and classical arguments can sometimes give non-trivial bounds for interesting sets H as small as $p^{1/4}$, but not much smaller. For H a multiplicative subgroup, the first bound of the form $\sum_{x \in H} \psi(x) \ll_\delta p^{-\delta}|H|$ with $\delta > 0$ and for $|H|$ significantly smaller than $p^{1/2}$ was obtained when $|H| \gg_\epsilon p^{3/7+\epsilon}$ (for all $\epsilon > 0$) by Shparlinskiĭ [14], and later refined to $|H| \gg_\epsilon p^{3/8+\epsilon}$ by Konyagin and Shparlinskiĭ (unpublished), for $|H| \gg_\epsilon p^{1/3+\epsilon}$ by Heath-Brown and Konyagin [12], and for $|H| \gg_\epsilon p^{1/4+\epsilon}$ by Konyagin [13]. An essential ingredient in these results are upper bounds on the number of $\mathbb{F}_p$-points on certain curves/varieties that significantly go beyond what the Weil bounds give.

In several recent articles Bourgain along with Chang, Glibichuk, and, Konyagin showed how to get non-trivial upper bounds for various interesting H that are much smaller, using completely different methods — the techniques of additive combinatorics. The aim of this note is to give an exposition of these ideas in the simplest case[1] by showing that there is significant cancellation in such exponential sums over small multiplicative subgroups H of the finite field $\mathbb{F}_p$.

Theorem 1.1. *Given $\alpha > 0$, there exists $\beta = \beta(\alpha) > 0$ such that if $|H| > p^\alpha$, and H is a multiplicative subgroup of $\mathbb{F}_p$, then*

$$(1.1) \qquad \sum_{x \in H} \psi(x) \ll p^{-\beta}|H|.$$

A proof of this result was first sketched by Bourgain and Konyagin in [10], and detailed proofs were subsequently given by Bourgain, Glibichuk, and Konyagin in [8]. This note is based on the arguments by Bourgain and Chang in [6], and is a somewhat streamlined version of notes from a lecture series given at KTH.

However, as alluded to above, the idea of using additive combinatorics is very versatile. For instance, in [2, 4] Bourgain showed that under certain circumstances it is enough to assume that H has a small multiplicative doubling set, i.e., that $|H \cdot H| < |H|^{1+\tau}$ for $\tau > 0$ small. In particular, one can take $H = \{g^t : t_0 \leq t \leq t_1\}$ as long as the multiplicative order of g modulo p and $t_1 - t_0$ are not too small, and thus it is also possible to non-trivially bound incomplete exponential sums over small (as well as large) multiplicative subgroups. Further, by suitably generalizing the sum-product theorem to subsets of $\mathbb{F}_p \times \mathbb{F}_p$ (some care is required since there are subsets of $\mathbb{F}_p \times \mathbb{F}_p$, e.g., any line passing through $(0,0)$, that violate a naive generalization of the sum-product theorem), Bourgain showed that there is considerable cancellation in sums of the form $\sum_{s_1=1}^t |\sum_{s_2=1}^t \psi(ag^{s_1} + bg^{s_1 s_2})|$ (consequently proving equidistribution for so-called Diffie–Hellman triples in $\mathbb{F}_p^3$) and in [3, 5] he obtained bounds for Mordell type exponential sums $\sum_{x=1}^p \psi\bigl(f(x)\bigr)$, where $f(x) = \sum_{i=1}^r a_i x^{k_i}$ is a sparse polynomial (under suitable conditions on the k_i's.) Moreover, in [6, 7] Bourgain and Chang obtained bounds on sums over multiplicative subgroups (and "almost subgroups") of general finite fields $\mathbb{F}_{p^n}$, respectively $\mathbb{Z}/q\mathbb{Z}$ where q is allowed to be composite, but with a bounded number of prime divisors.

---

[1]See Section 5 for an easy extension to the case of incomplete sums.

1.1. A brief outline of the argument.
Define an H-invariant probability measure μ_H on $\mathbb{F}_p$ by

$$\mu_H(x) := \begin{cases} 1/|H| & \text{if } x \in H, \\ 0 & \text{otherwise,} \end{cases}$$

and assume that (1.1) is violated, i.e., that there exists $\xi \in \mathbb{F}_p^\times$ for which

(1.2) $$\widehat{\mu}_H(\xi) = \sum_{x \in \mathbb{F}_p} \mu_H(x) \exp\left(\frac{2\pi i x \xi}{p}\right) > p^{-\beta}.$$

Let $\nu = \mu_H * \mu_H^-$, where $\mu_H^-(x) = \mu_H(-x)$, and let ν_k be the k-fold convolution of ν. Using (1.2), it is possible to show (see Proposition 4.4) that for some tiny η and k sufficiently large,

(1.3) $$\sum_{x,\xi \in \mathbb{F}_p} |\widehat{\nu}_k(\xi)|^2 |\widehat{\nu}_k(x\xi)|^2 \nu_k(x) > p^{-10\eta} \sum_{\xi \in \mathbb{F}_p} |\widehat{\nu}_k(\xi)|^2,$$

and that the support of $\widehat{\nu}_k$ is essentially contained in the set of "large Fourier coefficients" Λ_δ (cf. Proposition 4.2.) Now, $\widehat{\nu}_k$ being essentially supported on Λ_δ means that $\widehat{\nu}_k$ and $\widehat{\nu}_{2k}$ are "similar" (note that $\widehat{\nu}_{2k}(\xi) = \widehat{\nu}_k(\xi)^2$, and $\widehat{\nu}_k(\xi) \geq 0$ for all ξ), hence ν_k and $\nu_{2k} = \nu_k * \nu_k$ are also similar, and this might be seen as a form of statistical, or approximate, additive invariance for the measure ν_k. Further, by Parseval, (1.3) says that $\sum_{x,y \in \mathbb{F}_p} \nu_{2k}(y) \nu_{2k}(x^{-1}y) \nu_k(x) > p^{-10\eta} \sum_{x \in \mathbb{F}_p} \nu_k(x)^2$, which we may interpret as $\sum_{y \in \mathbb{F}_p} \nu_{2k}(y) \nu_{2k}(x^{-1}y)$ being correlated with ν_k, and this in turn might be seen as statistical multiplicative invariance. (Also see Remarks 4.2 and 4.3.) With S_1 being the set of points assigned large relative mass (i.e., those x for which $\nu_k(x)$ is close to $\|\nu_k\|_\infty$) as a starting point, these invariance properties can then be used to find a subset of S_1 with both small sum and product sets. More precisely, using (1.3), together with the Balog–Gowers–Szemerédi theorem (cf. Theorem 2.2) in multiplicative form, we can find a fairly large subset $S_3 \subset S_1$ with a small product set. Using the Balog–Gowers–Szemerédi theorem again, but in additive form, we then find a large subset $S_4 \subset S_3$ which has a small sum set. Now, since $S_4 \subset S_3$, S_4 also has a small product set, hence it contradicts the sum-product theorem (cf. Theorem 2.1.)

2. Some additive combinatorics results

We will need two essential ingredients from additive combinatorics. First we recall the sum-product theorem for subsets of $\mathbb{F}_p$, due to Bourgain, Katz and Tao [9] (for an expository note, see [11].)

Theorem 2.1. *For any $\epsilon > 0$ there exists $\delta = \delta(\epsilon)$ such that the following holds: If $A \subset \mathbb{F}_p$ is a subset for which $p^\epsilon < |A| < p^{1-\epsilon}$ then*

$$|A + A| + |A \cdot A| \gg |A|^{1+\delta}.$$

We will also need the following version of the Balog–Gowers–Szemerédi theorem (this version of Theorem BGS' in [6] is an immediate consequence of Theorem 5 in Balog's article herein [1]):

Theorem 2.2. *Let A and B be finite subsets of an additive abelian group, Z, and G be a subset of $A \times B$, and let $S = \{a + b : (a,b) \in G\}$. If $|A|, |B|, |S| \leq N$*

and $|G| \geq \alpha N^2$ then there is an $A' \subset A$ such that

(2.1) $\quad$ (i) $|A' + A'| \leq \dfrac{2^{37}}{\alpha^8} N,$ $\quad$ (ii) $|A'| \geq \dfrac{\alpha^4}{2^{15}} N.$

3. The main technical result

In this section we prove the key technical result (cf. [6, Proposition 2.1].):

Proposition 3.1. *Let μ be a probability measure on $\mathbb{F}_p$. If there exists a constant $\Delta \in (0, \frac{1}{2}]$ such that*

(3.1) $$\sum_{\xi, y \in \mathbb{F}_p} |\widehat{\mu}(\xi)|^2 |\widehat{\mu}(y\xi)|^2 \mu(y) > \Delta \sum_{\xi \in \mathbb{F}_p} |\widehat{\mu}(\xi)|^2,$$

and

(3.2) $$\mu(0), \sum_{x \in \mathbb{F}_p} \mu(x)^2 < \Delta/4$$

then there exist a subset $S \subset \mathbb{F}_p^\times$ such that

(3.3) $$\frac{\Delta^{254}}{2^{900}} p < |S| \sum_{\xi \in \mathbb{F}_p} |\widehat{\mu}(\xi)|^2 < \frac{8}{\Delta} p,$$

and

$$|S + S| + |S \cdot S| < \frac{2^{2729}}{\Delta^{768}} |S|.$$

To prove Proposition 3.1 we will construct a sequence of subsets $\mathbb{F}_p \supset S_1 \supset S_2 \supset S_3 \supset S_4$ such that $|S_i|/|S_{i+1}| = \Delta^{O(1)}$, where S_3 has a small product set and S_4 has a small sum set.

First let us recall some useful properties of the finite Fourier transform. For a given probability measure μ on $\mathbb{F}_p$ define its Fourier transform to be

$$\widehat{\mu}(\xi) := \sum_{x \in \mathbb{F}_p} \mu(x) \psi(x\xi),$$

so that $\overline{\widehat{\mu}(\xi)} = \widehat{\mu}(-\xi)$. With this normalization, Parseval's formula reads as

$$p \sum_{x \in \mathbb{F}_p} |\mu(x)|^2 = \sum_{\xi \in \mathbb{F}_p} |\widehat{\mu}(\xi)|^2.$$

As μ is a probability measure, we see that

$$\phi(x) := p(\mu * \mu^-)(x) = \sum_{\xi \in \mathbb{F}_p} |\widehat{\mu}(\xi)|^2 \psi(x\xi)$$

is ≥ 0 for all x. We will replace the middle term in (3.3) by $|S|\phi(0)$. Moreover,

$$\sum_{x \in \mathbb{F}_p} \phi(x) = p,$$

since $\mu * \mu^-$ is also a probability measure. From the Fourier expansion of ϕ, we have

(3.4) $$\max_{x \in \mathbb{F}_p} \phi(x) = \phi(0) = p \cdot (\mu * \mu^-)(0) = p \sum_x \mu(x)^2 \leq \Delta p/4$$

by (3.2).

3.1. Multiplicative stability. We obtain the following form of "statistical multiplicative stability."

Lemma 3.2. *If* (3.1) *and* (3.2) *hold then*

(3.5) $$\sum_{x \in \mathbb{F}_p} \sum_{y \in \mathbb{F}_p^\times} \phi(x)\phi(xy)\mu(y) > \frac{3}{4}\Delta p \phi(0).$$

PROOF. For y fixed, we have

$$\sum_{x \in \mathbb{F}_p} \phi(x)\phi(xy) = \sum_{\xi,\tau \in \mathbb{F}_p} |\widehat{\mu}(\xi)|^2 |\widehat{\mu}(\tau)|^2 \sum_{x \in \mathbb{F}_p} \psi(x\tau + xy\xi) = p \sum_{\xi \in \mathbb{F}_p} |\widehat{\mu}(\xi)|^2 |\widehat{\mu}(-y\xi)|^2.$$

Summing this over all $y \in \mathbb{F}_p^\times$, we see that the left hand side of (3.5) equals

$$p \sum_{y,\xi \in \mathbb{F}_p} |\widehat{\mu}(\xi)|^2 |\widehat{\mu}(-y\xi)|^2 \mu(y) - p \sum_{\xi \in \mathbb{F}_p} |\widehat{\mu}(\xi)|^2 |\widehat{\mu}(0)|^2 \mu(0)$$

$$\geq p \Delta \sum_{\xi \in \mathbb{F}_p} |\widehat{\mu}(\xi)|^2 - p(\Delta/4)|\widehat{\mu}(0)|^2 \sum_{\xi \in \mathbb{F}_p} |\widehat{\mu}(\xi)|^2$$

by (3.1) and (3.2), as $|\widehat{\mu}(-y\xi)|^2 = |\widehat{\mu}(y\xi)|^2$, which yields the result since $|\widehat{\mu}(0)|^2 \leq 1$. □

Remark 3.1. Note that

$$\sum_{x \in \mathbb{F}_p} \sum_{y \in \mathbb{F}_p^\times} \phi(x)\phi(xy)\mu(y) \leq \phi(0) \sum_{x,y \in \mathbb{F}_p} \phi(x)\mu(y) \leq p\phi(0).$$

In our applications, we shall take $\Delta = p^{-\epsilon}$, and for this choice of Δ, the lower bound (3.5) is fairly good.

As a starting point for a multiplicatively stable subset, we use the points which are assigned large measure by $\mu * \mu^-$.

Lemma 3.3. *If* (3.1) *and* (3.2) *hold and*

$$S_1 := \{x \in \mathbb{F}_p : \phi(x) > \tfrac{1}{8}\Delta\phi(0)\}$$

then

(3.6) $$\sum_{\substack{x \in S_1, y \in \mathbb{F}_p^\times \\ xy \in S_1}} \phi(x)\phi(xy)\mu(y) > \tfrac{1}{2}\Delta p \phi(0).$$

PROOF. We have

$$\sum_{\substack{x \in S_1, y \in \mathbb{F}_p^\times \\ xy \in S_1}} \geq \sum_{\substack{x \in \mathbb{F}_p \\ y \in \mathbb{F}_p^\times}} - \sum_{\substack{x \in \mathbb{F}_p \setminus S_1 \\ y \in \mathbb{F}_p^\times}} - \sum_{\substack{x \in \mathbb{F}_p, y \in \mathbb{F}_p^\times \\ xy \notin S_1}}.$$

By (3.5), the first term on the right-hand side is $> (\tfrac{3}{4})\Delta p \phi(0)$. The second term

$$\sum_{\substack{x \in \mathbb{F}_p \setminus S_1 \\ y \in \mathbb{F}_p^\times}} \phi(x)\phi(xy)\mu(y)$$

is, since $\phi(x) \leq \Delta\phi(0)/8$ for $x \notin S_1$, bounded by

$$\frac{\Delta\phi(0)}{8} \sum_{\substack{x \in \mathbb{F}_p \setminus S_1 \\ y \in \mathbb{F}_p^\times}} \phi(xy)\mu(y) \leq \frac{\Delta\phi(0)}{8} \sum_{y \in \mathbb{F}_p^\times} \mu(y) \sum_{x \in \mathbb{F}_p} \phi(xy) \leq \frac{\Delta p\phi(0)}{8}$$

since $\sum_{x \in \mathbb{F}_p} \phi(xy) = p$ for $y \neq 0$ and μ is a probability measure. Similarly, the third term is bounded by $\Delta p\phi(0)/8$, hence the left-hand side of (3.6) is $> \Delta p\phi(0)(\frac{3}{4} - \frac{1}{8} - \frac{1}{8}) \geq \Delta p\phi(0)/2$. $\square$

We proceed to estimate the size of S_1.

Lemma 3.4. *If* (3.1) *and* (3.2) *hold then*

$$(3.7) \qquad \frac{\Delta p}{2\phi(0)} < |S_1| < \frac{8p}{\Delta\phi(0)}.$$

Moreover, if we let

$$S_2 := S_1 \setminus \{0\} \subset \mathbb{F}_p^\times,$$

then $|S_2| \geq |S_1|/2$.

PROOF. For the lower bound, note that

$$(3.8) \qquad |S_1| = \sum_{y \in \mathbb{F}_p} |S_1|\mu(y) \geq \sum_{y \in \mathbb{F}_p^\times} |S_1 \cap y^{-1}S_1|\mu(y) = \sum_{\substack{x \in S_1, y \in \mathbb{F}_p^\times \\ xy \in S_1}} \mu(y)$$

$$\geq \frac{1}{\phi(0)^2} \sum_{\substack{x \in S_1, y \in \mathbb{F}_p^\times \\ xy \in S_1}} \phi(x)\phi(xy)\mu(y) > \frac{\Delta p}{2\phi(0)}$$

by (3.6), which is ≥ 2 by (3.4), so that $|S_2| \geq |S_1|/2$. For the upper bound, note that

$$|S_1| < \frac{8}{\Delta\phi(0)} \sum_{x \in S_1} \phi(x) \leq \frac{8}{\Delta\phi(0)} \sum_{x \in \mathbb{F}_p} \phi(x) = \frac{8p}{\Delta\phi(0)}. \qquad \square$$

To show that there are many y such that $|S_2 \cap y^{-1}S_2|$ is fairly large, we begin by giving a lower bound on the expected size of the intersection.

Lemma 3.5. *If* (3.1) *and* (3.2) *hold then*

$$(3.9) \qquad \sum_{y \in \mathbb{F}_p^\times} |S_2 \cap y^{-1}S_2|\mu(y) \geq \frac{\Delta p}{4\phi(0)}$$

PROOF. Since $S_2 \cap y^{-1}S_2 = (S_1 \cap y^{-1}S_1) \setminus \{0\}$ for all $y \in \mathbb{F}_p^\times$ we have

$$\sum_{y \in \mathbb{F}_p^\times} |S_2 \cap y^{-1}S_2|\mu(y) \geq \sum_{y \in \mathbb{F}_p^\times} |S_1 \cap y^{-1}S_1|\mu(y) - \sum_{y \in \mathbb{F}_p^\times} \mu(y)$$

$$> \frac{\Delta p}{2\phi(0)} - 1 \geq \frac{\Delta p}{4\phi(0)}$$

by the right-hand side of (3.8) and as $\sum_{y \in \mathbb{F}_p} \mu(y) = 1$, and then by (3.4). $\square$

In the next result we show that there are many y for which $|S_2 \cap y^{-1}S_2|$ is large:

Lemma 3.6. *If* (3.1) *and* (3.2) *hold and*

(3.10) $$T := \left\{ y \in \mathbb{F}_p^\times : |S_2 \cap y^{-1} S_2| > \frac{\Delta p}{8\phi(0)} \right\}$$

then

(3.11) $$|T| \geq \frac{\Delta^5}{2^{15}} |S_1|$$

PROOF.

(3.12) $$|S_2| \mu(T) = |S_2| \sum_{y \in T} \mu(y) \geq \sum_{y \in T} |S_2 \cap y^{-1} S_2| \mu(y)$$
$$= \sum_{y \in \mathbb{F}_p^\times} |S_2 \cap y^{-1} S_2| \mu(y) - \sum_{y \in \mathbb{F}_p^\times \setminus T} |S_2 \cap y^{-1} S_2| \mu(y)$$
$$\geq \frac{\Delta p}{8\phi(0)} > \frac{\Delta^2}{64} |S_2|$$

by (3.9) and from the definition of T, and then by (3.7) and the trivial bound $|S_2| \leq |S_1|$, so that $\mu(T) > \Delta^2/64$.

On the other hand, by Cauchy-Schwartz and Parseval's identity,

$$\mu(T) \leq |T|^{1/2} \left(\sum_{x \in T} \mu(x)^2 \right)^{1/2} \leq |T|^{1/2} \left(\frac{1}{p} \sum_{\xi \in \mathbb{F}_p} |\widehat{\mu}(\xi)|^2 \right)^{1/2} = \left(\frac{|T|\phi(0)}{p} \right)^{1/2},$$

so that $|T| \geq p\Delta^4/(2^{12}\phi(0)) > (\Delta^5/2^{15})|S_1|$, by (3.7). $\square$

Thus, by shrinking T if necessary, we have found a set T such that
$$(\Delta^5/2^{15})|S_2| \leq |T| \leq |S_2|$$
with the property that for all $y \in T$,

(3.13) $$|S_2 \cap y^{-1} S_2| > \frac{\Delta p}{8\phi(0)} > \frac{\Delta^2}{2^6} |S_1| \geq \frac{\Delta^2}{2^6} |S_2|$$

by (3.7).

Let $G := \{(x, y) : x \in S_2, y \in T, xy \in S_2\} \subset S_2 \times T \subset \mathbb{F}_p^\times \times \mathbb{F}_p^\times$. By (3.13), the number of x such that $(x, y) \in G$ is at least $2^{-6} \Delta^2 |S_2|$ for each $y \in T$. Therefore, since $|T| \geq 2^{-15} \Delta^5 |S_2|$, we find that
$$|G| \geq 2^{-6} \Delta^2 |S_2| \cdot 2^{-15} \Delta^5 |S_2| = (\Delta/8)^7 |S_2|^2.$$

By the definition of G we know that
$$\{st : (s, t) \in G\} \subset S_2;$$
so, with g a primitive root modulo p and defining $\log_{g,p}(s)$ to be the smallest integer $m \geq 0$ such that $g^m \equiv s \bmod p$, and by taking $A = \{\log_{g,p} s : s \in S_2\}$, $B = \{\log_{g,p} t : t \in T\}$ with $N = |S_2|$ and $\alpha = (\Delta/8)^7$ in Theorem 2.2, we obtain a subset A' of A, with $|A'| > (\Delta^{28}/2^{99})|A|$, for which
$$|A' + A'| \leq (2^{205}/\Delta^{56})N < (2^{304}/\Delta^{84})|A'|.$$

Therefore $S_3 = \{g^a : a \in A'\}$ is a subset of S_2 for which

(3.14) $$|S_3| > (\Delta^{28}/2^{100})|S_1|,$$

by Lemma 3.4, and
$$|S_3 \cdot S_3| \leq (2^{304}/\Delta^{84})|S_3|.$$

3.2. Additive stability. We finish the proof of Proposition 3.1 by finding a subset S_4 of S_3 with a small sum set. We first show that S_3 exhibits "statistical additive stability"; to do this we only need to use that $S_3 \subset S_1$, together with the definition of S_1.

Lemma 3.7. *If* (3.1) *and* (3.2) *hold then*

(3.15) $$\sum_{x_1,x_2 \in S_3} \phi(x_1 - x_2) > 2^{-6}\Delta^2\phi(0)|S_3|^2.$$

PROOF. Recalling that $\phi(x) = p(\mu * \mu^-)(x)$, we find, using the Cauchy–Schwarz inequality, that

$$\left(\frac{1}{p}\sum_{x \in S_3} \phi(x)\right)^2 = \left(\sum_{y \in \mathbb{F}_p} \mu(y) \sum_{x \in S_3} \mu(x+y)\right)^2 \leq \sum_{y \in \mathbb{F}_p} \mu(y)^2 \cdot \sum_{y \in \mathbb{F}_p}\left(\sum_{x \in S_3} \mu(x+y)\right)^2$$

$$= \frac{\phi(0)}{p} \sum_{x_1,x_2 \in S_3} \sum_{y \in \mathbb{F}_p} \mu(x_1+y)\mu(x_2+y) = \frac{\phi(0)}{p^2} \sum_{x_1,x_2 \in S_3} \phi(x_1 - x_2).$$

Now $\sum_{x \in S_3} \phi(x) > \frac{\Delta}{8}\phi(0)|S_3|$, since $S_3 \subset S_1$, and the lemma follows. □

To obtain an additively stable subset we will, as before, use Theorem 2.2. First, let

(3.16) $$S_0 := \{x \in \mathbb{F}_p : \phi(x) > 2^{-7}\Delta^2\phi(0)\}$$

Then
$$|S_0| \leq \frac{2^7}{\Delta^2\phi(0)} \sum_{x \in S_0} \phi(x) \leq \frac{2^7 p}{\Delta^2\phi(0)} \leq \frac{2^8}{\Delta^3}|S_1| < \frac{2^{108}}{\Delta^{31}}|S_3|$$

by (3.7) and then (3.14).

Using S_0, S_3 we can now define a fairly large graph G'.

Lemma 3.8. *If* (3.1) *and* (3.2) *hold then*
$$G' := \{(x_1, -x_2) \in S_3 \times (-S_3) : x_1 - x_2 \in S_0\} \subset S_3 \times (-S_3).$$
has at least $2^{-7}\Delta^2|S_3|^2$ *elements.*

PROOF. We have
$$|G'| \cdot \phi(0) \geq \sum_{(x_1,-x_2) \in G'} \phi(x_1 - x_2)$$
$$= \sum_{x_1,x_2 \in S_3} \phi(x_1 - x_2) - \sum_{(x_1,-x_2) \in S_3 \times (-S_3) \setminus G'} \phi(x_1 - x_2)$$
$$\geq 2^{-6}\Delta^2\phi(0)|S_3|^2 - 2^{-7}\Delta^2\phi(0)|S_3|^2$$

by (3.15) and (3.16), and the result follows. □

Since $\{x_1 - x_2 : (x_1, -x_2) \in G'\} \subset S_0$ we can apply Theorem 2.2 with $A = S_3$, $B = -S_3$, $G = G'$, $N = (2^{108}/\Delta^{31})|S_3|$ and $\alpha = \Delta^{64}/2^{223}$ to obtain a subset $S_4 \subset S_3$ with

(3.17) $$|S_4| > \frac{\Delta^{256}}{2^{907}}N = \frac{\Delta^{225}}{2^{799}}|S_3|$$

for which
$$|S_4 + S_4| < \frac{2^{1821}}{\Delta^{512}} N = \frac{2^{1929}}{\Delta^{543}}|S_3| < \frac{2^{2728}}{\Delta^{768}}|S_4|.$$
Moreover, since $S_4 \subset S_3$, we find that
$$|S_4 \cdot S_4| \leq |S_3 \cdot S_3| < (2^{304}/\Delta^{84})|S_3| < (2^{1103}/\Delta^{309})|S_4|.$$
Finally, by (3.7), then (3.17), (3.14), and Lemma 3.4, we have
$$\frac{8p}{\Delta\phi(0)} > |S_1| \geq |S_4| > \frac{\Delta^{225}}{2^{799}}|S_3| > \frac{\Delta^{253}}{2^{899}}|S_1| > \frac{\Delta^{254}}{2^{900}}\frac{p}{\phi(0)}.$$
Taking $S = S_4$ we have found a set with the desired properties.

4. Proof of Theorem 1.1

4.1. Preliminaries. Let μ be a given probability measure on $\mathbb{F}_p$. Recall that the Fourier transform of μ was defined to be $\widehat{\mu}(\xi) := \sum_{x \in \mathbb{F}_p} \mu(x)\psi(x\xi)$, and hence $\overline{\widehat{\mu}(\xi)} = \widehat{\mu}(-\xi)$. With this normalization, Parseval's formula reads as $p \sum_{x \in \mathbb{F}_p} |\mu(x)|^2 = \sum_{\xi \in \mathbb{F}_p} |\widehat{\mu}(\xi)|^2$. Moreover, if ν is another probability measure then
$$\sum_{x \in \mathbb{F}_p} \mu(x)\widehat{\nu}(x) = \sum_{\xi \in \mathbb{F}_p} \overline{\widehat{\mu}(\xi)}\nu(-\xi) = \sum_{\xi \in \mathbb{F}_p} \widehat{\mu}(-\xi)\nu(-\xi) = \sum_{\xi \in \mathbb{F}_p} \widehat{\mu}(\xi)\nu(\xi)$$

Let $\nu := \mu * \mu^-$, that is $\nu(x) = \sum_{y,z : y-z=x} \mu(y)\mu(z)$, so that $\nu(-x) = \nu(x)$ and $\widehat{\nu}(x) = |\widehat{\mu}(x)|^2$. If ν_k is the k-fold convolution of ν, that is
$$\nu_k(x) := \sum_{\substack{y_1,y_2,\ldots,y_k \in \mathbb{F}_p \\ y_1+y_2+\cdots+y_k=x}} \nu(y_1)\nu(y_2)\cdots\nu(y_k),$$
then $\widehat{\nu_k}(x) = |\widehat{\mu}(x)|^{2k} \geq 0$. Notice that
$$\nu(x) = \sum_{y,z : y-z=x} \mu(y)\mu(z) \leq \max_z \mu(z) \sum_y \mu(y) = \max_z \mu(z)$$
for all x; and similarly
$$\max_x \nu_k(x) \leq \max_z \mu(z) \quad \text{for all } k. \tag{4.1}$$

We have
$$\|\mu_H\|_2^2 = \sum_{x \in \mathbb{F}_p} |\mu_H(x)|^2 = 1/|H|.$$
Note that $\mu_H(hx) = \mu_H(x)$ for all $h \in H$, and so $\widehat{\mu}_H(hx) = \widehat{\mu}_H(x)$ for all $h \in H$, and $\nu_k(hx) = \nu_k(x)$ for all $h \in H$ and $k \geq 1$.

4.2. The set of large Fourier coefficients. Given $\delta > 0$, let
$$\Lambda_\delta := \{\xi \in \mathbb{F}_p : |\widehat{\mu}(\xi)| > p^{-\delta}\}$$
be the set of "large" Fourier coefficients of μ.

Lemma 4.1. *Suppose that $\mu = \mu_H$. We have*
$$|\Lambda_\delta| \leq p^{1+2\delta}/|H|.$$
Also if $|\widehat{\mu}_H(\xi)| > p^{-\delta}$ for some nonzero $\xi \in \mathbb{F}_p^\times$, then
$$|\Lambda_\delta| \geq |H|.$$

PROOF. For any measure μ on $\mathbb{F}_p$ we have

$$|\Lambda_\delta| \leq p^{2\delta} \sum_{\xi \in \Lambda_\delta} |\widehat{\mu}(\xi)|^2 \leq p^{2\delta} \sum_{\xi \in \mathbb{F}_p} |\widehat{\mu}(\xi)|^2 = p^{1+2\delta} \sum_{x \in \mathbb{F}_p} |\mu(x)|^2,$$

and the first result follows since this last sum equals $1/|H|$ for $\mu = \mu_H$. For the second result note that if $\xi \in \Lambda_\delta$ then $|\widehat{\mu}_H(h\xi)| = |\widehat{\mu}_H(\xi)| > p^{-\delta}$ for all $h \in H$, so that $h\xi \in \Lambda_\delta$ for all $h \in H$. □

We will now show that it is possible to find k, δ so that the support of $\widehat{\nu}_k$ is, in L^2-sense, essentially given by Λ_δ.

Proposition 4.2. *For any measure μ on $\mathbb{F}_p$, where $p \geq 3$, and any $\eta \geq 5/(p^3 \log p)$, there exists an integer $k \geq 4$ and*

$$\delta \in (0, \eta/k^2)$$

such that

(4.2) $$p^{-\eta}|\Lambda_\delta| \leq \sum_{\xi \in \mathbb{F}_p} |\widehat{\nu}_k(\xi)|^2 \leq p^\eta |\Lambda_\delta|$$

and, in particular,

(4.3) $$\sum_{\xi \in \mathbb{F}_p} |\widehat{\nu}_k(\xi)|^2 \leq p^{2\eta} \sum_{\xi \in \Lambda_\delta} |\widehat{\nu}_k(\xi)|^2.$$

PROOF. For any $k \in \mathbb{N}$ we have
(4.4)
$$\sum_{\xi \in \mathbb{F}_p} |\widehat{\nu}_k(\xi)|^2 = \sum_{\xi \in \Lambda_{1/k}} |\widehat{\nu}_k(\xi)|^2 + \sum_{\xi \notin \Lambda_{1/k}} |\widehat{\nu}_k(\xi)|^2 \leq |\Lambda_{1/k}| + p(p^{-1/k})^{4k} = |\Lambda_{1/k}| + 1/p^3$$

since each $\widehat{\nu}_k(\xi) \leq 1$.

We define a sequence of integers $k_0 = 4 < k_1 < \ldots$ where $k_{i+1} = [k_i^2/\eta] + 1$ for each $i \geq 0$, and let $\delta_i = 1/k_{i+1}$ for each i. Note that $k_i^2/\eta < k_{i+1} = 1/\delta_i$ so that $k_i \delta_i < \eta/k_i \leq \eta/4$. Since $\widehat{\nu}_{k_i}(\xi) = |\widehat{\mu}_H(\xi)|^{2k_i}$, we have

$$\sum_{\xi \in \Lambda_{\delta_i}} |\widehat{\nu}_{k_i}(\xi)|^2 > |\Lambda_{\delta_i}| \cdot p^{-4k_i \delta_i} \geq |\Lambda_{\delta_i}| \cdot p^{-\eta}.$$

We note that the lower bound in (4.2) follows from this, as well as (4.3), once we establish the upper bound in (4.2).

Now, there exists an integer $i \in [0, M]$, where $M = 2([1/\eta] + 1)$, such that $\sum_{\xi \in \mathbb{F}_p} |\widehat{\nu}_{k_i}(\xi)|^2 \leq p^\eta |\Lambda_{\delta_i}|$ else

$$p^\eta |\Lambda_{1/k_{i+1}}| = p^\eta |\Lambda_{\delta_i}| < \sum_{\xi \in \mathbb{F}_p} |\widehat{\nu}_{k_i}(\xi)|^2 \leq |\Lambda_{1/k_i}| + 1/p^3 \leq |\Lambda_{1/k_i}|(1 + 1/p^3)$$

for each i, by (4.4), and so

$$|\Lambda_{1/k_M}| < p^{-M\eta}|\Lambda_{1/k_0}|(1 + 1/p^3)^M \leq p^{1-M\eta}(1 + 1/p^3)^M \leq p^{-1}(1 + 1/p^3)^M < 1$$

since $M \leq \frac{1}{2} p^3 \log p$, which is untrue (as $0 \in \Lambda_{1/k}$ for all $k \in \mathbb{N}$).

We select $k = k_i$ and $\delta = \delta_i$. □

Remark 4.1. Note that the proof gives us $k \ll \exp\bigl(\exp(O(1/\eta))\bigr)$.

Remark 4.2. Since the support of $\widehat{\nu}_k$ is essentially given by Λ_δ, it is easy to see that the same holds for $\widehat{\nu}_{2k}$; we may interpret this as $\nu_k * \nu_k$ being "similar" to ν_k, and hence that ν_k is "approximately additively stable".

In the following key lemma, the H-invariance of μ_H, and hence of $\widehat{\nu}_k$, is essential.

Lemma 4.3. *For $\mu = \mu_H$ and all $\xi \in \mathbb{F}_p$, we have*
$$\widehat{\nu}_k(\xi)^{4k} \leq \sum_{x \in \mathbb{F}_p} \widehat{\nu}_k(x\xi)^2 \nu_k(x).$$

PROOF. The case $\xi = 0$ is immediate, hence we may assume that $\xi \neq 0$. Now, since $\widehat{\nu}_k(h\xi) = \widehat{\nu}_k(\xi)$ for all $h \in H$, we have
$$\widehat{\nu}_k(\xi)^2 = \sum_{x \in \mathbb{F}_p} \widehat{\nu}_k(x\xi)^2 \mu_H(x) = \sum_{x \in \mathbb{F}_p} \nu_{2k}(-x\xi^{-1}) \widehat{\mu}_H(x),$$
by Parseval's formula. Now note that if μ is any probability measure and $l \geq 1$, then $\sum_x \mu(x) f(x) \leq (\sum_x \mu(x) |f(x)|^l)^{1/l}$. Therefore the above gives
$$\widehat{\nu}_k(\xi)^{4k} \leq \sum_{x \in \mathbb{F}_p} \nu_{2k}(-x\xi^{-1}) |\widehat{\mu}_H(x)|^{2k} = \sum_{x \in \mathbb{F}_p} \nu_{2k}(-x\xi^{-1}) \widehat{\nu}_k(x)$$
since $|\widehat{\mu}_H(x)|^{2k} = \widehat{\nu}(x)^k = \widehat{\nu}_k(x)$ and, applying Parseval one more time, we obtain
$$\widehat{\nu}_k(\xi)^{4k} \leq \sum_{x \in \mathbb{F}_p} \widehat{\nu}_k(-x\xi)^2 \nu_k(-x) = \sum_{x \in \mathbb{F}_p} \widehat{\nu}_k(x\xi)^2 \nu_k(x). \qquad \square$$

We consequently obtain:

Proposition 4.4. *With k, η as in Proposition 4.2, we have*
$$p^{-10\eta} \sum_{\xi \in \mathbb{F}_p} \widehat{\nu}_k(\xi)^2 \leq \sum_{\xi \in \mathbb{F}_p} \sum_{x \in \mathbb{F}_p} \widehat{\nu}_k(\xi)^2 \widehat{\nu}_k(x\xi)^2 \nu_k(x).$$

PROOF. By Proposition 4.2, we have
$$p^{-2\eta} \sum_{\xi \in \mathbb{F}_p} \widehat{\nu}_k(\xi)^2 \leq \sum_{\xi \in \Lambda_\delta} \widehat{\nu}_k(\xi)^2 \leq p^{8k^2\delta} \sum_{\xi \in \Lambda_\delta} \widehat{\nu}_k(\xi)^{4k+2} \leq p^{8\eta} \sum_{\xi \in \mathbb{F}_p} \widehat{\nu}_k(\xi)^{4k+2}$$
which, by Lemma 4.3, is
$$\leq p^{8\eta} \sum_{\xi \in \mathbb{F}_p} \sum_{x \in \mathbb{F}_p} \widehat{\nu}_k(\xi)^2 \widehat{\nu}_k(x\xi)^2 \nu_k(x). \qquad \square$$

Remark 4.3. Since $\widehat{\nu}_k(x\xi) \leq 1$ and ν_k is a probability measure, we find that $\sum_{\xi, x \in \mathbb{F}_p} \widehat{\nu}_k(\xi)^2 \widehat{\nu}_k(x\xi)^2 \nu_k(x) \leq \sum_{\xi \in \mathbb{F}_p} \widehat{\nu}_k(\xi)^2$, so the lower bound on the double sum in Proposition 4.4 is quite good. Further, using Parseval on the two sums over ξ (ignoring the term $x = 0$) we find that $\sum_{y \in \mathbb{F}_p} \nu_{2k}(y) \nu_{2k}(yx^{-1})$, which we can interpret as a multiplicative translate of ν_{2k} with itself, is highly correlated with $\nu_k(x)$. Thus, the Proposition might be interpreted as a statement of "approximate multiplicative stability" of ν_k. (Since the essential support of $\widehat{\nu}_k$ is given by Λ_δ, the same holds for $\widehat{\nu}_{2k}$, so in some sense ν_k and ν_{2k} are "similar.")

To go from statistical additive/multiplicative stability to a subset that contradicts the sum-product Theorem, we will apply Proposition 3.1 with $\mu = \nu_k$ and $\Delta = p^{-10\eta}$ (and note that (4.1) implies (3.2) provided $1/|H| < \Delta/4$), and select δ and k as in Proposition 4.2. Assume that $|\widehat{\mu}_H(\xi)| > p^{-\delta}$ for some $\xi \in \mathbb{F}_p^\times$. We thus obtain a set S such that

$$|S+S| + |S \cdot S| < 2^{2729} p^{7680\eta} |S|.$$

Note that

$$p^{-\eta}|H| \leq p^{-\eta}|\Lambda_\delta| \leq \sum_{\xi \in \mathbb{F}_p} |\widehat{\nu}_k(\xi)|^2 \leq p^\eta |\Lambda_\delta| \leq p^{1+\eta+2\delta}/|H|$$

by (4.2) and Lemma 4.1, so that (3.3) gives, as $2\delta < \eta$,

$$\frac{1}{2^{900}} \frac{|H|}{p^{2542\eta}} < |S| < 8 \frac{p^{1+11\eta}}{|H|}.$$

Now select $\eta = \min\{\alpha/6000, \delta(\alpha/2)/8000\}$, so that the sum-product Theorem 2.1 is violated with $\epsilon = \alpha/2$ for p sufficiently large, and thus $|\widehat{\mu}_H(\xi)| \leq p^{-\delta}$ for all $\xi \in \mathbb{F}_p^\times$. The Theorem follows with $\beta = \delta \gg \exp(-\exp(C/\eta))$ for some constant $C > 0$.

5. Incomplete sums

The proof of Theorem 1.1 can fairly easily be extended to incomplete sums over multiplicative subgroups.

Theorem 5.1. *Let $g \in \mathbb{F}_p^\times$ have multiplicative order at least T, and let $H = \{g^t : 0 \leq t < T\}$. If $|H| = T > p^\alpha$, then*

$$\sum_{x \in H} \psi(x) \ll p^{-\beta}|H|.$$

Define $\mu_H, \widehat{\mu}_H, \nu_k, \Lambda_\delta$ etc. as before. To obtain a contradiction, we will assume that $|\widehat{\mu}_H(\xi_0)| > 2p^{-\delta}$ for some $\xi_0 \in \mathbb{F}_p^\times$.

We begin by showing that Λ_δ, the set of large Fourier coefficients, is almost of size $|H|$, and that $\widehat{\mu}$ is quite large on $\Lambda_\delta \cdot H_1$ for a fairly large subset $H_1 \subset H$.

Lemma 5.2. *Let*

$$H_1 := \{g^t : 0 \leq t < |H|p^{-\delta}/4\}.$$

If $|\widehat{\mu}(\xi_0)| > 2p^{-\delta}$ for some $\xi_0 \in \mathbb{F}_p^\times$, then

$$|\Lambda_\delta| \geq |H_1|.$$

Moreover, if $\xi \in \Lambda_\delta$ and $h \in H_1$, then

$$|\widehat{\mu}_H(h\xi)| > |\widehat{\mu}_H(\xi)|/2.$$

PROOF. For $l \in \mathbb{Z}$ such that $0 \leq l < T$, we have

$$\widehat{\mu}_H(g^l \xi) = \sum_{x \in \mathbb{F}_p} \psi(g^l \xi x) \mu_H(x) = \sum_{x \in \mathbb{F}_p} \psi(\xi x) \mu_H(g^{-l} x)$$

$$= \frac{1}{|H|} \sum_{x \in g^l H} \psi(\xi x) = \frac{1}{|H|} \left(\sum_{x \in H} \psi(\xi x) + 2\theta l \right)$$

for some θ such that $|\theta| \leq 1$. Thus, if $l < |H|p^{-\delta}/4$, then

(5.1) $$|\widehat{\mu}_H(g^l\xi)| > |\widehat{\mu}_H(\xi)| - p^{-\delta}/2.$$

In particular, if $h \in H_1$, then $|\widehat{\mu}_H(h\xi_0)| \geq |\widehat{\mu}_H(\xi_0)| - p^{-\delta}/2 > 2p^{-\delta} - p^{-\delta}/2 > p^{-\delta}$ and hence $|\Lambda_\delta| \geq |H_1|$. Finally, if $\xi \in \Lambda_\delta$ then $|\widehat{\mu}_H(\xi)| > p^{-\delta}$, so the second assertion follows from (5.1). $\square$

Lemma 5.3. *If $\xi \in \Lambda_\delta$, then*
$$\widehat{\nu}_k(\xi)^{4k} \leq 2^{8k^2+6k} p^{2k\delta} \sum_{x \in \mathbb{F}_p} \widehat{\nu}_k(h\xi)^2 \nu_k(x).$$

PROOF. If $\xi \in \Lambda_\delta$, then $|\widehat{\mu}_H(\xi h)| \geq |\widehat{\mu}_H(\xi)|/2$ for all $h \in H_1$. Hence
$$\widehat{\nu}_k(\xi)^2 \leq \frac{2^{4k}}{|H_1|} \sum_{h \in H_1} \widehat{\nu}_k(h\xi)^2 \leq \frac{2^{4k}|H|}{|H_1|} \sum_{x \in \mathbb{F}_p} \widehat{\nu}_k(h\xi)^2 \mu_H(x)$$
$$= 2^{4k+3} p^\delta \sum_{x \in \mathbb{F}_p} \widehat{\nu}_k(h\xi)^2 \mu_H(x)$$

since $|H|/|H_1| \leq 8p^\delta$. Thus, if $\xi \in \Lambda_\delta$, then
$$\widehat{\nu}_k(\xi)^{4k} \leq 2^{8k^2+6k} p^{2k\delta} \left(\sum_{x \in \mathbb{F}_p} \widehat{\nu}_k(h\xi)^2 \mu_H(x) \right)^{2k} \leq 2^{8k^2+6k} p^{2k\delta} \sum_{x \in \mathbb{F}_p} \widehat{\nu}_k(h\xi)^2 \nu_k(x)$$

by the same argument used in the proof of Lemma 4.3. $\square$

Proposition 5.4. *For p sufficiently large,*
$$p^{-11\eta} \sum_{\xi \in \mathbb{F}_p} \widehat{\nu}_k(\xi)^2 \leq \sum_{\xi, x \in \mathbb{F}_p} \widehat{\nu}_k(\xi)^2 \widehat{\nu}_k(\xi x)^2 \nu_k(x).$$

PROOF. Arguing as in the proof of Proposition 4.4 find that
$$p^{-2\eta} \sum_{\xi \in \mathbb{F}_p} \widehat{\nu}_k(\xi)^2 \leq \sum_{\xi \in \Lambda_\delta} \widehat{\nu}_k(\xi)^2 \leq p^{8k^2\delta} \sum_{\xi \in \Lambda_\delta} \widehat{\nu}_k(\xi)^{4k+2} \leq p^{8\eta} \sum_{\xi \in \Lambda_\delta} \widehat{\nu}_k(\xi)^{4k+2}$$

which, by Lemma 5.3 is
$$\leq p^{8\eta+2k\delta} 2^{8k^2+6k} \sum_{\xi \in \Lambda_\delta} \sum_{x \in \mathbb{F}_p} \widehat{\nu}_k(\xi)^2 \widehat{\nu}_k(\xi x)^2 \nu_k(x) \leq p^{9\eta} \sum_{x,\xi \in \mathbb{F}_p} \widehat{\nu}_k(\xi)^2 \widehat{\nu}_k(\xi x)^2 \nu_k(x). \quad \square$$

The rest of the proof is now essentially the same as the proof of Theorem 1.1.

Acknowledgment. It is my pleasure to thank John B. Friedlander and Andrew Granville for their encouragement, as well as many helpful comments and suggestions. I am also grateful to University of Toronto for its hospitality during my visit in April 2007, during which parts of this note were written up.

References

1. A. Balog, *Many additive quadruples*, in this volume.
2. J. Bourgain, *New bounds on exponential sums related to the Diffie–Hellman distributions*, C. R. Math. Acad. Sci. Paris **338** (2004), no. 11, 825–830.
3. ———, *Mordell type exponential sum estimates in fields of prime order*, C. R. Math. Acad. Sci. Paris **339** (2004), no. 5, 321–325.
4. ———, *Estimates on exponential sums related to the Diffie–Hellman distributions*, Geom. Funct. Anal. **15** (2005), no. 1, 1–34.

5. _____, *Mordell's exponential sum estimate revisited*, J. Amer. Math. Soc. **18** (2005), no. 2, 477–499.
6. J. Bourgain and M.-C. Chang, *Exponential sum estimates over subgroups and almost subgroups of $\mathbb{Z}_Q^*$, where Q is composite with few prime factors*, Geom. Funct. Anal. **16** (2006), no. 2, 327–366.
7. _____, *A Gauss sum estimate in arbitrary finite fields*, C. R. Math. Acad. Sci. Paris, to appear.
8. J. Bourgain, A. A. Glibichuk, and S. V. Konyagin, *Estimates for the number of sums and products and for exponential sums in fields of prime order*, J. London Math. Soc. (2) **73** (2006), 380–398.
9. J. Bourgain, N. Katz, and T. Tao, *A sum-product estimate in finite fields and their applications*, Geom. Funct. Anal. **14** (2004), no. 1, 27–57.
10. J. Bourgain and S. V. Konyagin, *Estimates for the number of sums and products and for exponential sums over subgroups in fields of prime order*, C. R. Math. Acad. Sci. Paris **337** (2003), no. 2, 75–80.
11. B. Green, *Sum-product estimates*, available at http://www.dpmms.cam.ac.uk/~bjg23/notes.html.
12. D. R. Heath-Brown and S. V. Konyagin, *New bounds for Gauss sums derived from kth powers, and for Heilbronn's exponential sum*, Q. J. Math. **51** (2000), no. 2, 221–235.
13. S. V. Konyagin, *Estimates for trigonometric sums over subgroups and for Gauss sums*, IV International Conference "Modern Problems of Number Theory and its Applications", Current Problems, Part III (Tula, 2001), Mosk. Gos. Univ. im. Lomonosova, Mekh.-Mat. Fak., Moscow, 2002, pp. 86–114 (Russian).
14. I. E. Shparlinskiĭ, *Estimates for Gauss sums*, Mat. Zametki **50** (1991), no. 1, 122–130.

DEPARTMENT OF MATHEMATICS, ROYAL INSTITUTE OF TECHNOLOGY, SE-100 44 STOCKHOLM, SWEDEN

E-mail address: kurlberg@math.kth.se

Montréal Notes on Quadratic Fourier Analysis

Ben Green

ABSTRACT. These are notes to accompany four lectures that I gave at the *School on additive combinatorics,* held in Montréal, Québec between March 30th and April 5th 2006.

My aim is to introduce "quadratic Fourier analysis" in so far as we understand it at the present time. Specifically, we will describe "quadratic objects" of various types and their relation to additive structures, particularly four-term arithmetic progressions.

I will focus on qualitative results, referring the reader to the literature for the many interesting quantitative questions in this theory. Thus these lectures have a distinctly "soft" flavor in many places.

Some of the notes cover unpublished work which is joint with Terence Tao. This will be published more formally at some future juncture.

1. Lecture 1

Topics to be covered:

- Introduction. The finite field philosophy.
- Review of notation and basic properties of the Fourier transform
- Counting 3- and 4-term arithmetic progressions using the Gowers U^2- and U^3-norms: generalized von Neumann theorems.
- Inverse theorem for the Gowers U^2-norm.
- The "quadratic" example for the Gowers U^3-norm.
- Brief revision of key results from additive combinatorics.

What is "quadratic Fourier analysis?" The aim of this series of lectures is to give a reasonably detailed answer to that question, at least in so far as is possible at the present time.

It would, however, be presumptuous to suppose that any reader would venture to the end of these notes in order to discover the meaning of the title, so we begin with a very brief introduction.

Fourier analysis, or "linear" Fourier analysis as we shall call it in these notes, is a multi-faceted subject. One rather small part of it is concerned with solving

---

2000 *Mathematics Subject Classification.* Primary 42B99; Secondary 11B25, 11N13.

The author is a Clay Research fellow. He is grateful for the support of the Clay Mathematics Institute, which enabled him to attend the activities in Montréal.

This is the final form of the paper.

©2007 American Mathematical Society

linear equations. Two examples of theorems which may be proven using some kind of study of the Fourier transform are

- (Chowla/van der Corput) *There are infinitely many 3-term arithmetic progressions of primes.*
- (Roth) *Let $\delta > 0$ be fixed. Then if $N > N_0(\delta)$ is sufficiently large, any subset $A \subseteq \{1, \ldots, N\}$ with size at least δN contains three distinct elements in arithmetic progression.*

Note that an arithmetic progression of length three is defined by a single linear equation $x_1 + x_3 = 2x_2$.

Standard Fourier analysis fails in many situations where we are interested in a pair of linear equations. The natural example here is a progression of length four, which is defined by the equations $x_1 + x_3 = 2x_2$, $x_2 + x_4 = 2x_3$. This is the situation where quadratic Fourier analysis is appropriate. Thus by developing the methods that we will talk about in these lectures, it is possible to prove

- (Green-Tao) *There are infinitely many 4-term arithmetic progressions of primes.*
- (Szemerédi) *Let $\delta > 0$ be fixed. Then if $N > N_0(\delta)$ is sufficiently large, any subset $A \subseteq \{1, \ldots, N\}$ with size at least δN contains three distinct elements in arithmetic progression.*

In fact we will not prove either of these theorems in this course, since we will be working in a model setting. A common theme in additive combinatorics is the consideration of *finite field models*. A full discussion may be found in [12], but the basic idea is as follows. For many problems in additive combinatorics one is interested in the interval $\{1, \ldots, N\}$. However, it is convenient to work in a group, and so one often uses various technical devices in order to place the problem at hand in $\mathbb{Z}/N\mathbb{Z}$. Once this is done, it is often easy to formulate an analogous question inside an arbitrary finite abelian group G. In most applications that we know of, this more general problem is scarcely harder to solve than in the specific case $G = \mathbb{Z}/N\mathbb{Z}$. However, there is a family of groups, namely the groups $\mathbb{F}_p^n$ where p is a small prime, in which it can be relatively easy to work. Techniques used to prove theorems in this setting can often be used to guide proof techniques in $\mathbb{Z}/N\mathbb{Z}$, which provide theorems of actual number theoretic interest.

In this series of lectures we will focus almost exclusively on the group $G = \mathbb{F}_5^n$. I am rather fond of the prime 5 since it is the smallest for which the notion of a 4-term arithmetic progression is sensible.

We will conclude with a discussion of the general case at the end, in as much detail as time permits. It turns out that the theory for $\mathbb{Z}/N\mathbb{Z}$ is surprisingly rich, and there are strong connections with the ergodic theory techniques that are discussed in the lectures of Bryna Kra in these proceedings.

Notation. Opinion seems to be converging in additive combinatorics about what constitutes the "standard" notation, and I will endeavor to keep to these norms. If X is any finite set and $f \colon X \to \mathbb{C}$ is any function then we write

$$\mathbb{E}_{x \in X} f(x) := |X|^{-1} \sum_{x \in X} f(x).$$

This means that it is often possible to avoid worrying about normalizing factors.

Unless specified otherwise, we will set $G := \mathbb{F}_5^n$ and write $N := |G| = 5^n$. Any character on G (that is, homomorphism $\gamma \colon G \to \mathbb{C}^\times$) has the form $x \mapsto \omega^{r^T x}$, where $\omega := e^{2\pi i/5}$ and $r \in \mathbb{F}_5^n$ is a vector. We write $\widehat{G}$ for $\mathbb{F}_5^n$ when considered as the group of characters in this way.

If $f \colon G \to \mathbb{C}$ is any function then we define its Fourier transform $\hat{f} \colon \widehat{G} \to \mathbb{C}$ by
$$\hat{f}(r) := \mathbb{E}_{x \in G} f(x) \omega^{r^T x}.$$
We distinguish the *trivial character* corresponding to $r = 0$, which takes the value 1 for all $x \in G$. If $f, g \colon G \to \mathbb{C}$ are two functions then we define the *convolution* $f * g \colon G \to \mathbb{C}$ by
$$(f * g)(x) := \mathbb{E}_{y \in G} f(x) g(y - x).$$
Note that when working on G we always use the *Haar measure* which assigns weight $|G|^{-1}$ to any $x \in G$. When working on $\widehat{G}$ we use the *counting measure* which assigns weight 1 to every $r \in \widehat{G}$. These measures are dual to one another, which in practice means that in formulas such as those in Lemma 1.1 below one can simply write $\mathbb{E}_{x \in G}$ and $\sum_{r \in \widehat{G}}$, and thereafter be untroubled by normalizing factors.

When we talk of L^p norms, these will always be taken with respect to the appropriate underlying measure. Thus
$$\|f\|_1 := \mathbb{E}_{x \in G} |f(x)|,$$
whereas
$$\|\hat{f}\|_4 := \left(\sum_{r \in \widehat{G}} |\hat{f}(r)|^4 \right)^{1/4}.$$

I will be assuming familiarity with the basic properties of the Fourier transform, which are all straightforward consequences of the orthogonality relations
$$\sum_r \omega^{r^T x} = \begin{cases} N & \text{if } x = 0 \\ 0 & \text{otherwise} \end{cases}$$
and
$$\mathbb{E}_{x \in G} \omega^{r^T x} = \begin{cases} 1 & \text{if } r = 0 \\ 0 & \text{otherwise.} \end{cases}$$

Lemma 1.1 (Basic properties of the Fourier transform). *Suppose that $f, g \colon G \to \mathbb{C}$ are any two functions. Then*

(1) *We have $\hat{f}(0) = \mathbb{E}_{x \in G} f(x)$. For any r we have $|\hat{f}(r)| \leq \|f\|_1$.*

(2) *(Parseval identity) We have*
$$\mathbb{E}_{x \in G} f(x) \bar{g}(x) = \sum_{r \in \widehat{G}} \hat{f}(r) \overline{\hat{g}(r)}.$$

In particular $\|f\|_2 = \|\hat{f}\|_2$.

(3) *(Inversion) We have*
$$f(x) = \sum_{r \in \widehat{G}} \hat{f}(r) \omega^{-r^T x}.$$

(4) *(Convolution) We have $(f * g)^\wedge = \hat{f} \hat{g}$.*

The last item here illustrates how we will denote the Fourier transform of expressions E for which it would be too cumbersome to write $\widehat{E}$.

Let us now start with the main business of these lectures. Let G be a finite abelian group with order N which is coprime to 6, and let $f_1, \ldots, f_4 \colon G \to [-1, 1]$ be functions. In these notes a central rôle will be played by the multilinear operators Λ_3 and Λ_4, defined by

$$\Lambda_3(f_1, f_2, f_3) := \mathbb{E}_{x,d}\, f_1(x) f_2(x+d) f_3(x+2d)$$

and

$$\Lambda_4(f_1, f_2, f_3, f_4) := \mathbb{E}_{x,d}\, f_1(x) f_2(x+d) f_3(x+2d) f_4(x+3d).$$

Thus Λ_3 counts the number of 3-term arithmetic progressions "along the f_i," whilst Λ_4 counts the number of 4-term progressions.[1]

When the functions f_i are characteristic functions, the operators Λ_3 and Λ_4 may be interpreted combinatorially.

Observation 1.2. Suppose that $f_i = 1_{A_i}$, where $A_i \subseteq G$ is a set. Then $\Lambda_3(1_{A_1}, 1_{A_2}, 1_{A_3})$ is equal to N^{-2} times the number of triples $(a_1, a_2, a_3) \in A_1 \times A_2 \times A_3$ which are in arithmetic progression. Similarly, $\Lambda_4(1_{A_1}, 1_{A_2}, 1_{A_3}, 1_{A_4})$ is equal to N^{-2} times the number of quadruples $(a_1, a_2, a_3, a_4) \in A_1 \times A_2 \times A_3 \times A_4$ which are in arithmetic progression.

There are certainly many situations in which one might be interested in counting the number of 3- or 4-term progressions inside a set. To do this, we normally proceed as follows. If $A \subseteq G$ is a set with size αN, then write $f_A := 1_A - \alpha$. This is called the *balanced function* of A, and it has expected value 0.

Lemma 1.3 (Balanced function decomposition). *Suppose that $A_1, \ldots, A_4 \subseteq G$, and that $|A_i| = \alpha_i N$. Then we have*

$$(1.1) \qquad \Lambda_3(1_{A_1}, 1_{A_2}, 1_{A_3}) = \alpha_1 \alpha_2 \alpha_3 + (\text{seven other terms}),$$

where each of the seven terms has the form $\Lambda_3(g_1, g_2, g_3)$ where each g_i is either f_{A_i} or α_i and at least one is equal to f_{A_i}. Similarly

$$(1.2) \qquad \Lambda_4(1_{A_1}, 1_{A_2}, 1_{A_3}, 1_{A_4}) = \alpha_1 \alpha_2 \alpha_3 \alpha_4 + (\text{fifteen other terms}),$$

where each of the fifteen terms has the form $\Lambda_4(g_1, g_2, g_3, g_4)$ where each g_i is either f_{A_i} or α_i and at least one is equal to f_{A_i}.

Let us specialize to the case $A_1 = A_2 = A_3 = A_4 = A$ for simplicity, and write $|A| = \alpha N$. What do we "expect" $\Lambda_3(1_A, 1_A, 1_A)$ and $\Lambda_4(1_A, 1_A, 1_A, 1_A)$ to be? It is not hard to see that for a "random" set A, generated by tossing a coin which comes up heads with probability α to decide whether each $x \in G$ lies in A, the expected value of $\Lambda_3(1_A, 1_A, 1_A)$ is approximately α^3, whilst the expected value of $\Lambda_4(1_A, 1_A, 1_A, 1_A)$ is approximately α^4. Note that these quantities are exactly the "main terms" in the expansions of Lemma 1.3. It is thus reasonable to suggest that the other seven terms in (1.1) measure some kind of "nonuniformity" of A relevant to 3-term progressions, whilst the fifteen terms in (1.2) do the same for 4-term progressions.

Let us make a preliminary definition.

---

[1] Whilst we will talk exclusively about 3- and 4-term arithmetic progressions, the reader should note that much of what we have to say may be adapted to more general problems where it is of interest to count the number of solutions to a linear equation, or to a pair of linear equations.

Definition 1.4 (Uniformity along progressions). Let $A \subseteq G$ be a set with $|A| = \alpha N$, and let $f_A := 1_A - \alpha$ be the balanced function of A. Let $\delta \in (0,1)$ be a parameter. Then we say that A exhibits δ-uniformity along 3-term progressions if whenever we have three functions $g_1, g_2, g_3 \to [-1,1]$, at least one of which is equal to f_A, then
$$|\Lambda_3(g_1, g_2, g_3)| \leq \delta.$$
We define nonunifomity along 4-term progressions similarly.

Remark. It is not, at first sight, obvious that there are *any* sets which are uniform along progressions.

Lemma 1.5. *Suppose that $A \subseteq G$ is a set with $|A| = \alpha N$. If A is δ-uniform along 3-term progressions, then*
$$|\Lambda_3(1_A, 1_A, 1_A) - \alpha^3| \leq 7\delta.$$
If A is δ-uniform along 4-term progressions, then
$$|\Lambda_4(1_A, 1_A, 1_A, 1_A) - \alpha^4| \leq 15\delta.$$

PROOF. Immediate consequence of Lemma 1.3. □

The following question will be a recurring theme of these lectures:

Question 1.6. Suppose that A is not δ-uniform along 3- or 4-term progressions. Can we say something "useful" about A?

Of course, the notion of "useful" is a subjective one. The reader may assume, however, that the mere failure of Definition 1.4 does not constitute "useful." We will see that if A is not uniform along 3-term progressions, then it exhibits "linear" behavior, whilst functions which are not uniform along 4-term progressions are somehow "quadratic."

Formulating, proving, and using statements of this type is our main goal in these notes.

Question 1.6 may be answered very satisfactorily using Fourier analysis. The key tool is the following simple lemma, whose proof is an amusing exercise using the basic properties of the Fourier transform.

Lemma 1.7. *Let $f_1, f_2, f_3 \colon G \to \mathbb{R}$ be any three functions. Then*
$$(1.3) \qquad \Lambda_3(f_1, f_2, f_3) = \sum_r \hat{f}_1(r) \hat{f}_2(-2r) \hat{f}_3(r)$$

Proposition 1.8 (Inverse result for 3-term progressions. I). *Suppose that A is not δ-uniform along 3-term progressions. Then $\|\hat{f}_A\|_\infty \geq \delta$, that is to say there is some $r \in \hat{G}$ such that $|\hat{f}_A(r)| \geq \delta$.*

PROOF. Suppose that
$$|\Lambda_3(g_1, g_2, f_A)| \geq \delta$$
for some functions $g_1, g_2 \colon G \to [-1,1]$ (the analysis of the other two cases, when $g_1 = f_A$ or $g_2 = f_A$, is more-or-less identical). We have, by Lemma 1.7 the formula
$$\Lambda_3(g_1, g_2, f_A) = \sum_r \hat{g}_1(r) \hat{g}_2(-2r) \hat{f}_A(r).$$

Thus by Cauchy–Schwarz and Parseval's identity we infer that

(1.4) $$\delta \leq |\sum_r \hat{g}_1(r)\hat{g}_2(-2r)\hat{f}_A(r)| \leq \|\hat{f}_A\|_\infty \|\hat{g}_1\|_2 \|\hat{g}_2\|_2 \leq \|\hat{f}_A\|_\infty.$$ □

This is a very clean result, but the method of proof (appealing to a formula in Fourier analysis) has not, so far, proved amenable to generalization. One way to generalize an argument is to first try and find a more longwinded, less natural looking approach and try and generalize that. We will describe such an approach now, though we hope that any reader looking back on this section later on will not consider it so unnatural. Note that the result is the same as Proposition 1.8, but the bound is slightly worse.

Proposition 1.9 (Inverse result for 3-term progressions. II). *Suppose that A is not δ-uniform along 3-term arithmetic progressions. Then $\|\widehat{f_A}\|_\infty \geq \delta^2$.*

PROOF. Let us first observe that
$$\Lambda_3(g_1, g_2, f_A) = \mathbb{E}_{y_1, y_2} g_1(-y_1) g_2(\tfrac{1}{2} y_2) f_A(y_1 + y_2).$$
This is a simple reparametrization. Applying the Cauchy–Schwarz inequality, we have
$$|\Lambda_3(g_1, g_2, f_A)|^2 \leq \mathbb{E}_{y_2} |\mathbb{E}_{y_1} g_1(-y_1) f_A(y_1 + y_2)|^2$$
$$= \mathbb{E}_{y_1, y_1', y_2} f_A(y_1 + y_2) f_A(y_1' + y_2) g_1(-y_1) g_1(-y_1').$$
Applying Cauchy–Schwarz again, we have

(1.5) $|\Lambda_3(g_1, g_2, f_A)|^4$
$$\leq \mathbb{E}_{y_1, y_1'} |\mathbb{E}_{y_2} f_A(y_1 + y_2) f_A(y_1' + y_2)|^2$$
$$= \mathbb{E}_{y_1, y_1', y_2, y_2'} f_A(y_1 + y_2) f_A(y_1' + y_2) f_A(y_1 + y_2') f_A(y_1' + y_2').$$

This last expression is called the (fourth power of) the *Gowers U^2-norm of f_A*. Thus we define

(1.6) $$\|f_A\|_{U^2}^4 := \mathbb{E}_{y_1, y_1', y_2, y_2'} f_A(y_1 + y_2) f_A(y_1' + y_2) f_A(y_1 + y_2') f_A(y_1' + y_2').$$

It is often useful to write this in the alternative form
$$\|f_A\|_{U^2}^4 = \mathbb{E}_{x, h_1, h_2} f_A(x) f_A(x + h_1) f_A(x + h_2) f_A(x + h_1 + h_2).$$
It is not hard to show that $\|\cdot\|_{U^2}$ is a norm using the Cauchy–Schwarz inequality several times. We will not make much use of this fact, and refer the reader to [9] for the proof.

Note that (1.5) implies that if $|\Lambda_3(g_1, g_2, f_A)| \geq \delta$ then $\|f_A\|_{U^2} \geq \delta$.

What now? Another way to see that $\|\cdot\|_{U^2}$ is a norm is to observe that
$$\|f\|_{U^2}^4 = \|f * f\|_2^2 = \|(f * f)^\wedge\|_2^2 = \|\hat{f}\|_4^4.$$
Thus if $\|f_A\|_{U^2} \geq \delta$ then we have
$$\delta^4 \leq \|\hat{f}_A\|_4^4 \leq \|\hat{f}_A\|_\infty^2 \|\hat{f}_A\|_2^2 \leq \|\hat{f}_A\|_\infty^2,$$
which concludes the proof in the case that $|\Lambda_3(g_1, g_2, f_A)| \geq \delta$. Again, the cases when $f_A = g_1$ or g_2 can be dealt with very similarly, and are left to the reader; the parametrizations leading to (1.5) must be modified slightly. □

At the moment, it is hard to see what has been gained here. To prove the result, we still had to fall back on a formula of Fourier analysis, and furthermore the bound we obtain is worse than that in Proposition 1.8.

We may summarize the argument in Proposition 1.9 as follows, giving the two distinct parts a name.

- (Generalized von Neumann theorem) *The operator Λ_3 is controlled by the Gowers U^2-norm. Specifically for any three functions $f_1, f_2, f_3 \colon G \to [-1,1]$ we have*

$$|\Lambda_3(f_1, f_2, f_3)| \leq \inf_{i=1,2,3} \|f_i\|_{U^2}.$$

- (Gowers inverse theorem) *If the Gowers U^2-norm of a function $f \colon G \to [-1,1]$ is large, f must have a large Fourier coefficient:*

$$\|f\|_{U^2} \geq \delta \implies \|\hat{f}\|_\infty \geq \delta^2.$$

We note that the Gowers inverse theorem is necessary and sufficient. Indeed if $\|\hat{f}\|_\infty \geq \delta$ then clearly $\|\hat{f}\|_4 \geq \delta$, and so of course $\|f\|_{U^2} \geq \delta$.

This division of labor into two parts turns out to be the natural way to proceed for Λ_4 (and higher operators). The first part of the argument (the definition of the Gowers norm and the Generalized von Neumann theorem) goes through somewhat straightforwardly. The second part (the Gowers inverse theorem) does not, since we do not know of a formula analogous to $\|f\|_{U^2} = \|\hat{f}\|_4$.

Definition 1.10 (Gowers U^3-norm). Let $f \colon G \to [-1,1]$ be a function. Then we define

$$\|f\|_{U^3}^8 := \mathbb{E}_{\substack{y_1,y_2,y_3 \\ y_1',y_2',y_3'}} f(y_1+y_2+y_3)f(y_1'+y_2+y_3)f(y_1+y_2'+y_3)f(y_1+y_2+y_3')$$
$$\times f(y_1'+y_2'+y_3)f(y_1'+y_2+y_3')f(y_1+y_2'+y_3')f(y_1'+y_2'+y_3')$$
$$= \mathbb{E}_{x,h_1,h_2,h_3 \in G} f(x)f(x+h_1)f(x+h_2)f(x+h_3)f(x+h_1+h_2)$$
$$\times f(x+h_1+h_3)f(x+h_2+h_3)f(x+h_1+h_2+h_3).$$

Note that this is a kind of sum of f over 3-dimensional parallelepipeds. We omit the proof that $\|f\|_{U^3}$ is actually a norm (see [9]).

Proposition 1.11 (Generalized von Neumann theorem for 4-term APs). *Let $f_1, \ldots, f_4 \colon G \to [-1,1]$ be any four functions. Then we have*

$$|\Lambda_4(f_1, \ldots, f_4)| \leq \inf_{i=1,\ldots,4} \|f_i\|_{U^3}.$$

In particular if A is not δ-uniform along four-term progressions then $\|f_A\|_{U^3} \geq \delta$.

PROOF. The idea is the same as in Proposition 1.9. Here, we find a suitable reparametization of $\Lambda_4(f_1, \ldots, f_4)$, and then apply the Cauchy–Schwarz inequality three times. A "suitable reparametization" turns out to be

(1.7) $\Lambda_4(f_1, f_2, f_3, f_4)$
$$= \mathbb{E}_{y_1,y_2,y_3 \in G} f_1(-\tfrac{1}{2}y_2 - 2y_3) f_2(\tfrac{1}{3}y_1 - y_3) f_3(\tfrac{2}{3}y_1 + \tfrac{1}{2}y_2) f_4(y_1 + y_2 + y_3).$$

For the rest of this section let $\mathbf{b}()$ denote any function bounded by 1. Different occurrences of $\mathbf{b}$ may denote different functions. The Cauchy–Schwarz inequality implies that

(1.8) $\quad |\mathbb{E}_{x \in X} \mathbb{E}_{y \in Y} \mathbf{b}(x) f(x,y)| \leq \left| \mathbb{E}_{x \in X} \mathbb{E}_{y^{(0)}, y^{(1)} \in Y} f(x, y^{(0)}) f(x, y^{(1)}) \right|^{1/2}.$

We apply this three times. At the first application we take $X := \{y_2, y_3\}$ and $Y = \{y_1\}$, and put the function f_1 inside the $\mathbf{b}()$ term. We now have variables $y_1^{(0)}$, $y_1^{(1)}$, y_2, y_3. Now set $X := \{y_1^{(0)}, y_1^{(1)}, y_3\}$, $Y := \{y_2\}$ and arrange for everything involving f_2 to be placed in the $\mathbf{b}()$ term. We now have variables $y_1^{(0)}$, $y_1^{(1)}$, $y_2^{(0)}$, $y_2^{(1)}$, y_3. For the final application of Cauchy–Schwarz set $X := \{y_1^{(0)}, y_1^{(1)}, y_2^{(0)}, y_2^{(1)}\}$ and $Y := \{y_3\}$, and arrange for everything involving f_3 to be placed in the $\mathbf{b}()$ term. Note that at this point we have eliminated everything involving f_1, f_2, f_3 and have

$$|\mathbb{E}_{y_1, y_2, y_3} f_1(-\tfrac{1}{2}y_2 - 2y_3) f_2(\tfrac{1}{3}y_1 - y_3) f_3(\tfrac{2}{3}y_1 + \tfrac{1}{2}y_2) f_4(y_1 + y_2 + y_3)|$$

$$\leq |\mathbb{E}_{y_1^{(0)}, y_1^{(1)}, y_2^{(0)}, y_2^{(1)}, y_3^{(0)}, y_3^{(1)}} f_4(y_1^{(0)} + y_2^{(0)} + y_3^{(0)})$$
$$\times f_4(y_1^{(1)} + y_2^{(0)} + y_3^{(0)}) \cdots f_4(y_1^{(1)} + y_2^{(1)} + y_3^{(1)})|^{1/8}.$$

The right-hand side here is precisely $\|f_4\|_{U^3}$.

To show that $\Lambda(f_1, f_2, f_3, f_4)$ is bounded by the other expressions $\|f_i\|_{U^3}$, one may proceed similarly. We leave the details to the reader. □

We now come to the central question of quadratic Fourier analysis: when is $\|f\|_{U^3}$ large? The first key observation is that the answer is not simply the same as for the U^2-norm.

Lemma 1.12 (Key example). *There is a function $f: G \to \mathbb{C}$ with $\|f\|_\infty \leq 1$ such that $\|f\|_{U^3} = 1$, but such that $\|\hat{f}\|_\infty \leq N^{-1/2}$.*

PROOF. Before embarking on the proof, we must remark that $\|\cdot\|_{U^3}$ has only been defined for real-valued functions thus far. To define it for complex-valued functions, one must take complex conjugates of the terms $f(x + h_1)$, $f(x + h_2)$, $f(x + h_3)$ and $f(x + h_1 + h_2 + h_3)$. The extension to complex-valued functions facilitates the discussion of examples, but is not otherwise essential in the theory. Keeping track of complex conjugates is rather a tedious affair, so will endeavor to work with real functions whenever possible.

Set $f(x) = \omega^{x^T x}$. We have

$$\|f\|_{U^3}^8 = \mathbb{E}_{x, h_1, h_2, h_3} \omega^{x^T x - (x+h_1)^T(x+h_1) - \cdots - (x+h_1+h_2+h_3)^T(x+h_1+h_2+h_3)} = 1.$$

This can be seen by intelligent direct computation (or even by naïve direct computation); the phase vanishes since it is essentially the third derivative of a quadratic.

To evaluate $\|\hat{f}\|_\infty$, observe that we have

$$\left| \mathbb{E}_{x \in G} \omega^{x^T x + r^T x} \right| = \left| \prod_{j=1}^n \mathbb{E}_{x_j \in \mathbb{F}_5} \omega^{x_j^2 + r_j x_j} \right| = 5^{-n/2}.$$

This concludes the proof of the lemma. □

We conclude this first lecture by stating three key results in additive combinatorics which we will need in the second lecture. These results will all be discussed and proved in other lectures in this school. In these results, $0 < c < 1 < C$ are absolute constants.

Proposition 1.13 (The Balog–Szemerédi–Gowers theorem). *Let G be an abelian group, and suppose that $A \subseteq G$ is a set with $|A| = n$. Suppose that there are at least δn^3 additive quadruples in A, that is to say solutions to $a_1 + a_2 = a_3 + a_4$. Then there is a subset $A' \subseteq A$ with $|A'| \geq c\delta^C |A|$ such that $|A' + A'| \leq C\delta^{-C}|A'|$.*

This result will be the subject of Antal Balog's lecture at the school.

Proposition 1.14 (Freĭman's theorem in finite fields). *Let p be a prime, and write $\mathbb{F}_p^n$ for the n-dimensional vector space over the finite field with p elements. Suppose that $A \subseteq \mathbb{F}_p^n$ is a set with $|A+A| \leq K|A|$. Then there is a subspace $H \leq \mathbb{F}_p^n$ such that $A \subseteq H$ and for which we have the bound $|H| \leq p^{CK^C}|A|$.*

This result will be discussed by Imre Ruzsa.

Exercises. For the reader wishing to familiarize herself with the Gowers norms, we offer a handful of exercises. Discussions pertinent to these exercises may be found in the papers [9, 17, 18].

(1) Let $k \geq 2$ be any integer, and define the Gowers U^k-norm by

$$(1.9) \qquad \|f\|_{U^k}^{2^k} := \mathbb{E}_{x,h_1,\ldots,h_k \in G} \prod_{\omega \in \{0,1\}^k} f(x + \omega \cdot h).$$

Show that $\|\cdot\|_{U^k}$ is a norm. (*Hint*: first define the *Gowers inner product* $\langle f_\omega \rangle_{\omega \in \{0,1\}^k}$ for 2^k functions $(f_\omega)_{\omega \in \{0,1\}^k}$ by modifying (1.9). Then use several applications of the Cauchy–Schwarz inequality to prove the *Gowers–Cauchy–Schwarz inequality*

$$|\langle f_\omega \rangle_{\omega \in \{0,1\}^k}| \leq \prod_\omega \|f_\omega\|_{U^k}.$$

Finally, use this to prove the triangle inequality for $\|f\|_{U^k}$).

(2) Prove that the Gowers U^k-norms are *nested*:

$$\|f\|_{U^2} \leq \|f\|_{U^3} \leq \cdots.$$

(3) By generalizing Lemma 1.12, show that the Gowers norms are *strictly* nested in the following strong sense. For any $k \geq 3$ there is $c_k > 0$ such that the following is true. For any N, there is a group G with $|G| \geq N$ and a function $f \colon G \to \mathbb{C}$ with $\|f\|_\infty = \|f\|_{U^k} = 1$ such that $\|f\|_{U^{k-1}} \ll N^{-c_k}$.

(4) We noted that the U^2 inverse theorem is an if and only if statement. That is, if f is a bounded function with $|\mathbb{E}_x f(x)\omega^{r^T x}| \geq \delta$ for some r then f has large U^2-norm. Prove this without using the fact that $\|f\|_{U^2} = \|\hat{f}\|_4$. (*Hint*: use the Gowers–Cauchy–Schwarz inequality of Exercise (1).)

(5) Let $G = \mathbb{F}_5^n$. Suppose that

$$|\mathbb{E}_{x \in G} f(x)\omega^{x^T Mx + r^t x}| \geq \delta.$$

for some matrix M and vector r. Prove that $\|f\|_{U^3} \geq \delta$. (*Hint*: apply the Gowers–Cauchy–Schwarz inequality again. You will need the generalization of $\|\cdot\|_{U^3}$ which covers complex-valued functions; this can be obtained by inserting appropriate complex conjugate symbols, as was discussed during the proof of Lemma 1.12.

(6) (Generalizing the generalized von Neumann theorem) Show that

$$|\Lambda_k(f_1, \ldots, f_k)| \leq \inf_{i=1,\ldots,k} \|f_i\|_{U^{k-2}}.$$

Further reading. This material was originally laid out in Gowers [9], though the notation was slightly different and (of course) the Gowers norms were not named as such! Various expositions of the material may be found in papers by one or both of Terry Tao and myself. See, for example, [17, 18].

A very general version of the generalized von Neumann theorem (linking systems of s equations in t unknowns to the U^{s+1} norm) may be found in our forthcoming paper [20], and an even more general version (applying to functions which are not necessarily bounded by 1) may be found in [21].

Analogues of much of the material in this lecture were discovered in ergodic theory about 20 years ago. For more on this fascinating connection, the lectures of Kra in these Proceedings are illuminating.

The Balog–Szemerédi–Gowers theorem was originally proved by Gowers [8], and is a quantitative version of the earlier result of Balog and Szemerédi [1] (see also Balog's article in these proceedings). A version with a good value of the exponent C may be found in [5]. This material is also covered in my notes [14]. The Plünnecke–Ruzsa inequality was obtained in [25] and afforded an elegant proof by Ruzsa in [26]. The original reference for Proposition 1.14 is the paper [28] by Imre Ruzsa. For self-contained notes on Plünnecke's inequality and Freĭman's theorem, see [13]. For a discussion of all of the material in this lecture (and indeed much of the material in the other lectures) see the book [34].

2. Lecture 2

Topics to be covered:

- The inverse theorem for the U^3-norm on $\mathbb{F}_5^n$.

Some notation. Let E, E' be real-valued expressions. We will write $E \gg_\delta E'$ to mean that there is some function $c(\delta) > 0$ such that $E \geq c(\delta) E'$. There is nothing particularly unusual about this notation, but one aspect of the manner in which we shall apply it is somewhat subtle. When we write, for example, "let $N \gg_\delta 1$," we mean "let $N \geq c(\delta)$, where $c \colon \mathbb{R}_+ \to \mathbb{R}_+$ is some function which may be chosen so that later arguments work." We do *not* (of course) mean that an arbitrary function c may be chosen.

We will also, on occasion, use the notation $O_\delta(1)$ to denote a finite quantity which depends only on δ.

We have deliberately chosen topics within the subject of quadratic Fourier analysis for which bounds are unimportant, since these are the topics most allied to the "infinitary" ideas which feature in the lectures of Kra and Tao in these proceedings. It is quite reasonable to think of there being just two types of quantity in these lectures: *finite* quantities which depend only on δ, and *infinite* quantities which depend on the size of $\mathbb{F}_5^n$.

Let us recall the main question we are trying to address.

Question 2.1 (Gowers inverse question). Suppose that $f \colon G \to [-1, 1]$ is a function and that $\|f\|_{U^3} \geq \delta$. What can we say about f?

It turns out to be *much* easier to address this question in a finite field setting such as $G = \mathbb{F}_5^n$. We showed in the exercises to Lecture 1 that if f correlates with a quadratic phase $\omega^{x^T M x + r^T x}$ then f has large U^3 norm. It turns out that the converse is also true, though this is much harder to prove and will be our main goal in this lecture.

Proposition 2.2 (Inverse theorem for the U^3-norm on $\mathbb{F}_5^n$). *Suppose that $f \colon G \to [-1, 1]$ is a function for which $\|f\|_{U^3} \geq \delta$. Then there is a matrix*

$M \in \mathfrak{M}_n(\mathbb{F}_5)$ and a vector $r \in \mathbb{F}_5^n$ so that
$$|\mathbb{E}_{x \in G} f(x) \omega^{x^T M x + r^T x}| \gg_\delta 1.$$

Remark. Write $E := \sup_{r,M} |\mathbb{E}_{x \in G} f(x) \omega^{x^T M x + r^T x}|$. It is not hard to check that the proof we give would allow one to replace $E \gg_\delta 1$ by some bound of the form $E \geq \exp(-C\delta^{-C})$. For our later application, we will merely need *some* lower bound of the form $E \gg_\delta 1$. There are other applications where bounds are important—see the *further reading* at the end of this lecture for a discussion.

To prove Proposition 2.2 we will essentially follow the approach of Gowers [8]. We will, however, employ a slight twist which is essentially due to Samorodnitsky [29].

Definition 2.3 (Derivatives). Suppose that $f \colon G \to \mathbb{R}$ is a function. Then for any $h \in G$ we define the function $\Delta(f;h)$ by
$$\Delta(f;h)(x) := f(x) f(x-h).$$

Remark. It is convenient, though perhaps slightly mystifying, to give the name "derivative" to this construction. If we extended the definition to complex-valued functions by setting $\Delta(f;h)(x) = f(x)\overline{f(x-h)}$ and applied it with $f(x) = e^{2\pi i \phi(x)}$, the mystery might be reduced somewhat as the phase ϕ is indeed being differentiated.

Proposition 2.4 (Samorodnitsky's identity). Let $f \colon G \to \mathbb{R}$ be any function. Then we have

(2.1) $$\sum_{r_1+r_2=r_3+r_4} \mathbb{E}_{h_1+h_2=h_3+h_4} |\Delta(f;h_1)^\wedge(r_1)|^2 \cdots |\Delta(f;h_4)^\wedge(r_4)|^2 = \mathbb{E}_h \|\Delta(f;h)^\wedge\|_8^8.$$

PROOF. The idea of the proof is simple: we show that both sides are equal to

(2.2) $$\sum_{(c_1,\ldots,c_8,c'_1,\ldots,c'_8) \in \mathcal{C}} f(c_1) \cdots f(c_8) f(c'_1) \ldots f(c'_8),$$

where the sum is over all configurations $\mathcal{C}$ with
$$c_1 + \cdots + c_4 = c_5 + \cdots + c_8$$
and
$$c'_1 - c_1 = \cdots = c'_8 - c_8.$$

To show that the right-hand side of (2.1) is equal to (2.2) is the easier of the two tasks to accomplish. One notes that
$$\|\Delta(f;h)^\wedge\|_8^8 = \mathbb{E}_x |\Delta(f;h) * \Delta(f;h) * \Delta(f;x) * \Delta(f;h)(x)|^2,$$
by Parseval's identity and the fact that $(f*g)^\wedge = \hat{f}\hat{g}$. That the expectation of this over h is equal to (2.2) follows by expansion.

To prove that the left-hand side of (2.1) is equal to (2.2), it is convenient to introduce some notation. If $\psi \colon \widehat{G} \to \mathbb{C}$ is a function then we define $\psi^\vee \colon G \to \mathbb{C}$ by
$$\psi^\vee(x) := \sum_{r \in \widehat{G}} \psi(r) \omega^{-r^T x}.$$
Note that the inversion formula is equivalent to

(2.3) $$(\hat{f})^\vee = f.$$

If $\psi, \phi: \widehat{G} \to \mathbb{C}$ are two functions then we define
$$\psi * \phi(r) := \sum_{s \in \widehat{G}} \psi(s)\phi(r-s)$$
and note the formula
$$(\psi * \phi)^\vee = \psi^\vee \phi^\vee.$$
It follows from these facts and Parseval's identity that for any four functions $g_1, \ldots, g_4 : G \to \mathbb{C}$ we have
$$(2.4) \quad \sum_{r_1+r_2=r_3+r_4} \widehat{g_1}(r_1)\widehat{g_2}(r_2)\overline{\widehat{g_3}(r_3)\widehat{g_4}(r_4)} = \sum_r \widehat{g_1}*\widehat{g_2}(r)\overline{\widehat{g_3}*\widehat{g_4}(r)}$$
$$= \mathbb{E}_x \, g_1(x)g_2(x)\overline{g_3(x)g_4(x)}.$$
We apply this with
$$g_i = \Delta(f; h_i) * \Delta(f; h_i)^\circ,$$
where we have defined $f^\circ(x) := \overline{f(-x)}$. Noting that $(f^\circ)^\wedge = \overline{\widehat{f}}$, we see that
$$\widehat{g_i}(r) = |\Delta(f; h_i)^\wedge(r)|^2.$$
Substituting into (2.4), we see that the left-hand side of (2.1) is equal to
$$\mathbb{E}_{h_1+h_2=h_3+h_4} \mathbb{E}_x \prod_{i=1}^4 \Delta(f; h_i) * \Delta(f; h_i)^\circ(x).$$
Expanding out, we recover (2.2) once more. $\square$

Using this identity, we can prove the following crucial result, which provides the first link between functions f with large U^3-norm and quadratic phases. It states that the derivatives $\Delta(f; h)$ obey a sort of weak linearity property.

Proposition 2.5 (Gowers). *Let $f: G \to [-1, 1]$ be a function, and suppose that $\|f\|_{U^3} \geq \delta$. Suppose that $|G| \gg_\delta 1$. Then there is a function $\phi: G \to \widehat{G}$ such that*
(1) $|\Delta(f; h)^\wedge(\phi(h))| \gg_\delta 1$ *for all $h \in S$, where $|S| \gg_\delta |G|$;*
(2) *There are $\gg_\delta |G|^3$ quadruples $(s_1, s_2, s_3, s_4) \in S^4$ such that $s_1 + s_2 = s_3 + s_4$ and $\phi(s_1) + \phi(s_2) = \phi(s_3) + \phi(s_4)$.*

PROOF. Set $N := |G|$. One may easily check that
$$\|f\|_{U^3}^8 = \mathbb{E}_h \|\Delta(f; h)\|_{U^2}^4.$$
Recalling that the U^2-norm is the L^4 norm of the Fourier transform, we thus have
$$\|f\|_{U^3}^8 = \mathbb{E}_h \|\Delta(f; h)^\wedge\|_4^4.$$
Now Hölder's inequality and Parseval's identity imply that for any h we have
$$\|\Delta(f; h)^\wedge\|_4^4 \leq \|\Delta(f; h)^\wedge\|_2^{4/3} \|\Delta(f; h)^\wedge\|_8^{8/3} \leq \|\Delta(f; h)^\wedge\|_8^{8/3}.$$
Another application of Hölder yields
$$\mathbb{E}_h \|\Delta(f; h)^\wedge\|_8^{8/3} \leq \left(\mathbb{E}_h \|\Delta(f; h)^\wedge\|_8^8\right)^{1/3}.$$
Combining these observations, we conclude that
$$\mathbb{E}_h \|\Delta(f; h)^\wedge\|_8^8 \geq \delta^{24}.$$

Samorodnitsky's identity then allows us to conclude that

$$(2.5) \qquad \sum_{r_1+r_2=r_3+r_4} \mathbb{E}_{h_1+h_2=h_3+h_4} |\Delta(f;h_1)^\wedge(r_1)|^2 \cdots |\Delta(f;h_4)^\wedge(r_4)|^2 \geq \delta^{24}.$$

To each $h \in G$, we associate the set $\Phi(h)$ of characters r for which $|\Delta(f;h)^\wedge(r)| \geq \delta^{50}$. It is immediate from Parseval's identity that $|\Phi(h)| \leq \delta^{-100}$ for all h. Now the contribution to (2.5) from those h_i, r_i for which $r_1 \notin \Phi(h_1)$ (say) is bounded by

$$\delta^{100} \sum_{r_2,r_3,r_4} \mathbb{E}_{h_2,h_3,h_4} |\Delta(f;h_2)^\wedge(r_2)|^2 |\Delta(f;h_3)^\wedge(r_3)|^2 |\Delta(f;h_4)^\wedge(r_4)|^2 \leq \delta^{100}.$$

It follows that

$$\sum_{r_1+r_2=r_3+r_4} \mathbb{E}_{h_1+h_2=h_3+h_4} 1_{r_1 \in \Phi(h_1)} |\Delta(f;h_1)^\wedge(r_1)|^2 \cdots 1_{r_4 \in \Phi(h_4)} |\Delta(f;h_4)^\wedge(r_4)|^2 \geq \delta^{24}/2,$$

and so in particular there are at least $\delta^{24} N^3/2$ additive octuples $(h_1, r_1, \ldots, h_4, r_4)$ such that $h_1 + h_2 = h_3 + h_4$, $r_1 + r_2 = r_3 + r_4$ and $r_i \in \Phi(h_i)$ for $i = 1, \ldots, 4$. We say that an octuple is *proper* if $h_1, \ldots, h_4$ are all distinct. The number of our additive octuples which fail to be proper is clearly $\ll_\delta N^2$ and hence, since N is so large, at least $\delta^{24} N^3/4$ of them *are* proper.

Let S be the set of all h for which $\Phi(h) \neq \varnothing$. It is easy to see that $|S| \gg_\delta |G|$, since otherwise there could not be enough additive octuples. For each $h \in S$, pick an element $\phi(h)$ uniformly at random from $\Phi(h)$, and suppose that these choices are independent for different h. For each proper additive octuple $(h_1, r_1, \ldots, h_4, r_4)$, the probability that it *fits* ϕ, that is to say that $r_i = \phi(h_i)$ for $i = 1, 2, 3, 4$, is precisely $1/|\Phi(h_1)| \cdots |\Phi(h_4)|$. This is $\gg_\delta 1$. It follows that the expected number of additive octuples which fit ϕ is $\gg_\delta |G|^3$. In particular there is some specific choice of ϕ for which $\gg |G|^3$ additive octuples fit ϕ.

It takes a few seconds to realize that we have, in fact, proved the result. Indeed, an octuple which fits ϕ is precisely an additive quadruple of points $h_1, \ldots, h_4$ such that $\phi(h_1) + \phi(h_2) = \phi(h_3) + \phi(h_4)$ and $\phi(h_i) \in \Phi(h_i)$, that is to say $|\Delta(f,h)^\wedge(\phi(h))| \geq \delta^{50}$. $\square$

We have made a crucial step: assuming that $\|f\|_{U^3}$ was large, we deduced that the derivative of f has a certain weak linearity property. We must now work with this property and make it somewhat stronger.

Proposition 2.6 (From weak linearity to linearity). *Suppose that $\phi: G \to \widehat{G}$ is a function with the property in Proposition 2.5(2), that is to say there is some set $S \subseteq G$ with $|S| \gg_\delta |G|$ such that there are $\gg_\delta |G|^3$ additive quadruples (s_1, s_2, s_3, s_4) such that $s_1 + s_2 = s_3 + s_4$ and $\phi(s_1) + \phi(s_2) = \phi(s_3) + \phi(s_4)$. Then there is some linear function $\psi(x) = Mx + b$, where $M \in \mathfrak{M}_n(\mathbb{F}_5)$ and $b \in \mathbb{F}_5^n$, such that $\phi(x) = \psi(x)$ for $\gg_\delta |G|$ values of $x \in S$.*

PROOF. The first step is to observe that the conclusion of Proposition 2.5 may be rephrased using the *graph*

$$\Gamma := \{(h, \phi(h)) : h \in S\},$$

which is a subset of $G \times \widehat{G}$. Statement (2) of Proposition 2.5 is just the same as saying that Γ has $\gg_\delta |G|^3$ additive quadruples. It follows from the Balog–

Szemerédi–Gowers theorem that there is a subset $\Gamma' \subseteq \Gamma$ with
$$|\Gamma'| \gg_\delta |\Gamma| \gg_\delta |G|$$
and
$$|\Gamma' + \Gamma'| \ll_\delta |\Gamma'|.$$
Define $S' \subseteq S$ by
$$\Gamma' := \{(h, \phi(h)) : h \in S'\},$$
and note that
$$|S'| \gg_\delta |G|.$$
Now we may identify $G \times \widehat{G}$ with $\mathbb{F}_5^n \times \mathbb{F}_5^n$ and hence with $\mathbb{F}_5^{2n}$. From Ruzsa's finite field analogue of Freĭman's theorem, it follows that there is some subspace $H \leq \mathbb{F}_5^n \times \mathbb{F}_5^n$,

(2.6) $$|H| \ll_\delta |G|,$$

such that $\Gamma' \subseteq H$.

Consider the map $\pi : H \to G$ onto the first factor. The image of this linear map contains S', and so from (2.6) and the lower bound for $|S'|$ we see that
$$\dim_{\mathbb{F}_5} \ker \pi \ll_\delta 1.$$
It follows that we may foliate H into $\ll_\delta 1$ cosets of some subspace H', such that π is injective on each of these cosets. By averaging, we see that there is some x such that
$$|(x + H') \cap \Gamma'| \gg_\delta |G|.$$
Set $\Gamma'' := (x + H') \cap \Gamma'$, and define $S'' \subseteq S'$ accordingly. Then $\pi|_{x+H'}$ is an affine isomorphism onto its image V, which means that there is an affine linear map $\psi : V \to \widehat{G}$ such that $(s'', \psi(s'')) \in \Gamma''$ for all $s'' \in S''$, that is to say $\psi(s'') = \phi(s'')$ for all $s'' \in S''$. □

Let us put this last result together with Proposition 2.5.

Corollary 2.7 (Linearity of the derivative). *Suppose that $f : G \to [-1, 1]$ is a function with $\|f\|_{U^3} \geq \delta$. Suppose that $|G| \gg_\delta 1$. Then there is some $M \in \mathfrak{M}_n(\mathbb{F}_5)$ and some $b \in \mathbb{F}_5^n$ such that*
$$\mathbb{E}_h |\Delta(f;h)^\wedge(Mh + b)|^2 \gg_\delta 1.$$

PROOF. Recall that ϕ is defined for $h \in S$, where
$$|S| \gg_\delta |G|$$
and that it has the property that
$$|\Delta(f;h)^\wedge(\phi(h))| \gg_\delta 1$$
for all $h \in S$. We proved in Proposition 2.6 that there is an affine linear function $\psi(h) = Mh + b$ such that $\phi(h) = \psi(h)$ for all $h \in S''$, where $|S''| \gg_\delta |G|$. The corollary follows immediately. □

Corollary 2.7 shows that the derivative of a function f with large U^3 norm correlates with a linear function. Recall that our aim is to show that f correlates with a quadratic function $x \mapsto \omega^{x^T M x + r^T x}$. This latter function does have a linear derivative, but this derivative is *symmetric*. For that reason we need the following lemma, which states that the matrix M in Corollary 2.7 is automatically nearly symmetric.

Lemma 2.8 (Symmetry argument). *Suppose that $f\colon G \to [-1,1]$ is a function, that $M \in \mathfrak{M}_n(\mathbb{F}_5)$, and that $b \in \mathbb{F}_5^n$. Suppose that*

$$\mathbb{E}_h |\Delta(f;h)^{\wedge}(Mh+b)|^2 \gg_\delta 1.$$

Then M is approximately symmetric in the sense that

$$\operatorname{rk}(M - M^T) \ll_\delta 1.$$

PROOF. Write $D = M - M^T$. Expanding the assumption gives

$$\mathbb{E}_{x,y,h} f(x)f(x-h)f(y)f(y-h)\omega^{(x-y)^T Mh + (x-y)^T b} \gg_\delta 1,$$

Making the substitution $z = x + y - h$, this becomes

$$\mathbb{E}_{x,y,z} f(x)f(z-x)f(y)f(z-y)\omega^{(x-y)^T M(x+y-z) + (x-y)^T b} \gg_\delta 1,$$

which can be written

$$\mathbb{E}_z \mathbb{E}_x \Delta'(f;z)(x)\omega^{x^T M(x-z) + x^T b} \mathbb{E}_y \Delta'(f;z)(y) \omega^{-y^T M(y-z) - y^T b} \omega^{x^T Dy} \gg_\delta 1.$$

Here, we have written

$$\Delta'(f;z)(t) := f(t)f(z-t).$$

Writing

$$g_z(x) := \Delta'(f;z)(x)\omega^{x^T M(x-z) + x^T b},$$

we have

$$\mathbb{E}_z \mathbb{E}_{x,y} g_z(x)\overline{g_z(y)}\omega^{x^T Dy} \gg_\delta 1.$$

Averaging over z, we see that there is some function $g\colon G \to \mathbb{C}$ with $\|g\|_\infty \leq 1$ such that

$$|\mathbb{E}_x g(x)\overline{g(y)}\omega^{x^T Dy}| \gg_\delta 1,$$

that is to say

$$|\mathbb{E}_x g(x)\widehat{g}(Dx)| \gg_\delta 1.$$

This implies that

$$\mathbb{E}_x |\widehat{g}(Dx)| \gg_\delta 1,$$

and so in particular there are $\gg_\delta |G|$ values of x such that $|\widehat{g}(Dx)| \gg_\delta 1$. However we know from Parseval's identity that the number of r such that $|\widehat{g}(r)| \gg_\delta 1$ is $\ll_\delta 1$. Thus there is some set $S \subseteq \mathbb{F}_5^n$ with $|S| \gg_\delta |G|$ and $|D(S)| \ll_\delta 1$. This implies that

$$|\ker(D)| \gg_\delta |G|,$$

which immediately implies the result. $\square$

We have shown that if $\|f\|_{U^3}$ is large then the derivative of f correlates with a symmetric linear form. To complete the proof of Proposition 2.2, we must "integrate" this statement and show that f correlates with a quadratic. We give this integration now.

PROOF OF PROPOSITION 2.2. From Corollary 2.7 and Lemma 2.8, we know that

(2.7) $$\mathbb{E}_h |\Delta(f;h)^{\wedge}(Mh+b)|^2 \gg_\delta 1,$$

where

$$\operatorname{rk}(M - M^T) \ll_\delta 1.$$

Write $M_{\text{sym}} := \frac{1}{2}(M + M^T)$, and let $V := \ker(M - M^T)$. For each $t \in G$ there is some b_t such that we have
$$Mh + b = M_{\text{sym}}h + b_t.$$
for all $h \in V + t$. By a trivial averaging argument and the fact that $\text{codim}(V) \ll_\delta 1$, we may find a t such that
$$\mathbb{E}_h \, 1_{h \in V+t} |\Delta(f;h)^\wedge(Mh + b)|^2 \gg_\delta 1.$$
This of course implies that
$$\mathbb{E}_h \, 1_{h \in V+t} |\Delta(f;h)^\wedge(M_{\text{sym}}h + b_t)|^2 \gg_\delta 1,$$
and hence by positivity that
$$\mathbb{E}_h \, |\Delta(f;h)^\wedge(M_{\text{sym}}h + b_t)|^2 \gg_\delta 1.$$
By redefining M to be M_{sym} and b to be b_t, it follows that we may assume in (2.7) that M is symmetric.

Expanding out (2.7) we obtain
$$\mathbb{E}_{h,x,y} f(x)f(x-h)f(y)f(y-h)\omega^{h^T M(x-y) + b^T(x-y)} \gg_\delta 1.$$
Substituting $y := x - k$, we obtain
$$\mathbb{E}_{h,x,k} f(x)f(x-h)f(x-k)f(x-h-k)\omega^{h^T Mk + b^T k} \gg_\delta 1.$$
Using the identity
$$x^T Mx - (x-h)^T M(x-h) - (x-k)^T M(x-k) + (x-h-k)^T M(x-h-k) = 2h^T Mk,$$
this may be written as
$$(2.8) \qquad \mathbb{E}_{h,x,k} \, g_1(x)g_2(x-h)g_3(x-k)g_4(x-h-k) \gg_\delta 1,$$
where
$$g_1(x) := f(x)\omega^{(x^T Mx)/2}, \qquad g_2(x) := f(x)\omega^{-(x^T Mx)/2 - b^T x},$$
$$g_3(x) := f(x)\omega^{-(x^T Mx)/2}, \qquad g_4(x) := f(x)\omega^{(x^T Mx)/2 - b^T x}.$$
Note that the functions g_2, g_3, g_4 are bounded by 1; this is, in fact, the only property of them that we shall use.

Now the left-hand side of (2.8) may be rewritten using the Fourier transform as
$$\sum_r \widehat{g_1}(r)\widehat{g_2}(-r)\widehat{g_3}(-r)\widehat{g_4}(r).$$
It follows immediately from Hölder's inequality that
$$\|\widehat{g_1}\|_4 \gg_\delta 1,$$
which, since $\|\widehat{g_1}\|_2 \leq 1$, implies that
$$\|\widehat{g_1}\|_\infty \gg_\delta 1,$$
that is to say there is some $r \in \mathbb{F}_5^n$ such that
$$|\mathbb{E}_x f(x)\omega^{(x^T Mx)/2 + r^T x}| \gg_\delta 1.$$
This, at last, completes the proof of Proposition 2.2. $\square$

Remark. In going from (2.8) to the end of the proof, what we have really done is apply the Gowers–Cauchy–Schwarz inequality (cf. the exercises following Lecture 1) and the inverse theorem for the U^2-norm.

2.1. Further reading.

The original argument of Gowers is in [8]. This took place in the group $G = \mathbb{Z}/N\mathbb{Z}$, not in a finite field model, and did not quite give a necessary and sufficient inverse theorem for the U^3-norm. It was instead shown that if $f\colon \mathbb{Z}/N\mathbb{Z} \to [-1,1]$ has large U^3-norm then f correlates with a quadratic polynomial on some subprogression of length a power of N. This is a "local" statement, and as such is much weaker than having large U^3-norm, which is "global", i.e., involves averaging over the whole group G.

To get an inverse theorem, one extra ingredient must be added to Gowers' work. This is the symmetry argument, Lemma 2.8. It was first given in [18]. That paper gives an inverse theorem for the U^3-norm in any finite abelian group of odd order. To even state the result is somewhat complicated, and we defer a discussion until we have thoroughly examined the finite field case. An inverse theorem for the U^3-norm in $\mathbb{F}_2^n$ was given by Samorodnitsky [29], using the method we have described but with a slight twist to enable him to handle characteristic 2. It is very likely that a combination of his methods and ours would allow one to prove an inverse theorem in *any* finite abelian G, but to my knowledge no-one has yet undertaken this task.

As we remarked, one may replace our $\gg_\delta 1$ notation with more precise bounds, ending up with a version of Proposition 2.2 with a function of the form $\exp(-C\delta^{-C})$ on the right-hand side. It would be of great interest to know whether this could be improved, perhaps even to $c\delta^C$. This would follow from the so-called Polynomial Freĭman–Ruzsa conjecture, the finite field version of which is discussed in [12].

The strongest known inverse result for the U^3 norm on $\mathbb{F}_5^n$ is the following, proved in [18].

Proposition 2.9 (Inverse theorem for the U^3-norm on $\mathbb{F}_5^n$. II). *Suppose that $f\colon \mathbb{F}_5^n \to [-1,1]$ is a function for which $\|f\|_{U^3} \geq \delta$. Then there exists a subspace $H \leq \mathbb{F}_5^n$ with $\mathrm{codim}(H) \leq C\delta^{-C}$, together with a system of quadratic forms $r_y^T x + x^T M_y x$ indexed by the cosets $y + H$ of H, such that*

$$\mathbb{E}_y \,|\, \mathbb{E}_{x \in y+H} f(x) \omega^{x^T M_y x + r_y^T x} | \geq c\delta^C.$$

Note that the amount of correlation is $c\delta^C$ rather than $\exp(-C\delta^{-C})$, but one must pass to a coset of a subspace of somewhat large codimension.

The proof of this result is rather longer than that of Proposition 2.2, and involves a good deal more machinery (Bogolyubov's method and Freĭman homomorphisms). This stronger result is necessary for certain applications, for example in our paper [19] in which it is shown that $r_4(\mathbb{F}_5^n) \ll N(\log N)^{-c}$.

3. Lecture 3

Topics to be covered:
- Quadratic factors
- The energy increment lemma
- The idea of approximating a function by projecting onto a low-complexity factor
- The Koopman–von Neumann decomposition
- The arithmetic regularity decomposition

Our main effort so far has been devoted to proving a result of the form "if $\|f\|_{U^3}$ is large then f has a large quadratic Fourier coefficient."

In this section we turn to a discussion of how this kind of information can be useful to us. There are many instances in additive combinatorics where study of a single Fourier coefficient is fruitful. However there are many other occasions on which it is beneficial to consider *several* Fourier coefficients of f, say the set of large Fourier coefficients of f. We must develop analogues of this theory in the quadratic setting.

From now on, matrices $M \in \mathfrak{M}_n(\mathbb{F}_5)$ will only appear in quadratic forms $x^T M x$. Thus from this point onwards it is natural to adopt the convention that *all matrices are symmetric*. We note that a (slightly) more high-brow approach to the whole theory, avoiding the use of bases, appears in our paper [19].

The following simple lemma will be used over and over again.

Lemma 3.1 (Gauss sums). *Suppose that M is symmetric and that* $\operatorname{rk} M = d$. *Then for any $r \in G$ we have*

$$|\mathbb{E}_{x \in G}\, \omega^{x^T M x + r^T x}| \leq 5^{-d/2}.$$

If $r = 0$ then equality occurs.

PROOF. Squaring, we obtain

$$|\mathbb{E}_{x \in G}\, \omega^{x^T M x + r^T x}|^2 = \mathbb{E}_h\, \omega^{h^T M h + r^T h} \mathbb{E}_x\, \omega^{2 h^T M x}$$
$$\leq \mathbb{E}_h\, |\mathbb{E}_x\, \omega^{2 h^T M x}|.$$

The inner sum is zero unless $h \in \ker(M)$. This occurs with probability 5^{-r}, and so we do indeed get

$$|\mathbb{E}_{x \in G}\, \omega^{x^T M x + r^T x}|^2 \leq 5^{-d}.$$

If $r = 0$ then the phase $\omega^{h^T M h + r^T h}$ is actually equal to 1 when $h \in \ker(M)$, and so equality occurs. $\square$

Using this lemma, we may highlight one of the immediate difficulties with formulating "quadratic Fourier analysis."

Lemma 3.2 (Profusion of large QFCs). *Let $f: \mathbb{F}_5^n \to [-1,1]$ be a function. Then there at most δ^{-2} values of r for which*

$$|\hat{f}(r)| = |\mathbb{E}_{x \in \mathbb{F}_5^n} f(x) \omega^{r^T x}| \geq \delta.$$

However, the number of pairs (M, r) such that

$$|\mathbb{E}_{x \in \mathbb{F}_5^n} f(x) \omega^{x^T M x + r^T x}| \geq \delta$$

need not be bounded in terms of δ.

PROOF. The first statement, which is included for comparison with the classical setting, is immediate from Parseval's identity. To illustrate the second, one may consider a function as simple as $f(x) \equiv 1$. For any symmetric matrix M with $\operatorname{rk}(M) \leq \log_5(1/\delta)$, we have

$$|\mathbb{E}_{x \in \mathbb{F}_5^n} f(x) \omega^{x^T M x}| \geq \delta.$$

The number of such matrices is not bounded in terms of δ. $\square$

This lemma suggests that we should perhaps only consider QFCs as "essentially different" if they are not too close in rank. This turns out to be a useful idea, and we will return to it later when we are in a position to formulate it properly.

As we said there are many arguments (e.g., [6, 10, 23, 30]) where one considers the set of δ-large Fourier coefficients

$$\operatorname{Spec}_\delta(f) := \{r \in \mathbb{F}_5^n : |\hat{f}(r)| \geq \delta\}.$$

Without going into details of the applications, let us describe a useful way to think about the way this construction is often used.

Definition 3.3 (Factors). Let $\phi_1, \ldots, \phi_k \colon \mathbb{F}_5^n \to \mathbb{F}_5$ be any functions. These functions describe a σ-algebra $\mathcal{B}$ on $\mathbb{F}_5^n$, the atoms of which are sets (of which there are at most 5^k) of the form $\{x : \phi_1(x) = c_1, \ldots, \phi_k(x) = c_k\}$. If $f \colon \mathbb{F}_5^n \to \mathbb{C}$ is a function then we often consider the *conditional expectation* $\mathbb{E}(f \mid \mathcal{B})$. Note that $\mathbb{E}(f \mid \mathcal{B})(x)$ is just the average of f over the atom $\mathcal{B}(x)$ which contains x. We will usually refer to σ-algebras arising in this way as *factors*, by analogy with ergodic theory. We say that a factor $\mathcal{B}'$ *refines* $\mathcal{B}$ if every atom of $\mathcal{B}'$ is contained in an atom of $\mathcal{B}$. Thus $\mathcal{B}'$ is at least as fine a partition of $\mathbb{F}_5^n$ as $\mathcal{B}$ is.

Definition 3.4 (Linear factors). Suppose that $r_1, \ldots, r_k \in \mathbb{F}_5^n$. Then the σ-algebra $\mathcal{B}$ whose atoms are the sets $\{x : r_i^T x = c_i, i = 1, \ldots, k\}$ is called a *linear factor* of complexity at most k.

Proposition 3.5 (Linear Koopman–von Neumann decomposition). *Let $f \colon \mathbb{F}_5^n \to [-1, 1]$ be a function and let $\delta > 0$ be a parameter. Then there is a linear factor $\mathcal{B}$ of complexity at most $4\delta^{-4}$ such that*

$$f = f_1 + f_2,$$

where

$$f_1 := \mathbb{E}(f \mid \mathcal{B})$$

and

$$\|f_2\|_{U^2} \leq \delta.$$

Remark. The Koopman–von Neumann theorem may be described in words as "any bounded function is the sum of a "low complexity" function formed by projecting onto a linear factor, and a "uniform" function which is small in U^2.

PROOF. The proof we give uses Fourier analysis, and does not generalize to give a result for the U^3-norm. We include it to justify the fact that this is a proposition which encodes the notion of "taking all the large Fourier coefficients of f."

Write $\eta := \delta^2/2$. Let $S := \operatorname{Spec}_\eta(f)$: note that by Parseval's identity we have $|S| \leq 4\delta^{-4}$. Let $H = S^\perp$ be the annihilator of f and write μ_H for the Haar measure on H, that is to say $\mu_H := 1_H / \mathbb{E} 1_H$. Define $f_1 := f * \mu_H$ and $f_2 := f - f * \mu_H$. It is not hard to see that $f_1 = \mathbb{E}(f \mid \mathcal{B})$, where $\mathcal{B}$ is the factor defined by the linear functions $r^T x$, $r \in S$. To conclude the proof, we only need check that $\|\hat{f}_2\|_\infty$ is small. To that end, we have

$$|\hat{f}_2(r)| = |\hat{f}(r)||1 - \widehat{\mu_H}(r)|.$$

If $r \in \operatorname{Spec}_\eta(f)$ then $\widehat{\mu_H}(r) = 1$, and so $\hat{f}_2(r) = 0$. If $r \notin \operatorname{Spec}_\eta(f)$ then by definition we have $|\hat{f}(r)| \leq \eta$, and so $|\hat{f}_2(r)| \leq 2\eta$ in this case. It follows that $\|\hat{f}_2\|_\infty \leq 2\eta$, and thus by the inverse theorem for the U^2-norm we have $\|f_2\|_{U^2} \leq \sqrt{2\eta}$. The result follows. □

Definition 3.6 (Quadratic factors). Let $r_1, \ldots, r_{d_1} \in \mathbb{F}_5^n$ be vectors, and let $M_1, \ldots, M_{d_2} \in \mathfrak{M}_n(\mathbb{F}_5)$ be symmetric matrices. We write $\mathcal{B}_1$ for the linear factor generated by the $r_j^T x$. Write $\mathcal{B}_2$ for the σ-algebra generated by the functions $r_j^T x$ and the pure quadratic functions $x^T M_j x$. Clearly $\mathcal{B}_2$ refines $\mathcal{B}_1$. We call the pair $(\mathcal{B}_1, \mathcal{B}_2)$ a (homogeneous) quadratic factor of complexity (d_1, d_2).

Proposition 3.7 (Quadratic Koopman–von Neumann decomposition). Let $(\mathcal{B}_1^{(0)}, \mathcal{B}_2^{(0)})$ be a quadratic factor with complexity at most $(d_1^{(0)}, d_2^{(0)})$. Let $f \colon \mathbb{F}_5^n \to [-1, 1]$ be a function and let $\delta > 0$ be a parameter. Then there is a quadratic factor $(\mathcal{B}_1, \mathcal{B}_2)$ of complexity at most $(d_1^{(0)} + O_\delta(1), d_2^{(0)} + O_\delta(1))$ which refines $(\mathcal{B}_1^{(0)}, \mathcal{B}_2^{(0)})$, and such that

$$f = f_1 + f_2,$$

where

$$f_1 := \mathbb{E}(f \mid \mathcal{B}_2)$$

and

$$\|f_2\|_{U^3} \leq \delta.$$

Remark. For applications in which bounds are unimportant, it is better to apply the arithmetic regularity lemma which we will give later. A version of the Koopman–von Neumann theorem with reasonable bounds is the key tool in [19]. In that application we take $(\mathcal{B}_1^{(0)}, \mathcal{B}_2^{(0)})$ to be the trivial factor.

The key to proving the Koopman–von Neumann decomposition lies in the following result.

Lemma 3.8 (Energy increment). Let $(\mathcal{B}_1, \mathcal{B}_2)$ be a quadratic factor of complexity at most (d_1, d_2), and let $f \colon \mathbb{F}_5^n \to [-1, 1]$ be a function such that

$$\|f - \mathbb{E}(f \mid \mathcal{B}_2)\|_{U^3} \geq \delta.$$

Then exists a refinement $(\mathcal{B}_1', \mathcal{B}_2')$ of $(\mathcal{B}_1, \mathcal{B}_2)$ of complexity at most $(d_1 + 1, d_2 + 1)$ such that we have the energy increment

(3.1) $$\|\mathbb{E}(f \mid \mathcal{B}_2')\|_2^2 \geq \|\mathbb{E}(f \mid \mathcal{B}_2)\|_2^2 + c(\delta),$$

where $c \colon (0, 1) \to \mathbb{R}_+$ is some nondecreasing function of δ.

PROOF. The function $g := f - \mathbb{E}(f \mid \mathcal{B}_2)$ is certainly bounded by 2, so we may apply the inverse theorem for the U^3-norm (Proposition 2.2) to conclude that there is a quadratic $x^T M x + r^T x$ so that

(3.2) $$|\mathbb{E}_x g(x) \omega^{x^T M x + r^T x}| \geq c(\delta).$$

We may clearly assume that $c \colon (0, 1) \to \mathbb{R}_+$ is a nondecreasing function. The linear part $r^T x$ and the pure quadratic part $x^T M x$ of this quadratic together induce a quadratic factor $(\widetilde{\mathcal{B}}_1, \widetilde{\mathcal{B}}_2)$ of complexity $(1, 1)$.

Now since $x^T M x + r^T x$ is $\widetilde{\mathcal{B}}_2$-measurable, it is clear that

$$\mathbb{E}_x g(x) \omega^{x^T M x + r^T x} = \mathbb{E}_x \mathbb{E}(g \mid \widetilde{\mathcal{B}}_2)(x) \omega^{x^T M x + r^T x},$$

In particular, (3.2) implies that

(3.3) $$\|\mathbb{E}(g \mid \widetilde{\mathcal{B}}_2)\|_1 \geq c(\delta).$$

Now define $\mathcal{B}'_1 := \mathcal{B}_1 \vee \widetilde{\mathcal{B}}_1$ and $\mathcal{B}'_2 := \mathcal{B}_2 \vee \widetilde{\mathcal{B}}_2$. Again, the meaning of this is the obvious one; simply intersect all the atoms of $\mathcal{B}_i$ with those of $\widetilde{\mathcal{B}}_i$. It is clear that $(\mathcal{B}'_1, \mathcal{B}'_2)$ is a quadratic factor of complexity at most $(d_1 + 1, d_2 + 1)$.

It remains to establish the energy increment (3.1). A key tool is

Pythagoras' theorem. *Suppose that $\mathcal{B}$, $\mathcal{B}'$ are two σ-algebras on $\mathbb{F}_5^n$ such that $\mathcal{B}'$ refines $\mathcal{B}$. Let $f: \mathbb{F}_5^n \to [-1,1]$ be any function. Then*
$$\|\mathbb{E}(f \mid \mathcal{B}')\|_2^2 = \|\mathbb{E}(f \mid \mathcal{B})\|_2^2 + \|\mathbb{E}(f \mid \mathcal{B}') - \mathbb{E}(f \mid \mathcal{B})\|_2^2.$$

Now we have the chain of inequalities
$$\|\mathbb{E}(f \mid \mathcal{B}'_2)\|_2^2 - \|\mathbb{E}(f \mid \mathcal{B}_2)\|_2^2 = \|\mathbb{E}(f \mid \mathcal{B}'_2) - \mathbb{E}(f \mid \mathcal{B}_2)\|_2^2 = \|\mathbb{E}(g \mid \mathcal{B}'_2)\|_2^2$$
$$\geq \|\mathbb{E}(g \mid \widetilde{\mathcal{B}}_2)\|_2^2 \geq \|\mathbb{E}(g \mid \widetilde{\mathcal{B}}_2)\|_1^2 \geq c(\delta).$$

The justification of these five equalities and inequalities uses respectively Pythagoras' theorem, the fact that $\mathcal{B}'_2$ refines $\mathcal{B}_2$, Pythagoras' theorem together with the fact that $\mathcal{B}'_2$ refines $\widetilde{\mathcal{B}}_2$, the Cauchy–Schwarz inequality, and (3.3). □

PROOF OF PROPOSITION 3.7. Start with $(\mathcal{B}_1, \mathcal{B}_2) = (\mathcal{B}_1^{(0)}, \mathcal{B}_2^{(0)})$. If
$$\|f - \mathbb{E}(f \mid \mathcal{B}_2)\|_{U^3} \leq \delta \tag{3.4}$$
then STOP. Otherwise, we may apply Lemma 3.8 to extend $(\mathcal{B}_1, \mathcal{B}_2)$ to a quadratic factor with complexity incremented by at most $(1,1)$ and the energy $\|\mathbb{E}(f \mid \mathcal{B}_2)\|_2^2$ incremented by at least $c(\delta)$. If (3.4) holds then STOP, otherwise repeat the process. Since f is bounded, the energy $\|\mathbb{E}(f \mid \mathcal{B}_2)\|_2^2$ lies in the interval $[0,1]$. Since $c: (0,1) \to \mathbb{R}_+$ is nondecreasing, we cannot iterate the above procedure more than $1/c(\delta)$ times before we STOP. The claim follows. □

We will not give an application of the Koopman–von Neumann decomposition, since the interesting applications require quantitative versions of the result (cf. [19]). The result has a significant shortcoming, which is that the uniformity parameter δ need not be small in terms of the complexity of $(\mathcal{B}_1, \mathcal{B}_2)$. For such situations there is another type of decomposition, which we call the *arithmetic regularity lemma* because of an analogy with Szemerédi's regularity lemma in graph theory. We note that any use of this type of decomposition necessarily results in terrible "tower-type" bounds: see for example [7, 11]. As we have stated, however, bounds are not our concern in these lectures.

Proposition 3.9 (Arithmetic regularity lemma for U^3). *Let $\delta > 0$ be a parameter, and let $\omega: \mathbb{R}_+ \to \mathbb{R}_+$ be an arbitrary growth function[2] (which may depend on δ). Suppose that $n > n_0(\omega, \delta)$ is sufficiently large, and let $f: \mathbb{F}_5^n \to [-1,1]$ be a function. Let $(\mathcal{B}_1^{(0)}, \mathcal{B}_2^{(0)})$ be a quadratic factor of complexity $(d_1^{(0)}, d_2^{(0)})$. Then there is $C = C(\delta, \omega, d_1^{(0)}, d_2^{(0)})$ and a quadratic factor $(\mathcal{B}_1, \mathcal{B}_2)$ which refines $(\mathcal{B}_1^{(0)}, \mathcal{B}_2^{(0)})$ and has complexity at most (d,d), $d \leq C$, together with a decomposition*
$$f = f_1 + f_2 + f_3,$$
where
$$f_1 := \mathbb{E}(f \mid \mathcal{B}_2), \quad \|f_2\|_2 \leq \delta \quad \text{and} \quad \|f_3\|_{U^3} \leq 1/\omega(d).$$

---

[2] The use of arbitrary growth functions really does put us in the domain of "discrete analogues of infinitary mathematics." The arithmetic regularity lemma is indeed very close in spirit to the main result of the ergodic-theoretic paper [2].

PROOF. Apply the Koopman–von Neumann theorem iteratively, with parameters δ_i, $i = 1, 2, \ldots$, to obtain quadratic factors $(\mathcal{B}_1^{(i)}, \mathcal{B}_2^{(i)})$ with complexities at most (C_i, C_i) such that

- $(\mathcal{B}_1^{(i)}, \mathcal{B}_2^{(i)})$ is a refinement of $(\mathcal{B}_1^{(i-1)}, \mathcal{B}_2^{(i-1)})$;
- $\|f - \mathbb{E}(f \mid \mathcal{B}_2^{(i)})\|_{U^3} \leq \delta_i$;
- C_i is bounded above in terms of C_{i-1} and δ_i.

Choose the sequence of δ_is such that $\delta_{i+1} \leq 1/\omega(C_i)$ for all i. Since C_i is bounded above by a quantity depending only on $\delta_1, \ldots, \delta_i$, this is certainly possible.

Now the energies $\|\mathbb{E}(f \mid \mathcal{B}_2^{(i)})\|_2^2$ are nondecreasing, and are all bounded by 1. By the pigeonhole principle there is therefore some $i \leq \lceil \delta^{-2} \rceil$ such that

$$\|\mathbb{E}(f \mid \mathcal{B}_2^{(i+1)})\|_2^2 - \|\mathbb{E}(f \mid \mathcal{B}_2^{(i)})\|_2^2 \leq \delta^2.$$

For such an i, we may take for our decomposition

$$f_1 := \mathbb{E}(f \mid \mathcal{B}_2^{(i)}), \quad f_2 := \mathbb{E}(f \mid \mathcal{B}_2^{(i+1)}) - \mathbb{E}(f \mid \mathcal{B}_2^{(i)}), \quad f_3 := f - \mathbb{E}(f \mid \mathcal{B}_2^{(i+1)}).$$

It follows from Pythagoras' theorem that $\|f_2\|_2 \leq \delta$, as required. □

What is the point of the Koopman–von Neumann and arithmetic regularity results, say for the U^3-norm? The answer is that they often reduce the study of general functions (say from the point of view of counting 4-term arithmetic progressions) to the study of projections $\mathbb{E}(f \mid \mathcal{B})$ onto "low-complexity" quadratic factors. This, however, is of little consequence unless we can study those supposedly simple objects.

Definition 3.10 (Rank of quadratic factors). Suppose that $(\mathcal{B}_1, \mathcal{B}_2)$ is a quadratic factor of complexity (d_1, d_2), being defined by d_1 linear forms $r_1^T x, \ldots, r_{d_1}^T x$ and d_2 pure quadratics $x^T M_1 x, \ldots, x^T M_{d_2} x$. We say that $(\mathcal{B}_1, \mathcal{B}_2)$ has rank at least r if

$$\mathrm{rk}(\lambda_1 M_1 + \cdots + \lambda_{d_2} M_{d_2}) \geq r$$

whenever $\lambda_1, \ldots, \lambda_{d_2}$ are elements of $\mathbb{F}_5$, not all zero.

When we are not concerned with bounds, it turns out that we may assume our quadratic factors have exceedingly large rank. We will see in the next lecture that factors with high rank are much easier to handle than factors with small rank.

Lemma 3.11 (Making factors high-rank). *Let $\omega : \mathbb{R}_+ \to \mathbb{R}_+$ be an arbitrary growth function. Then there is another function $\tau = \tau_\omega$ with the following property. Let $(\mathcal{B}_1, \mathcal{B}_2)$ be a quadratic factor with complexity at most (d_1, d_2). Then there is a refinement $(\mathcal{B}'_1, \mathcal{B}'_2)$ of $(\mathcal{B}_1, \mathcal{B}_2)$ with complexity at most (d'_1, d_2), where $d'_1 \leq \tau(d_1, d_2)$, which has rank at least $\omega(d'_1 + d'_2)$.*

PROOF. Suppose as usual that $(\mathcal{B}_1, \mathcal{B}_2)$ is described by d_1 linear functions $r_1^T x, \ldots, r_{d_1}^T x$ and d_2 "pure quadratics" $x^T M_1 x, \ldots, x^T M_{d_2} x$. Suppose that $(\mathcal{B}_1, \mathcal{B}_2)$ does not have rank at least $\omega(d_1 + d_2)$. Then there is some relation

$$\mathrm{rk}(\lambda_1 M_1 + \cdots + \lambda_{d_2} M_{d_2}) \leq \omega(d),$$

where we may assume without loss of generality that $\lambda_{d_2} = 1$. Let $s_1, \ldots, s_k$, $k \leq \omega(d)$ be a basis for $\ker(U)^\perp$, where $U := \lambda_1 M_1 + \cdots + \lambda_{d_2} M_{d_2}$, and let $(\mathcal{B}_1^\dagger, \mathcal{B}_2^\dagger)$ be the homogeneous quadratic factor defined by the linear forms $r_1^T x, \ldots, r_{d_1}^T x, s_1^T x, \ldots, s_k^T x$ and the quadratic forms $x^T M_1 x, \ldots, x^T M_{d_2 - 1} x$. It has complexity bounded by $(d_1^\dagger, d_2 - 1)$, where $d_1^\dagger \leq d + \omega(d_1 + d_2)$. The value of $x^T M_{d_2} x$ is determined by the values of the $x^T M_i x$, $i = 1, \ldots, d_2 - 1$ together with the value of $x^T U x$. This in turn is determined by the coset of $\ker(U)$ that x lies in, and hence by $s_1^T x, \ldots, s_k^T x$. It follows that $(\mathcal{B}_1^\dagger, \mathcal{B}_2^\dagger)$ refines $(\mathcal{B}_1, \mathcal{B}_2)$.

Now we ask whether $(\mathcal{B}_1^\dagger, \mathcal{B}_2^\dagger)$ has rank no more than $\omega(d_1^\dagger + d_2)$. If so, we refine again, obtaining a new factor $(\mathcal{B}_1^{\dagger\dagger}, \mathcal{B}_2^{\dagger\dagger})$ with complexity bounded by at most $(d_1^\dagger + \omega(d_1^\dagger + d_2), d_2 - 2)$. This procedure can last no more than d_2 steps, however, since at each stage the number of pure quadratic phases is reduced by one. We may take $(\mathcal{B}_1', \mathcal{B}_2')$ to be the factor that we have when the procedure terminates. $\square$

Proposition 3.12 (Arithmetic regularity lemma for U^3. II). Let $\delta > 0$ be a parameter, and let $\omega_1, \omega_2 \colon \mathbb{R}_+ \to \mathbb{R}_+$ be arbitrary growth functions (which may depend on δ). Let $n > n_0(\delta, \omega_1, \omega_2)$ be sufficiently large, and let $f \colon \mathbb{F}_5^n \to [-1, 1]$ be a function. Let $(\mathcal{B}_1^{(0)}, \mathcal{B}_2^{(0)})$ be a quadratic factor of complexity $(d_1^{(0)}, d_2^{(0)})$. Then there is a quadratic factor $(\mathcal{B}_1, \mathcal{B}_2)$ with the following properties:

(1) $(\mathcal{B}_1, \mathcal{B}_2)$ refines $(\mathcal{B}_1^{(0)}, \mathcal{B}_2^{(0)})$;
(2) The complexity of $(\mathcal{B}_1, \mathcal{B}_2)$ is at most (d_1, d_2), where

$$d_1, d_2 \leq C(\delta, \omega_1, \omega_2, d_1^{(0)}, d_2^{(0)}),$$

for some fixed function C;
(3) The rank of $(\mathcal{B}_1, \mathcal{B}_2)$ is at least $\omega_1(d_1 + d_2)$;
(4) There is a decomposition $f = f_1 + f_2 + f_3$, where

$$f_1 := \mathbb{E}(f \mid \mathcal{B}_2), \quad \|f_2\|_2 \leq \delta \quad \text{and} \quad \|f_3\|_{U^3} \leq 1/\omega_2(d_1 + d_2).$$

Remark. The formulation is very similar to that in Proposition 3.9, but we now insist that the factor $(\mathcal{B}_1, \mathcal{B}_2)$ be homogeneous, and also include a condition on its rank. The statement of Proposition 3.12 will look complicated at first sight, but there is nothing much to be scared of. As always with complicated propositions, it is as well to attempt to formulate what has been proved in a somewhat looser, wordier way. Here is an attempt:

Let f be any function on $\mathbb{F}_5^n$. Then, up to an error which is small in L^2, we may write f as a sum of a function which is measurable with respect to a bounded complexity quadratic factor, plus an error which is miniscule in $\|\cdot\|_{U^3}$. Furthermore we may insist that the rank of the quadratic factor is huge in comparison to its complexity.

PROOF. Apply Proposition 3.9 to get a factor $(\mathcal{B}_1, \mathcal{B}_2)$ refining $(\mathcal{B}_1^{(0)}, \mathcal{B}_2^{(0)})$, and a decomposition $f = f_1 + f_2 + f_3$ such that $f_1 = \mathbb{E}(f \mid \mathcal{B}_2)$, $\|f_2\|_2 \leq \delta/2$ and $\|f_3\|_{U^3} \leq 1/\omega_2(\tau(d_1, d_2) + d_2)$, where (d_1, d_2) is an upper bound for the complexity of $(\mathcal{B}_1, \mathcal{B}_2)$ and $\tau = \tau_{\omega_1}$ is the function appearing in Lemma 3.11. Using that lemma, we may refine $(\mathcal{B}_1, \mathcal{B}_2)$ to a quadratic factor $(\mathcal{B}_1', \mathcal{B}_2')$ with complexity at most (d_1', d_2'), where $d_1' \leq \tau(d_1, d_2)$ and $d_2' \leq d_2$, and with rank at least $\omega_1(d_1' + d_2')$. Define a new decomposition $f = f_1' + f_2' + f_3'$, where

$$f_1' := \mathbb{E}(f \mid \mathcal{B}_2'), \quad f_2' := f_2 + \mathbb{E}(f \mid \mathcal{B}_2) - \mathbb{E}(f \mid \mathcal{B}_2')$$

and $f_3' = f_3$. Either this has the desired properties, or else we have

$$\|\mathbb{E}(f \mid \mathcal{B}_2) - \mathbb{E}(f \mid \mathcal{B}_2')\|_2 \geq \delta/2.$$

By Pythagoras' theorem this leads to the energy increment
$$\|\mathbb{E}(f \mid \mathcal{B}'_2)\|_2^2 \geq \|\mathbb{E}(f \mid \mathcal{B}_2)\|_2^2 + \delta^2/4. \tag{3.5}$$
In this eventuality we apply Proposition 3.9 again, initializing with $(\mathcal{B}_1^{(0)}, \mathcal{B}_2^{(0)}) := (\mathcal{B}'_1, \mathcal{B}'_2)$. In view of the energy increment (3.5), we can only repeat this $\lceil 4/\delta^2 \rceil$ times before we reach a decomposition with the properties we desire. □

Further reading. There is a wealth of directions to go in. Results of Koopman–von Neumann type go back, implicitly, a long way. The *name* was first given, by Tao and I, to a result in our paper [17] on primes in AP. That result was somewhat different to the results here, but the method of proof (the energy increment strategy) is the same.

The arithmetic regularity lemma for the U^3-norm will be the subject of a forthcoming paper by Tao and I [22]. There is, of course, an analogous result for U^2-norm, and this was implicit in Bourgain [3]. The proof there used the Fourier transform rather than the energy-increment strategy. A substantially more difficult (!) proof of the same result was given 15 years later by me [11]; a number of applications were given there. The energy-increment proof of Proposition 3.9 seems at the moment to be the "right" way to think about these issues, and is essentially the approach taken in [32].

There are connections with regularity results for graphs and hypergraphs, the first result of this type being Szemerédi's regularity lemma [31]. There are also parallels with results in ergodic theory such as [2]. Perhaps it is best to refer the reader to the lectures by Kra and Tao at this school. The ICM article by Tao [33] has many references and would represent a fine place to begin further investigations.

4. Lecture 4

Topics to be covered

- Working on a quadratic factor; the configuration space.
- A theorem on progressions of length 4: an example of how to put all the ingredients together.

Our aim in this lecture is to prove the following theorem by using the machinery we have developed. Recall that we are writing $N := 5^n$.

Theorem 4.1 (G.–Tao). *Let $\alpha, \epsilon > 0$ be real numbers. Then there is an $n_0 = n_0(\alpha, \epsilon)$ with the following property. Suppose that $n > n_0(\alpha, \epsilon)$, and that $A \subseteq \mathbb{F}_5^n$ is a set with density α. Then there is some $d \neq 0$ such that A contains at least $(\alpha^4 - \epsilon)N$ four-term arithmetic progressions with common difference d.*

Remarks. It is easy to see that one cannot replace α^4 by anything larger, by considering a random set of density α. This theorem has, as a consequence, a version of Szemerédi's theorem for progressions of length four in finite fields, namely $r_4(\mathbb{F}_5^n) = o(N)$. The theorem is a finite field version of a conjecture of Bergelson, Host and Kra. Rather bizarrely at first sight, this result does not generalize to progressions longer than four.

Now in the last lecture we worked rather hard in order to show that, in various senses, the study of an arbitrary function $f : \mathbb{F}_5^n \to [-1, 1]$ can be reduced to the study of a $\mathcal{B}_2$-measurable function $\mathbb{E}(f \mid \mathcal{B}_2)$, where $(\mathcal{B}_1, \mathcal{B}_2)$ is a quadratic factor with "bounded complexity" and high rank. To make use of this, we need to be able to understand $\mathcal{B}_2$-measurable functions. At the very least, we are going to want to know about the size of the atoms in $\mathcal{B}_2$ and, for any four atoms, the number of four-term progressions spanned by those atoms. It turns out that the "high-rank" assumption allows us to simply compute these quantities using Fourier analysis.

Suppose, throughout this lecture, that $(\mathcal{B}_1, \mathcal{B}_2)$ is a quadratic factor defined by d_1 linear forms $r_j^T x$ and d_2 pure quadratics $x^T M_j x$. (Recall that $\mathcal{B}_1$ is the σ-algebra generated by the linear functions, and $\mathcal{B}_2$ is the σ-algebra generated by the linear *and* quadratic functions.) We will always suppose (as we clearly may) that the vectors r_j are linearly independent.

To understand $\mathcal{B}_2$-measurable functions, that is to say functions which are constant on atoms of $\mathcal{B}_2$ (or alternatively functions which have the form $\mathbb{E}(f \mid \mathcal{B}_2)$), it is helpful to work in *configuration space* $\mathbb{F}_5^{d_1} \times \mathbb{F}_5^{d_2}$. We write $\Gamma \colon \mathbb{F}_5^n \to \mathbb{F}_5^{d_1}$ and $\Phi \colon \mathbb{F}_5^n \to \mathbb{F}_5^{d_2}$ for the maps defined by $\Gamma(x) := (r_1^T x, \ldots, r_{d_1}^T x)$ and $\Phi(x) := (x^T M_1 x, \ldots, x^T M_{d_2} x)$.

Lemma 4.2 (Size of atoms). *Suppose that $(\mathcal{B}_1, \mathcal{B}_2)$ has rank at least r. Let $(a, b) \in \mathbb{F}_5^{d_1} \times \mathbb{F}_5^{d_2}$. Then the probability that a randomly chosen $x \in \mathbb{F}_5^n$ has $\Gamma(x) = a$ and $\Phi(x) = b$ is $5^{-d_1 - d_2} + O(5^{-r/2})$.*

Remark. In this lemma and the next, the probabilistic language is present only to avoid normalizing factors of $N = 5^n$. This is really a statement about the *number* of x with $\Gamma(x) = a$, $\Phi(x) = b$.

PROOF. The quantity in question is given by
$$5^{-d_1-d_2} \mathbb{E}_x \prod_{i=1}^{d_1} \bigg(\sum_{\mu_i \in \mathbb{F}_5} \omega^{\mu_i(r_i^T x - a_i)} \bigg) \prod_{j=1}^{d_2} \bigg(\sum_{\lambda_j \in \mathbb{F}_5} \omega^{\lambda_j (x^T M_j x - b_j)} \bigg),$$
which rearranges as
$$(4.1) \quad 5^{-d_1 - d_2} \sum_{\mu_i, \lambda_j} \omega^{-\lambda_1 b_1 - \cdots - \lambda_{d_2} b_{d_2} - \mu_1 a_1 - \cdots - \mu_{d_1} a_{d_1}}$$
$$\times \mathbb{E}_x \omega^{x^T (\lambda_1 M_1 + \cdots + \lambda_{d_2} M_{d_2}) x + (\mu_1 r_1 + \cdots + \mu_{d_1} r_{d_1})^T x}.$$

Now the rank of $(\mathcal{B}_1, \mathcal{B}_2)$ is at least r, which means that
$$\mathrm{rk}(\lambda_1 M_1 + \cdots + \lambda_{d_2} M_{d_2}) \geq r.$$
In view of the Gauss sum estimate, Lemma 3.1, this means that every term in (4.1) in which the λ_i are not all zero is bounded by $5^{-d_1 - d_2 - r/2}$. Of the terms with $\lambda_1 = \cdots = \lambda_{d_2} = 0$, the linear independence of the r_i guarantees that the only term which does not vanish is that with $\mu_1 = \cdots = \mu_{d_1} = 0$. The result follows immediately. $\square$

Lemma 4.3 (4-term progressions). *Suppose that $(\mathcal{B}_1, \mathcal{B}_2)$ has rank at least r. Suppose that $(a^{(1)}, b^{(1)}), \ldots, (a^{(4)}, b^{(4)}) \in \mathbb{F}_5^{d_1} \times \mathbb{F}_5^{d_2}$. Suppose that a 4-term progression $(x, x+d, x+2d, x+3d) \in (\mathbb{F}_5^n)^4$ is chosen at random. If*

(4.2) $\qquad a^{(1)}, a^{(2)}, a^{(3)}, a^{(4)}$ *are in arithmetic progression*

and

(4.3) $\qquad\qquad b^{(1)} - 3b^{(2)} + 3b^{(3)} + b^{(4)} = 0$

then the probability that $\Gamma(x + id) = a^{(i)}$, $\Phi(x + id) = b^{(i)}$ for $i = 1, 2, 3, 4$ is $5^{-2d_1 - 3d_2} + O(5^{-r/2})$. Otherwise, it is zero.

PROOF. The important thing to appreciate here is that four elements in different atoms of $\mathcal{B}_2$ can only lie in arithmetic progression if the two constraints (4.2) and (4.3) are satisfied. Furthermore these are the only relevant constraints, in that if they are satisfied (and if the factor $(\mathcal{B}_1, \mathcal{B}_2)$ has large rank) then we can accurately count the number of four-term progressions involving those atoms.

The necessity of the constraints (4.2) and (4.3) is easy. If $(x, x+d, x+2d, x+3d)$ is an arithmetic progression, we need only observe that $\Gamma(x)$, $\Gamma(x+d)$, $\Gamma(x+2d)$, $\Gamma(x+3d)$ are also in arithmetic progression, and that $\Phi(x) - 3\Phi(x+d) + 3\Phi(x+2d) - \Phi(x+3d) = 0$.

To obtain the statement about probability, we proceed in the same manner as in Lemma 4.2. The notation here is, however, somewhat fearsome. We start with the observation that the probability in question is

$$5^{-4d_1-4d_2} \mathbb{E}_{x,d} \prod_{l=1}^{4} \prod_{i=1}^{d_1} \left(\sum_{\mu_i^{(l)} \in \mathbb{F}_5} \omega^{\mu_i^{(l)}(r_i^T(x+ld)-a_i^{(l)})} \right)$$
$$\times \prod_{j=1}^{d_2} \left(\sum_{\lambda_j^{(l)} \in \mathbb{F}_5} \omega^{\lambda_j^{(l)}((x+ld)^T M_j(x+ld)-b_j^{(l)})} \right),$$

and then swap the order of summation to rearrange as

(4.4) $$5^{-4d_1-4d_2} \sum_{\mu_i^{(l)}, \lambda_j^{(l)} \in \mathbb{F}_5} \mathbb{E}_{x,d} \, \omega^{x^T Px + 2x^T Qd + d^T Rd + u^T x + v^T d - w},$$

where

$$P = P(\lambda) = \sum_{j=1}^{d_2} (\lambda_j^{(1)} + \lambda_j^{(2)} + \lambda_j^{(3)} + \lambda_j^{(4)}) M_j,$$

$$Q = Q(\lambda) = \sum_{j=1}^{d_2} (\lambda_j^{(1)} + 2\lambda_j^{(2)} + 3\lambda_j^{(3)} + 4\lambda_j^{(4)}) M_j,$$

$$R = R(\lambda) = \sum_{j=1}^{d_2} (\lambda_j^{(1)} + 4\lambda_j^{(2)} + 9\lambda_j^{(3)} + 16\lambda_j^{(4)}) M_j,$$

$$u = u(\mu) = \sum_{i=1}^{d_1} (\mu_i^{(1)} + \mu_i^{(2)} + \mu_i^{(3)} + \mu_i^{(4)}) r_i,$$

$$v = v(\mu) = \sum_{i=1}^{d_1} (\mu_i^{(1)} + 2\mu_i^{(2)} + 3\mu_i^{(3)} + 4\mu_i^{(4)}) r_i$$

and

$$w = w(\mu, \lambda) = \sum_{l=1}^{4} \sum_{i=1}^{d_1} \mu_i^{(l)} a_i^{(l)} + \sum_{l=1}^{4} \sum_{j=1}^{d_2} \lambda_j^{(l)} b_j^{(l)}.$$

We use Lemma 3.1 repeatedly. By fixing either x or d, we see that the inner sum in (4.4) (that is, the expectation over x, d) is $O(5^{-r/2})$ unless

(4.5) $$\lambda_j^{(1)} + \lambda_j^{(2)} + \lambda_j^{(3)} + \lambda_j^{(4)} = \lambda_j^{(1)} + 4\lambda_j^{(2)} + 9\lambda_j^{(3)} + 16\lambda_j^{(4)} = 0,$$

in which case certainly $P = R = 0$. In this case, the inner sum is a rather purer-looking

(4.6) $$\mathbb{E}_{x,d}\,\omega^{x^T Q d + u^T x + v^T d - w}.$$

For fixed d, this is zero unless $Qd + u = 0$. If $\lambda_j^{(1)} + 2\lambda_j^{(2)} + 3\lambda_j^{(3)} + 4\lambda_j^{(4)} \neq 0$ then, since $\mathrm{rk}(Q) \geq r$, this cannot happen for more than 5^{-r} of all d, and (4.6) is bounded by 5^{-r}. If on the other hand

(4.7) $$\lambda_j^{(1)} + 2\lambda_j^{(2)} + 3\lambda_j^{(3)} + 4\lambda_j^{(4)} = 0$$

then (4.6) further reduces to

$$\mathbb{E}_{x,d}\,\omega^{u^T x + v^T d - w},$$

which clearly vanishes unless

(4.8) $$\mu_i^{(1)} + \mu_i^{(2)} + \mu_i^{(3)} + \mu_i^{(4)} = \mu_i^{(1)} + 2\mu_i^{(2)} + 3\mu_i^{(3)} + 4\mu_i^{(4)} = 0.$$

We have shown that the inner sum in (4.4) is $O(5^{-r/2})$ unless the five linear conditions (4.5), (4.7), (4.8) are satisfied. The total contribution to (4.4) from cases where one of these five conditions is not satisfied is therefore $O(5^{-r/2})$. The total contribution from cases when the five conditions *are* satisfied is

$$5^{-4d_1 - 4d_2} \sum_{l=1}^{4} \sum_{\mu_i^{(l)}, \lambda_j^{(l)}} \omega^{-w(\mu, \lambda)}.$$

Since the $a^{(i)}$ are in arithmetic progression and the $b^{(i)}$ satisfy $b^{(1)} - 3b^{(2)} + 3b^{(3)} - b^{(4)}$, it is easy to check that $w(\mu, \lambda) = 0$ when the five conditions are satisfied. It remains only to note that, of the $5^{4d_1 + 4d_2}$ choices for μ, λ, the five conditions are satisfied for $5^{2d_1 + d_2}$ of them. $\square$

If $f: \mathbb{F}_5^n \to \mathbb{C}$ is a $\mathcal{B}$-measurable function then we write $\mathbf{f}: \mathbb{F}_5^{d_1} \times \mathbb{F}_5^{d_2} \to \mathbb{C}$ for the function which satisfies

$$f(x) = \mathbf{f}\big(\Gamma(x), \Phi(x)\big)$$

for all $x \in \mathbb{F}_5^n$. We will adopt this convention of using bold letters to denote functions on configuration space for the rest of these lectures without further comment.

We are now in a position to prove Theorem 4.1.

PROOF OF THEOREM 4.1. Recall that $A \subseteq \mathbb{F}_5^n$ is a set with density α. Apply Proposition 3.12 to find a quadratic factor $(\mathcal{B}_1, \mathcal{B}_2)$ with complexity (d_1, d_2), $d_i \leq d_0(\alpha, \epsilon)$ and rank r satisfying (say)

$$r \geq 100(\log(1/\epsilon) + \log(1/\alpha) + d_1 + d_2)$$

together with a decomposition $1_A = f_1 + f_2 + f_3$ such that $f_1 = \mathbb{E}(1_A \mid \mathcal{B}_2)$, $\|f_2\|_2 \leq \delta$ and $\|f_3\|_{U^3} \leq 1/\omega(d_1 + d_2)$. The parameter δ and the growth function ω will be specified as the proof unfolds, but will depend only on α and ϵ.

Let $r_1^T x, \ldots, r_{d_1}^T x$ be the linear functions involved in $\mathcal{B}_1$, and then let $H := \langle r_1, \ldots, r_{d_1} \rangle^T$. Let 1_H be the characteristic function of H, and let μ_H be the normalized measure on H, thus $\mu_H := 1_H / \mathbb{E}\, 1_H$. We are going to prove that

(4.9) $$\mathbb{E}_{x,d}\, 1_A(x) 1_A(x+d) 1_A(x+2d) 1_A(x+3d) \mu_H(d) \geq \alpha^4 - \epsilon,$$

which clearly implies the theorem (for some $d \in H$). To do this, we split the left-hand side of (4.9) into 81 parts by substituting $1_A = f_1 + f_2 + f_3$.

Claim 1. *The contribution from any of the 65 terms which contain f_2 is no more than $\epsilon/200$.*

PROOF. Suppose that the term is

(4.10) $\quad \mathbb{E}_{x,d}\, g_1(x)g_2(x+d)g_3(x+2d)g_4(x+3d)\mu_H(d),$

where $g_1 = f_2$ (the proofs of the other cases are very similar). Set $F(x) := \mathbb{E}_d\, g_2(x+d)g_3(x+2d)g_4(x+3d)\mu_H(d)$, and observe that $\|F\|_\infty \leq 1$. It follows that

$$|\mathbb{E}_{x,d}\, g_1(x)g_2(x+d)g_3(x+2d)g_4(x+3d)\mu_H(d)| \leq |\mathbb{E}_x\, g_1(x)F(x)| \leq \|f_2\|_1 \leq \|f_2\|_2.$$

This proves the claim provided that $\delta \leq \epsilon/200$. □

Claim 2. *The contribution from any of the 65 terms which contain f_3 is no more than $\epsilon/200$.*

PROOF. Suppose that the term is

(4.11) $\quad \mathbb{E}_{x,d}\, g_1(x)g_2(x+d)g_3(x+2d)g_4(x+3d)\mu_H(d),$

where $g_1 = f_3$ (the proofs of the other cases are very similar). We have

$$1_H(d) = \sum_t 1_{t+H}(x+2d)1_{t+H}(x+d),$$

where the sum is over all cosets $t+H$ of H in $\mathbb{F}_5^n$. By the generalized von Neumann theorem (Proposition 1.11), we have

$$|\mathbb{E}_{x,d}\, g_1(x)g_2(x+d)1_{t+H}(x+d)g_3(x+2d)1_{t+H}(x+2d)g_4(x+3d)| \leq \|f_3\|_{U^3}$$
$$\leq 1/\omega(d_1+d_2)$$

for each t. It follows that (4.11) is no more than $5^{2d_1}/\omega(d_1+d_2)$, which proves the claim provided that $\omega(t) \geq 5^{t+4}/\epsilon$. □

Remarks. Note carefully that for Claim 2 to follow we required the regularity parameter $\omega(t)$ to be *exponential* in t, rather than (say) polynomial. This is why the full arithmetic regularity lemma is required, rather than just the Koopman–von Neumann theorem.

These two claims account for 80 of the 81 terms into which we have decomposed the left-hand side of (4.9). To finish the argument, it suffices to show that

(4.12) $\quad \mathbb{E}_{x,d}\, f_1(x)f_1(x+d)f_1(x+2d)f_1(x+3d)\mu_H(d) \geq \alpha^4 - \epsilon/2.$

Now f_1 is (by definition) constant on atoms of $\mathcal{B}_2$. Recall that these atoms are indexed by the configuration space $\mathbb{F}_5^{d_1} \times \mathbb{F}_5^{d_2}$, and that we write $\mathbf{f}_1(a,b)$ for the value of f_1 on the atom indexed by (a,b).

Claim 3. *We have*

(4.13) $\quad \mathbb{E}_{(a,b)\in \mathbb{F}_5^{d_1}\times \mathbb{F}_5^{d_2}}\, \mathbf{f}_1(a,b) = \alpha\bigl(1 + O(5^{2d_1+2d_2-r/2})\bigr).$

PROOF. Note that the result would be trivial (and would hold without the O-term) if all the atoms of $\mathcal{B}_2$ had *exactly* the same size. Now recall that Lemma 4.2 gives an approximate version of this statement. We leave the slightly tedious details to the reader. □

Claim 4. *We have*

$$\mathbb{E}_{x,d} f_1(x)f_1(x+d)f_1(x+2d)f_1(x+3d)\mu_H(d)$$
$$= \mathbb{E}_{\substack{a\in\mathbb{F}_5^{d_1}, b^{(1)},\ldots,b^{(4)}\in\mathbb{F}_5^{d_2} \\ b^{(1)}-3b^{(2)}+3b^{(3)}-b^{(4)}=0}} \mathbf{f}_1(a,b^{(1)})\mathbf{f}_1(a,b^{(2)})\mathbf{f}_1(a,b^{(3)})\mathbf{f}_1(a,b^{(4)}) + O(5^{2d_2+3d_2-r/2}).$$

PROOF. Condition on the quadruple $(a^{(1)}, b^{(1)}), \ldots, (a^{(4)}, b^{(4)})$ of atoms containing $(x, x+d, x+2d, x+3d)$. The constraint that $d \in H$ is equivalent to $a^{(1)} = a^{(2)} = a^{(3)} = a^{(4)} = a$, say. By Lemma 4.3, we must also have $b^{(1)} - 3b^{(2)} + 3b^{(3)} - b^{(4)} = 0$. Invoking that same lemma, we have

$$\mathbb{E}_{x,d} f_1(x)f_1(x+d)f_1(x+2d)f_1(x+3d)1_H(d)$$
$$= \left(5^{-2d_1-3d_2} + O(5^{-r/2})\right) \sum_{a\in\mathbb{F}_5^{d_1}} \sum_{\substack{b^{(1)},\ldots,b^{(4)}\in\mathbb{F}_5^{d_2} \\ b^{(1)}-3b^{(2)}+3b^{(3)}-b^{(4)}=0}} \mathbf{f}_1(a,b^{(1)})\mathbf{f}_1(a,b^{(2)})\mathbf{f}_1(a,b^{(3)})\mathbf{f}_1(a,b^{(4)}).$$

Normalizing, we obtain the stated result. $\square$

Now the rank r was chosen very large ($r > 100(\log(1/\epsilon) + \log(1/\alpha) + d_1 + d_2)$). All we need do to establish (4.12), then, is prove the inequality

$$(4.14) \quad \mathbb{E}_{\substack{a\in\mathbb{F}_5^{d_1}, b^{(1)},\ldots,b^{(4)}\in\mathbb{F}_5^{d_2} \\ b^{(1)}-3b^{(2)}+3b^{(3)}-b^{(4)}=0}} \mathbf{f}_1(a,b^{(1)})\mathbf{f}_1(a,b^{(2)})\mathbf{f}_1(a,b^{(3)})\mathbf{f}_1(a,b^{(4)}) \geq \left(\mathbb{E}_{(a,b)\in\mathbb{F}_5^{d_1}\times\mathbb{F}_5^{d_2}} \mathbf{f}_1(a,b)\right)^4.$$

Noting that the left-hand side is

$$\mathbb{E}_{a\in\mathbb{F}_5^{d_1}} \mathbb{E}_{x\in\mathbb{F}_5^{d_2}} \left(\mathbb{E}_{\substack{b,b'\in\mathbb{F}_5^{d_2} \\ b-3b'=x}} \mathbf{f}_1(a,b)\mathbf{f}_1(a,b')\right)^2,$$

this follows from two applications of the Cauchy–Schwarz inequality.

Alternatively, it is amusing to give an interpretation in terms of the Fourier transform. The left-hand side of (4.14) is

$$(4.15) \quad \mathbb{E}_{a\in\mathbb{F}_5^{d_1}} \sum_{r\in\widehat{\mathbb{F}_5^{d_2}}} |\widetilde{\mathbf{f}_1}(a,r)|^2 |\widetilde{\mathbf{f}_1}(a,-3r)|^2.$$

In this expression the tilde denotes Fourier transform in the second variable, which was called b in (4.14).

A lower bound for (4.15) comes from ignoring all terms except those with $r = 0$, yielding

$$\mathbb{E}_{a\in\mathbb{F}_5^n} |\widetilde{\mathbf{f}_1}(a,0)|^4 = \mathbb{E}_{a\in\mathbb{F}_5^n} |\mathbb{E}_{b\in\mathbb{F}_5^{d_2}} \mathbf{f}_1(a,b)|^4.$$

The result now follows from Hölder's inequality. $\square$

A more interesting application of these partial Fourier transforms may be found in [19].

5. Lecture 5

Topics to be covered
- An introduction to the theory on $\mathbb{Z}/N\mathbb{Z}$.

For simplicity I will assume that N is a large prime.

I am only scheduled to give four lectures at the school. These notes are here for two reasons: firstly, it is possible that I will finish the material from the first four lectures early. More importantly, it is the theory on the group $\mathbb{Z}/N\mathbb{Z}$ that is of most interest for applications in number theory, and it would be remiss of me to not at least point the reader in directions where she may learn more.

Note that the theory on $\mathbb{Z}/N\mathbb{Z}$ is actually rather richer than for an arbitrary abelian group G, because we have been able to pursue analogies with ergodic theory. This is concerned with $\mathbb{Z}$-actions, and $\mathbb{Z}/N\mathbb{Z}$ is the finite abelian group which most closely models $\mathbb{Z}$.

One way of motivating the theory is to try and take what we know for $\mathbb{F}_5^n$ and attempt to adapt it to $\mathbb{Z}/N\mathbb{Z}$. Let us note that the basic definitions of Gowers norms and the basic generalized von Neumann theorems of Lecture 1 go over essentially unchanged to $\mathbb{Z}/N\mathbb{Z}$. The first stumbling block comes at the point where we ask for a conjectural analogue of Proposition 2.2. A first guess might be:

Conjecture 5.1. *Suppose that $f\colon \mathbb{Z}/N\mathbb{Z} \to [-1,1]$ is a function with $\|f\|_{U^3} \geq \delta$. Then there are $r, s \in \mathbb{Z}/N\mathbb{Z}$ such that*

$$\left| \mathbb{E}_{x \in \mathbb{Z}/N\mathbb{Z}} f(x) e\left(\frac{rx^2 + sx}{N} \right) \right| \gg_\delta 1.$$

Remark. As usual in analytic number theory we have written $e(\theta) := e^{2\pi i \theta}$.

It turns out that this conjecture is false. One example of a function on $\mathbb{Z}/N\mathbb{Z}$ which has a large U^3-norm, but does not correlate with any quadratic form $e(rx^2 + sx/N)$, is a quadratic $e(\theta x^2)$ where $\theta \not\approx r/N$. Such a quadratic is most naturally defined on $\mathbb{Z}$, but by restricting its domain to $\{1, \ldots, N\}$ one obtains a function which can be defined on $\mathbb{Z}/N\mathbb{Z}$. Another example is a "bracket quadratic" such as $e(\theta_1 x \{\theta_2 x\})$, where $\{t\}$ denotes the fractional part of t. The second of these counterexamples is somehow more serious, but it is also rather harder to see that this rather exotic function *does* provide a counterexample to Conjecture 5.1. For a brief discussion see [15, §6], and for more detail see [18].

If the only obvious generalization of Proposition 2.2 is wrong, how should we proceed? It turns out that a hint is given to us by the quantitatively stronger form of the inverse theorem for the U^3-norm on $\mathbb{F}_5^n$, namely Proposition 2.9. We are not concerned with quantitative issues here, so let us state a weak consequence of that result. This is actually a trivial consequence of Proposition 2.2, too.

Proposition 5.2 (Inverse result for U^3-norm on $\mathbb{F}_5^n$. III). *Suppose that $f\colon \mathbb{F}_5^n \to [-1,1]$ is a function with $\|f\|_{U^3} \geq \delta$. Then there is a subspace $H \leq \mathbb{F}_5^n$ with $\operatorname{codim} H \ll_\delta 1$, a matrix $M \in \mathfrak{M}_n(\mathbb{F}_5)$ and a vector $r \in \mathbb{F}_5^n$ such that*

$$\left| \mathbb{E}_x f(x) 1_H(x) \omega^{x^T M x + r^T x} \right| \gg_\delta 1.$$

Remark. It is not too hard to show that this is *equivalent* to Proposition 2.2: we leave this as an exercise to the reader.

Let us try and generalize this result. There are two objects which do not obviously transfer to $\mathbb{Z}/N\mathbb{Z}$: the notion of *subspace*, and (implicitly) the notion of *quadratic form*. It turns out that the second notion can be sensibly formulated for functions defined on any set.

Definition 5.3 (Quadratic forms). Let S be a set in some abelian group, and let $\psi: S \to \mathbb{R}/\mathbb{Z}$ be a function. We say that ψ is a quadratic form if the second derivative
$$\psi''(h_1, h_2) := \psi(x+h_1+h_2) - \psi(x+h_1) - \psi(x+h_2) + \psi(x)$$
is well-defined, that is to say if this definition does not depend on x whenever $x, x+h_1, x+h_2, x+h_1+h_2 \in S$.

Whilst the notion of subspace is rather vacuous in $\mathbb{Z}/N\mathbb{Z}$, there is a plentiful supply of *approximate* subspaces. These are more usually called Bohr sets.

Definition 5.4 (Approximate subspaces/Bohr sets). Let $R = \{r_1, \ldots, r_k\} \subseteq \mathbb{Z}/N\mathbb{Z}$ and let $\epsilon > 0$. Then we write
$$B(R, \epsilon) := \{x \in \mathbb{Z}/N\mathbb{Z} : |e(rx/N) - 1| \leq \epsilon\}.$$
This is called the Bohr set with width ϵ corresponding to frequency set R.

The set R should actually be thought of as a set of characters on $\mathbb{Z}/N\mathbb{Z}$, each value r corresponding to the character $x \mapsto e(rx/N)$. Once thought of in this way, it is easy to see how Bohr sets can be defined on any finite abelian group G. Bohr sets on $\mathbb{F}_5^n$ do not depend very seriously on the width parameter ϵ, and certainly for $\epsilon < 1/10$ (say) they are just vector subspaces.

There is a lot to say about Bohr sets, and much information may be found in [34]. See also [12], where there is a discussion of the place of Bohr sets in the transition from finite field models to $\mathbb{Z}/N\mathbb{Z}$ in various settings. We caution the reader that there are certain technicalities associated with the study of Bohr sets in additive combinatorics, most particularly the need to consider *regular* Bohr sets (ones that "behave well at the edges"). In this brief overview we will say nothing more about these technicalities, other than that most of them were overcome in a seminal paper of Bourgain [4].

To return to the point, we may now state Theorem 2.7(i) of [18], which is an inverse theorem for the U^3-norm on $\mathbb{Z}/N\mathbb{Z}$. In the light of the above discussion, the reader will see that it is a natural generalization of Proposition 5.2.

Proposition 5.5 (Inverse theorem for the U^3-norm on $\mathbb{Z}/N\mathbb{Z}$. I). *Suppose that $f: \mathbb{Z}/N\mathbb{Z} \to [-1,1]$ is a function and that $\|f\|_{U^3} \geq \delta$. Then there is a set $R \subseteq \mathbb{Z}/N\mathbb{Z}$, $|R| \ll_\delta 1$, a parameter $\epsilon \gg_\delta 1$ such that the Bohr set $B := B(R, \epsilon)$ is regular, some $y \in \mathbb{Z}/N\mathbb{Z}$ and a quadratic form $\psi: y + B \to \mathbb{R}/\mathbb{Z}$ such that*

(5.1) $$\left|\mathbb{E}_x f(x) 1_{y+B}(x) e(\psi(x))\right| \gg_\delta 1.$$

It turns out that result is necessary and sufficient, that is to say if (5.1) is satisfied then $\|f\|_{U^3}$ is large. See [18, Theorem 2.7(ii)] (note that this is the only point at which the regularity of $B(R, \epsilon)$ is relevant). This is, at first sight, a very unsatisfactory state of affairs: we have a theorem which gives a necessary and sufficient condition for a natural problem which interests us, yet the theorem is somewhat inelegant and difficult to state.

Our subject being in some sense an extension of the work of Hardy and Littlewood, one should perhaps recall at this point Hardy's view that there is "no permanent place in the world for ugly mathematics."

With this in mind we observe that although Proposition 5.5 is necessary and sufficient, it need not be the *only* necessary and sufficient condition. In what follows we will be rather vague. Write $\mathcal{Q} = \mathcal{Q}(\delta)$ for the collection of all "quadratic obstructions" of the form $1_{y+B}(x)e(\psi(x))$, where B, ψ are as above. Any other collection $\mathcal{Q}'$ with the property that anything in $\mathcal{Q}$ is approximately a linear combination of elements in $\mathcal{Q}'$, and vice versa, will also be a necessary and sufficient collection of quadratic obstructions for $\mathbb{Z}/N\mathbb{Z}$.

It turns out that there is a very natural choice for $\mathcal{Q}'$, the collection of 2-*step nilsequences*. The idea that we should look at these objects came to us from ergodic theory — there will be much more on this in the lectures of Bryna Kra at the school.

Let G be a connected, simply-connected 2-step nilpotent Lie group over $\mathbb{R}$ and let $\Gamma \leq G$ be a discrete, cocompact submanifold. The quotient G/Γ is called a 2-step nilmanifold. For the sake of illustration, we recommend that the reader take

$$G := \begin{pmatrix} 1 & \mathbb{R} & \mathbb{R} \\ 0 & 1 & \mathbb{R} \\ 0 & 0 & 1 \end{pmatrix}, \quad \Gamma := \begin{pmatrix} 1 & \mathbb{Z} & \mathbb{Z} \\ 0 & 1 & \mathbb{Z} \\ 0 & 0 & 1 \end{pmatrix},$$

in which case G/Γ is a 3-dimensional compact manifold called the *Heisenberg nilmanifold*.

Let $g \in G$ and $x \in G/\Gamma$ be arbitrary. The element g induces a continuous map $T_g \colon G/\Gamma \to G/\Gamma$ by multiplication on the left. Any sequence of the form $\bigl(F(T_g^n \cdot x)\bigr)_{n \in \mathbb{N}}$, where $F \colon G/\Gamma \to [-1, 1]$ is continuous, is called a 2-step nilsequence. It turns out that the collection of 2-step nilsequences can play the rôle of $\mathcal{Q}'$ as discussed above. The following is proved in [18, Theorem 12.8].

Proposition 5.6 (Inverse theorem for the U^3-norm on $\mathbb{Z}/N\mathbb{Z}$. II). *Let $f \colon \mathbb{Z}/N\mathbb{Z} \to [-1, 1]$ be a function, and suppose that $\|f\|_{U^3} \geq \delta$. Then there is a 2-step nilsequence $\bigl(F(T_g^n \cdot x)\bigr)_{n \in \mathbb{N}}$ with complexity $\ll_\delta 1$ such that*

$$|\mathbb{E}_{n \leq N} f(n) F(T_g^n \cdot x)| \gg_\delta 1.$$

If, conversely, f correlates with a 2-step nilsequence of bounded complexity then the $\|\cdot\|_{U^3}$-norm of f is large.

We have not defined the *complexity* of a nilsequence. It is some number associated to $\bigl(F(T_g^n \cdot x)\bigr)_{n \in \mathbb{N}}$, which bounds both the dimension of the underlying nilmanifold G/Γ, and also the Lipschitz constant of F with respect to some sensible metric. There is no canonical way of defining the complexity, but this is not important for the theory.

We do not attempt to explain why this collection $\mathcal{Q}'$ of 2-step nilsequences is "equivalent" to the collection $\mathcal{Q}$ used in Proposition 5.5. Detailed technical discussions may be found in [18, 20]. A short calculation involving the Heisenberg example, showing how a 2-step nilsequence on it resembles a quadratic form on a Bohr set, is given in [15].

References

1. A. Balog and E. Szemerédi, *A statistical theorem of set addition*, Combinatorica **14** (1994), no. 3, 263–268.
2. V. Bergelson, Host B., and B. Kra, *Multiple recurrence and nilsequences*, Invent. Math. **160** (2005), no. 2, 261–303.
3. J. Bourgain, *A Szemerédi-type theorem for sets of positive density in $\mathbb{R}^k$*, Israel J. Math. **54** (1986), no. 3, 307–316.
4. _____, *On triples in arithmetic progression*, Geom. Funct. Anal. **9** (1999), no. 5, 968–984.
5. M. C. Chang, *On problems of Erdős and Rudin*, J. Funct. Anal. **207** (2004), no. 2, 444–460.

6. E. Croot, *The minimal number of 3-term arithmetic progressions modulo a prime converges to a limit*, Canad. Math. Bull., to appear.
7. W. T. Gowers, *Lower bounds of tower type for Szemerédi's uniformity lemma*, Geom. Funct. Anal. **7** (1997), no. 2, 322–337.
8. _____, *A new proof of Szemerédi's theorem for progressions of length four*, Geom. Funct. Anal. **8** (1998), no. 3, 529–551.
9. _____, *A new proof of Szemerédi's theorem*, Geom. Funct. Anal. **11** (2001), no. 3, 465–588.
10. B. J. Green, *Roth's theorem in the primes*, Ann. of Math. (2) **161** (2005), no. 3, 1609–1636.
11. _____, *A Szemerédi-type regularity lemma in abelian groups, with applications*, Geom. Funct. Anal. **15** (2005), no. 2, 340–376.
12. _____, *Finite field models in additive combinatorics*, Surveys in Combinatorics 2005 (Durham, 2005) (B. S. Webb, ed.), London Math. Soc. Lecture Note Ser., vol. 327, Cambridge Univ. Press, Cambridge, 2005, pp. 1–27.
13. _____, *Edinburgh–MIT lecture notes on Freĭman's theorem*, available at http://www.dpmms.cam.ac.uk/~bjg23.
14. _____, *Notes on the Bourgain–Katz–Tao theorem*, available at http://www.dpmms.cam.ac.uk/~bjg23.
15. _____, *Generalising the Hardy–Littlewood method for primes*, Proceedings of the International Congress of Mathematicians, Vol. 2 (Madrid, 2006).
16. B. J. Green and S. Konyagin, *On the Littlewood problem modulo a prime*, Canad. J. Math., to appear.
17. B. J. Green and T. C. Tao, *The primes contain arbitrarily long arithmetic progressions*, Ann. of Math. (2), to appear.
18. _____, *An inverse theorem for the Gowers U^3-norm, with applications*, Proc. Edinb. Math. Soc. (2), to appear.
19. _____, *A new bound for Szemerédi's theorem in finite field geometries, for progressions of length 4*, preprint.
20. _____, *Quadratic uniformity of the Möbius function*, Ann. Inst. Fourier (Grenoble), to appear.
21. _____, *Linear equations in primes*, Ann. of Math. (2), to appear.
22. _____, *Arithmetic regularity lemmas*, to be written.
23. D. R. Heath-Brown, *Integer sets containing no arithmetic progressions*, J. London Math. Soc. (2) **35** (1987), no. 3, 385–394.
24. B. Host and B. Kra, *Nonconventional ergodic averages and nilmanifolds*, Ann. of Math. (2) **161** (2005), no. 1, 397–488.
25. H. Plünnecke, *Eigenschaften und Abschätzungen von Wirkingsfunktionen*, BMwF-GMD-22, Gesellschaft für Mathematik und Datenverarbeitung, Bonn, 1969.
26. I. Z. Ruzsa, *An application of graph theory to additive number theory*, Sci. Ser. A Math. Sci. (N.S.) **3** (1989), 97–109.
27. _____, *Generalized arithmetical progressions and sumsets*, Acta Math. Acad. Sci. Hungar. **65** (1994), no. 4, 379–388.
28. _____, *An analog of Freĭman's theorem in groups*, Structure Theory of Set Addition (J.-M. Deshouillers, Landreau B., and A. A. Yudin, eds.), Astérisque, vol. 258, Soc. Math. France, Paris, 1999, pp. 323–326.
29. A. Samorodnitsky, *Low-degree tests at large distances*, available at arXiv:math.CO/0604353.
30. E. Szemerédi, *Integer sets containing no arithmetic progressions*, Acta Math. Acad. Sci. Hungar. **56** (1990), no. 1-2, 155–158.
31. E. Szemerédi, *Regular partitions of graphs*, Problémes combinatoires et théorie des graphes (Orsay, 1976), Colloq. Internat. CNRS, vol. 260, CNRS, Paris, 1978, pp. 399–401.

32. T. C. Tao, *A quantitative ergodic theory proof of Szemerédi's theorem*, Electron. J. Combin., to appear.
33. _____, *The dichotomy between structure and randomness, arithmetic progressions, and the primes*, Proceedings of the International Congress of Mathematicians. Vol. 1 (Madrid, 2006), to appear.
34. T. C. Tao and V. H. Vu, *Additive combinatorics*, Cambridge Stud. Adv, Math, vol. 105, Cambridge Univ. Press, Cambridge, 2006.
35. T. Ziegler, *Universal characteristic factors and Furstenberg averages*, J. Amer. Math. Soc. **20** (2007), no. 1, 53–97.

CENTRE FOR MATHEMATICAL SCIENCES, WILBERFORCE ROAD, CAMBRIDGE CB3 0WA, ENGLAND

E-mail address: `b.j.green@dpmms.cam.ac.uk`

Ergodic Methods in Additive Combinatorics

Bryna Kra

ABSTRACT. Shortly after Szemerédi's proof that a set of positive upper density contains arbitrarily long arithmetic progressions, Furstenberg gave a new proof of this theorem using ergodic theory. This gave rise to the field of combinatorial ergodic theory, in which problems motivated by additive combinatorics are addressed with ergodic theory. Combinatorial ergodic theory has since produced combinatorial results, some of which have yet to be obtained by other means, and has also given a deeper understanding of the structure of measure preserving systems. We outline the ergodic theory background needed to understand these results, with an emphasis on recent developments in ergodic theory and the relation to recent developments in additive combinatorics.

These notes are based on four lectures given during the School on Additive Combinatorics at the Centre de recherches mathématiques, Montreal in April, 2006. The talks were aimed at an audience without background in ergodic theory. No attempt is made to include complete proofs of all statements and often the reader is referred to the original sources. Many of the proofs included are classic, included as an indication of which ingredients play a role in the developments of the past ten years.

1. Combinatorics to ergodic theory

1.1. Szemerédi's theorem. Answering a long standing conjecture of Erdős and Turán [11], Szemerédi [54] showed that a set $E \subset \mathbb{Z}$ with positive upper density[1] contains arbitrarily long arithmetic progressions. Soon thereafter, Furstenberg [16] gave a new proof of Szemerédi's Theorem using ergodic theory, and this has lead to the rich field of combinatorial ergodic theory. Before describing some of the results in this subject, we motivate the use of ergodic theory for studying combinatorial problems.

We start with the finite formulation of Szemerédi's theorem:

Theorem 1.1 (Szemerédi [54]). *Given $\delta > 0$ and $k \in \mathbb{N}$, there is a function $N(\delta, k)$ such that if $N > N(\delta, k)$ and $E \subset \{1, \ldots, N\}$ is a subset with $|E| \geq \delta N$, then E contains an arithmetic progression of length k.*

1991 *Mathematics Subject Classification.* Primary 37A30; Secondary 11B25.
Key words and phrases. Ergodic theory, additive combinatorics.
The author was supported in part by NSF Grant DMS-#0555250.
This is the final form of the paper.

[1] Given a set $E \subset \mathbb{Z}$, its *upper density* $d^*(E)$ is defined by $d^*(E) = \limsup_{N\to\infty} |E \cap \{1,\ldots,N\}|/N$.

It is clear that this statement immediately implies the first formulation of Szemerédi's theorem, and a compactness argument gives the converse implication.

1.2. Translation to a probability system. Starting with Szemerédi's theorem, one gains insight into the intersection of sufficiently many sets with positive measure in an arbitrary probability system.[2] Note that $N(\delta, k)$ denotes the quantity in Theorem 1.1.

Corollary 1.2. *Let $\delta > 0$, $k \in \mathbb{N}$, $(X, \mathcal{X}, \mu)$ be a probability space and $A_1, \ldots, A_N \in \mathcal{X}$ with $\mu(A_i) \geq \delta$ for $i = 1, \ldots, N$. If $N > N(\delta, k)$, then there exist $a, d \in \mathbb{N}$ such that*

$$A_a \cap A_{a+d} \cap A_{a+2d} \cap \cdots \cap A_{a+kd} \neq \varnothing.$$

PROOF. For $A \subset \mathcal{X}$, let $\mathbf{1}_A(x)$ denote the characteristic function of A (meaning that $\mathbf{1}_A(x)$ is 1 for $x \in A$ and is 0 otherwise). Let $N > N(\delta, k)$. Then

$$\int_X \frac{1}{N} \sum_{n=0}^{N-1} \mathbf{1}_{A_n} \, d\mu \geq \delta.$$

Thus there exists $x \in X$ such that

$$\frac{1}{N} \sum_{n=0}^{N-1} \mathbf{1}_{A_n}(x) \geq \delta.$$

Then $E = \{n : x \in A_n\}$ satisfies $|E| \geq \delta N$, and so Szemerédi's theorem implies that E contains an arithmetic progression of length k. By the definition of E, we have a sequence of sets with the desired property. □

1.3. Measure preserving systems. A *probability measure preserving system* is a quadruple $(X, \mathcal{X}, \mu, T)$, where $(X, \mathcal{X}, \mu)$ is a probability space and $T \colon X \to X$ is a bijective, measurable, measure preserving transformation. This means that for all $A \in \mathcal{X}$, $T^{-1}A \in \mathcal{X}$ and

$$\mu(T^{-1}A) = \mu(A).$$

In general, we refer to a probability measure preserving system as a *system*.

Without loss of generality, we can place several simplifying assumptions on our systems. We assume that $\mathcal{X}$ is countably generated; thus for $1 \leq p < \infty$, $L^p(\mu)$ is separable. We implicitly assume that all sets and functions are measurable with respect to the appropriate σ-algebra, even when this is not explicitly stated. Equality between sets or functions is always meant up to sets of measure 0.

---

[2]A *σ-algebra* is a collection $\mathcal{X}$ of subsets of X satisfying: (i) $X \in \mathcal{X}$ (ii) for any $A \in \mathcal{X}$ we also have $X \setminus A \in \mathcal{X}$ (iii) for any countable collection $A_n \in \mathcal{X}$, we also have $\bigcup_{n=1}^{\infty} A_n \in \mathcal{X}$. A σ-algebra is endowed with operations $\bigvee$, $\bigwedge$, and c, which correspond to union, intersection, and taking complements. By a *probability system*, we mean a triple $(X, \mathcal{X}, \mu)$ where X is a measure space, $\mathcal{X}$ is a σ-algebra of measurable subsets of X, and μ is a probability measure. In general, we use the convention of denoting the σ-algebra $\mathcal{X}$ by the associated calligraphic version of the measure space X.

1.4. Furstenberg multiple recurrence.
In a system, one can use Szemerédi's theorem to derive a bit more information about intersections of sets. If $(X, \mathcal{X}, \mu, T)$ is a system and $A \in \mathcal{X}$ with $\mu(A) \geq \delta > 0$, then
$$A, T^{-1}A, T^{-2}A, \ldots, T^{-n}A, \ldots$$
are all sets of the same measure, and so all have measure $\geq \delta$. Applying Corollary 1.2 to this sequence of sets, we have the existence of $a, d \in \mathbb{N}$ with
$$T^{-a}A \cap T^{-(a+d)}A \cap T^{-(a+2d)} \cap \cdots \cap T^{-(a+kd)}A \neq \varnothing.$$

Furthermore, the measure of this intersection must be positive. If not, we could remove from A a subset of measure zero containing all the intersections and obtain a subset of measure at least δ without this property. In this way, starting with Szemerédi's Theorem, we have derived Furstenberg's multiple recurrence theorem:

Theorem 1.3 (Furstenberg [16]). *Let $(X, \mathcal{X}, \mu, T)$ be a system and let $A \in \mathcal{X}$ with $\mu(A) > 0$. Then for any $k \geq 1$, there exists $n \in \mathbb{N}$ such that*
$$(1.1) \qquad \mu\bigl(A \cap T^{-n}A \cap T^{-2n}A \cap \cdots \cap T^{-kn}A\bigr) > 0.$$

2. Ergodic theory to combinatorics

2.1. Strong form of multiple recurrence.
We have seen that Furstenberg multiple recurrence can be easily derived from Szemerédi's theorem. More interesting is the converse implication, showing that one can use ergodic theory to prove regularity properties of subsets of the integers, and in particular derive Szemerédi's theorem. This is what Furstenberg did in his landmark paper [16], and the techniques introduced in this paper have been used subsequently to deduce other patterns in subsets of integers with positive upper density. (See Section 9.) Moreover, Furstenberg's proof lead to new questions within ergodic theory, about the structure of measure preserving systems. In turn, this finer analysis of measure preserving systems has had implications in additive combinatorics. We return to these questions in Section 3.

Furstenberg's approach to Szemerédi's theorem has two major components. The first is proving a certain recurrence statement in ergodic theory, like that of Theorem 1.3. The second is showing that this statement implies a corresponding statement about subsets of the integers. We now make this more precise.

To use ergodic theory to show that some intersection of sets has positive measure, it is natural to average the expression under consideration. This leads us to the strong form of Furstenberg's multiple recurrence:

Theorem 2.1 (Furstenberg [16]). *Let $(X, \mathcal{X}, \mu, T)$ be a system and let $A \in \mathcal{X}$ with $\mu(A) > 0$. Then for any $k \geq 1$,*
$$(2.1) \qquad \liminf_{N \to \infty} \frac{1}{N} \sum_{n=0}^{N-1} \mu(A \cap T^{-n}A \cap T^{-2n}A \cap \cdots \cap T^{-kn}A)$$
is positive.

In particular, this implies the existence of infinitely many $n \in \mathbb{N}$ such that the intersection in (1.1) is positive and Theorem 1.3 follows. In Section 3, we discuss how to prove Theorem 2.1.

2.2. The correspondence principle.
The second major component in Furstenberg's proof is using this multiple recurrence statement to derive a statement about integers, such as Szemerédi's theorem. This is the content of Furstenberg's correspondence principle:

Theorem 2.2 (Furstenberg [16,17]). *Let $E \subset \mathbb{Z}$ have positive upper density. There exist a system $(X, \mathcal{X}, \mu, T)$ and a set $A \in \mathcal{X}$ with $\mu(A) = d^*(E)$ such that*

$$\mu(T^{-m_1}A \cap \cdots \cap T^{-m_k}A) \leq d^*\big((E+m_1) \cap \cdots \cap (E+m_k)\big)$$

for all $k \in \mathbb{N}$ and all $m_1, \ldots, m_k \in \mathbb{Z}$.

PROOF. Let $X = \{0,1\}^\mathbb{Z}$ be endowed with the product topology and the shift map T given by $Tx(n) = x(n+1)$ for all $n \in \mathbb{Z}$. A point of X is thus a sequence $\mathbf{x} = \{x(n)\}_{n \in \mathbb{Z}}$, and the distance between two points $\mathbf{x} = \{x(n)\}_{n \in \mathbb{Z}}, \mathbf{y} = \{y(n)\}_{n \in \mathbb{Z}}$ is defined to be 0 if $\mathbf{x} = \mathbf{y}$ and to be 2^{-k} if $\mathbf{x} \neq \mathbf{y}$ and $k = \min\{|n| : x(n) \neq y(n)\}$. Define $\mathbf{a} = \{a(n)\}_{n \in \mathbb{Z}} \in \{0,1\}^\mathbb{Z}$ by

$$a(n) = \begin{cases} 1 & \text{if } n \in E \\ 0 & \text{otherwise} \end{cases}$$

and let $A = \{\mathbf{x} \in X : x(0) = 1\}$. Thus A is a clopen (closed and open) set.

The set $A \in \mathcal{X}$ plays the same role as the set $E \subset \mathbb{Z}$: for all $n \in \mathbb{Z}$,

$$T^n \mathbf{a} \in A \text{ if and only if } n \in E.$$

By definition of $d^*(E)$, there exist sequences $\{M_i\}$ and $\{N_i\}$ of integers with $N_i \to \infty$ such that

$$\lim_{i \to \infty} \frac{1}{N_i} \big| E \cap [M_i, M_i + N_i - 1] \big| \to d^*(E).$$

It follows that

$$\lim_{i \to \infty} \frac{1}{N_i} \sum_{n=M_i}^{M_i+N_i-1} \mathbf{1}_A(T^n\mathbf{a}) = \lim_{i \to \infty} \frac{1}{N_i} \sum_{n=M_i}^{M_i+N_i-1} \mathbf{1}_E(n) = d^*(E).$$

Let $\mathcal{C}$ be the countable algebra generated by cylinder sets, meaning sets that are defined by specifying finitely many coordinates of each element and leaving the others free. We can define an additive measure μ on $\mathcal{C}$ by

$$\mu(B) = \lim_{i \to \infty} \frac{1}{N_i} \sum_{n=M_i}^{M_i+N_i-1} \mathbf{1}_B(T^n\mathbf{a}),$$

where we pass, if necessary, to subsequences $\{N_i\}, \{M_i\}$ such that this limit exists for all $B \in \mathcal{C}$. (Note that $\mathcal{C}$ is countable and so by diagonalization we can arrange it such that this limit exists for all elements of $\mathcal{C}$.)

We can extend the additive measure to a σ-additive measure μ on all Borel sets $\mathcal{X}$ in X, which is exactly the σ-algebra generated by $\mathcal{C}$. Then μ is an invariant measure, meaning that for all $B \in \mathcal{C}$,

$$\mu(T^{-1}B) = \lim_{i \to \infty} \frac{1}{N_i} \sum_{n=M_i}^{M_i+N_i-1} \mathbf{1}_B(T^{n-1}\mathbf{a}) = \mu(B).$$

Furthermore,
$$\mu(A) = \lim_{i \to \infty} \frac{1}{N_i} \sum_{n=M_i}^{M_i+N_i-1} \mathbf{1}_A(T^n \mathbf{a}) = d^*(E).$$

If $m_1, \ldots, m_k \in \mathbb{Z}$, then the set $T^{-m_1}A \cap \cdots \cap T^{-m_k}A$ is a clopen set, its indicator function is continuous, and

$$\mu(T^{-m_1}A \cap \cdots \cap T^{-m_k}A) = \lim_{i \to \infty} \frac{1}{N_i} \sum_{n=M_i}^{M_i+N_i-1} \mathbf{1}_{T^{-m_1}A \cap \cdots \cap T^{-m_k}A}(T^n \mathbf{a})$$

$$= \lim_{i \to \infty} \frac{1}{N_i} \sum_{n=M_i}^{M_i+N_i-1} \mathbf{1}_{(E+m_1) \cap \cdots \cap (E+m_k)}(n)$$

$$\leq d^*\big((E+m_1) \cap \cdots \cap (E+m_k)\big). \qquad \square$$

We use this to deduce Szemerédi's theorem from Theorem 1.3. As in the proof of the correspondence principle, define $\mathbf{a} \in \{0,1\}^{\mathbb{Z}}$ by

$$a(n) = \begin{cases} 1 & \text{if } n \in E \\ 0 & \text{otherwise,} \end{cases}$$

and set $A = \{\mathbf{x} \in \{0,1\}^{\mathbb{Z}} : x(0) = 1\}$. Thus $T^n \mathbf{a} \in A$ if and only if $n \in E$.
By Theorem 1.3, there exists $n \in \mathbb{N}$ such that

$$\mu(A \cap T^{-n}A \cap T^{-2n}A \cap \cdots \cap T^{-kn}A) > 0.$$

Therefore for some $m \in \mathbb{N}$, $T^m \mathbf{a}$ enters this multiple intersection and so

$$a(m) = a(m+n) = a(m+2n) = \cdots = a(m+kn) = 1.$$

But this means that

$$m, m+n, m+2n, \ldots, m+kn \in E$$

and so we have found an arithmetic progression of length $k+1$ in E.

3. Convergence of multiple ergodic averages

3.1. Convergence along arithmetic progressions.
Furstenberg's multiple recurrence theorem left open the question of the existence of the limit in (2.1). More generally, one can ask if given a system $(X, \mathcal{X}, \mu, T)$ and $f_1, f_2, \ldots, f_k \in L^{\infty}(\mu)$, does

$$(3.1) \qquad \lim_{N \to \infty} \frac{1}{N} \sum_{n=0}^{N-1} f_1(T^n x) \cdot f_2(T^{2n} x) \cdot \cdots \cdot f_k(T^{kn} x)$$

exist? Moreover, we can ask in what sense (in $L^2(\mu)$ or pointwise) does this limit exist, and if it does exist, what can be said about the limit? Setting each function f_i to be the indicator function of a measurable set A, we are back in the context of Furstenberg's theorem.

For $k = 1$, existence of the limit in $L^2(\mu)$ is the mean ergodic theorem of von Neumann. In Section 4.2, we give a proof of this statement. For $k = 2$, existence of the limit in $L^2(\mu)$ was proven by Furstenberg [16] as part of his proof of Szemerédi's theorem. Furthermore, in the same paper he showed the existence of the limit in $L^2(\mu)$ in a weak mixing system for arbitrary k; we define weak mixing in Section 5.5 and outline the proof for this case.

For $k \geq 3$, the proof of existence of the limit in 3.1 requires a more subtle understanding of measure preserving systems, and we begin discussing this case in Section 5.8. Under some technical hypotheses, the existence of the limit in $L^2(\mu)$ for $k = 3$ was first proven by Conze and Lesigne (see [8, 9]), then by Furstenberg and Weiss [22], and in the general case by Host and Kra [32]. More generally, we showed the existence of the limit for all $k \in \mathbb{N}$:

Theorem 3.1 (Host and Kra [34]). *Let $(X, \mathcal{X}, \mu, T)$ be a system, let $k \in \mathbb{N}$, and let $f_1, f_2, \ldots, f_k \in L^\infty(\mu)$. Then the averages*

$$\frac{1}{N} \sum_{n=0}^{N-1} f_1(T^n x) \cdot f_2(T^{2n} x) \cdot \ldots \cdot f_k(T^{kn} x)$$

converge in $L^2(\mu)$ as $N \to \infty$.

Such a convergence result for a finite system is trivial. For example, if $X = \mathbb{Z}/N\mathbb{Z}$, then $\mathcal{X}$ consists of all partitions of X and μ is the uniform probability measure, meaning that the measure of a set is proportional to the cardinality of the set. The transformation T is given by $Tx = x + 1 \mod N$. It is then trivial to check the convergence of the average in (3.1). However, although the ergodic theory is trivial in this case, there are common themes to be explored. Throughout these notes, an effort is made to highlight the connection with recent advances in additive combinatorics (see [39] for more on this connection). Of particular interest is the role played by nilpotent groups, and homogeneous spaces of nilpotent groups, in the proof of the ergodic statement.

Much of the present notes is devoted to understanding the ingredients in the proof of Theorem 3.1, and the role of nilpotent groups in this proof. Other expository accounts of this proof can be found in [31, 40]. In this context, 2-step nilpotent groups first appeared in the work of Conze-Lesigne in their proof of convergence for $k = 3$, and a $(k-1)$-step nilpotent group plays a similar role in convergence for the average in (3.1). Nilpotent groups also play some role in the combinatorial setup, and this has been recently verified by Green and Tao (see [26–28]) for progressions of length 4 (which corresponds to the case $k = 3$ in (3.1)). For more on this connection, see the lecture notes of Ben Green in this volume.

3.2. Other results. Using ergodic theory, other patterns have been shown to exist in sets of positive upper density and we discuss these results in Section 9. We briefly summarize these results. A striking example is the theorem of Bergelson and Leibman [6] showing the existence of polynomial patterns in such sets. Analogous to the linear average corresponding to arithmetic progressions, existence of the associated polynomial averages was shown in [35, 45]. One can also average along "cubes"; existence of these averages and a corresponding combinatorial statement was shown in [34]. For commuting transformations, little is known and these partial results are summarized in Section 9.1. An explicit formula for the limit in (3.1) was given by Ziegler [56], who also has recently given a second proof [57] of Theorem 3.1.

4. Single convergence (the case $k = 1$)

4.1. Poincaré recurrence. The case $k = 1$ in Furstenberg's multiple recurrence (Theorem 1.3) is Poincaré recurrence:

Theorem 4.1 (Poincaré [49]). *If $(X, \mathcal{X}, \mu, T)$ is a system and $A \in \mathcal{X}$ with $\mu(A) > 0$, then there exist infinitely many $n \in \mathbb{N}$ such that $\mu(A \cap T^{-n}A) > 0$.*

PROOF. Let $F = \{x \in A : T^{-n}x \notin A \text{ for all } n \geq 1\}$. Thus $F \cap T^{-n}F = \varnothing$ for all $n \geq 1$, and so for all integers $n \neq m$,

$$T^{-m}A \cap T^{-n}A = \varnothing.$$

In particular, $F, T^{-1}F, T^{-2}F, \ldots$ are all pairwise disjoint sets and each set in this sequence has measure equal to $\mu(F)$. If $\mu(F) > 0$, then

$$\mu\left(\bigcup_{n \geq 0} T^{-n}F\right) = \sum_{n \geq 0} \mu(F) = \infty,$$

a contradiction of μ being a probability measure.
Therefore $\mu(F) = 0$ and the statement is proven. $\square$

In fact the same proof shows a bit more: by a simple modification of the definition of F, we have that μ-almost every $x \in A$ returns to A infinitely often.

4.2. The von Neumann ergodic theorem. Although the proof of Poincaré recurrence is simple, unfortunately there seems to be no way to generalize it to show multiple recurrence. In order to find a method that generalizes for multiple recurrence, we prove a stronger statement than Poincaré recurrence, taking the average of the expression under consideration and showing that the lim inf of this average is positive. It is not any harder (for $k = 1$ only!) to show that the limit of this average exists (and is positive). This is the content of the von Neumann mean ergodic theorem. We first give the statement in a general Hilbert space:

Theorem 4.2 (von Neumann [55]). *If U is an isometry of a Hilbert space $\mathcal{H}$ and P is orthogonal projection onto the U-invariant subspace $\mathcal{I} = \{f \in \mathcal{H} : Uf = f\}$, then for all $f \in \mathcal{H}$,*

$$(4.1) \qquad \lim_{N \to \infty} \frac{1}{N} \sum_{n=0}^{N-1} U^n f = Pf.$$

Thus the case $k = 1$ in Theorem 3.1 is an immediate corollary of the von Neumann ergodic theorem.

PROOF. If $f \in \mathcal{I}$, then

$$\frac{1}{N} \sum_{n=0}^{N-1} U^n f = f$$

for all $N \in \mathbb{N}$ and so obviously the average converges to f. On the other hand, if $f = g - Ug$ for some $g \in \mathcal{H}$, then

$$\sum_{n=0}^{N-1} U^n f = g - U^N g$$

and so the average converges to 0 as $N \to \infty$. Set $\mathcal{J} = \{g - Ug : g \in \mathcal{H}\}$. If $f_k \in \mathcal{J}$ and $f_k \to f \in \overline{\mathcal{J}}$, then

$$\left\| \frac{1}{N} \sum_{n=0}^{N-1} U^n f \right\| \leq \left\| \frac{1}{N} \sum_{n=0}^{N-1} U^n (f - f_k) \right\| + \left\| \frac{1}{N} \sum_{n=0}^{N-1} U^n (f_k) \right\|$$

$$\leq \left\| \frac{1}{N} \sum_{n=0}^{N-1} U^n \right\| \cdot \| f - f_k \| + \left\| \frac{1}{N} \sum_{n=0}^{N-1} U^n (f_k) \right\|.$$

Thus for $f \in \overline{\mathcal{J}}$, the average $(1/N) \sum_{n=0}^{N-1} U^n f$ also converges to 0 as $N \to \infty$.

We now show that an arbitrary $f \in \mathcal{H}$ can be written as a combination of functions which exhibit these behaviors, meaning that any $f \in \mathcal{H}$ can be written as $f = f_1 + f_2$ for some $f_1 \in \mathcal{I}$ and $f_2 \in \overline{\mathcal{J}}$. If $h \in \mathcal{J}^\perp$, then for all $g \in \mathcal{H}$,

$$0 = \langle h, g - Ug \rangle = \langle h, g \rangle - \langle h, Ug \rangle = \langle h, g \rangle - \langle U^* h, g \rangle = \langle h - U^* h, y \rangle$$

and so $h = U^* h$ and $h = Uh$. Conversely, reversing the steps we have that if $h \in \mathcal{I}$, then $h \in \mathcal{J}^\perp$.

Since $\overline{\mathcal{J}}^\perp = \mathcal{J}^\perp$, we have that

$$\mathcal{H} = \mathcal{I} \oplus \overline{\mathcal{J}}.$$

Thus writing $f = f_1 + f_2$ with $f_1 \in \mathcal{I}$ and $f_2 \in \overline{\mathcal{J}}$, we have

$$\frac{1}{N} \sum_{n=0}^{N-1} U^n f = \frac{1}{N} \sum_{n=0}^{N-1} U^n f_1 + \frac{1}{N} \sum_{n=0}^{N-1} U^n f_2.$$

As $N \to \infty$, the first sum converges to the identity and the second sum to 0. □

The idea behind the proof of von Neumann's Theorem is simple: decompose an arbitrary function into two pieces and then show that the limit exists for each of these pieces. This sort of decomposition is used (in some sense) in Furstenberg's proof of Theorem 2.1, the original proof of Szemerédi's theorem, the convergence result of Theorem 3.1, and in the recent results of Green and Tao on patterns in the prime numbers.

Under a mild hypothesis on the system, we have an explicit formula for the limit (4.1). Let $(X, \mathcal{X}, \mu, T)$ be a system. A subset $A \subset X$ is said to be *invariant* if $T^{-1} A = A$. The invariant sets form a sub-σ-algebra $\mathcal{I}$ of $\mathcal{X}$. The system $(X, \mathcal{X}, \mu, T)$ is said to be *ergodic* if $\mathcal{I}$ is trivial, meaning that every invariant set has either measure 0 or measure 1.

A measure preserving transformation $T \colon X \to X$ defines a linear operator $U_T \colon L^2(\mu) \to L^2(\mu)$ by

$$(U_T f)(x) = f(Tx).$$

It is easy to check that the operator U_T is a unitary operator (meaning its adjoint is equal to its inverse). In a standard abuse of notation, we use the same letter to denote the operator and the transformation, writing $Tf(x) = f(Tx)$ instead of the more cumbersome $U_T f(x) = f(Tx)$.

Applying von Neumann's ergodic theorem in a measure preserving system, we have:

Corollary 4.3. *If $(X, \mathcal{X}, \mu, T)$ is a system and $f \in L^2(\mu)$, then*

$$\frac{1}{N} \sum_{n=0}^{N-1} f(T^n x)$$

converges in $L^2(\mu)$, as $N \to \infty$, to a T-invariant function $\tilde{f}$. If the system is ergodic, then the limit is the constant function $\int f \, d\mu$.

Let $(X, \mathcal{X}, \mu, T)$ be an ergodic system and let $A, B \in \mathcal{X}$. Taking $f = \mathbf{1}_A$ in Corollary 4.3 and integrating with respect to μ over a set B, we have that

$$\lim_{N \to \infty} \frac{1}{N} \sum_{n=0}^{N-1} \int_B \mathbf{1}_A(T^n x) \, d\mu(x) = \int_B \left(\int \mathbf{1}_A(y) \, d\mu(y) \right) d\mu(x).$$

Rewriting this, we have

$$\lim_{N \to \infty} \frac{1}{N} \sum_{n=0}^{N-1} \mu(A \cap T^{-n} B) = \mu(A) \mu(B).$$

In fact, one can check that this condition holds for all $A, B \in \mathcal{X}$ if and only if the system is ergodic.

As already discussed, convergence in the case of the finite system $\mathbb{Z}/N\mathbb{Z}$ with the transformation of adding $1 \mod N$ is trivial. Furthermore this system is ergodic. More generally, any permutation on $\mathbb{Z}/N\mathbb{Z}$ can be expressed as a product of disjoint cyclic permutations. These permutations are the "indecomposable" invariant subsets of an arbitrary transformation on $\mathbb{Z}/N\mathbb{Z}$ and the restriction of the transformation to one of these subsets is ergodic.

This idea of dividing a space into indecomposable components generalizes: an arbitrary measure preserving system can be decomposed into, perhaps continuously many, indecomposable components, and these are exactly the ergodic ones. Using this *ergodic decomposition* (see, for example, [10]), instead of working with an arbitrary system, we reduce most of the recurrence and convergence questions we consider here to the same problem in an ergodic system.

5. Double convergence (the case $k = 2$)

5.1. A model for double convergence. We now turn to the case of $k = 2$ in Theorem 3.1, and study convergence of the double average

(5.1) $$\frac{1}{N} \sum_{n=0}^{N-1} f_1(T^n x) \cdot f_2(T^{2n} x)$$

for bounded functions f_1 and f_2. Our goal is to explain how a simple class of systems, the rotations, suffice to understand convergence for the double average.

First we explicitly define what is meant by a rotation. Let G be a compact abelian group, with Borel σ-algebra $\mathcal{B}$, Haar measure m, and fix some $\alpha \in G$. Define $T \colon G \to G$ by

$$Tx = x + \alpha.$$

The system $(G, \mathcal{B}, m, T)$ is called a *group rotation*. It is ergodic if and only if $\mathbb{Z}\alpha$ is dense in G. For example, when X is the circle $\mathbb{T} = \mathbb{R}/\mathbb{Z}$ and $\alpha \notin \mathbb{Q}$, the rotation by α is ergodic.

The double average is the simplest example of a *nonconventional* ergodic average: even for an ergodic system, the limit is not necessarily constant. This sort of behavior does not occur for the single average of von Neumann's theorem, where we have seen that the limit is constant in an ergodic system. Even for the simple example of an an ergodic rotation, the limit of the double average is not constant:

Example 5.1. Let $X = \mathbb{T}$, with Borel σ-algebra and Haar measure, and let $T\colon X \to X$ be the rotation $Tx = x + \alpha \mod 1$. Setting $f_1(x) = \exp(4\pi i x)$ and $f_2(x) = \exp(-2\pi i x)$, then for all $n \in \mathbb{N}$,

$$f_1(T^n x) \cdot f_2(T^{2n} x) = \overline{f_2(x)}.$$

In particular, the double average (5.1) for these functions converges to a nonconstant function.

More generally, if $\alpha \notin \mathbb{Q}$ and $f_1, f_2 \in L^\infty(\mu)$, the double average converges to

$$\int_\mathbb{T} f_1(x+t) \cdot f_2(x+2t) \, dt.$$

We shall see that Fourier analysis suffices to understand this average. By taking both functions to be the indicator function of a set with positive measure and integrating over this set, we then have that Fourier analysis suffices for the study of arithmetic progressions of length 3. This gives a complete proof of Roth's Theorem via ergodic theory. Later we shall see that more powerful methods are needed to understand the average along longer progressions. In a similar vein, rotations are the model for an ergodic average with 3 terms, but are not sufficient for more terms. We introduce some terminology to make these notions more precise.

5.2. Factors. For the remainder of this section, we assume that $(X, \mathcal{X}, \mu, T)$ is an ergodic system.

A *factor* of a system $(X, \mathcal{X}, \mu, T)$ can be defined in one of several equivalent ways. It is a T-invariant sub-σ-algebra $\mathcal{Y}$ of $\mathcal{X}$. A second characterization is that a factor is a system $(Y, \mathcal{Y}, \nu, S)$ and a measurable map $\pi\colon X \to Y$, the *factor map*, such that $\mu \circ \pi^{-1} = \nu$ and $S \circ \pi = \pi \circ T$ for μ-almost every $x \in X$. A third characterization is that a factor is a T-invariant subalgebra $\mathcal{F}$ of $L^\infty(\mu)$. One can check that the first two definitions agree by identifying $\mathcal{Y}$ with $\pi^{-1}(\mathcal{Y})$, and that the first and third agree by identifying $\mathcal{F}$ with $L^\infty(\mathcal{Y})$. When any of these conditions holds, we say that Y, or the appropriate sub-σ-algebra, is a factor of X and write $\pi\colon X \to Y$ for the factor map. We usually make use of a slight (and standard) abuse of notation, using the same letter T to denote both the transformation in the original system and the transformation in the factor system. If the factor map $\pi\colon X \to Y$ is also injective, we say that the two systems $(X, \mathcal{X}, \mu, T)$ and $(Y, \mathcal{Y}, \nu, S)$ are *isomorphic*.

For example, if $(X, \mathcal{X}, \mu, T)$ and $(Y, \mathcal{Y}, \nu, S)$ are systems, then each is a factor of the product system $(X \times Y, \mathcal{X} \times \mathcal{Y}, \mu \times \nu, T \times S)$ and the associated factor map for each is projection onto the appropriate coordinate.

A more interesting example can be given in the system $X = \mathbb{T} \times \mathbb{T}$, with Borel σ-algebra and Haar measure, and transformation $T\colon X \to X$ given by

$$T(x, y) = (x + \alpha, y + x).$$

Then $\mathbb{T}$ with the rotation $x \mapsto x + \alpha$ is a factor of X.

5.3. Conditional expectation.
If $\mathcal{Y}$ is a T-invariant sub-σ-algebra of $\mathcal{X}$ and $f \in L^2(\mu)$, the *conditional expectation* $\mathbb{E}(f \mid \mathcal{Y})$ *of f with respect to $\mathcal{Y}$* is the function on Y defined by $\mathbb{E}(f \mid Y) \circ \pi = \mathbb{E}(f \mid \mathcal{Y})$. It is characterized as the $\mathcal{Y}$-measurable function on X such that
$$\int_X f(x) \cdot g(\pi(x)) \, d\mu(x) = \int_Y \mathbb{E}(f \mid \mathcal{Y})(y) \cdot g(y) \, d\nu(y)$$
for all $g \in L^\infty(\nu)$ and it satisfies the identities
$$\int \mathbb{E}(f \mid \mathcal{Y}) \, d\mu = \int f \, d\mu$$
and
$$T\mathbb{E}(f \mid \mathcal{Y}) = \mathbb{E}(Tf \mid \mathcal{Y}).$$

As an example, take $X = \mathbb{T} \times \mathbb{T}$ endowed with the transformation $(x, y) \mapsto (x + \alpha, y + x)$. We have a factor $Z = \mathbb{T}$ endowed with the map $x \mapsto x + \alpha$. Considering $f(x, y) = \exp(x) + \exp(y)$, we have that $\mathbb{E}(f \mid \mathcal{Z}) = \exp(x)$. The factor σ-algebra $\mathcal{Z}$ is the σ-algebra of sets that depend only on the x coordinate.

5.4. Characteristic factors.
For $f_1, \ldots, f_k \in L^\infty(\mu)$, we are interested in convergence in $L^2(\mu)$ of:
$$(5.2) \qquad \frac{1}{N} \sum_{n=0}^{N-1} T^n f_1 \cdot T^{2n} f_2 \cdot \ldots \cdot T^{kn} f_k.$$

Instead of working with the whole system $(X, \mathcal{X}, \mu, T)$, it turns out that it is easier to find some factor of the system that characterizes this average, meaning find some well chosen factor such that we can prove convergence of the average in this factor and this convergence suffices to understand convergence of the same average in the original system. This motivates the following definition.

A factor Y of X is *characteristic* for the average (5.2) if the difference between (5.2) and
$$\frac{1}{N} \sum_{n=0}^{N-1} T^n \mathbb{E}(f_1 \mid \mathcal{Y}) \cdot T^{2n} \mathbb{E}(f_2 \mid \mathcal{Y}) \cdot \ldots \cdot T^{kn} \mathbb{E}(f_k \mid \mathcal{Y})$$
(the same average with $\mathbb{E}(f_i \mid \mathcal{Y})$ substituted for f_i for $i = 1, 2, \ldots, k$) converges to 0 in $L^2(\mu)$ as $N \to \infty$. Rewriting the average (5.2) in terms of $f_i - \mathbb{E}(f_i \mid \mathcal{Y})$ for $i = 1, 2, \ldots, k$, it follows that the factor Y is characteristic for the average (5.2) if and only if the average in (5.2) converges to 0 as $N \to \infty$ when $\mathbb{E}(f_i \mid \mathcal{Y}) = 0$ for some $i \in \{1, 2, \ldots, k\}$.

The idea of a characteristic factor is that the limiting behavior of the average under study can be reduced to that of a factor of the system. We have already seen an example of a characteristic factor in the von Neumann ergodic theorem: the trivial factor, consisting only of the constants, is characteristic. (Recall that we have assumed that the system is ergodic.)

By definition, the whole system is always a characteristic factor. Of course nothing is gained by using such a characteristic factor, and the notion only becomes useful when we can find a characteristic factor that has useful geometric and/or algebraic properties. A very short outline of the proof of convergence of the average (5.2) is as follows: find a characteristic factor that has sufficient structure so as to allow one to prove convergence. We return to this idea later.

The definition of a characteristic factor can be extended for any other average under consideration, with the obvious changes: the limit remains unchanged when each function is replaced by its conditional expectation on this factor. This notion has been implicit in the literature since Furstenberg's proof of Szemerédi's theorem, but the terminology we now use was only introduced more recently in [22].

5.5. Weak mixing systems. The system $(X, \mathcal{X}, \mu, T)$ is *weak mixing* if for all $A, B \in \mathcal{X}$,

$$\lim_{N \to \infty} \frac{1}{N} \sum_{n=0}^{N-1} |\mu(T^{-n}A \cap B) - \mu(A)\mu(B)| = 0.$$

Any weak mixing system is ergodic, and the example of an irrational circle rotation shows that converse does not hold. There are many equivalent formulations of weak mixing, and we give a few (see, for example [10]):

Proposition 5.2. *Let $(X, \mathcal{X}, \mu, T)$ be a system. The following are equivalent:*

(1) $(X, \mathcal{X}, \mu, T)$ *is weak mixing.*
(2) *There exists $J \subset \mathbb{N}$ of density zero such that for all $A, B \in \mathcal{X}$*

$$\mu(T^{-n}A \cap B) \to \mu(A)\mu(B) \quad as \ n \to \infty \ and \ n \notin J.$$

(3) *For all $A, B, C \in \mathcal{X}$ with $\mu(A)\mu(B)\mu(C) > 0$, there exists $n \in \mathbb{N}$ such that*

$$\mu(A \cap T^{-n}B)\mu(A \cap T^{-n}C) > 0.$$

(4) *The system $(X \times X, \mathcal{X} \times \mathcal{X}, \mu \times \mu, T \times T)$ is ergodic.*

Any system exhibiting rotational behavior (for example a rotation on a circle, or a system with a nontrivial circle rotation as a factor) is not weak mixing. We have already seen in Example 5.1 that weak mixing, or lack thereof, has an effect on multiple averages. We give a second example to highlight this effect:

Example 5.3. Suppose that $X = X_1 \cup X_2 \cup X_3$ with $T(X_1) = X_2$, $T(X_2) = X_3$ and $T(X_3) = X_1$, and further suppose that T^3 restricted to X_i, for $i = 1, 2, 3$, is weak mixing. For the double average

$$\frac{1}{N} \sum_{n=0}^{N-1} f_1(T^n x) \cdot f_2(T^{2n} x),$$

where $f_1, f_2 \in L^\infty(\mu)$, if $x \in X_1$, this average converges to

$$\frac{1}{3} \left(\int_{X_1} f_1 \, d\mu \int_{X_1} f_2 \, d\mu + \int_{X_2} f_1 \, d\mu \int_{X_3} f_2 \, d\mu + \int_{X_3} f_1 \, d\mu \int_{X_2} f_2 \, d\mu \right).$$

A similar expression with obvious changes holds for $x \in X_2$ or $x \in X_3$.

The main point is that in both Example 5.1 and in Example 5.3 (for the double average) the limit depends on the rotational behavior of the system. Example 5.3 lacks weak mixing and so has a nontrivial rotation factor. We now formalize this notion.

5.6. Kronecker factor.

The *Kronecker factor* $(Z_1, \mathcal{Z}_1, m, T)$ of $(X, \mathcal{X}, \mu, T)$ is the sub-σ-algebra of $\mathcal{X}$ spanned by the eigenfunctions. (Recall that there is a unitary operator U_T associated to the measure preserving transformation T. By eigenfunctions, we refer to the eigenfunctions of this unitary operator.) A classical result is that the Kronecker factor can be given the structure of a group rotation:

Theorem 5.4 (Halmos and von Neumann [30]). *The Kronecker factor of a system is isomorphic to a system $(Z_1, \mathcal{Z}_1, m, T)$, where Z_1 is a compact abelian group, $\mathcal{Z}_1$ is its Borel σ-algebra, m is the Haar measure, and $Tx = x + \alpha$ for some fixed $\alpha \in Z_1$.*

We use $\pi_1 \colon X \to Z_1$ to denote the factor map from the system $(X, \mathcal{X}, \mu, T)$ to its Kronecker factor $(Z_1, \mathcal{Z}_1, m, T)$. Then any eigenfunction f of X takes the form

$$f(x) = c\gamma(\pi_1(x)),$$

where c is a constant and $\gamma \in \widehat{Z_1}$ is a character of Z_1.

We give two examples of Kronecker factors:

Example 5.5. If $X = \mathbb{T} \times \mathbb{T}$, $\alpha \in \mathbb{T}$, and $T \colon X \to X$ is the map

$$T(x, y) = (x + \alpha, y + x),$$

then the rotation $x \mapsto x + \alpha$ on $\mathbb{T}$ is the Kronecker factor of X. It corresponds to the pure point spectrum. (The spectrum in the orthogonal complement of the Kronecker factor is countable Lebesgue.)

Example 5.6. If $X = \mathbb{T}^3$, $\alpha \in \mathbb{T}$, and $T \colon X \to X$ is the map

$$T(x, y, z) = (x + \alpha, y + x, z + y),$$

then again the rotation $x \mapsto x + \alpha$ on $\mathbb{T}$ is the Kronecker factor of X. This example has the same pure point spectrum as the first example, but the system in the first example is a factor of the system in the second example.

The Kronecker factor can be used to give another characterization of weak mixing:

Theorem 5.7 (Koopman and von Neumann [38]). *A system is not weak mixing if and only if it has a nontrivial factor which is a rotation on a compact abelian group.*

The largest of these factors is exactly the Kronecker factor.

5.7. Convergence for $k = 2$.

If we take into account the rotational behavior in a system, meaning the existence of a nontrivial Kronecker factor, then we can understand the limit of the double average

(5.3) $$\frac{1}{N} \sum_{n=0}^{N-1} T^n f_1 \cdot T^{2n} f_2.$$

An obvious constraint is that for μ-almost every x, the triple $(x, T^n x, T^{2n} x)$ projects to an arithmetic progression in the Kronecker factor $\mathcal{Z}_1$. Assuming that the Kronecker factor is a circle with rotation by some α, we can think of each point in the progression $(x, T^n x, T^{2n} x)$ as located on the fiber above the corresponding point in the progression $(z, z + \alpha, z + 2\alpha)$:

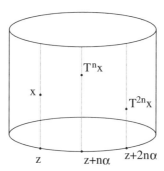

Furstenberg proved that this obvious restriction is the only restriction, showing that to prove convergence of the double average, one can assume that the system is an ergodic rotation on a compact abelian group:

Theorem 5.8 (Furstenberg [16]). *If $(X, \mathcal{X}, \mu, T)$ is an ergodic system, $(Z_1, \mathcal{Z}_1, m, T)$ is its Kronecker factor, and $f_1, f_2, \in L^\infty(\mu)$, then the limit*

$$\left\| \frac{1}{N} \sum_{n=0}^{N-1} T^n f_1 \cdot T^{2n} f_2 - \frac{1}{N} \sum_{n=0}^{N-1} T^n \mathbb{E}(f_1 \mid \mathcal{Z}_1) \cdot T^{2n} \mathbb{E}(f_2 \mid \mathcal{Z}_1) \right\|_{L^2(\mu)}$$

tends to 0 as $N \to \infty$.

In our terminology, this theorem can be quickly summarized: the Kronecker factor is characteristic for the double average. To prove the theorem, we use a standard trick for averaging, which is an iterated use of a variation of the van der Corput lemma on differences. (See [41] for uses of the van der Corput Lemma in number theory and [2] for its introduction to uses in ergodic theory.)

Lemma 5.9 (van der Corput). *Let $\{u_n\}$ be a sequence in a Hilbert space with $\|u_n\| \leq 1$ for all $n \in \mathbb{N}$. For $h \in \mathbb{N}$, set*

$$\gamma_h = \limsup_{N \to \infty} \left| \frac{1}{N} \sum_{n=0}^{N-1} \langle u_{n+h}, u_n \rangle \right|.$$

Then

$$\limsup_{N \to \infty} \left\| \frac{1}{N} \sum_{n=0}^{N-1} u_n \right\|^2 \leq \limsup_{H \to \infty} \frac{1}{H} \sum_{h=0}^{H-1} \gamma_h.$$

PROOF. Given $\varepsilon > 0$ and $H \in \mathbb{N}$, for N sufficiently large we have that

$$\left| \frac{1}{N} \sum_{n=0}^{N-1} u_n - \frac{1}{N} \frac{1}{H} \sum_{n=0}^{N-1} \sum_{h=0}^{H-1} u_{n+h} \right| < \varepsilon.$$

By convexity,

$$\left\| \frac{1}{N} \sum_{n=0}^{N-1} \frac{1}{H} \sum_{h=0}^{H-1} u_{n+h} \right\|^2 \leq \frac{1}{N} \sum_{n=0}^{N-1} \left\| \frac{1}{H} \sum_{h=0}^{H-1} u_{n+h} \right\|^2$$

$$= \frac{1}{N} \frac{1}{H^2} \sum_{n=0}^{N-1} \sum_{h_1, h_2=0}^{H-1} \langle u_{n+h_1}, u_{n+h_2} \rangle$$

and this approaches
$$\frac{1}{H^2} \sum_{h_1, h_2}^{H-1} \gamma_{h_1 - h_2}$$
as $N \to \infty$. But the assumption implies that this approaches 0 as $H \to \infty$. □

We now use this in the proof of Furstenberg's theorem:

PROOF OF THEOREM 5.8. By replacing f_1 by $f_1 - \mathbb{E}(f_1 \mid \mathcal{Z}_1)$ and f_2 by $f_2 - \mathbb{E}(f_2 \mid \mathcal{Z}_1)$, it suffices to show that if some f_i, for $i = 1, 2$ satisfies $\mathbb{E}(f_i \mid \mathcal{Z}_1) = 0$, then the doubule average converges to 0. Without loss, we assume that $\mathbb{E}(f \mid \mathcal{Z}_1) = 0$.

Set $u_n = T^n f_1 \cdot T^{2n} f_2$. Then
$$\langle u_n, u_{n+h} \rangle = \int T^n f_1 \cdot T^{2n} f_2 \cdot T^{n+h} \overline{f_1} \cdot T^{2n+2h} \overline{f_2} \, d\mu$$
$$= \int (f_1 \cdot T^h \overline{f_1}) \cdot T^n (f_2 \cdot T^{2h} \overline{f_2}) \, d\mu.$$

Thus
$$\frac{1}{N} \sum_{n=0}^{N-1} \langle u_n, u_{n+h} \rangle = \left(\int f_1 \cdot T^h \overline{f_1} \, d\mu \right) \frac{1}{N} \sum_{n=0}^{N-1} T^n (f_2 \cdot T^{2h} \overline{f_2}) \, d\mu.$$

By the von Neumann ergodic theorem (Theorem 4.2) applied to the second term, the limit
$$\gamma_h = \lim_{N \to \infty} \frac{1}{N} \sum_{n=0}^{N-1} \langle u_n, u_{n+h} \rangle$$
exists. Moreover, it is equal to

(5.4) $$\gamma_h = \int f_1 \cdot T^h \overline{f_1} \cdot \mathbb{P}(f_2 \cdot T^{2h} \overline{f_2}) \, d\mu,$$

where $\mathbb{P}$ is projection onto the T-invariant functions of $L^2(\mu)$. Since T is ergodic, $\mathbb{P}$ is projection onto the constant functions. But since $\mathbb{E}(f_1 \mid \mathcal{Z}_1) = 0$, f_1 is orthogonal to the constant functions and so by averaging over h, we have that
$$\lim_{H \to \infty} \frac{1}{H} \sum_{h=0}^{H-1} \gamma_h = 0.$$

By the van der Corput lemma, it follows that the double average also converges to 0. □

Furstenberg used a similar argument combined with induction to show that in a weak mixing system, the average (5.2) converges to the product of the integrals in $L^2(\mu)$ for all $k \geq 1$. This is one of the (simpler) steps in the proof of the Furstenberg's multiple recurrence theorem (Theorem 1.3) and gives a proof of multiple recurrence for weakly mixing systems. However, much more is needed to prove Theorem 1.3 in an arbitrary system; this is carried out by showing that for any function, the average along arithmetic progressions can be decomposed into two pieces, one of which has a generalized weak mixing property and the other of which is rigid in some sense. We have already seen a simple example of such a decomposition, in the proof of the von Neumann ergodic theorem. Some sort of decomposition is behind all of the multiple recurrence and convergence results we discuss.

We now return to showing that a set of integers with positive upper density contains arithmetic progressions of length three (Roth's theorem). By Furstenberg's correspondence principle it suffices to show double recurrence:

Theorem 5.10 (Theorem 1.3 for $k = 2$). *Let $(X, \mathcal{X}, \mu, T)$ be an ergodic system, and let $A \in \mathcal{X}$ with $\mu(A) > 0$. There exists $n \in \mathbb{N}$ with*
$$\mu(A \cap T^{-n} A \cap T^{-2n} A) > 0.$$

PROOF. Let $f = \mathbf{1}_A$. Then
$$\mu(A \cap T^{-n} A \cap T^{-2n} A) = \int f \cdot T^n f \cdot T^{2n} f \, d\mu.$$

It suffices to show that
$$\limsup_{N \to \infty} \frac{1}{N} \sum_{n=0}^{N-1} \int f \cdot T^n f \cdot T^{2n} f \, d\mu$$

is positive. However, we will show the stronger statement that the limit exists and is positive, rather than just the lim sup is positive.[3]

By Theorem 5.8, the limiting behavior of the double average $(1/N) \sum_{n=0}^{N-1} T^n f \cdot T^{2n} f$ is unchanged if f is replaced by $\mathbb{E}(f \mid \mathcal{Z}_1)$. Multiplying by f and integrating, it thus suffices to show that

(5.5) $$\lim_{N \to \infty} \frac{1}{N} \sum_{n=0}^{N-1} \int f \cdot T^n \mathbb{E}(f \mid \mathcal{Z}_1) \cdot T^{2n} \mathbb{E}(f \mid \mathcal{Z}_1) \, d\mu$$

exists and is positive. Since $\mathcal{Z}_1$ is T-invariant, $T^n \mathbb{E}(f \mid \mathcal{Z}_1) \cdot T^{2n} \mathbb{E}(f \mid \mathcal{Z}_1)$ is measurable with respect to $\mathcal{Z}_1$ and so we can replace (5.5) by

$$\lim_{N \to \infty} \frac{1}{N} \sum_{n=0}^{N-1} \int \mathbb{E}(f \mid \mathcal{Z}_1) \cdot T^n \mathbb{E}(f \mid \mathcal{Z}_1) \cdot T^{2n} \mathbb{E}(f \mid \mathcal{Z}_1) \, d\mu.$$

This means that we can assume that the first term is also measurable with respect to the Kronecker factor, and so we can assume that f is a nonnegative function that is measurable with respect to the Kronecker. Thus the system X can be assumed to be Z_1 and the transformation T is rotation by some irrational α. Thus it suffices to show that

$$\lim_{N \to \infty} \frac{1}{N} \sum_{n=0}^{N-1} \int_{Z_1} f(s) \cdot f(s + n\alpha) \cdot f(s + 2n\alpha) \, dm(s)$$

exists and is positive. But the convergence of this last expression is immediate using Fourier analysis. Since $\{n\alpha\}$ is equidistributed in Z_1, this limit approaches

(5.6) $$\iint_{Z_1 \times Z_1} f(s) \cdot f(s + t) \cdot f(s + 2t) \, dm(s) \, dm(t).$$

---

[3]In Furstenberg's proof of Szemerédi's theorem via Theorem 1.3, he showed that the analogous lim sup for $k \geq 2$ is positive and only showed the existence of the associated limit for $k = 2$. The positivity of the lim sup suffices for proving Szemerédi's theorem. As we are interested in the existence of the limit for $k > 2$ and the finer combinatorial information that can be gleaned from this, we prove the deeper statement here.

But
$$\lim_{t\to 0}\int_{Z_1} f(s)\cdot f(s+t)\cdot f(s+2t)\,dm(s) = \int_{Z_1} f(s)^3\,dm(s),$$
which is clearly positive. In particular, the double integral in (5.6) is positive. □

In the proof we have actually proven a stronger statement than needed to obtain Roth's theorem: we have shown the existence of the limit of the double average in $L^2(\mu)$. Letting $\tilde{f} = \mathbb{E}(f \mid \mathcal{Z}_1)$ for $f \in L^\infty(\mu)$, we have shown that the double average (5.3) converges to
$$\int_{Z_1} \tilde{f}_1(\pi_1(x)+s)\cdot \tilde{f}_2(\pi_1(x)+2s)\,dm(s).$$

More generally, the same sort of argument can be used to show that in a weak mixing system, the Kronecker factor is characteristic for the averages (3.1) for all $k \geq 1$, meaning that to prove convergence of these averages in a weak mixing system it suffices to assume that the system is a Kronecker system. Using Fourier analysis, one then gets convergence of the averages (3.1) for weak mixing systems.

5.8. Multiple averages. We want to carry out similar analysis for the multiple averages
$$\frac{1}{N}\sum_{n=0}^{N-1} T^n f_1 \cdot T^{2n} f_2 \cdot \ldots \cdot T^{kn} f_k$$
and show the existence of the limit in $L^2(\mu)$ as $N \to \infty$. In his proof of Szemerédi's theorem in [16] and subsequent proofs of Szemerédi's theorem via ergodic theory such as [21], the approach of Section 5.7 is not the one used for $k \geq 3$. Namely, they do not show the existence of the limit and then analyze the limit itself to show it is positive. A weaker statement is proved, only giving that the lim inf of (2.1) is positive. We will not discuss the intricate structure theorem and induction needed to prove this.

Already to prove convergence for $k = 3$, one needs to consider more than just rotational behavior.

Example 5.11. Given a system $(X, \mathcal{X}, \mu, T)$, let $F(Tx) = f(x)F(x)$, where
$$f(Tx) = \lambda f(x) \quad \text{and} \quad |\lambda| = 1.$$
Then
$$F(T^n x) = f(x)f(Tx)\cdots f(T^{n-1}x)F(x) = \lambda^{n(n-1)/2}\big(f(x)\big)^n F(x)$$
and so
$$F(x) = \big(F(T^n x)\big)^3 \big(F(T^{2n} x)\big)^{-3} F(T^{3n} x).$$
This means that there is some relation among
$$(x, T^n x, T^{2n} x, T^{3n} x)$$
that does not arise from the Kronecker factor.

One can construct more complicated examples (see Furstenberg [18]) that show that even such generalized eigenfunctions do not suffice for determining the limiting behavior for $k = 3$. More precisely, the factor corresponding to generalized eigenfunctions (the *Abramov factor*) is not characteristic for the average (3.1) with $k = 3$.

To understand the triple average, one needs to take into account systems more complicated than Kronecker and Abramov systems. The simplest such example is a 2-step nilsystem (the use of this terminology will be clarified later):

Example 5.12. Let $X = \mathbb{T} \times \mathbb{T}$, with Borel σ-algebra, and Haar measure. Fix $\alpha \in \mathbb{T}$ and define $T \colon X \to X$ by

$$T(x,y) = (x+\alpha, y+x)$$

The system is ergodic if and only if $\alpha \notin \mathbb{Q}$.

The system is not isomorphic to a group rotation, as can be seen by defining $f(x,y) = e(y) = \exp(2\pi i y)$. Then for all $n \in \mathbb{Z}$,

$$T^n(x,y) = \left(x + n\alpha, y + nx + \frac{n(n-1)}{2}\alpha\right)$$

and so

$$f(T^n(x,y)) = e(y)e(nx)e\left(\frac{n(n-1)}{2}\alpha\right).$$

Quadratic expressions like these do not arise from a rotation on a group.

6. The structure theorem

6.1. Major steps in the proof of Theorem 3.1. In broad terms, there are four major steps in the proof of Theorem 3.1.

For each $k \in \mathbb{N}$, we inductively define a seminorm $|||\cdot|||_k$ that controls the asymptotic behavior of the average. More precisely, we show that if $|f_1| \leq 1, \ldots, |f_k| \leq 1$, then

$$(6.1) \qquad \limsup_{N \to \infty} \left\| \frac{1}{N} \sum_{n=o}^{N-1} T^n f_1 \cdot T^{2n} f_2 \cdot \cdots \cdot T^{kn} f_k \right\|_{L^2(\mu)} \leq \min_{1 \leq j \leq k} |||f_j|||_k.$$

Using these seminorms, we define factors Z_k of X such that for $f \in L^\infty(\mu)$,

$$\mathbb{E}(f \mid Z_{k-1}) = 0 \text{ if and only if } |||f|||_k = 0.$$

It follows from (6.1) that the factor Z_{k-1} is characteristic for the average (3.1).

The bulk of the work is then to give a "geometric" description of these factors. This description is in terms of nilpotent groups, and more precisely we show that the dynamics of translations on homogeneous spaces of a nilpotent Lie group determines the limiting behavior of these averages. This is the content of the Structure Theorem, explained in Section 6.2. (A more detailed expository version of this is given in Host [31]; for full details, see [34].)

Finally, we show convergence for these particular types of systems.

Roughly speaking, this same outline applies to other convergence results we consider in the sequel, such as averages along polynomial times, averages along cubes, or averages for commuting transformations. For each average, we find a characteristic factor that can be described in geometric terms, allowing us to prove convergence in the characteristic factor.

6.2. The role of nilsystems.
We have already seen that the limit behavior of the double average is controlled by group rotations, meaning the Kronecker factor is characteristic for this average. Furthermore, we have seen that something more is needed to control the limit behavior of the triple average. Our goal here is to explain how the multiple averages of (3.1), and some more general averages, are controlled by nilsystems. We start with some terminology.

Let G be a group. If $g, h \in G$, let $[g, h] = g^{-1}h^{-1}gh$ denote the commutator of g and h. If $A, B \subset G$, we write $[A, B]$ for the subgroup of G spanned by $\{[a, b] : a \in A, b \in B\}$. The lower central series

$$G = G_1 \supset G_2 \supset \cdots \supset G_j \supset G_{j+1} \supset \cdots$$

of G is defined inductively, by setting $G_1 = G$ and $G_{j+1} = [G, G_j]$ for $j \geq 1$. We say that G is *k-step nilpotent* if $G_{k+1} = \{1\}$.

If G is a k-step nilpotent Lie group and Γ is a discrete cocompact subgroup, the compact manifold $X = G/\Gamma$ is a *k-step nilmanifold*.

The group G acts naturally on X by left translation: if $a \in G$ and $x \in X$, the translation T_a by a is given by $T_a(x\Gamma) = (ax)\Gamma$. There is a unique Borel probability measure μ (the *Haar measure*) on X that is invariant under this action. We let $\mathcal{G}/\Gamma$ denote the associated Borel σ-algebra on G/Γ. Fixing an element $a \in G$, the system $(G/\Gamma, \mathcal{G}/\Gamma, T_a, \mu)$ is a *k-step nilsystem* and T_a is a *nilrotation*.

The system $(X, \mathcal{X}, \mu, T)$ is an *inverse limit* of a sequence of factors $\{(X_j, \mathcal{X}_j, \mu_j, T)\}$ if $\{\mathcal{X}_j\}_{j \in \mathbb{N}}$ is an increasing sequence of T-invariant sub-σ-algebras such that $\bigvee_{j \in \mathbb{N}} \mathcal{X}_j = \mathcal{X}$ up to null sets.[4] If each system $(X_j, \mathcal{X}_j, \mu_j, T)$ is isomorphic to a k-step nilsystem, then $(X, \mathcal{X}, \mu, T)$ is an *inverse limit of k-step nilsystems*.

Proving convergence of the averages (3.1) is only possible if one has a good description of some characteristic factor for these averages. This is the content of the Structure Theorem:

Theorem 6.1 (Host and Kra [34]). *There exists a characteristic factor for the averages (3.1) which is isomorphic to an inverse limit of $(k-1)$-step nilsystems.*

The advantage of reducing to nilsystems is that convergence of the averages under study is much easier in nilsystems. This is further discussed in Section 8.2.

6.3. Examples of nilsystems.
We give two examples of nilsystems that illustrate their general properties.

Example 6.2. Let $G = \mathbb{Z} \times \mathbb{T} \times \mathbb{T}$ with multiplication given by

$$(k, x, y) * (k', x', y') = (k + k', x + x' \pmod 1, y + y' + 2kx' \pmod 1).$$

The commutator subgroup of G is $\{0\} \times \{0\} \times \mathbb{T}$, and G is 2-step nilpotent. The subgroup $\Gamma = \mathbb{Z} \times \{0\} \times \{0\}$ is discrete and cocompact, and thus $X = G/\Gamma$ is a nilmanifold. Let $\mathcal{X}$ denote the Borel σ-algebra and let μ denote Haar measure on X. Fix some irrational $\alpha \in \mathbb{T}$, let $a = (1, \alpha, \alpha)$, and let $T : X \to X$ be translation by a. Then (X, μ, T) is a 2-step nilsystem.

The Kronecker factor of X is $\mathbb{T}$ with rotation by α. Identifying X with $\mathbb{T}^2$ via the map $(k, x, y) \mapsto (x, y)$, the transformation T takes on the familiar form of a

---

[4]Recall that if $\mathcal{X}_1$ and $\mathcal{X}_1$ are sub-σ-algebras of $\mathcal{X}$, then $\mathcal{X}_1 \bigvee \mathcal{X}_2$ denotes the smallest sub-σ-algebra of $\mathcal{X}$ containing both $\mathcal{X}_1$ and $\mathcal{X}_2$. Thus $\mathcal{X}_1 \bigvee \mathcal{X}_2$ consists of all sets which are unions of sets of the form the form $A \cap B$ for $A \in \mathcal{X}_1$ and $B \in \mathcal{X}_2$.

skew transformation:
$$T(x,y) = (x+\alpha, y+2x+\alpha).$$
This system is ergodic if and only if $\alpha \notin \mathbb{Q}$: for $x, y \in X$ and $n \in \mathbb{Z}$,
$$T^n(x,y) = (x+n\alpha, y+2nx+n^2\alpha)$$
and equidistribution of the sequence $\{T^n(x,y)\}$ is equivalent to ergodicity.

Example 6.3. Let G be the Heisenberg group $\mathbb{R} \times \mathbb{R} \times \mathbb{R}$ with multiplication given by
$$(x,y,z) * (x',y',z') = (x+x', y+y', z+z'+xy').$$
Then G is a 2-step nilpotent Lie group. The subgroup $\Gamma = \mathbb{Z} \times \mathbb{Z} \times \mathbb{Z}$ is discrete and cocompact and so $X = G/\Gamma$ is a nilmanifold. Letting T be the translation by $a = (a_1, a_2, a_3) \in G$ where a_1, a_2 are independent over $\mathbb{Q}$ and $a_3 \in \mathbb{R}$, and taking $\mathcal{X}$ to be the Borel σ-algebra and μ to be the Haar measure, we have that $(X, \mathcal{X}, \mu, T)$ is a nilsystem. The system is ergodic if and only if a_1, a_2 are independent over $\mathbb{Q}$.

The compact abelian group $G/G_2\Gamma$ is isomorphic to $\mathbb{T}^2$ and the rotation on $\mathbb{T}^2$ by (a_1, a_2) is ergodic (again for a_1, a_2 independent over $\mathbb{Q}$). The Kronecker factor of X is the factor induced by functions on x_1, x_2. The system $(X, \mathcal{X}, \mu, T)$ is (uniquely) ergodic.

The dynamics of the first example gives rise to quadratic sequences, such as $\{n^2\alpha\}$, and the dynamics of the second example gives rise to generalized quadratic sequences such as $\{\lfloor n\alpha \rfloor n\beta\}$.

6.4. Motivation for nilpotent groups. The content of the Structure Theorem is that nilpotent groups, or more precisely the dynamics of a translation on the homogeneous space of a nilpotent Lie group, control the limiting behavior of the averages along arithmetic progressions. We give some motivation as to why nilpotent groups arise.

If G is an abelian group, then
$$\{(g, gz, gz^2, \ldots, gz^n) : g, z \in G\}$$
is a subgroup of G^n. However, this does not hold if G is not abelian. To make these arithmetic progressions into a group, one must take into account the commutators. This is the content of the following theorem, proven in different contexts by Hall [29], Petresco [48], Lazard [42], Leibman [43]. (Recall that $G_i = [G, G_i]$ denotes the ith entry in the lower central series of G.)

Theorem 6.4. *If G is a group, then for any $x, y \in G$, there exist $z \in G$ and $w_i \in G_i$ such that*
$$(x, x^2, x^3, \ldots, x^n) \times (y, y^2, y^3, \ldots, y^n)$$
$$= \left(z, z^2 w_1, z^3 w_1^3 w_2, \ldots, z^{\binom{n}{1}} w_1^{\binom{n}{2}} w_2^{\binom{n}{3}} \ldots w_{n-1}^{\binom{n}{n}}\right).$$
Furthermore, these expressions form a group.

If G is a group, a *geometric progression* is a sequence of the form
$$g, gz, gz^2 w_1, gz^3 w_1^3 w_2, \ldots, gz^{\binom{n}{1}} w_1^{\binom{n}{2}} \ldots w_{n-1}^{\binom{n}{n}}, \ldots$$
where $g, z \in G$ and $w_i \in G_i$.

Thus if G is abelian, g and z determine the whole sequence. On the other hand, if G is k-step nilpotent with $k < n$, the first k terms determine the whole sequence. (This holds because each w_i appears first in the ith term of the sequence and with exponent 1, and so it is completely determined, and for $i > k$, each w_i is trivial.)

Similarly, if $(G/\Gamma, \mathcal{G}/\Gamma, \mu, T_a)$ is a k-step nilsystem and
$$x_1 = g_1\Gamma, \quad x_2 = g_2\Gamma, \ldots, \quad x_k = g_k\Gamma, \ldots, \quad x_n = g_n\Gamma$$
is a geometric progression in G/Γ, then the first k terms determine the rest. Thus in a k-step nilsystem, $a^{k+1}x\Gamma$ is a function of the first k terms $ax\Gamma, a^2x\Gamma, \ldots, a^kx\Gamma$.

This means that the $(k+1)$st term $T^{(k+1)n}x$ in an arithmetic progression $T^nx, \ldots, T^{kn}x$ is constrained by the first k terms. More interestingly, the converse also holds: in an arbitrary system $(X, \mathcal{X}, \mu, T)$, any k-step nilpotent factor places a constraint on $(x, T^nx, T^{2n}x, \ldots, T^{kn}x)$.

7. Building characteristic factors

The material in this and the next section is based on [34] and the reader is referred to [34] for full proofs. To describe characteristic factors for the averages (3.1), for each $k \in \mathbb{N}$ we define a seminorm and use it to define these factors. We start by defining certain measures that are then used to define the seminorms. Throughout this section, we assume that $(X, \mathcal{X}, \mu, T)$ is an ergodic system.

7.1. Definition of the measures. Let $X^{[k]} = X^{2^k}$ and define $T^{[k]}\colon X^{[k]} \to X^{[k]}$ by $T^{[k]} = T \times \cdots \times T$ (taken 2^k times).

We write a point $\mathbf{x} \in X^{[k]}$ as $\mathbf{x} = (x_\epsilon : \epsilon \in \{0,1\}^k)$ and make the natural identification of $X^{[k+1]}$ with $X^{[k]} \times X^{[k]}$, writing $\mathbf{x} = (\mathbf{x}', \mathbf{x}'')$ for a point of $X^{[k+1]}$, with $\mathbf{x}', \mathbf{x}'' \in X^{[k]}$.

By induction, we define a measure $\mu^{[k]}$ on $X^{[k]}$ invariant under $T^{[k]}$. Set $\mu^{[0]} := \mu$. Let $\mathcal{I}^{[k]}$ be the invariant σ-algebra of $(X^{[k]}, \mathcal{X}^{[k]}, \mu^{[k]}, T^{[k]})$. (Note that this system is not necessarily ergodic.) Then $\mu^{[k+1]}$ is defined to be the *relatively independent joining* of $\mu^{[k]}$ with itself over $\mathcal{I}^{[k]}$, meaning that if F and G are bounded functions on $X^{[k]}$,

$$(7.1) \quad \int_{X^{[k+1]}} F(\mathbf{x}') \cdot G(\mathbf{x}'') \, d\mu^{[k+1]}(\mathbf{x})$$
$$= \int_{X^{[k]}} \mathbb{E}(F \mid \mathcal{I}^{[k]})(\mathbf{y}) \cdot \mathbb{E}(G \mid \mathcal{I}^{[k]})(\mathbf{y}) \, d\mu^{[k]}(\mathbf{y}).$$

Since $(X, \mathcal{X}, \mu, T)$ is assumed to be ergodic, $\mathcal{I}^{[0]}$ is trivial and $\mu^{[1]} = \mu \times \mu$. If the system is weak mixing, then for all $k \geq 1$, $\mu^{[k]}$ is the product measure $\mu \times \mu \times \cdots \times \mu$, taken 2^k times.

7.2. Symmetries of the measures. Writing a point $\mathbf{x} \in X^{[k]}$ as
$$\mathbf{x} = (x_\epsilon \colon \epsilon \in \{0,1\}^k),$$
we identify the indexing set $\{0,1\}^k$ of this point with the vertices of the Euclidean cube.

An isometry σ of $\{0,1\}^k$ induces a map $\sigma_*\colon X^{[k]} \to X^{[k]}$ by *permuting the coordinates*:
$$(\sigma_*(\mathbf{x}))_\epsilon = x_{\sigma(\epsilon)}.$$

For example, from the diagonal symmetries for $k=2$, we have the permutations
$$(x_{00}, x_{01}, x_{10}, x_{11}) \mapsto (x_{00}, x_{10}, x_{01}, x_{11})$$
$$(x_{00}, x_{01}, x_{10}, x_{11}) \mapsto (x_{11}, x_{01}, x_{10}, x_{00}).$$

By induction, the measures are invariant under permutations:

Lemma 7.1. *For each $k \in \mathbb{N}$, the measure $\mu^{[k]}$ is invariant under all permutations of coordinates arising from isometries of the unit Euclidean cube.*

7.3. Defining seminorms. For each $k \in \mathbb{N}$, we define a seminorm on $L^\infty(\mu)$ by setting
$$|||f|||_k^{2^k} = \int_{X^{[k]}} \prod_{\epsilon \in \{0,1\}^k} f(x_\epsilon) \, d\mu^{[k]}(\mathbf{x}).$$

By definition of the measure $\mu^{[k]}$, this integral is equal to
$$\int_{X^{[k-1]}} \mathbb{E}\left(\prod_{\epsilon \in \{0,1\}^{k-1}} f(x_\epsilon) \mid \mathcal{I}^{[k-1]}\right)^2 d\mu^{[k-1]}$$

and so in particular it is nonnegative.

From the symmetries of the measure $\mu^{[k]}$ (Lemma 7.1), we have a version of the Cauchy–Schwarz inequality for the seminorms, referred to as a Cauchy–Schwarz–Gowers inequality:

Lemma 7.2. *For $\epsilon \in \{0,1\}^k$, let $f_\epsilon \in L^\infty(\mu)$. Then*
$$\left| \int \prod_{\epsilon \in \{0,1\}^k} f_\epsilon(x_\epsilon) \, d\mu^{[k]}(\mathbf{x}) \right| \leq \prod_{\epsilon \in \{0,1\}^k} |||f_\epsilon|||_k.$$

As a corollary, the map $f \mapsto |||f|||_k$ is subadditive (meaning that $|||f+g|||_k \leq |||f|||_k + |||g|||_k$ for all $f, g \in L^\infty(\mu)$) and so:

Corollary 7.3. *For every $k \in \mathbb{N}$, $|||\cdot|||_k$ is a seminorm on $L^\infty(\mu)$.*

Since the system $(X, \mathcal{X}, \mu, T)$ is ergodic, the σ-algebra $\mathcal{I}^{[0]}$ is trivial, $\mu^{[1]} = \mu \times \mu$ and $|||f|||_1 = |\int f \, d\mu|$. By induction,
$$|||f|||_1 \leq |||f|||_2 \leq \cdots \leq |||f|||_k \leq \cdots \leq \|f\|_\infty.$$

If the system is weak mixing, then $|||f|||_k = |||f|||_1$ for all $k \in \mathbb{N}$.

By induction and the ergodic theorem, we have a second presentation of these seminorms:

Lemma 7.4. *For every $k \geq 1$,*
$$|||f|||_{k+1}^{2^{k+1}} = \lim_{N \to \infty} \frac{1}{N} \sum_{n=0}^{N-1} |||f \cdot T^n f|||_k^{2^k}.$$

7.4. Seminorms control the averages (3.1). The seminorms $|||\cdot|||_k$ control the averages along arithmetic progressions:

Lemma 7.5. *Assume that $(X, \mathcal{X}, \mu, T)$ is ergodic and let $k \in \mathbb{N}$. If $\|f_1\|_\infty, \ldots, \|f_k\|_\infty \leq 1$, then*
$$\limsup_{N \to \infty} \left\| \frac{1}{N} \sum_{n=0}^{N-1} T^n f_1 \cdot T^{2n} f_2 \cdot \cdots \cdot T^{kn} f_k \right\|_{L^2(\mu)} \leq \min_{\ell=1,\ldots,k} |||f_\ell|||_k.$$

PROOF. We proceed by induction on k. For $k = 1$, by the ergodic theorem
$$\lim_{N\to\infty}\left\|\frac{1}{N}\sum_{n=0}^{N-1}T^n f_1\right\|_{L^2(\mu)} = \left|\int f_1\,\mathrm{d}\mu\right| = \|\!|f_1|\!\|_1.$$
Assume it holds for $k \geq 1$. Let $f_1, f_2, \ldots, f_{k+1} \in L^\infty(\mu)$ with $\|f_j\|_\infty \leq 1$ for $j = 1, 2, \ldots, k+1$ and define $u_n = T^n f_1 \cdot T^{2n} f_2 \cdots T^{(k+1)n} f_{k+1}$. Assume that $\ell \in \{2, 3, \ldots, k+1\}$ (the case $\ell = 1$ is similar). Then
$$\left|\frac{1}{N}\sum_{n=0}^{N-1}\langle u_{n+h}, u_n\rangle\right| = \left|\int (f_1 \cdot T^h f_1)\frac{1}{N}\sum_{n=0}^{N-1}\prod_{j=2}^{k+1}T^{(j-1)n}(f_j \cdot T^{jh}f_j)\,\mathrm{d}\mu\right|$$
$$\leq \|f_1 \cdot T^h f_1\|_{L^2(\mu)}\left\|\frac{1}{N}\sum_{n=0}^{N-1}\prod_{j=2}^{k+1}T^{(j-1)n}(f_j \cdot T^{jh}f_j)\right\|_{L^2(\mu)}.$$
Set
$$\gamma_h = \limsup_{N\to\infty}\left|\frac{1}{N}\sum_{n=0}^{N-1}\langle u_{n+h}, u_n\rangle\right|.$$
Then by the inductive hypothesis, with f_{j-1} replaced by $f_j \cdot T^{jh}f_j$ for $j = 2, 3, \ldots, k+1$, we have that
$$\gamma_h \leq \ell \cdot \|\!|f_\ell \cdot T^{\ell h}f_\ell|\!\|_k.$$
Thus
$$\frac{1}{H}\sum_{h=0}^{H-1}\gamma_h \leq \ell^2 \frac{1}{\ell H}\sum_{n=0}^{\ell H-1}\|\!|f_\ell \cdot T^n f_\ell|\!\|_k$$
and the statement follows from the van der Corput lemma (Lemma 5.9) and the definition of the seminorm $\|\!|\cdot|\!\|_{k+1}$. □

7.5. The Kronecker factor, revisited ($k = 2$).
We have seen two presentations of the Kronecker factor $(Z_1, \mathcal{Z}_1, m, T)$: it is the largest abelian group rotation factor and it is the sub-σ-algebra of $\mathcal{X}$ that gives rise to all eigenfunctions. Another equivalent formulation is that it is the smallest sub-σ-algebra of $\mathcal{X}$ such that all invariant functions of $(X \times X, \mathcal{X} \times \mathcal{X}, \mu \times \mu, T \times T)$ are measurable with respect to $\mathcal{Z}_1 \times \mathcal{Z}_1$. Recall that $\pi_1 \colon X \to Z_1$ denotes the factor map.

We give an explicit description of the measure $\mu^{[2]}$, and thus give yet another description of the Kronecker factor. For $f \in L^\infty(\mu)$, write $\tilde f = \mathbb{E}(f \mid \mathcal{Z}_1)$.

For $s \in Z_1$ and $f_0, f_1 \in L^\infty(\mu)$, we define a probability measure μ_s on $X \times X$ by
$$\int_{X\times X} f_0(x_0)f_1(x_1)\,\mathrm{d}\mu_s(x_0, x_1) := \int_{Z_1}\tilde f_0(z)\tilde f_1(z+s)\,\mathrm{d}m(z).$$
This measure is $T \times T$-invariant and the ergodic decomposition of $\mu \times \mu$ under $T \times T$ is given by
$$\mu \times \mu = \int_{Z_1}\mu_s\,\mathrm{d}m(s).$$
Thus for m-almost every $s \in Z_1$, the system $(X \times X, \mathcal{X} \times \mathcal{X}, \mu_s, T \times T)$ is ergodic and
$$\mu^{[2]} = \int_{Z_1}\mu_s \times \mu_s\,\mathrm{d}m(s).$$

More generally, if f_ϵ, $\epsilon \in \{0,1\}^2$, are measurable functions on X, then

$$\int_{X^{[2]}} f_{00} \otimes f_{01} \otimes f_{10} \otimes f_{11} \, d\mu^{[2]}$$
$$= \int_{Z_1^3} \tilde{f}_{00}(z) \cdot \tilde{f}_{01}(z+s) \cdot \tilde{f}_{10}(z+t) \cdot \tilde{f}_{11}(z+s+t) \, dm(z) \, dm(s) \, dm(t).$$

It follows immediately that:

$$|||f|||_2^4 := \int f \otimes f \otimes f \otimes f \, d\mu^{[2]}$$
$$= \int_{Z_1^3} \tilde{f}(z) \cdot \tilde{f}(z+s) \cdot \tilde{f}(z+t) \cdot \tilde{f}(z+s+t) \, dm(z) \, dm(s) \, dm(t).$$

As a corollary, $|||f|||_2$ is the ℓ^4-norm of the Fourier Transform of $\tilde{f}$ and the factor Z_1, defined by $|||f|||_2 = 0$ if and only if $\mathbb{E}(f \mid \mathcal{Z}_1) = 0$ for $f \in L^\infty(\mu)$, is the Kronecker factor of $(X, \mathcal{X}, \mu, T)$.

7.6. Factors for all $k \geq 1$. Using these seminorms, we define factors $Z_k = Z_k(X)$ for $k \geq 1$ of X that generalize the relation between the Kronecker factor Z_1 and the second seminorm $|||\cdot|||_2$. We define $\mathcal{Z}_k$ as follows: for $f \in L^\infty(\mu)$, $\mathbb{E}(f \mid \mathcal{Z}_k) = 0$ if and only if $|||f|||_{k+1} = 0$. We let Z_k denote the associated factor. That this does define a factor needs proof and to further explain this and the definition, we start by describing some geometric properties of the measures $\mu^{[k]}$.

Indexing $X^{[k]}$ by the coordinates $\{0,1\}^k$ of the Euclidean cube, it is natural to use geometric terms like *side*, *edge*, *vertex* for subsets of $\{0,1\}^k$. For example, Figure 1 illustrates the point $\mathbf{x} \in X^{[3]}$ with the side $\alpha = \{010, 011, 110, 111\}$:

Let $\alpha \subset \{0,1\}^k$ be a side. The *side transformation* $T_\alpha^{[k]}$ of $X^{[k]}$ is defined by:

$$\left(T_\alpha^{[k]} \mathbf{x}\right)_\epsilon = \begin{cases} Tx_\epsilon & \text{if } \epsilon \in \alpha; \\ x_\epsilon & \text{otherwise.} \end{cases}$$

We can represent the transformation T_α associated to the side $\{010, 011, 110, 111\}$ by Figure 2.

Since permutations of coordinates leave the measure $\mu^{[k]}$ invariant and act transitively on the sides, we have:

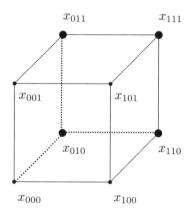

FIGURE 1.

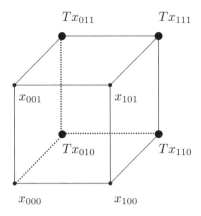

FIGURE 2.

Lemma 7.6. *For all $k \in \mathbb{N}$, the measure $\mu^{[k]}$ is invariant under the side transformations.*

We now view $X^{[k]}$ in a different way, identifying $X^{[k]} = X \times X^{2^k-1}$. A point $\mathbf{x} \in X^{[k]}$ is now written as

$$\mathbf{x} = (x_{\mathbf{0}}, \tilde{x}) \quad \text{where } \tilde{x} \in X^{2^k-1},\ x_{\mathbf{0}} \in X, \text{ and } \mathbf{0} = (00\ldots 0) \in \{0,1\}^k.$$

Although the $\mathbf{0}$ coordinate has been singled out and seems to play a particular role, it follows from the symmetries of the measure $\mu^{[k]}$ (Lemma 7.1) that any other coordinate could have been used instead.

If $\alpha \subset \{0,1\}^k$ is a side that does not contain $\mathbf{0}$ (there are k such sides), the transformation $T_\alpha^{[k]}$ leaves the coordinate $\mathbf{0}$ invariant. It follows from induction and the definition of the measure $\mu^{[k]}$ that:

Proposition 7.7. *Let $k \in \mathbb{N}$. If $B \subset X^{2^k-1}$, there exists $A \subset X$ with*

(7.2) $$\mathbf{1}_A(x_{\mathbf{0}}) = \mathbf{1}_B(\tilde{x}) \quad \text{for almost all } \mathbf{x} = (x_{\mathbf{0}}, \tilde{x}) \in X^{[k]}$$

if and only if $X \times B$ is invariant under the k transformations $T_\alpha^{[k]}$ arising from the k sides α not containing $\mathbf{0}$.

This means that the subsets $A \subset X$ such that there exists $B \subset X^{2^k-1}$ satisfying (7.2) form an invariant sub-σ-algebra $\mathcal{Z}_{k-1} = \mathcal{Z}_{k-1}(X)$ of $\mathcal{X}$. We define $Z_{k-1} = Z_{k-1}(X)$ to be the associated factor. Thus $\mathcal{Z}_{k-1}(X)$ is defined to be the sub-σ-algebra of sets $A \subset X$ such that (7.2) holds for some set $B \subset X^{2^k-1}$.

We give some properties of the factors:

Proposition 7.8. (1) *For every bounded function f on X,*

$$|||f|||_k = 0 \text{ if and only if } \mathbb{E}(f \mid \mathcal{Z}_{k-1}) = 0.$$

(2) *For bounded functions f_ϵ, $\epsilon \in \{0,1\}^k$, on X,*

$$\int \prod_{\epsilon \in \{0,1\}^k} f_\epsilon(x_\epsilon)\, d\mu^{[k]}(\mathbf{x}) = \int \prod_{\epsilon \in \{0,1\}^k} \mathbb{E}(f_\epsilon \mid \mathcal{Z}_{k-1})(x_\epsilon)\, d\mu^{[k]}(\mathbf{x}).$$

Furthermore, $\mathcal{Z}_{k-1}$ is the smallest sub-σ-algebra of $\mathcal{X}$ with this property.

(3) The invariant sets of $(X^{[k]}, \mathcal{X}^{[k]}, \mu^{[k]}, T^{[k]})$ are measurable with respect to $\mathcal{Z}_k^{[k]}$. Furthermore, $\mathcal{Z}_k$ is the smallest sub-σ-algebra of $\mathcal{X}$ with this property.

The proof of this proposition relies on showing a similar formula to that used (in (7.1)) to define the measures $\mu^{[k]}$, but with respect to the new identification separating the 0 coordinate from the $2^k - 1$ others. Namely, for bounded functions f on X and F on X^{2^k-1},

$$\int_{X^{[k]}} f(x_0) \cdot F(\tilde{x}) \, d\mu^{[k]}(\mathbf{x}) = \int_{X^{[k-1]}} \mathbb{E}(f \mid \mathcal{Z}_{k-1}) \cdot \mathbb{E}(F \mid \mathcal{Z}_{k-1}) \, d\mu^{[k-1]}.$$

The given properties then follow using induction and the symmetries of the measures.

We have already seen that Z_0 is the trivial factor and Z_1 is the Kronecker factor. More generally, the sequence of factors is increasing:

$$Z_0 \leftarrow Z_1 \leftarrow \cdots \leftarrow Z_k \leftarrow Z_{k+1} \leftarrow \cdots \leftarrow X.$$

If X is weak mixing, then $Z_k(X)$ is the trivial factor for every k.

An immediate consequence of Lemma 7.5 and the definition of the factors is that the factor Z_{k-1} is characteristic for the average along arithmetic progressions:

Proposition 7.9. *For all $k \geq 1$, the factor Z_{k-1} is characteristic for the convergence of the averages*

$$\frac{1}{N} \sum_{n=0}^{N-1} T^n f_1 \cdot T^{2n} f_2 \cdot \cdots \cdot T^{kn} f_k.$$

This means that in order to understand the long term behavior of the multiple average along a k-term arithmetic progression, it suffices to assume that the space itself is Z_k. In particular, once we show that the factor Z_k has some useful structure (and this is the content of the Structure Theorem of [34], Theorem 8.1, discussed in Section 8), we are able to prove the existence of the limit of the average along arithmetic progressions. Proposition 7.9 would be meaningless if we were not able to explicitly describe the structure of Z_k in some way other than the abstract definition already given, and then use that description to prove convergence.

8. Structure theorem

8.1. Systems of order k. For $k \geq 0$, an ergodic system X is said to be of *order k* if $Z_k(X) = X$. This means that $\|\!\|\!\|\cdot\|\!\|\!\|_{k+1}$ is a norm on $L^\infty(\mu)$.

Given an ergodic system $(X, \mathcal{X}, \mu, T)$, $Z_k(X)$ is a system of order k, since $Z_k(Z_k(X)) = Z_k(X)$. The unique system of order zero is the trivial system, and a system of order 1 is an ergodic rotation. By definition, if a system is of order k, then it is also of order k' for any $k' > k$.

By Proposition 7.9, to show convergence of

$$\frac{1}{N} \sum_{n=0}^{N-1} T^n f_1 \cdot T^{2n} f_2 \cdot \cdots \cdot T^{kn} f_k$$

in an arbitrary system, it suffices to assume that each function is defined on the factor Z_{k-1}. But since $Z_{k-1}(X)$ is a system of order k, it suffices to prove convergence of this average for systems of order $k - 1$.

In this language, the Structure Theorem becomes:

Theorem 8.1 (Host and Kra [34]). *A system of order k is the inverse limit of a sequence of k-step nilsystems.*

Before turning to the proof of the Structure Theorem, we show convergence for the average along arithmetic progressions in a nilsystem. Combining this convergence with Theorem 8.1 completes the proof of Theorem 3.1.

8.2. Convergence on a nilmanifold.
Using general properties of nilmanifolds (see Furstenberg [15] and Parry [47]), Lesigne [46] showed for connected group G and Leibman [44] showed in the general case, convergence in a nilsystem:

Theorem 8.2. *If $(X = G/\Gamma, \mathcal{G}/\Gamma, \mu, T)$ is a nilsystem and f is a continuous function on X, then*
$$\frac{1}{N}\sum_{n=0}^{N-1} f(T^n x)$$
converges for every $x \in X$.

(See also Ratner [50] and Shah [53] for related convergence results.)

As a corollary, we have convergence in $L^2(\mu)$ for the average along arithmetic progressions in a nilmanifold:

Corollary 8.3. *If $(X = G/\Gamma, \mathcal{G}/\Gamma, \mu, T)$ is a nilsystem, $k \in \mathbb{N}$, and $f_1, f_2, \ldots, f_k \in L^\infty(\mu)$, then*
$$\lim_{N\to\infty} \frac{1}{N}\sum_{n=0}^{N-1} T^n f_1 \cdot T^{2n} f_2 \cdot \ldots \cdot T^{kn} f_k$$
exists in $L^2(\mu)$.

PROOF. By density, we can assume that the functions are continuous. By assumption, G^k is a nilpotent Lie group, Γ^k is a discrete cocompact subgroup and $X^k = G^k/\Gamma^k$ is a nilmanifold. Let
$$s = (t, t^2, \ldots, t^k) \in G^k$$
and let $S\colon X^k \to X^k$ be the translation by s, meaning that
$$S = T \times T^2 \times \cdots \times T^k.$$
We apply Theorem 8.2 to (X^k, S) with the continuous function
$$F(x_1, x_2, \ldots, x_k) = f_1(x_1) f_2(x_2) \ldots f_k(x_k)$$
at the point $y = (x, x, \ldots, x)$ and so the averages converge everywhere. □

Thus Theorem 3.1 holds in a nilsystem, and we are left with proving the Structure Theorem.

8.3. A group of transformations.
To each ergodic system, we associate a group of measure preserving transformations. The general approach is to show that for sufficiently many systems of order k, this group is a nilpotent Lie group. The bulk of the work is to then show that this group acts transitively on the system. Thus the system can be given the structure of a nilmanifold and the Structure Theorem (Theorem 8.1) follows.

Most proofs are sketched or omitted completely, and the reader is referred to [34] for the details.

Let $(X, \mathcal{X}, \mu, T)$ be an ergodic system. If $S\colon X \to X$ and $\alpha \subset \{0,1\}^k$, define $S_\alpha^{[k]}\colon X^{[k]} \to X^{[k]}$ by:

$$\left(S_\alpha^{[k]}\mathbf{x}\right)_\epsilon = \begin{cases} Sx_\epsilon & \text{if } \epsilon \in \alpha; \\ x_\epsilon & \text{otherwise.} \end{cases}$$

Let $\mathcal{G} = \mathcal{G}(X)$ be the group of transformations $S\colon X \to X$ such that for all $k \in \mathbb{N}$ and all sides $\alpha \subset \{0,1\}^k$, the measure $\mu^{[k]}$ is invariant under $S_\alpha^{[k]}$.

Some properties of this group are immediate. By symmetry, it suffices to consider one side. By definition, $T \in \mathcal{G}$, and if $ST = TS$ then we also have that $S \in \mathcal{G}$. If $S \in \mathcal{G}$ and $k \in N$, then $\mu^{[k]}$ is invariant under $S^{[k]}\colon X^{[k]} \to X^{[k]}$. Furthermore, $S^{[k]}E = E$ for every $E \in \mathcal{I}^{[k]}$.

By induction, the invariance of the measure $\mu^{[k]}$ under the side transformations, and commutator relations, we have:

Proposition 8.4. *If X is a system of order k, then $\mathcal{G}(X)$ is a k-step nilpotent group.*

8.4. Proof of the Structure Theorem.
We proceed by induction. By the inductive assumption, we can assume that we are given a system $(X, \mathcal{X}, \mu, T)$ of order k. We have a factor $(Y, \mathcal{Y}, \nu, T)$, where $Y = Z_{k-1}(X)$ and $\pi\colon X \to Y$ is the factor map. Furthermore, Y is an inverse limit of a sequence of $(k-1)$-step nilsystems

$$Y = \varprojlim Y_i; \quad Y_i = G_i/\Gamma_i.$$

We want to show that X is an inverse limit of k-step nilsystems.

We have already shown that if f_ϵ, $\epsilon \in \{0,1\}^k$, are bounded functions on X, then

$$\int \prod_{\epsilon \in \{0,1\}^k} f_\epsilon(x_\epsilon)\, d\mu^{[k]}(\mathbf{x}) = \int \prod_{\epsilon \in \{0,1\}^k} \mathbb{E}(f_\epsilon \mid \mathcal{Y})(x_\epsilon)\, d\mu^{[k]}(\mathbf{x}).$$

In particular, for $f \in L^\infty(\mu)$,

$$|\!|\!|f|\!|\!|_k = 0 \text{ if and only if } \mathbb{E}(f \mid \mathcal{Y}) = 0.$$

Furthermore, X does not admit a strict sub-σ-algebra $\mathcal{Z}$ such that all invariant sets of $(X^{[k]}, \mu^{[k]}, T^{[k]})$ are measurable with respect to $\mathcal{Z}^{[k]}$. Recall also that the system $(X^{[k]}, \mu^{[k]}, T^{[k]})$ is defined as a relatively independent joining.

In [16], Furstenberg described the invariant σ-algebra for an arbitrary relatively independent joining. It follows that X is an *isometric extension* of Y, meaning that $X = Y \times H/K$ where H is a compact group and K is a closed subgroup, $\mu = \nu \times m$, where m is the Haar measure of H/K, and the transformation T is given by

$$T(y, u) = (Ty, \rho(y) \cdot u)$$

for some map $\rho\colon Y \to H$. (Note that we are making a slight, but standard, abuse of notation in using the same letter T to denote both the transformation in X and Y.)

Lemma 8.5. *For every $h \in H$, the transformation $(y, u) \mapsto (y, h \cdot u)$ of X belongs to the center of $\mathcal{G}(X)$.*

Thus H is abelian. We can substitute H/K for H, and we use additive notation for H.

We therefore have more information: X is an *abelian extension* of Y, meaning that $X = Y \times H$ for some compact abelian group H, $\mu = \nu \times m$, where m is the

Haar measure of H, and the transformation T is given by $T(y, u) = (Ty, u + \rho(y))$ for some map $\rho\colon Y \to H$. We call ρ the *cocycle* defining the extension.

Furthermore, we show that the cocycle defining this extension has a particular form, given by a particular functional equation:

Proposition 8.6. *If $(X, \mathcal{X}, \mu, T)$ is a system of order k and $(Y, \mathcal{Y}, \nu, T) = Z_{k-1}(X)$, then X is an abelian extension of Y via a compact group H and for the cocycle ρ defining this extension, there exists a map $\Phi\colon Y^{[k]} \to H$ such that*

$$(8.1) \qquad \sum_{\epsilon \in \{0,1\}^k} (-1)^{\epsilon_1 + \cdots + \epsilon_k} \rho(y_\epsilon) = \Phi(T^{[k]}\mathbf{y}) - \Phi(\mathbf{y})$$

for $\nu^{[k]}$-a.e. $\mathbf{y} \in Y^{[k]}$.

We can make a few more assumptions on our system. Namely, by induction we can deduce that H is connected. Since every connected compact abelian group H is an inverse limit of a sequence of tori, we can further reduce to the case that $H = \mathbb{T}^d$.

8.5. The case $k = 2$ (The Conze–Lesigne equation).

We maintain notation of the preceding section and review what this means for the case $k = 2$. By assumption, we have that $(Y, \mathcal{Y}, \nu, T)$ is a system of order 1, meaning it is a group rotation. The measure $\nu^{[2]}$ is the Haar measure of the subgroup

$$\{(y, y + s, y + t, y + s + t) : y, s, t \in Y\}$$

of Y^4. The functional equation of Proposition 8.6 is: there exists $\Phi\colon Y^3 \to \mathbb{T}^d$ with

$$\rho(y) - \rho(y + s) - \rho(y + t) + \rho(y + s + t) = \Phi(y + 1, s, t) - \Phi(y, s, t)$$

It follows that for every $s \in Y$, there exists $\phi_s\colon Y \to \mathbb{T}^d$ and $c_s \in \mathbb{T}^d$ satisfying the *Conze–Lesigne equation* (see [9]):

(CL) $\qquad \rho(y) - \rho(y + s) = \phi_s(y + 1) - \phi_s(y) + c_s.$

The group $\mathcal{G}(X)$ associated to the system is the group of transformations of $X = Y \times \mathbb{T}^d$ of the form

$$(y, h) \mapsto (y + s, h + \phi_s(y))$$

where s and ϕ_s satisfy (CL).

8.6. Structure theorem in general.

We give a short outline of the steps needed to complete the proof of the Structure Theorem for $k \geq 3$. We have that $Y = Z_{k-1}(X)$ is a system of order $k - 1$, $X = Y \times \mathbb{T}^d$, $T(y, h) = (Ty, h + \rho(y))$, and $\rho\colon Y \to \mathbb{T}^d$ satisfies the functional equation (8.1). By the induction hypothesis $Y = \varprojlim Y_i$ where each $Y_i = G_i/\Gamma_i$ is a $(k-1)$-step nilsystem.

We first show that the cocycle ρ is *cohomologous* to a cocycle measurable with respect to $\mathcal{Y}_i$ for some i, meaning that the difference between the two cocycles is a coboundary. This reduces us to the case that ρ is measurable with respect to some $\mathcal{Y}_i$, and so we can assume that $Y = \mathcal{Y}_i$ for some i. Thus Y is a $(k-1)$-step nilsystem and we can assume that $Y = G/\Gamma$ with $G = \mathcal{G}(Y)$.

We then use the functional equation (8.1) to lift every transformation $S \in G$ to a transformation of X belonging to $\mathcal{G}(X)$. Starting with the case $S \in G_{k-1}$, we move up the lower central series of G. Lastly we show that we obtain sufficiently many elements of the group $\mathcal{G}(X)$ in this way.

8.7. Relations to the finite case.
The seminorms $|||\cdot|||_k$ play the same role that the Gowers norms play in Gowers's proof [23] of Szemerédi's theorem and in Green and Tao's proof [25] that the primes contain arbitrarily long arithmetic progressions. We let U_k denote the k-th Gowers norm. For the finite system $\mathbb{Z}/N\mathbb{Z}$, $|||f|||_k = \|f\|_{U_k}$. Furthermore, $\|\cdot\|_{U_k}$ is a norm, not only a seminorm. The analog of Lemma 7.5 is that if $\|f_0\|_\infty, \|f_1\|_\infty, \ldots, \|f_k\|_\infty \leq 1$, then there exists some constant $C_k > 0$ such that

$$\left|\mathbb{E}\big(f_0(x)f_1(x+y)\ldots f_k(x+ky) \mid x,y \in \mathbb{Z}/p\mathbb{Z}\big)\right| \leq C_k \min_{0 \leq j \leq k} \|f_j\|_{U_k}.$$

Other parts of the program are not as easy to translate to the finite setting. Consider defining a factor of the system using the seminorms. If p is prime, then $\mathbb{Z}/p\mathbb{Z}$ has no nontrivial factor and so there is no factor of $\mathbb{Z}/p\mathbb{Z}$ playing the role of the factor Z_k, meaning there is no factor with

$$\mathbb{E}(f \mid Z_k) = 0 \text{ if and only if } \|f\|_{U_k} = 0.$$

Instead, the corresponding results have a different flavor: if $\|f\|_{U_k}$ is large in some sense, then f has large conditional expectation on some (noninvariant) σ-algebra or it has large correlation with a function of some particular class. Although we have a complete characterization of the seminorms $|||\cdot|||_k$ (and so also of the factors Z_k) in terms of nilmanifolds, there are only partial combinatorial characterizations in this direction (see [26–28]).

9. Other patterns

9.1. Commuting transformations.
Ergodic theory has been used to detect other patterns that occur in sets of positive upper density, using Furstenberg's correspondence principle and an appropriately chosen strengthening of Furstenberg multiple recurrence. A first example is for commuting transformations:

Theorem 9.1 (Furstenberg and Katznelson [19]). *Let $(X, \mathcal{X}, \mu)$ be a probability measure space, let $k \geq 1$ be an integer, and assume that $T_j \colon X \to X$ are commuting measure preserving transformations for $j = 1, 2, \ldots, k$. Then for all $A \in \mathcal{X}$ with $\mu(A) > 0$, there exist infinitely many $n \in \mathbb{N}$ such that*

(9.1) $$\mu(A \cap T_1^{-n}A \cap T_2^{-n}A \cap \cdots \cap T_k^{-n}A) > 0.$$

(In [20], Furstenberg and Katznelson proved a strengthening of this result, showing that one can place some restrictions on the choice of n; we do not discuss these "IP" versions of this theorem or the theorems given in the sequel.) Via correspondence, a multidimensional version of Szemerédi's theorem follows: if $E \subset \mathbb{Z}^r$ has positive upper density and $F \subset \mathbb{Z}^r$ is a finite subset, then there exist $z \in \mathbb{Z}^r$ and $n \in \mathbb{N}$ such that $z + nF \subset E$.

Again, this theorem is proven by showing that the associated liminf of the average of the quantity in Equation (9.1) is positive. And again, it is natural to ask whether the limit

$$\lim_{N \to \infty} \frac{1}{N} \sum_{n=0}^{N-1} \mu(A \cap T_1^{-n}A \cap \cdots \cap T_k^{-n}A)$$

exists in $L^2(\mu)$ for commuting maps $T_1, \ldots, T_k$. Only partial results are known. For $k = 2$, Conze and Lesigne ([8, 9]) proved convergence. For $k \geq 3$, the only known results rely on strong hypotheses of ergodicity:

Theorem 9.2 (Frantzikinakis and Kra [13]). *Let $k \in \mathbb{N}$ and assume that $T_1, T_2, \ldots, T_k$ are commuting invertible ergodic measure preserving transformations of a measure space $(X, \mathcal{X}, \mu)$ such that $T_i T_j^{-1}$ is ergodic for all $i, j \in \{1, 2, \ldots, k\}$ with $i \neq j$. If $f_1, f_2, \ldots, f_k \in L^\infty(\mu)$ the averages,*

$$\frac{1}{N} \sum_{n=0}^{N-1} T_1^n f_1 \cdot T_2^n f_2 \cdot \cdots \cdot T_k^n f_k$$

converge in $L^2(\mu)$ as $N \to \infty$.

The idea is to prove an analog of Lemma 7.5 for commuting transformations, thus reducing the problem to working in a nilsystem. The factors Z_k that are characteristic for averages along arithmetic progressions are also characteristic for these particular averages of commuting transformations. Without the strong hypotheses of ergodicity, this no longer holds and the general case remains open.

9.2. Averages along cubes. Another type of average is along k-dimensional cubes, the natural objects that arise in the definition of the seminorms. For example, a 2-dimensional cube is an expression of the form:

$$f(x) f(T^m x) f(T^n x) f(T^{m+n} x).$$

In [4], Bergelson showed the existence in $L^2(\mu)$ of

$$\lim_{N \to \infty} \frac{1}{N^2} \sum_{n,m=0}^{N-1} T^n f_1 \cdot T^m f_2 \cdot T^{n+m} f_3,$$

where $f_1, f_2, f_3 \in L^\infty(\mu)$. Similarly, one can define a 3-dimensional cube:

$$f_1(T^m x) f_2(T^n x) f_3(T^{m+n} x) f_4(T^p x) f_5(T^{m+p} x) f_6(T^{n+p} x) f_7(T^{m+n+p} x)$$

and existence of the limit of the average of this expression $L^2(\mu)$ for bounded functions $f_1, f_2, \ldots, f_7$ was shown in [33].

More generally, this theorem holds for cubes of $2^k - 1$ functions. Recalling the notation of Section 7, we have for $\epsilon = \epsilon_1 \ldots \epsilon_k \in \{0,1\}^k$ and $\mathbf{n} = (n_1, \ldots, n_k) \in \mathbb{Z}^k$,

$$\epsilon \cdot \mathbf{n} = \epsilon_1 n_1 + \epsilon_2 n_2 + \cdots + \epsilon_k n_k,$$

and $\mathbf{0}$ denotes the element $00\ldots0$ of $\{0,1\}^k$. We have:

Theorem 9.3 (Host and Kra [34]). *Let $(X, \mathcal{X}, \mu, T)$ be a system, let $k \geq 1$ be an integer, and let f_ϵ, $\epsilon \in \{0,1\}^k \setminus \{\mathbf{0}\}$, be $2^k - 1$ bounded functions on X. Then the averages*

$$\frac{1}{N^k} \cdot \sum_{\mathbf{n} \in [0, N-1]^k} \prod_{\substack{\epsilon \in \{0,1\}^k \\ \epsilon \neq \mathbf{0}}} T^{\epsilon \cdot \mathbf{n}} f_\epsilon$$

converge in $L^2(\mu)$ as $N \to \infty$.

The same result holds for translated averages, meaning the average for $\mathbf{n} \in [M_1, N_1] \times \cdots \times [M_k, N_k]$, as $N_1 - M_1, \ldots, N_k - M_k \to \infty$.

By Furstenberg's correspondence principle, this translates to a combinatorial statement. A subset $E \subset \mathbb{Z}$ is *syndetic* if $\mathbb{Z}$ can be covered by finitely many translates of E. In other words, there exists $N > 0$ such that every interval of size N contains at least one element of E. (Thus it is natural to refer to a syndetic set

in the integers as a set with *bounded gaps*.) More generally, $E \subset \mathbb{Z}^k$ is *syndetic* if there exists an integer $N > 0$ such that
$$E \cap \big([M_1, M_1 + N] \times \cdots \times [M_k, M_k + N]\big) \neq \varnothing$$
for all $M_1, \ldots, M_k \in \mathbb{Z}$.

Restricting Theorem 9.3 to indicator functions, the limit of the averages
$$\prod_{i=1}^k \frac{1}{N_i - M_i} \cdot \sum_{n_1 \in [M_1, N_1], \ldots, n_k \in [M_k, N_k]} \mu\bigg(\bigcap_{\epsilon \in \{0,1\}^k} T^{\epsilon \cdot \mathbf{n}} A\bigg)$$
exists and is greater than or equal to $\mu(A)^{2^k}$ when $N_1 - M_1, \ldots, N_k - M_k \to \infty$. Thus for every $\varepsilon > 0$,
$$\bigg\{\mathbf{n} \in \mathbb{Z}^k : \mu\bigg(\bigcap_{\epsilon \in \{0,1\}^k} T^{\epsilon \cdot \mathbf{n}} A\bigg) > \mu(A)^{2^k} - \varepsilon\bigg\}$$
of $\mathbb{Z}^k$ is syndetic.

By the correspondence principle, we have that if $E \subset \mathbb{Z}^k$ has upper density $d^*(E) > \delta > 0$ and $k \in \mathbb{N}$, then
$$\bigg\{\mathbf{n} \in \mathbb{Z}^k : d^*\bigg(\bigcap_{\epsilon \in \{0,1\}^k} (E + \epsilon \cdot \mathbf{n})\bigg) \geq \delta^{2^k}\bigg\}$$
is syndetic.

9.3. Polynomial patterns. In a different direction, one can restrict the iterates arising in Furstenberg's multiple recurrence. A natural choice is polynomial iterates, and the corresponding combinatorial statement is that a set of integers with positive upper density contains elements who differ by a polynomial:

Theorem 9.4 (Sárközy [51], Furstenberg [17]). *If $E \subset \mathbb{N}$ has positive upper density and $p \colon \mathbb{Z} \to \mathbb{Z}$ is a polynomial with $p(0) = 0$, then there exist $x, y \in E$ and $n \in \mathbb{N}$ such that $x - y = p(n)$.*

As for arithmetic progressions, Furstenberg's proof relies on the correspondence principle and an averaging theorem:

Theorem 9.5 (Furstenberg [17]). *Let $(X, \mathcal{X}, \mu, T)$ be a system, let $A \in \mathcal{X}$ with $\mu(A) > 0$ and let $p \colon \mathbb{Z} \to \mathbb{Z}$ be a polynomial with $p(0) = 0$. Then*
$$\liminf_{N \to \infty} \frac{1}{N} \sum_{n=0}^{N-1} \mu(A \cap T^{-p(n)} A) > 0.$$

The multiple polynomial recurrence theorem, simultaneously generalizing this single polynomial result and Furstenberg's multiple recurrence, was proven by Bergelson and Leibman:

Theorem 9.6 (Bergelson and Leibman [6]). *Let $(X, \mathcal{X}, \mu, T)$ be a system, let $A \in \mathcal{X}$ with $\mu(A) > 0$, and let $k \in \mathbb{N}$. If $p_1, p_2, \ldots, p_k \colon \mathbb{Z} \to \mathbb{Z}$ are polynomials with $p_j(0) = 0$ for $j = 1, \ldots, k$, then*

(9.2) $$\liminf_{N \to \infty} \frac{1}{N} \sum_{n=0}^{N-1} \mu\big(A \cap T^{-p_1(n)} A \cap \cdots \cap T^{-p_k(n)} A\big) > 0.$$

By the correspondence principle, one immediately deduces a polynomial Szemerédi theorem: if $E \subset \mathbb{Z}$ has positive upper density, then it contains arbitrary polynomial patterns, meaning there exists $n \in \mathbb{N}$ such that
$$x, x + p_1(n), x + p_2(n), \ldots, x + p_k(n) \in E.$$
(More generally, Bergelson and Leibman proved a version of Theorem 9.6 for commuting transformations, with a multidimensional polynomial Szemerédi theorem as a corollary.)

Again, it is natural to ask whether the lim inf in (9.2) is actually a limit. A first result in this direction was given by Furstenberg and Weiss [22], who proved convergence in $L^2(\mu)$ of
$$\frac{1}{N}\sum_{n=0}^{N-1} T^{n^2} f_1 \cdot T^n f_2$$
and
$$\frac{1}{N}\sum_{n=0}^{N-1} T^{n^2} f_1 \cdot T^{n^2+n} f_2$$
for bounded functions f_1, f_2.

The proof of convergence for general polynomial averages uses the technology of the seminorms, reducing to the same characteristic factors Z_k that can be described using nilsystems, as for averages along arithmetic progressions:

Theorem 9.7 (Host and Kra [35], Leibman [45]). *Let $(X, \mathcal{X}, \mu, T)$ be a system, $k \in \mathbb{N}$, and $f_1, f_2, \ldots, f_k \in L^\infty(\mu)$. Then for any polynomials $p_1, p_2, \ldots, p_k \colon \mathbb{Z} \to \mathbb{Z}$, the averages*
$$\frac{1}{N}\sum_{n=0}^{N-1} T^{p_1(n)} f_1 \cdot T^{p_2(n)} f_2 \cdot \ldots \cdot T^{p_k(n)} f_k$$
converge in $L^2(\mu)$.

Recently, Johnson [36] has shown that under strong ergodicity conditions, similar to those in Theorem 9.2, one can generalize this and prove $L^2(\mu)$-convergence of the polynomial averages for commuting transformations:
$$\frac{1}{N}\sum_{n=0}^{N-1} T_1^{p_1(n)} f_1 \cdot T_2^{p_2(n)} f_2 \cdot \ldots \cdot T_k^{p_k(n)} f_k$$
for $f_1, f_2, \ldots, f_k \in L^\infty(\mu)$.

For a *totally ergodic system* (meaning that T^n is ergodic for all $n \in \mathbb{N}$), Furstenberg and Weiss showed a stronger result, giving an explicit and simple formula for the limit:
$$\frac{1}{N}\sum_{n=0}^{N-1} T^n f_1 \cdot T^{n^2} f_2 \to \int f_1 \, d\mu \cdot \int f_2 \, d\mu$$
in $L^2(\mu)$.

Bergelson [3] asked whether the same result holds for k polynomials of different degrees, meaning that the limit of the polynomial average for a totally ergodic system is the product integrals. In [12], we show that the answer is yes under a more general condition. A family of polynomials $p_1, p_2, \ldots, p_k \colon \mathbb{Z} \to \mathbb{Z}$ is *rationally independent* if the polynomials $\{1, p_1, p_2, \ldots, p_k\}$ are linearly independent over the rationals. We show:

Theorem 9.8 (Frantzikinakis and Kra [12]). *Let $(X, \mathcal{X}, \mu, T)$ be a totally ergodic system, let $k \geq 1$ be an integer, and assume that $p_1, p_2 \ldots, p_k \colon \mathbb{Z} \to \mathbb{Z}$ are rationally independent polynomials. If $f_1, f_2, \ldots, f_k \in L^\infty(\mu)$,*

$$\lim_{N \to \infty} \left\| \frac{1}{N} \sum_{n=0}^{N-1} T^{p_1(n)} f_1 \cdot T^{p_2(n)} f_2 \cdot \ldots \cdot T^{p_k(n)} f_k - \prod_{i=1}^{k} \int f_i \, d\mu \right\|_{L^2(\mu)} = 0.$$

As a corollary, if $(X, \mathcal{X}, \mu, T)$ is totally ergodic, $\{1, p_1, \ldots, p_k\}$ are rationally independent polynomials taking on integer values on the integers, and $A_0, A_1, \ldots, A_k \in \mathcal{X}$ with $\mu(A_i) > 0$, $i = 0, \ldots, k$, then

$$\mu(A_0 \cap T^{-p_1(n)} A_1 \cap \cdots \cap T^{-p_k(n)} A_k) > 0$$

for some $n \in \mathbb{N}$. Thus in a totally ergodic system, one can strengthen Bergelson and Leibman's multiple polynomial recurrence theorem, allowing the sets A_i to be distinct, and allowing the polynomials p_i to have nonzero constant term. It is not clear whether or not this has a combinatorial interpretation.

10. Strengthening Poincaré recurrence

10.1. Khintchine recurrence.
Poincaré recurrence states that a set of positive measure returns to intersect itself infinitely often. One way to strengthen this is to ask that the set return to itself often with 'large' intersection. Khintchine made this notion precise, showing that large self intersection occurs on a syndetic set:

Theorem 10.1 (Khintchine [37]). *Let $(X, \mathcal{X}, \mu, T)$ be a system, let $A \in \mathcal{X}$ have $\mu(A) > 0$, and let $\varepsilon > 0$. Then*

$$\{n \in \mathbb{Z} : \mu(A \cap T^n A) > \mu(A)^2 - \varepsilon\}$$

is syndetic.

It is natural to ask for a simultaneous generalization of Furstenberg multiple recurrence and Khintchine recurrence. More precisely, if $(X, \mathcal{X}, \mu, T)$ is a system, $A \in \mathcal{X}$ has positive measure, $k \in \mathbb{N}$, and $\varepsilon > 0$, is the set

$$\{n \in \mathbb{Z} \colon \mu(A \cap T^n A \cap \cdots \cap T^{kn} A) > \mu(A)^{k+1} - \varepsilon\}$$

syndetic?

Furstenberg multiple recurrence implies that there exists some constant $c = c(\mu(A)) > 0$ such that

$$\{n \in \mathbb{Z} \colon \mu(A \cap T^n A \cap \cdots \cap T^{kn} A) > c\}$$

is syndetic. But to generalize Khintchine recurrence, one needs $c = \mu(A)^{k+1}$. It turns out that the answer depends on the length k of the arithmetic progression.

Theorem 10.2 (Bergelson, Host and Kra [5]). *Let $(X, \mathcal{X}, \mu, T)$ be an ergodic system and let $A \in \mathcal{X}$. Then for every $\varepsilon > 0$, the sets*

$$\{n \in \mathbb{Z} : \mu(A \cap T^n A \cap T^{2n} A) > \mu(A)^3 - \varepsilon\}$$

and

$$\{n \in \mathbb{Z} : \mu(A \cap T^n A \cap T^{2n} A \cap T^{3n} A) > \mu(A)^4 - \varepsilon\}$$

are syndetic.

Furthermore, this result fails on average, meaning that the average of the left hand side expressions is not necessarily greater than $\mu(A)^3 - \varepsilon$ or $\mu(A)^4 - \varepsilon$, respectively.

On the other hand, based on an example of Ruzsa contained in the appendix of [5], we have:

Theorem 10.3 (Bergelson, Host and Kra [5]). *There exists an ergodic system $(X, \mathcal{X}, \mu, T)$ and for all $\ell \in \mathbb{N}$ there exists a set $A = A(\ell) \in \mathcal{X}$ with $\mu(A) > 0$ such that*
$$\mu(A \cap T^n A \cap T^{2n} A \cap T^{3n} A \cap T^{4n} A) \leq \mu(A)^\ell / 2$$
for every integer $n \neq 0$.

We now briefly outline the major ingredients in the proofs of these theorems.

10.2. Positive ergodic results. We start with the ergodic results needed to prove Theorem 10.2. Fix an integer $k \geq 1$, an ergodic system $(X, \mathcal{X}, \mu, T)$, and $A \in \mathcal{X}$ with $\mu(A) > 0$. The key ingredient is the study of the *multicorrelation sequence*
$$\mu(A \cap T^n A \cap T^{2n} A \cap \cdots \cap T^{kn} A).$$
More generally, for a real valued function $f \in L^\infty(\mu)$, we consider the *multicorrelation sequence*
$$I_f(k, n) := \int f \cdot T^n f \cdot T^{2n} f \cdot \, \cdots \, \cdot T^{kn} f \, \mathrm{d}\mu(x).$$

When $k = 1$, Herglotz's theorem implies that the correlation sequence $I_f(1, n)$ is the Fourier transform of some positive measure $\sigma = \sigma_f$ on the torus $\mathbb{T}$:
$$I_f(1, n) = \widehat{\sigma}(n) := \int_\mathbb{T} e^{2\pi i n t} \, \mathrm{d}\sigma(t).$$

Decomposing the measure σ into its continuous part σ^c and its discrete part σ^d, can write the multicorrelation sequence $I_f(1, n)$ as the sum of two sequences
$$I_f(1, n) = \widehat{\sigma^c}(n) + \widehat{\sigma^d}(n).$$
The sequence $\{\widehat{\sigma^c}(n)\}$ *tends to 0 in density*, meaning that

(10.1) $$\lim_{N \to \infty} \sup_{M \in \mathbb{Z}} \frac{1}{M} \sum_{n=M}^{M+N-1} |\widehat{\sigma^c}(n)| = 0.$$

Equivalently, for any $\varepsilon > 0$, the upper Banach density[5] of the set $\{n \in \mathbb{Z} : |\widehat{\sigma^c}(n)| > \varepsilon\}$ is zero. The sequence $\{\widehat{\sigma^d}(n)\}$ is *almost periodic*, meaning that there exists a compact abelian group G, a continuous real valued function ϕ on G, and $a \in G$ such that $\widehat{\sigma^d}(n) = \phi(a^n)$ for all n.

A compact abelian group can be approximated by a compact abelian Lie group. Thus any almost periodic sequence can be uniformly approximated by an almost periodic sequence arising from a compact abelian Lie group.

In general, however, for higher k the answer is more complicated. We find a similar decomposition for the multicorrelation sequences $I_f(k, n)$ for $k \geq 2$. The notion of an almost periodic sequence is replaced by that of a *nilsequence*: for an

---

[5]The *upper Banach density* $\bar{d}(E)$ of a set $E \subset \mathbb{Z}$ is defined by $\bar{d}(e) = \lim_{N \to \infty} \sum_{M \in \mathbb{Z}} (1/N) |E \cap [M, M+N-1]|$.

integer $k \geq 2$, a k-step nilmanifold $X = G/\Gamma$, a continuous real (or complex) valued function ϕ on G, $a \in G$, and $e \in X$, the sequence $\{\phi(a^n \cdot e)\}$ is called a *basic k-step nilsequence*. A k-step nilsequence is a uniform limit of basic k-step nilsequences.

It follows that a 1-step nilsequence is the same as an almost periodic sequence. An inverse limit of compact abelian Lie groups is a compact group. However an inverse limit of k-step nilmanifolds is not, in general, the homogeneous space of some locally compact group, and so for higher k, the decomposition result must take into account the uniform limits of basic nilsequences. We have:

Theorem 10.4 (Bergelson, Host and Kra [5]). *Let $(X, \mathcal{X}, \mu, T)$ be an ergodic system, $f \in L^\infty(\mu)$ and $k \geq 1$ an integer. The sequence $\{I_f(k, n)\}$ is the sum of a sequence tending to zero in density and a k-step nilsequence.*

Due to the connections between the use of the seminorms in ergodic theory and the Gowers uniformity norms in additive combinatorics, it is natural that nilsequences also have a role to play on the combinatorial side. Recently, Green and Tao (see [26–28]) have adapted the idea of a nilsequence to combinatorics, and this plays a role in the asymptotics for the number of arithmetic progressions of length 4 in the primes. Ben Green's notes in this volume have more on this connection.

Finally, we explain how Theorem 10.4 can be used to prove Theorem 10.2. Let $\{a_n\}_{n \in \mathbb{Z}}$ be a bounded sequence of real numbers. The *syndetic supremum* of this sequence is defined to be

$$\sup\{c \in \mathbb{R} : \{n \in \mathbb{Z} : a_n > c\} \text{ is syndetic}\}.$$

Every nilsequence $\{a_n\}$ is uniformly recurrent.[6] In particular, if $S = \sup(a_n)$ and $\varepsilon > 0$, then $\{n \in \mathbb{Z} : a_n \geq S - \varepsilon\}$ is syndetic.

If $\{a_n\}$ and $\{b_n\}$ are two sequences of real numbers such that $a_n - b_n$ tends to 0 in density (in the sense of definition (10.1)), then the two sequences have the same syndetic supremum. Therefore the syndetic supremums of the sequences

$$\{\mu(A \cap T^n A \cap T^{2n} A)\}$$

and

$$\{\mu(A \cap T^n A \cap T^{2n} A \cap T^{3n} A)\}$$

are equal to the supremum of the associated nilsequences, and we are reduced to showing that they are greater than or equal to $\mu(A)^3$ and $\mu(A)^4$, respectively.

10.3. Nonergodic counterexample. Ergodicity is not needed for Khintchine's theorem, but is essential for Theorem 10.2:

Theorem 10.5 (Bergelson, Host, and Kra [5]). *There exists a (nonergodic) system $(X, \mathcal{X}, \mu, T)$, and for every $\ell \in \mathbb{N}$ there exists $A \in \mathcal{X}$ with $\mu(A) > 0$ such that*

$$\mu(A \cap T^n A \cap T^{2n} A) \leq \frac{1}{2}\mu(A)^\ell.$$

for integer $n \neq 0$.

Actually there exists a set A of arbitrarily small positive measure with

$$\mu(A \cap T^n A \cap T^{2n} A) \leq \mu(A)^{-c \log(\mu(A))}$$

for every integer $n \neq 0$ and for some positive universal constant c.

---

[6]A sequence $\{a_n\}$ of real numbers is said to be *uniformly recurrence* if for all $\varepsilon > 0$ and all $h \in \mathbb{N}$, the set $\{n : |a_{n+h} - a_n| < \varepsilon\}$ is syndetic

The proof is based on Behrend's construction of a set containing no arithmetic progression of length 3:

Theorem 10.6 (Behrend [1]). *For all $L \in \mathbb{N}$, there exists a subset $E \subset \{0, 1, \ldots, L-1\}$ having more than $L\exp(-c\sqrt{\log L})$ elements that does not contain any nontrivial arithmetic progression of length 3.*

PROOF (OF THEOREM 10.5). Let $X = \mathbb{T} \times \mathbb{T}$, with Haar measure $\mu = m \times m$ and transformation $T\colon X \to X$ given by $T(x,y) = (x, y+x)$.

Let $E \subset \{0, 1, \ldots, L-1\}$, not containing any nontrivial arithmetic progression of length 3. Define
$$B = \bigcup_{j \in E} \left[\frac{j}{2L}, \frac{j}{2L} + \frac{1}{4L}\right),$$
which we consider as a subset of the torus and $A = \mathbb{T} \times B$.

For every integer $n \neq 0$, we have $T^n(x, y) = (x, y+nx)$ and
$$\mu(A \cap T^n A \cap T^{2n} A) = \iint_{\mathbb{T} \times \mathbb{T}} \mathbf{1}_B(y)\mathbf{1}_B(y+nx)\mathbf{1}_B(y+2nx)\,dm(y)\,dm(x)$$
$$= \iint_{\mathbb{T} \times \mathbb{T}} \mathbf{1}_B(y)\mathbf{1}_B(y+x)\mathbf{1}_B(y+2x)\,dm(y)\,dm(x).$$

Bounding this integral, we have that:
$$\mu(A \cap T^n A \cap T^{2n} A) = \iint_{\mathbb{T} \times \mathbb{T}} \mathbf{1}_B(y)\mathbf{1}_B(y+x)\mathbf{1}_B(y+2x)\,dm(x)\,dm(y)$$
$$\leq \frac{m(B)}{4L}.$$

By Behrend's theorem, we can choose the set E with cardinality on the order of $L\exp(-c\sqrt{\log L})$. Choosing L sufficiently large, a simple computation gives the statement. □

For longer arithmetic progressions, the counterexample of Theorem 10.3 is based on a construction of Ruzsa. When P is a nonconstant integer polynomial of degree ≤ 2, the subset
$$\{P(0), P(1), P(2), P(3), P(4)\}$$
of $\mathbb{Z}$ is called a *quadratic configuration of* 5 *terms*, written QC5 for short.

Any QC5 contains at least 3 distinct elements. An arithmetic progression of length 5 is a QC5, corresponding to a polynomial of degree 1.

Theorem 10.7 (Ruzsa [5]). *For all $L \in \mathbb{N}$, there exists a subset $E \subset \{0, 1, \ldots, L-1\}$ having more than $L\exp(-c\sqrt{\log L})$ elements that does not contain any QC5.*

Based on this, we show:

Theorem 10.8 (Bergelson, Host and Kra [5]). *There exists an ergodic system $(X, \mathcal{X}, \mu, T)$ and, for every $\ell \in \mathbb{N}$, there exists $A \in \mathcal{X}$ with $\mu(A) > 0$ such that*
$$\mu(A \cap T^n A \cap T^{2n} A \cap T^{3n} A \cap T^{4n} A) \leq \frac{1}{2}\mu(A)^\ell$$
for every integer $n \neq 0$.

Once again, the proof gives the estimate $\mu(A)^{-c\log(\mu(A))}$, for some constant $c > 0$.

The construction again involves a simple example: $\mathbb{T}$ is the torus with Haar measure m, $X = \mathbb{T} \times \mathbb{T}$, and $\mu = m \times m$. Let $\alpha \in \mathbb{T}$ be irrational and let $T\colon X \to X$ be
$$T(x,y) = (x + \alpha, y + 2x + \alpha).$$

Combinatorially this example becomes: for all $k \in \mathbb{N}$, there exists $\delta > 0$ such that for infinitely many integers N, there is a subset $A \subset \{1,\ldots,N\}$ with $|A| \geq \delta N$ that contains no more than $\frac{1}{2}\delta^k N$ arithmetic progressions of length ≥ 5 with the same difference.

10.4. Combinatorial consequences. Via a slight modification of the correspondence principle, each of these results translates to a combinatorial statement. For $\varepsilon > 0$ and $E \subset \mathbb{Z}$ with positive upper Banach density (see the definition in Footnote 5), consider the set

(10.2) $\quad \{n \in \mathbb{Z}\colon \bar{d}(E \cap (E+n) \cap (E+2n) \cap \cdots \cap (E+kn)) \geq \bar{d}(E^{k+1}) - \varepsilon\}.$

From Theorems 10.2 and 10.3, for $k = 2$ and for $k = 3$, this set is syndetic, while for $k \geq 4$ there exists a set of integers E with positive upper Banach density such that the set in (10.2) is empty.

We can refine this a bit further. Recall the notation from Szemerédi's theorem: for every $\delta > 0$ and $k \in \mathbb{N}$, there exists $N(\delta, k)$ such that for all $N > N(\delta, k)$, every subset of $\{1,\ldots,N\}$ with at least δN elements contains an arithmetic progression of length k.

For an arithmetic progression $\{a, a+s, \ldots, a+(k-1)s\}$, s is the *difference* of the progression. Write $\lfloor x \rfloor$ for integer part of x. ¿From Szemerédi's theorem, we can deduce that every subset E of $\{1,\ldots,N\}$ with at least δN elements contains at least $\lfloor cN^2 \rfloor$ arithmetic progressions of length k, where $c = c(k, \delta) > 0$ is a constant. Therefore the set E contains at least $\lfloor c(k,\delta)N \rfloor$ progressions of length k with the same difference.

The ergodic results of Theorem 10.2 give some improvement for $k = 3$ and $k = 4$ (see [5] for the precise statement). For $k = 3$, this was strengthened by Green:

Theorem 10.9 (Green [24]). *For all $\delta, \varepsilon > 0$, there exists $N_0(\delta, \varepsilon)$ such that for all $N > N_0(\delta, \varepsilon)$ and any $E \subset \{1,\ldots,N\}$ with $|E| \geq \delta N$, E contains at least $(1-\varepsilon)\delta^3 N$ arithmetic progressions of length 3 with the same difference.*

On the other hand, the similar bound for longer progressions with length $k \geq 5$ does not hold. The proof in [5], based on an example of Rusza, does not use ergodic theory. We show that for all $k \in \mathbb{N}$, there exists $\delta > 0$ such that for infinitely many N, there exists a subset E of $\{1,\ldots,N\}$ with $|E| \geq \delta N$ that contains no more than $\frac{1}{2}\delta^k N$ arithmetic progressions of length ≥ 5 with the same step.

10.5. Polynomial averages. One can ask whether similar lower bounds hold for the polynomial averages. For independent polynomials, using the fact that the characteristic factor is the Kronecker factor, we can show:

Theorem 10.10 (Frantzikinakis and Kra [14]). *Let $k \in \mathbb{N}$, $(X, \mathcal{X}, \mu, T)$ be a system, $A \in \mathcal{X}$, and let $p_1, p_2, \ldots, p_k\colon \mathbb{Z} \to \mathbb{Z}$ be rationally independent polynomials*

with $p_i(0) = 0$ for $i = 1, 2, \ldots, k$. Then for every $\varepsilon > 0$, the set

$$\left\{n \in \mathbb{Z} : \mu(A \cap T^{p_1(n)}A \cap T^{p_2(n)} \cap \cdots \cap T^{p_k(n)}A) > \mu(A)^{k+1} - \varepsilon\right\}$$

is syndetic.

Once again, this result fails on average.

Via correspondence, analogous to the results of Section 10.4, we have that for $E \subset \mathbb{Z}$ and rationally independent polynomials $p_1, p_2, \ldots, p_k \colon \mathbb{Z} \to \mathbb{Z}$ with $p_i(0) = 0$ for $i = 1, 2, \ldots, k$, then for all $\varepsilon > 0$, the set

$$\left\{n \in \mathbb{Z} \colon \bar{d}\left(E \cap \left(E + p_1(n)\right) \cap \cdots \cap \left(E + p_k(n)\right)\right) \geq \bar{d}(E)^{k+1} - \varepsilon\right\}$$

is syndetic.

Moreover, in [14] we strengthen this and show that there are many configurations with the same n giving the differences: if $p_1, p_2, \ldots, p_k \colon \mathbb{Z} \to \mathbb{Z}$ are rationally independent polynomials with $p_i(0) = 0$ for $i = 1, 2, \ldots, k$, then for all $\delta, \varepsilon > 0$, there exists $N(\delta, \varepsilon)$ such that for all $N > N(\delta, \varepsilon)$ and any subset $E \subset \{1, \ldots, N\}$ with $|E| \geq \delta N$ contains at least $(1 - \varepsilon)\delta^{k+1}N$ configurations of the form

$$\{x, x + p_1(n), x + p_2(n), \ldots, x + p_k(n)\}$$

for a fixed $n \in \mathbb{N}$.

References

1. F. A. Behrend, *On sets of integers which contain no three terms in arithmetic progression*, Proc. Nat. Acad. Sci. U.S.A. **32** (1946), 331–332.
2. V. Bergelson, *Weakly mixing PET*, Ergodic Theory Dynam. Systems **7** (1987), no. 3, 337–349.
3. _____, *Ergodic Ramsey theory — an update*, Ergodic Theory of $\mathbb{Z}^d$-Actions (Warwick, 1993–1994) (M. Pollicott and K. Schmidt, eds.), London Math. Soc. Lecture Note Ser., vol. 228, Cambridge Univ. Press, Cambridge, 1996.
4. _____, *The multifarious Poincaré recurrence theorem*, Descriptive Set Theory and Dynamical Systems (Marseille-Luminy, 1996) (M. Foreman, A. S. Kechris, A. Louveau, and B. Weiss, eds.), London Math. Soc. Lecture Note Ser., vol. 277, Cambridge Univ. Press, Cambridge, 2000, pp. 31–57.
5. V. Bergelson, Host B., and B. Kra, *Multiple recurrence and nilsequences*, Invent. Math. **160** (2005), no. 2, 261–303.
6. V. Bergelson and A. Leibman, *Polynomial extensions of van der Waerden's and Szemerédi's theorems*, J. Amer. Math. Soc. **9** (1996), no. 3, 725–753.
7. J. Bourgain, *On the maximal ergodic theorem for certain subsets of the positive integers*, Israel J. Math. **61** (1988), no. 1, 39–72.
8. J.-P. Conze and E. Lesigne, *Sur un théorème ergodique pour des mesures diagonales*, Probabilités, Publ. Inst. Rech. Math. Rennes, vol. 1987-1, Univ. Rennes I, Rennes, 1987, pp. 1–31.
9. _____, *Sur un théorème ergodique pour des mesures diagonales*, C. R. Acad. Sci. Paris Sér. I Math. **306** (1988), no. 12, 491–493.
10. I. P. Cornfeld, S. V. Fomin, and Ya. G. Sinaĭ, *Ergodic theory*, Grundlehren Math. Wiss., vol. 245, Springer, New York, 1982.
11. P. Erdős and P. Turán, *On some sequences of integers*, J. London Math. Soc. **11** (1936), 261–264.
12. N. Frantzikinakis and B. Kra, *Polynomial averages converge to the product of the integrals*, Israel J. Math. **148** (2005), 267–276.
13. _____, *Convergence of multiple ergodic averages for some commuting transformations*, Ergodic Theory Dynam. Systems **25** (2005), no. 3, 799–809.
14. _____, *Ergodic averages for independent polynomials and applications*, J. London Math. Soc. (2) **74** (2006), no. 1, 131–142.

15. H. Furstenberg, *Strict ergodicity and transformations of the torus*, Amer. J. Math. **83** (1961), 573–601.
16. _____, *Ergodic behavior of diagonal measures and a theorem of Szemerédi on arithmetic progressions*, J. Analyse Math. **31** (1977), 204–256.
17. _____, *Recurrence in ergodic theory and combinatorial number theory*, M. B. Porter Lectures, Princeton Univ. Press, Princeton, NJ, 1981.
18. _____, *Nonconventional ergodic averages*, The legacy of John von Neumann (Hempstead, NY, 1988), Proc. Sympos. Pure Math., vol. 50, Amer. Math. Soc., Providence, RI, 1990, pp. 43–56.
19. H. Furstenberg and Y. Katznelson, *An ergodic Szemerédi theorem for commuting transformation*, J. Analyse Math. **34** (1978), 275–291.
20. _____, *An ergodic Szemerédi theorem for IP-systems and combinatorial theory*, J. Analyse Math. **45** (1985), 117–268.
21. H. Furstenberg, Y. Katznelson, and D. Ornstein, *The ergodic theoretical proof of Szemerédi's theorem*, Bull. Amer. Math. Soc. (N.S.) **7** (1982), no. 3, 527–552.
22. H. Furstenberg and B. Weiss, *A mean ergodic theorem for $(1/N) \sum_{n=1}^{N} f(T^n x)g(T^{n^2} x)$*, Convergence in Ergodic Theory and Probability (Columbus, OH, 1993), Ohio State Univ. Math. Res. Inst. Publ., vol. 5, de Gruyter, Berlin, 1996, pp. 193–227.
23. W. T. Gowers, *A new proof of Szemerédi's theorem*, Geom. Funct. Anal. **11** (2001), no. 3, 465–588.
24. B. J. Green, *A Szemerédi-type regularity lemma in abelian groups, with applications*, Geom. Funct. Anal. **15** (2005), no. 2, 340–376.
25. B. J. Green and T. C. Tao, *The primes contain arbitrarily long arithmetic progressions*, Ann. of Math. (2), to appear.
26. _____, *An inverse theorem for the Gowers U^3-norm, with applications*, Proc. Edinb. Math. Soc. (2), to appear.
27. _____, *Quadratic uniformity of the Möbius function*, preprint.
28. _____, *Linear equations in primes*, preprint.
29. P. Hall, *A contribution to the theory of groups of prime-power order*, Proc. London Math. Soc. **36** (1933), 29–95.
30. P. R. Halmos and J. von Neumann, *Operator methods in classical mechanics. II*, Ann. of Math. (2) **43** (1942), 332–350.
31. B. Host, *Convergence of multiple ergodic averages*, School on Information and Randomness (Santiago, 2004), to appear.
32. B. Host and B. Kra, *Convergence of Conze–Lesigne averages*, Ergodic Theory Dynam. Systems **21** (2001), no. 2, 493–509.
33. _____, *Averaging along cubes*, Modern Dynamical Systems and Applications, Cambridge Univ. Press, Cambridge, 2004, pp. 123–144.
34. _____, *Nonconventional ergodic averages and nilmanifolds*, Ann. of Math. (2) **161** (2005), no. 1, 397–488.
35. _____, *Convergence of polynomial ergodic averages*, Israel J. Math. **149** (2005), 1–19.
36. M. Johnson, *Convergence of polynomial ergodic averages of several variables for some commuting transformations*, preprint.
37. A. Y. Khintchine, *Eine Verschärfung des Poincaréschen "Wiederkehrsatzes"*, Compositio Math. **1** (1934), 177–179.
38. B. O. Koopman and J. von Neumann, *Dynamical systems of continuous spectra*, Proc. Nat. Acad. Sci. U.S.A. **18** (1932), 255–263.
39. B. Kra, *The Green–Tao theorem on arithmetic progressions in the primes: an ergodic point of view*, Bull. Amer. Math. Soc. (N.S.) **43** (2006), no. 1, 3–23.
40. _____, *From combinatorics to ergodic theory and back again*, Proceedings of the International Congress of Mathematicians, Vol. 3 (Madrid, 2006).
41. L. Kuipers and N. Niederreiter, *Uniform distribution of sequences*, Pure Appl. Math., John Wiley & Sons, New York, 1974.
42. M. Lazard, *Sur certaines suites d'éléments dans les groupes libres et leurs extensions*, C. R. Acad. Sci. Paris **236** (1953), 36–38.
43. A. Leibman, *Polynomial sequences in groups*, J. Algebra **201** (1998), no. 1, 189–206.
44. _____, *Pointwise convergence of ergodic averages for polynomial sequences of translations on a nilmanifold*, Ergodic Theory Dynam. Systems **25** (2005), no. 1, 201–213.

45. _____, *Convergence of multiple ergodic averages along polynomials of several variables*, Israel J. Math. **146** (2005), 303–316.
46. E. Lesigne, *Sur une nil-variété, les parties minimales associeée à une translation sont uniquement ergodiques*, Ergodic Theory Dynam. Systems **11** (1991), no. 2, 379–391.
47. W. Parry, *Ergodic properties of affine transformations and flows on nilmanifolds*, Amer. J. Math. **91** (1969), 757–771.
48. J. Petresco, *Sur les commutateurs*, Math. Z. **61** (1954), 348–356.
49. H. Poincaré, *Les méthodes nouvelles de la mécanique céleste*. I, Gathiers-Villars, Paris, 1892; II, 1893; III, 1899.
50. M. Ratner, *On Raghunathan's measure conjecture*, Ann. of Math. (2) **134** (1991), no. 3, 545–607.
51. A. Sárközy, *On difference sets of sequences of integers*. I, Acta Math. Acad. Sci. Hungar. **31** (1978), no. 1-2, 125–149.
52. _____, *On difference sets of sequences of integers*. III, Acta Math. Acad. Sci. Hungar. **31** (1978), no. 3-4, 355–386.
53. N. Shah, *Invariant measures and orbit closures on homogeneous spaces for actions of subgroups*, Lie Groups and Ergodic Theory (Mumbai, 996), Tata Inst. Fund. Res., Bombay, 1998.
54. E. Szemerédi, *On sets of integers containing no k elements in arithmetic progression*, Acta Arith. **27** (1975), 199–245.
55. J. von Neumann, *Proof of the quasi-ergodic hypothesis*, Proc. Nat. Acad. Sci. U.S.A. **18** (1932), 70–82.
56. T. Ziegler, *A non-conventional ergodic theorem for a nilsystem*, Ergodic Theory Dynam. Systems **25** (2005), no. 4, 1357–1370.
57. _____, *Universal characteristic factors and Furstenberg averages*, J. Amer. Math. Soc. **20** (2007), no. 1, 53–97.

DEPARTMENT OF MATHEMATICS, NORTHWESTERN UNIVERSITY, 2033 SHERIDAN ROAD, EVANSTON, IL 60208-2730, USA

E-mail address: kra@math.northwestern.edu

The Ergodic and Combinatorial Approaches to Szemerédi's Theorem

Terence Tao

ABSTRACT. A famous theorem of Szemerédi asserts that any set of integers of positive upper density will contain arbitrarily long arithmetic progressions. In its full generality, we know of four types of arguments that can prove this theorem: the original combinatorial (and graph-theoretical) approach of Szemerédi, the ergodic theory approach of Furstenberg, the Fourier-analytic approach of Gowers, and the hypergraph approach of Nagle–Rödl–Schacht–Skokan and Gowers. In this lecture series we introduce the first, second and fourth approaches, though we will not delve into the full details of any of them. One of the themes of these lectures is the strong similarity of ideas between these approaches, despite the fact that they initially seem rather different.

1. Introduction

These lecture notes will be centred upon the following fundamental theorem of Szemerédi:

Theorem 1.1 (Szemerédi's theorem [41]**).** *Let $A \subset \mathbb{Z}$ be a subset of the integers of positive upper density, thus $\limsup_{N \to \infty} |A \cap [-N, N]|/(2N + 1) > 0$. (Here and in the sequel, we use $|B|$ to denote the cardinality of a finite set B.) Then A contains arbitrarily long arithmetic progressions.*

This theorem is rather striking, because it assumes almost nothing on the given set A—other than that it is large—and concludes that A is necessarily structured in the sense that it contains arithmetic progressions of any given length k. This is a property special to arithmetic progressions (and a few other related patterns). Consider for instance the question asking whether a set A of positive density must contain a triplet of the form $\{x, y, x + y\}$. (Compare with the triplet $\{x, y, (x + y)/2\}$, which is an arithmetic progression of length three.) It is then clear that the odd numbers, which are certainly a set of positive upper density, do not contain such triples (see however Theorem 6.1 below). Or for another

---

2000 *Mathematics Subject Classification.* 05C55, 05C65, 05C75, 11B25, 11N13, 37A45, 37B20.

Key words and phrases. Szemerédi's theorem, van der Waerden's theorem, ergodic theory, graph theory, hypergraph theory.

The author is supported by a grant from the Packard Foundation.

This is the final form of the paper.

example, consider whether a set of positive upper density must contain a pair $\{x, x+2\}$. The multiples of 3 provide an immediate counterexample. (This is basically why the methods from [25] can leverage Szemerédi's theorem to show that the primes contain arbitrarily long arithmetic progressions, but are currently unable to make any progress whatsoever on the twin prime conjecture.) But the arithmetic progressions seem to be substantially more "indestructable" than these other types of patterns, in that they seem to occur in any large set A no matter how one tries to rearrange A to eliminate all the progressions.

We have contrasted Szemerédi's theorem with some negative results where the selected pattern need not occur. Now let us give the opposite contrast, in which it becomes *very easy* to find a pattern of a certain type in a set. Here is a basic example (a special case of a result of Hilbert):

Proposition 1.2. *Let $A \subset \mathbb{Z}$ have positive upper density. Then A contains infinitely many "parallelograms" $\{x, x+a, x+b, x+a+b\}$ where $a, b \neq 0$.*

Note that if we could just set $a = b$ in these parallelograms then we could find infinitely many arithmetic progressions of length three. Alas, things are not so easy, and while progressions are certainly intimately related to parallelograms (and more generally to higher-dimensional parallelopipeds, for which an analogue of Proposition 1.2 can be easily located), the existence of the latter does not instantly imply the existence of the former without substantial additional effort. For example, one can easily modify Proposition 1.2 to locate, for any $k \geq 1$, infinitely many parallelopipeds of the form $\{p + \sum_{i \in A} x_i : A \subset \{1, \ldots, k\}\}$ in the primes $\{2, 3, 5, \ldots\}$, where p is a prime and $x_1, \ldots, x_k > 0$ are positive integers, but this appears to be of no help whatsoever in locating long arithmetic progressions in the primes (one would need to somehow force all the x_i to be equal, which does not seem easily accomplishable).

PROOF. Since A has positive upper density, we can find a $\delta > 0$ and arbitrarily large integers N such that
$$|A \cap [-N, N]| \geq \delta N.$$
Now consider the collection of all differences $x - y$, where x, y are distinct elements of $A \cap [-N, N]$. On the one hand, there are $\delta N(\delta N - 1)$ possible pairs (x, y) that can generate such a difference. On the other hand, these differences range from $-2N$ to $2N$, and thus have at most $4N$ possible values. For N sufficiently large, $\delta N(\delta N - 1) > 4N$, and hence by the pigeonhole principle we can find distinct pairs $(x, y), (x', y')$ with $x, y, x', y' \in A \cap [-N, N]$ and $x - y = x' - y' \neq 0$. This generates a parallelogram. A simple modification of this argument (which we leave to the reader) in fact generates infinitely many such parallelograms. □

The above argument in fact yields a very large number of parallelograms; if $|A \cap [-N, N]| \geq \delta N$, then $A \cap [-N, N]$ in fact contain $\gg \delta^4 N^3$ parallelograms $\{x, x+a, x+b, x+a+b\}$. This should be compared against the total number of parallelograms in $[-N, N]$, which is comparable (up to multiplicative constants) to N^3. Thus the density of parallelograms in $A \cap [-N, N]$ differs only at most by a power of $1/\delta$ from the density of $A \cap [-N, N]$ itself. If arithmetic progressions behaved similarly, one would expect a set A in $[-N, N]$ of density δ to contain $\gg \delta^{C_k} N^2$ arithmetic progressions of a fixed length k. While this is trivially true for $k = 2$, it fails even for $k = 3$:

Proposition 1.3 (Behrend example [2]). *Let $0 < \delta \ll 1$ and $N \geq 1$. Then there exists a subset $A \subset \{1,\ldots,N\}$ of density $|A|/N \gg \delta$ which contains no more than $\delta^{c\log(1/\delta)} N^2$ arithmetic progressions $\{n, n+r, n+2r\}$ of length three, where $c > 0$ is an absolute constant.*

PROOF. The basic idea is to exploit the fact that convex sets in $\mathbb{R}^d$, such as spheres, do not contain arithmetic progressions of length three. The main challenge is then to somehow "embed" $\mathbb{R}^d$ into the interval $\{1,\ldots,N\}$. To do this, let $M, d \geq 1$ be chosen later, and let $\phi\colon \{1,\ldots,N\} \to \{0,\ldots,M-1\}^d$ denote the partial base M map

$$\phi(n) := (\lfloor n/M^i \rfloor \bmod M)_{i=0}^{d-1}$$

where $\lfloor x \rfloor$ is the greatest integer less than x, and $n \bmod M$ is the remainder of n when divided by M. We then pick an integer R between 1 and dM^2 uniformly at random, and let $B_R \subset \{0,\ldots,\lfloor M/10 \rfloor\}^d$ be the set

$$B_R := \{(x_1,\ldots,x_d) \in \{0,\ldots,\lfloor M/10\rfloor\}^d : x_1^2 + \cdots + x_d^2 = R\}$$

and then let $A_R := \phi^{-1}(B_R) \subset \{1,\ldots,N\}$ be the preimage of B_R. The set B_R is contained in a sphere and thus contains no arithmetic progressions of length three, other than the trivial ones $\{x,x,x\}$. Because there is no "carrying" when manipulating base M expansions with digits in $\{0,\ldots,\lfloor M/10\rfloor\}$, we thus conclude that A_R only contains an arithmetic progression $(n, n+r, n+2r)$ when r is a multiple of M^d. This shows that the number of progressions in A_R is at most $O(M^{-d} N^2)$. On the other hand, whenever $\phi(n) \in \{0,\ldots,M/10\}^d$, then n has a probability $1/dM^2$ of lying in A_R. Thus we have a lower bound

$$|A_R| \gg \frac{1}{dM^2} 10^{-d}.$$

If we set $d := c\log(1/\delta)$ and $M := \delta^c$ for some small constants $c > 0$ we obtain the claim. $\square$

This example shows that one cannot hope to prove Szemerédi's theorem by an argument as simple as that used to prove Proposition 1.2, as such simple arguments invariably give polynomial type bounds. Remarkably, this 60-year old bound of Behrend is still the best known (apart from the issue of optimising the constant c).

Another reason why Szemerédi's theorem is difficult is that it already implies the much simpler, but still nontrivial, theorem of van der Waerden:

Theorem 1.4 (van der Waerden's theorem [47]). *Suppose that the integers $\mathbb{Z}$ are partitioned into finitely many colour classes. Then one of the colour classes contains arbitrarily long arithmetic progressions.*

Indeed, from the pigeonhole principle one of the colour classes would have positive density, which by Szemerédi's theorem gives infinitely long progressions. The converse deduction is far more difficult; while certain proofs of Szemerédi's theorem do indeed use van der Waerden's theorem as a component (e.g., [41, 45], and Section 8 below), many more additional arguments are also needed.

While van der Waerden's theorem is not terribly difficult to prove (we give a proof in the next section), it already yields some nontrivial consequences. Here is one simple one:

Proposition 1.5 (Quadratic recurrence). *Let α be a real number and $\varepsilon > 0$. Then one has $\|\alpha r^2\|_{\mathbb{R}/\mathbb{Z}} < \varepsilon$ for infinitely many integers r, where $\|x\|_{\mathbb{R}/\mathbb{Z}}$ denotes the distance from x to the nearest integer.*

PROOF. Partition the unit circle $\mathbb{R}/\mathbb{Z}$ into finitely many intervals I of diameter $\leq \varepsilon/4$. Each interval I induces a colour class $\{n \in \mathbb{N} : \alpha n^2/2 \bmod 1 \in I\}$ on the integers $\mathbb{Z}$. (This is a basic example of a *structured* colouring; we will see the dichotomy between structure and randomness repeatedly in the sequel.) By van der Waerden's theorem, one of these classes contains progressions of length 3 with arbitrarily large spacing r, thus for each such r there is an n for which
$$\alpha n^2/2, \alpha(n+r)^2/2, \alpha(n+2r)^2/2 \in I.$$
The claim now follows from the identity
$$\alpha n^2/2 - 2\alpha(n+r)^2/2 + \alpha(n+2r)^2/2 = \alpha r^2. \qquad \square$$

A modification of the argument lets one also handle higher powers αr^k. More general polynomials (with more than one monomial, but with vanishing constant term) can also be handled, although the argument is more difficult. This simple example already demonstrates however that the number-theoretic question of the distribution of the fractional parts of polynomials is already encoded to some extent within Szemerédi's or van der Waerden's theorem.

Szemerédi's theorem has many further important extensions and generalisations which we will not discuss here (see for instance Bryna Kra's lectures for some of these). Instead, we will focus on two of the main approaches to proving Szemerédi's theorem in its full generality, namely the ergodic theory approach of Furstenberg and the combinatorial approach of Rödl and coauthors, as well as Gowers. We will also sketch in very vague terms the original combinatorial approach of Szemerédi. We will however not discuss the important Fourier-analytic approach, though, despite the many connections between that approach and the ones given here; see Ben Green's lectures for a detailed treatment of the Fourier-analytic method. The combinatorial and ergodic approaches may seem rather different at first glance, but we will try to emphasise the many similarities between them. In particular, both approaches are based around a *structure theorem*, which asserts that a general object (such as a subset A of the integers) can be somehow split into a "structured" component (which has low complexity, is somehow "compact," and has high self-correlation) and a "pseudorandom" component (which has high complexity, is somehow "mixing," and has negligible self-correlation). One then has to manipulate the structured and pseudorandom components in completely different ways to establish the result.

2. Prelude: van der Waerden's theorem

Before we plunge into proofs of Szemerédi's theorem, let us first study the much simpler model case of van der Waerden's theorem. This theorem has both a simple combinatorial proof and a simple dynamical proof; while these proofs do not easily scale up to proving Szemerédi's theorem, the comparison between the two is already illustrative.

We begin with the combinatorial proof. There are three key ideas in the argument (known as a *colour focusing argument*). The first is to induct on the length of the progression. The second is to establish an intermediate type of pattern between a progression of length k and a progression of length $k+1$, which one might call a "polychromatic fan." The third is a concatenation of colours trick in order to leverage the induction hypothesis on progressions of length k, which allows one to move from one fan to the next.

We need some notation. We use $a + [0, k) \cdot r$ to denote the arithmetic progression $a, a+r, \ldots, a+(k-1)r$.

Definition 2.1. Let $\mathbf{c}\colon \{1, \ldots, N\} \to \{1, \ldots, m\}$ be a colouring, let $k \geq 1$, $d \geq 0$, and $a \in \{1, \ldots, N\}$. We define a *fan of radius k, degree d, and base point a* to be a d-tuple $(a + [0, k) \cdot r_1, \ldots, a + [0, k) \cdot r_d)$ of progressions in $\{1, \ldots, N\}$ with $r_1, \ldots, r_d > 0$. We refer to the progressions $a + [1, k) \cdot r_i$, $1 \leq i \leq d$ as the *spokes* of the fan. We say that a fan is *polychromatic* if its base point and its d spokes are all monochromatic with distinct colours. In other words, there exist distinct colours $c_0, c_1, \ldots, c_d \in \{1, \ldots, m\}$ such that $\mathbf{c}(a) = c_0$, and $\mathbf{c}(a + jr_i) = c_i$ for all $1 \leq i \leq d$ and $1 \leq j \leq k$.

Theorem 2.2 (van der Waerden again). *Let $k, m \geq 1$. Then there exists N such that any m-colouring of $\{1, \ldots, N\}$ contains a monochromatic progression of length k.*

It is clear that this implies Theorem 1.4; the converse implication can also be obtained by a simple compactness argument which we leave as an exercise to the reader.

PROOF. We induct on k. The base case $k = 1$ is trivial, so suppose $k \geq 2$ and the claim has already been proven for $k - 1$.

We now claim inductively that for all $d \geq 0$ there exists a positive integer N such that any m-colouring of $\{1, \ldots, N\}$ contains either a monochromatic progression of length k, or a polychromatic fan of radius k and degree d. The base case $d = 0$ is trivial; as soon as we prove the claim for $d = m$ we are done, as it is impossible in an m-colouring for a polychromatic fan to have degree larger than or equal to m.

Assume now that $d > 1$ and the claim has already been proven for $d - 1$. We define $N = 4kN_1N_2$, where N_1 and N_2 are sufficiently large and will be chosen later. Let $\mathbf{c}\colon \{1, \ldots, N\} \to \{1, \ldots, m\}$ be an m-colouring of $\{1, \ldots, N\}$. Then for any $b \in \{1, \ldots, N_2\}$, the set $\{bkN_1 + 1, \ldots, bkN_1 + N_1\}$ is a subset of $\{1, \ldots, N\}$ of cardinality N_1. Applying the inductive hypothesis, we see (if N_1 is large enough) that $\{bkN_1 + 1, \ldots, bkN_1 + N_1\}$ contains either a monochromatic progression of length k, or a polychromatic fan of radius k and degree $d - 1$. If there is at least one b in which the former case applies, we are done, so suppose that the latter case applies for every b. This implies that for every $b \in \{1, \ldots, N_2\}$ there exist $a(b), r_1(b), \ldots, r_{d-1}(b) \in \{1, \ldots, N_1\}$ and distinct colours $c_0(b), \ldots, c_{d-1}(b) \in \{1, \ldots, m\}$ such that $\mathbf{c}(bkN_1 + a(b)) = c_0(b)$ and $\mathbf{c}(bkN_1 + a(b) + jr_i(b)) = c_i(b)$ for all $1 \leq j \leq k - 1$ and $1 \leq i \leq d - 1$. In particular the map $b \mapsto (a(b), r_1(b), \ldots, r_{d-1}(b), c_0(b), \ldots, c_{d-1}(b))$ is a colouring of $\{1, \ldots, N_2\}$ by $m^d N_1^d$ colours (which we may enumerate as $\{1, \ldots, m^d N_1^d\}$ in some arbitrary fashion). Thus (if N_2 is large enough) there exists a monochromatic arithmetic progression $b + [0, k-1) \cdot s$ of length $k - 1$ in $\{1, \ldots, N_2\}$, with some colour $(a, r_1, \ldots, r_{d-1}, c_0, \ldots, c_{d-1})$. We may assume without loss of generality that s is negative since we can simply reverse the progression if s is positive.

Now we use an algebraic trick (similar to Cantor's famous diagonalization trick) which will convert a progression of identical fans into a new fan of one higher degree, the base points of the original fans being used to form the additional spoke of the new fan. Introduce the base point $b_0 := (b-s)kN_1 + a$, which lies in $\{1, \ldots, N\}$ by construction of N, and consider the fan

$$\big(b_0 + [0,k) \cdot skN_1, b_0 + [0,k) \cdot (skN_1 + r_1), \ldots, b_0 + [0,k) \cdot (skN_1 + r_{d-1})\big)$$

of radius k, degree d, and base point b_0. We observe that all the spokes of this fan are monochromatic. For the first spoke this is because

$$\mathbf{c}(b_0 + jskN_1) = \mathbf{c}\big((b + (j-1)s)kN_1 + a\big) = c_0(b + (j-1)s) = c_0$$

for all $1 \leq j \leq k-1$ and for the remaining spokes this is because

$$\mathbf{c}\big(b_0 + j(skN_1 + r_t)\big) = \mathbf{c}\big((b+(j-1)s)kN_1 + a + jr_t\big) = c_t(b+(j-1)s) = c_t$$

for all $1 \leq j \leq k-1$, $1 \leq t \leq d-1$. If the base point b_0 has the same colour as one of the spokes, then we have found a monochromatic progression of length k; if the base point b_0 has distinct colour to all of the spokes, we have found a polychromatic fan of radius k and degree d. In either case we have verified the inductive claim, and the proof is complete. □

Now let us give the dynamical proof. van der Waerden's theorem follows from the following abstract topological statement. Define a *topological dynamical system* to be a pair (X, T) where X is a compact nonempty topological space and $T \colon X \to X$ is a homeomorphism.[1]

Theorem 2.3 (Topological multiple recurrence theorem [15]). *Let (X, T) be a topological dynamical system. Then for any open cover $(V_\alpha)_{\alpha \in A}$ of X and $k \geq 2$, at least one of the sets in the cover contains a subset of the form $T^{[0,k) \cdot r} x := \{x, T^r x, \ldots, T^{(k-1)r} x\}$ for some $x \in X$ and $r > 0$. (We shall refer to such sets as* progressions of length k.)

PROOF OF VAN DER WAERDEN ASSUMING THEOREM 2.3. Let $\mathbf{c} \colon \mathbb{Z} \to \{1, \ldots, m\}$ be an m-colouring of the integers. We can identify $\mathbf{c}$ with a point $x_{\mathbf{c}} := (\mathbf{c}(n))_{n \in \mathbb{Z}}$ in the discrete infinite product space $\{1, \ldots, m\}^{\mathbb{Z}}$. Since each $\{1, \ldots, m\}$ is a compact topological space with the discrete topology, so is $\{1, \ldots, m\}^{\mathbb{Z}}$. The shift operator $T \colon \{1, \ldots, m\}^{\mathbb{Z}} \to \{1, \ldots, m\}^{\mathbb{Z}}$ defined by $T((x_n)_{n \in \mathbb{Z}}) := (x_{n-1})_{n \in \mathbb{Z}}$ is a homeomorphism. Let X be the closure of the orbit $\{T^n x_{\mathbf{c}} : n \in \mathbb{Z}\}$, then X is also compact, and is invariant under T, thus (X, T) is a topological dynamical system. We cover X by the open sets $V_i := \{(x_n)_{n \in \mathbb{Z}} : x_0 = i\}$ for $i = 1, \ldots, m$; by Theorem 2.3, one of these open sets, say V_i, contains a subset of the form $T^{[0,k) \cdot r} x$ for some $x \in X$ and $r > 0$. Since X is the closure of the orbit $\{T^n x_{\mathbf{c}} : n \in \mathbb{Z}\}$, we see from the open-ness of V_i and the continuity of T that V_i must in fact contain a set of the form $T^{[0,k) \cdot r} T^n x_{\mathbf{c}}$. But this implies that the progression $-n - [0,k) \cdot r$ is monochromatic with colour i, and the claim follows. □

Conversely, it is not difficult to deduce Theorem 2.3 from van der Waerden's theorem, so the two are totally equivalent. One can view this equivalence as an instance of a *correspondence principle* between colouring theorems and topological dynamics theorems. By invoking this correspondence principle one leaves the realm

---

[1] As it turns out, T only needs to be a continuous map rather than a homeomorphism, but we retain the homeomorphism property for some minor technical simplifications. It is also common to require X to be a metric space rather than a topological one but this does not make a major difference in the argument.

of number theory and enters the infinitary realm of abstract topology. However, a key advantage of doing this is that we can now manipulate a new object, namely the compact topological space X. Indeed, the proof proceeds by first proving the claim for a particularly simple class of such X, the *minimal* spaces X, and then extending to general X. This strategy can of course also be applied directly on the integers, without appeal to the correspondence principle, but it becomes somewhat less intuitive when doing so (we invite the reader to try it!).

The space X encodes in some sense all the "finite complexity, translation-invariant" information that is contained in the colouring **c**. For instance, if **c** is such that one never sees a red integer immediately after a blue integer, this fact will be picked up in X (which will be disjoint from the set $\{(x_n)_{n \in \mathbb{Z}} : x_0 \text{ blue}, x_1 \text{ red}\}$). The correspondence principle asserts that a colouring theorem can be derived purely by exploiting such information.

Definition 2.4 (Minimal topological dynamical system). A topological dynamical system (X, T) is said to be *minimal* if it does not contain any proper subsystem, i.e., there does not exist $\varnothing \subsetneq Y \subsetneq X$ which is closed with $TY = Y$.

Example 2.5. Consider the torus $X = \mathbb{R}/\mathbb{Z}$ with the doubling map $Tx := 2x$. Then the torus is not minimal, but it contains the minimal system $\{0\}$, the minimal system $\{\frac{1}{3}, \frac{2}{3}\}$, and many other minimal systems. On the other hand, the same torus with an irrational shift $Tx := x + \alpha$ for $\alpha \notin \mathbb{Q}$ is minimal. Minimality can be viewed as somewhat analogous to ergodicity in measure-preserving dynamical systems.

Lemma 2.6. *Every topological dynamical system contains at least one minimal topological dynamical subsystem.*

PROOF. Observe that the intersection of any totally ordered chain of topological dynamical systems is again a topological dynamical system (the nonemptiness of such an intersection follows from the finite intersection property of compact spaces). The claim now follows from Zorn's lemma. □

In light of this lemma, we see that in order to prove Theorem 2.3 it suffices to do so for minimal systems. One advantage of working with minimal systems is the following.

Lemma 2.7. *Let (X, T) be a minimal dynamical system, and let V be a nonempty open subset in X. Then X can be covered by finitely many shifts $T^n V$ of V.*

PROOF. If the shifts $T^n V$ do not cover X, then the complement $X \setminus \bigcup_{n \in \mathbb{Z}} T^n V$ is a proper closed invariant subset of X, contradicting minimality. Thus the $T^n V$ cover X, and the claim follows from compactness. □

Remark 2.8. There is a notion of a *minimal colouring* of the integers that corresponds to a minimal system; informally speaking, a minimal colouring is one that does not "strictly contain" any other colouring, in the sense that the set of finite blocks (i.e., the colour class of a finite consecutive string of integers) of the latter colouring is a proper subset of the set of finite blocks of the former colouring. For instance, the colouring which makes the even numbers red and odd numbers blue is minimal, but if one changes finitely many of these colours then the colouring is no longer minimal. This lemma then asserts that in a minimal colouring, any block that does appear in that colouring, in fact appears *syndetically* (the gaps between each appearance are bounded). Minimal colourings may be considered "maximally structured," in that all the finite blocks that appear in the sequence, appear for a "good reason." The opposite extreme to minimal colourings are *pseudorandom colourings*, in which every finite block of colours appears at least once in the sequence (so X is all of $\{1, \ldots, k\}^{\mathbb{Z}}$).

Now we can prove Theorem 2.3 for minimal dynamical systems. We induct on k. The $k = 1$ case is trivial; now suppose that $k \geq 2$ and the claim has already been proven for $k - 1$, thus given any open cover of X, one of the open sets contains a progression of length $k - 1$. Combining this with Lemma 2.7 (and the trivial observation that the shift of a progression is again a progression), we obtain

Corollary 2.9. *Let (X, T) be a minimal dynamical system, and let V be a nonempty open subset in X. Then V contains a progression of length $k - 1$.*

Now we can build fans again.

Definition 2.10. Let (X, T) be a minimal dynamical system, let $(V_\alpha)_{\alpha \in A}$ be an open cover of X, let $d \geq 0$, and $x \in X$. We define a *fan of radius k, degree d, and base point x* to be a d-tuple $(T^{[0,k) \cdot r_1} x, \ldots, T^{[0,k) \cdot r_d} x)$ of progressions of length k with $r_1, \ldots, r_d > 0$, and refer to the progressions $a + [1, k) \cdot r_i$, $1 \leq i \leq d$ as the *spokes* of the fan. We say that a fan is *polychromatic* if its base point and its d spokes each lie in a distinct element of the cover. In other words, there exist distinct $\alpha_0, \ldots, \alpha_d \in A$ such that $x \in A_{\alpha_0}$ and $T^{jr_i} x \in A_{\alpha_i}$ for all $1 \leq i \leq d$ and $1 \leq j \leq k$.

To prove Theorem 2.3 it now suffices to show

Proposition 2.11. *Let (X, T) be a minimal dynamical system, and let $(V_\alpha)_{\alpha \in A}$ be an open cover of X. Then for any $d \geq 0$ either there exists at least one polychromatic fan of radius k and degree d, or at least one of the sets in the open cover contains a progression of length k.*

Indeed, by compactness we can make the open cover finite, and the above proposition leads to the desired result by taking d large enough.

PROOF. The base case $d = 0$ is trivial. Assume now that $d \geq 1$ and the claim has already been proven for $d - 1$. If one of the V_α contains a progression of length k we are done, so we may assume that we have found a polychromatic fan $(T^{[0,k) \cdot r_1} x, \ldots, T^{[0,k) \cdot r_{d-1}} x)$ of degree $d - 1$, thus there exist distinct $\alpha_0, \ldots, \alpha_{d-1} \in A$ such that $x \in A_{\alpha_0}$ and $T^{jr_i} x \in A_{\alpha_i}$ for all $1 \leq i \leq d - 1$ and $1 \leq j \leq k$. Since the T^{jr_i} are continuous, we can thus find a neighbourhood V of x in A_{α_0} such that $T^{jr_i} V \subset A_{\alpha_i}$ for all $1 \leq i \leq d - 1$ and $1 \leq j \leq k$. By Corollary 2.9 V contains a progression of length $k - 1$, say $T^{[1,k) \cdot r_0} y$. Thus we see that $T^{jr_0} y \in A_{\alpha_0}$ for $1 \leq j \leq k$, and $T^{j(r_0 + r_i)} y \in A_{\alpha_i}$ for $1 \leq j \leq k$ and $1 \leq i \leq d - 1$. The point y itself lies in an open set A_α. If α equals one of the $\alpha_0, \alpha_1, \ldots, \alpha_d$, then V_α contains a progression of length k; if α is distinct from $\alpha_0, \alpha_1, \ldots, \alpha_{d-1}$, we have a polychromatic fan of degree d. The claim follows. □

As one can see, the topological dynamics proof contains the same core arithmetical ideas as the combinatorial proof (namely, that a progression of fans can be converted to either a longer progression, or a fan of one higher degree) but the argument is somewhat cleaner as one does not have to keep track of superfluous parameters such as N. For the particular purpose of proving van der Waerden's theorem, the additional overhead in the dynamical proof makes the total argument longer than the combinatorial proof, but for more complicated colouring theorems the dynamical proofs tend to eventually be somewhat shorter and conceptually clearer than the combinatorial proofs, which are often burdened with substantial notation. The dynamical proofs seem to rely quite heavily on infinitary tools such as Tychonoff's theorem and Zorn's lemma, though one can reduce the dependence on these tools by making the argument more "quantitative" (of course, if one removes the infinitary framework completely, one ultimately ends up at an argument which is more or less just some reworking of the combinatorial argument).

3. Shelah's argument

Let us now present another proof of van der Waerden's theorem, due to Shelah [39]; it gives slightly better bounds by avoiding inductive arguments which massively increase the number of colours in play. This argument in fact proves a much stronger theorem, namely the *Hales-Jewett theorem*, but we shall content ourselves with a slightly less general result in order to avoid a certain amount of notation.

Definition 3.1 (Cubes). A *cube* of dimension d and length k is any set of integers of the form
$$a + [0,k)^d \cdot v = \{a + n_1 v_1 + \cdots + n_d v_d : 0 \leq n_1, \ldots, n_d \leq k\}$$
where $a \in \mathbb{Z}$ and $v = (v_1, \ldots, v_d)$ is a d-tuple of positive integers, with the property that all the elements $a + n_1 v_1 + \cdots + n_d v_d$ are distinct.

Cubes are a special case of *generalised arithmetic progressions*, which play an important role in this subject.

Theorem 3.2 (Hales–Jewett theorem [26]). *Let Q be a cube of dimension d and length k which is coloured into m colour classes. If $j \geq 1$, and d is sufficiently large depending on k, m, j, then Q contains a monochromatic subcube Q' of dimension j and length k.*

Note that the interval $\{1, \ldots, k^d\}$ can be viewed as a proper cube of dimension d and length k. As such, we see that the van der Waerden theorem follows from the $j = 1$ case of the Hales–Jewett theorem. (The original proof of this theorem proceeded by a colour focusing argument that directly generalised that used to prove van der Waerden's theorem, and we leave it as an exercise.)

Shelah's proof of this theorem proceeds by an induction on the length k. The $k = 1$ case is trivial, so suppose that $k \geq 1$ and that the theorem has already been proven for $k - 1$. Let us call a subcube
(3.1) $$Q' = \{a + n_1 v_1 + \cdots + n_d v_d : 0 \leq n_1, \ldots, n_d \leq k\}$$
of Q *weakly monochromatic* if whenever one of the $n_1, \ldots, n_d$ is swapped from $k-1$ to k or vice versa, the colour of the element of Q' is unchanged. It will suffice to show

Theorem 3.3 (Hales–Jewett theorem, first inductive step). *Let Q be a cube of dimension d and length k which is coloured into m colour classes. If $j \geq 1$, and d is sufficiently large depending on k, m, j, then Q contains a weakly monochromatic subcube Q' of dimension j and length k.*

To prove Theorem 3.2, one may first without loss of generality "stretch" the cube Q by making each v_i enormously large compared with the previous v_{i-1}. This allows us to eliminate certain "exotic" subcubes which would cause some technicalities later on. Then, we let J be a large integer depending on k, m, j to be chosen later. If d is large enough depending on k, m, J, then by Theorem 3.3 we can find a weakly monochromatic subcube Q' of Q of dimension J and length k. We contract each of the edges by 1 (deleting all the vertices where one of the n_i is equal to k) to create a subcube Q'' of Q of dimension J and length $k-1$. By the induction hypothesis, we see that if J is large enough then Q'' will in turn contain a monochromatic cube Q''' of dimension j and length $k-1$. Since Q' was weakly monochromatic, one can verify that Q''' extends back to a monochromatic cube Q'''' of dimension j and length k, which is contained in Q, and the claim follows.

It remains to prove Theorem 3.3. Let us modify the notion of weakly monochromatic somewhat. Let us call the subcube (3.1) *i-weakly monochromatic* for some $0 \leq i \leq d$ if whenever one of the $n_1, \ldots, n_i$ is swapped from $k-1$ to k or vice versa, the colour of the element of Q' is unchanged. It will suffice to show

Theorem 3.4 (Hales–Jewett theorem, second inductive step). *Let Q be a cube of dimension d and length k which is coloured into m colour classes which is already i-weakly monochromatic for some $i \geq 0$. If $j \geq i+1$, and d is sufficiently large depending on k, m, j, i, then Q contains a $i+1$-weakly monochromatic subcube Q' of dimension j and length k.*

Indeed, by iterating Theorem 3.4 in i we see that for d large enough depending on k, m, j, i, Q will contain an i-weakly monochromatic subcube of dimension j and length k (the case $i = 0$ is trivial); setting $i = j$ we obtain Theorem 3.3.

It remains to prove Theorem 3.4. As a warmup (and because we need the result to prove the general case) let us first give a simple special case of this theorem.

Lemma 3.5 (Hales–Jewett theorem, trivial case). *Let Q be a cube of dimension d and length k which is coloured into m colour classes. If $d \geq m+1$, then Q contains a 1-weakly monochromatic subcube Q' of dimension 1 and length k.*

PROOF. Write
$$Q = \{a + n_1 v_1 + \cdots + n_d v_d : 0 \leq n_1, \ldots, n_d \leq k\}$$
and consider the $m+1$ elements of Q of the form
$$a + (k-1)v_1 + \cdots + (k-1)v_s + kv_{s+1} + \cdots + kv_{m+1}$$
where s ranges from 1 to $m+1$. By the pigeonhole principle two of these have the same colour, thus we have $1 \leq s < s' \leq m+1$ such that the (1-dimensional, length k) subcube
$$\{a + (k-1)v_1 + \cdots + (k-1)v_s + n(v_{s+1} + \cdots + v_{s'}) + kv_{s'+1} \cdots + kv_{m+1} : 1 \leq n \leq k\}$$
is 1-weakly monochromatic, and the claim follows. □

Now we can prove Theorem 3.4 and hence the Hales–Jewett theorem. The main idea is to recast the cube Q, not as an i-weakly monochromatic m-coloured cube of dimension d and length k, but rather as an $m^{k^{j-1}}$-coloured cube of dimension $d - j + 1$ and length k. More precisely, let us write
$$Q = \{a + n_1 v_1 + \cdots + n_d v_d : 0 \leq n_1, \ldots, n_d \leq k\}$$

and consider now the modified cube of dimension $d - j + 1$ and length k

$$\widetilde{Q} := \{a + n_j v_j + \cdots + n_d v_d : 0 \leq n_j, \ldots, n_d \leq k\}.$$

Note that each element $x \in \widetilde{Q}$ is associated to k^{j-1} elements of Q, namely

$$\{x + n_1 v_1 + \cdots + n_{j-1} v_{j-1}\}.$$

Each of these elements has m colours, and so we can naturally associate an $m^{k^{j-1}}$-colouring of $\widetilde{Q}$. If d (and hence $d-j+1$) is large enough, we can apply Theorem 3.5 and find a 1-weakly monochromatic subcube $\widetilde{Q}'$ of dimension 1 and length k in $\widetilde{Q}$. It is easy to verify that this in turn induces a $i+1$-weakly monochromatic subcube Q' of dimension j and length k in Q, and we are done.

4. The Furstenberg correspondence principle

In a previous section, we saw how van der Waerden's theorem was shown to be equivalent to a recurrence theorem in topological dynamics. Similarly, Szemerédi's theorem is equivalent to a recurrence theorem in measure-preserving dynamics.

Definition 4.1. A *measure-preserving system* $(X, \mathcal{B}, \mu, T)$, is a probability space $(X, \mathcal{B}, \mu)$, where $\mathcal{B}$ is a σ-algebra of events on X, $\mu \colon \mathcal{B} \to [0,1]$ is a probability measure (thus μ is countably additive with $\mu(X) = 1$), and the *shift map* $T \colon X \to X$ is a bijection which is bi-measurable (thus $T^n \colon \mathcal{B} \to \mathcal{B}$ for all $n \in \mathbb{Z}$) and probability preserving (thus $\mu(T^n E) = \mu(E)$ for all $E \in \mathcal{B}$ and $n \in \mathbb{Z}$).

Example 4.2 (Circle shift). Take X to be the circle $\mathbb{R}/\mathbb{Z}$ with the Borel σ-algebra $\mathcal{B}$, the uniform probability measure μ, and the shift $T \colon x \mapsto x + \alpha$ where $\alpha \in \mathbb{R}$. Thus $T^n E = E + n\alpha$ for any $E \in \mathcal{B}$. This system is to recurrence theorems as *quasiperiodic sets*, such as the *Bohr set* $\{n \in \mathbb{Z} : \|n\alpha\|_{\mathbb{R}/\mathbb{Z}} \leq \theta\}$, is to Szemerédi's theorem — it is an extreme example of a *structured set*.

Example 4.3 (Finite systems). Take X to be a finite set, and let $\mathcal{B}$ be the σ-algebra generated by some partition $X = A_1 \cup \cdots \cup A_n$ of X into nonempty sets $A_1, \ldots, A_n$ (these sets are known as "atoms"). Thus a set is measurable in $\mathcal{B}$ if and only if it is the finite union of atoms. We take μ to be the uniform measure, thus $\mu(E) := |E|/|X|$ for all $E \in \mathcal{B}$. The shift map $T \colon X \to X$ is then a permutation on X, with the property that it maps atoms to atoms. Note that if two atoms have different sizes, it will be impossible for the shift map (or any power of the shift map) to take one to the other. If one assumes that the shift map is ergodic (we will define this later), this forces all the atoms to have the same size. The finite case is not the case of interest in recurrence theorems, but it does serve as a useful toy model that illustrates many of the basic concepts in the proofs without many of the technicalities. Finite systems have a counterpart in Szemerédi's theorem as *periodic sets* — which are trivial for the purpose of demonstrating existence of arithmetic progressions, but still serve as an important illustrative special case for certain components of the proof of Szemerédi's theorem.

Remark 4.4. The shift T induces an action $n \mapsto T^n$ of the additive integer group $\mathbb{Z}$ on X. One can also study actions of other groups; for instance, actions of $\mathbb{Z}^2$ are described by a pair S, T of commuting bimeasurable probability preserving transformations.

Given any measure-preserving system $(X, \mathcal{B}, \mu, T)$, a set E, and a point $x \in X$, we can define the recurrence set $A = A_{x,E} \subset \mathbb{Z}$ of integers by the formula

(4.1) $$A_{x,E} := \{n \in \mathbb{Z} : T^n x \in E\}.$$

This is a way of identifying sets E in a system with sets A in the integers. Similarly, given a function $f : X \to \mathbb{R}$ on the system, and an $x \in X$, we can define an associated sequence $F = F_{x,f} : \mathbb{Z} \to \mathbb{R}$ by the formula

(4.2) $$F_{x,f}(n) := f(T^n x).$$

This correspondence between sets and functions on the system, and sets and functions on the integers, underlies the Furstenberg correspondence principle. In particular, it allows one to equate Szemerédi's theorem—which is a theorem on the integers—to the following theorem on measure-preserving systems.

Theorem 4.5 (Furstenberg multiple recurrence theorem [11]). *Let $(X, \mathcal{B}, \mu, T)$ be a measure-preserving system. Then for any set $E \in \mathcal{B}$ of positive measure $\mu(E) > 0$ and any $k \geq 1$, we have*

$$\liminf_{N \to \infty} \mathbb{E}_{1 \leq r \leq N} \mu(E \cap T^r E \cap \cdots \cap T^{(k-1)r} E) > 0$$

where we use the averaging notation $\mathbb{E}_{1 \leq r \leq N} f(r) := (1/N) \sum_{r=1}^{N} f(r)$.

Remark 4.6. The $k = 1$ case is trivial. The $k = 2$ case follows easily from the pigeonhole principle and is known as the *Poincaré recurrence theorem*. The $k = 3$ case can be handled by spectral theory (i.e., Fourier analysis). However the general k case is significantly harder. It is known that the limit inferior on the left can actually be replaced by a limit (i.e., the limit exists), but this is even harder to establish (see Bryna Kra's lectures).

As one consequence of this theorem, we see that every set in X of positive measure contains arbitrarily long progressions. This should be contrasted with Theorem 2.3, which can easily be shown to be a special case of Theorem 4.5.

The *Furstenberg correspondence principle* asserts an equivalence between results such as Szemerédi's theorem in combinatorial number theory, and recurrence theorems in ergodic theory. Let us first show how the recurrence theorem implies Szemerédi's theorem.

PROOF OF SZEMERÉDI'S THEOREM ASSUMING THEOREM 4.5. This shall be analogous to the topological correspondence principle, in which we shifted the colouring function **c** around and took closures to create the dynamical system $X \subset \{1, \ldots, m\}^{\mathbb{Z}}$. This time we shift a set A around and take weak limits to create the measure-preserving system $X \subset \{0, 1\}^{\mathbb{Z}}$. One can view this as "inverting" the correspondence (4.1); whereas (4.1) starts with a set in a system and turns it into a set of integers, here we need to do things the other way around.

More precisely, suppose for contradiction that Szemerédi's theorem fails. Then there exists a $k \geq 1$, a set $A \subset \mathbb{Z}$ without progressions of length k, and a sequence N_i of integers going to infinity such that $\liminf_{i \to \infty} |A \cap [-N_i, N_i]|/(2N_i + 1) > 0$. Now for each i, consider the random set

$$A_i := A + x_i$$

where x_i is an integer chosen at random from $[-N_i, N_i]$. As the subsets of $\mathbb{Z}$ can be identified with elements of $X := \{0, 1\}^{\mathbb{Z}}$, we can think of A_i as a random variable taking values in X. More precisely, if we let $\mathcal{B}$ be the Borel σ-algebra of X, we can identify A_i with a probability measure μ_i on X (it is the average of $2N_i + 1$ Dirac masses). Now X is a separable compact Hausdorff space, and so the probability measures are weakly sequentially compact. This means that (after passing to a subsequence of i if necessary), the μ_i converge to another probability measure μ in the weak sense, thus

$$\lim_{i \to \infty} \int_X f \, d\mu_i = \int_X f \, d\mu$$

for any continuous function f on X. In particular, if we let[2] $E := \{(x_n)_{n \in \mathbb{Z}} \in \{0,1\}^{\mathbb{Z}} : x_n = 1\}$, then since E is both open and closed,

$$\lim_{i \to \infty} \mu_i(E) = \mu(E).$$

But a computation shows

$$\mu_i(E) = \frac{|A \cap [-N_i, N_i]|}{2N_i + 1}$$

and hence $\mu(E) > 0$. Similarly, if $T \colon X \to X$ is the shift operator $T(x_n)_{n \in \mathbb{Z}} := (x_{n-1})_{n \in \mathbb{Z}}$, then a brief computation shows that

$$\lim_{i \to \infty} \mu_i(TE) - \mu_i(E) = 0$$

and more generally

$$\lim_{i \to \infty} \mu_i(TF) - \mu_i(F) = 0$$

whenever F is a finite Boolean combination of E and its shifts. This means that

$$\mu(TF) = \mu(F)$$

for all such F, and then by the Kolmogorov extension theorem we see that μ is in fact shift-invariant. Finally, since A contains no arithmetic progressions of length k, we see that

$$\mu_i(E \cap T^r E \cap \cdots \cap T^{(k-1)r} E) = 0$$

for any $r > 0$, and hence on taking limits

$$\mu(E \cap T^r E \cap \cdots \cap T^{(k-1)r} E) = 0.$$

These facts together contradict the Furstenberg recurrence theorem, and we are done. $\square$

One can easily show that the Szemerédi theorem and the Furstenberg recurrence theorem are equivalent to slightly stronger versions of themselves. For instance, Furstenberg's multiple recurrence theorem generalises to

Theorem 4.7 (Furstenberg multiple recurrence theorem, again). *Let $(X, \mathcal{B}, \mu, T)$ be a measure-preserving system. Then for any bounded measurable function $f \colon X \to [0,1]$ with $\int_X f \, d\mu > 0$ and any $k \geq 1$, we have*

(4.3) $$\liminf_{N \to \infty} \mathbb{E}_{1 \leq r \leq N} \int_X f T^r f \cdots T^{(k-1)r} f \, d\mu > 0$$

where $T^r f := f \circ T^{-r}$ is the translation of f by r.

---

[2]This is the correct choice of E if one wants to invert the equivalence (4.1). Indeed, identifying A with a point in X, we see that $A = E_{A,E}$, $A_i = E_{A_i,E}$, and so forth.

This follows simply because if $\int_X f \, d\mu > 0$, then we have the pointwise bound $f \geq c 1_E$ for some $c > 0$ and some set E of positive measure, where 1_E is the indicator function of E. In a similar spirit, Szemerédi's theorem has the following quantitative formulation:

Theorem 4.8 (Szemerédi's theorem, again). *Let $\mathbb{Z}/N\mathbb{Z}$ is a cyclic group. Then for any bounded function $f : \mathbb{Z}/N\mathbb{Z} \to [0,1]$ with $\mathbb{E}_{n \in \mathbb{Z}/N\mathbb{Z}} f(n) \geq \delta > 0$ and any $k \geq 1$, we have*

$$\mathbb{E}_{n,r \in \mathbb{Z}/N\mathbb{Z}} f(n) T^r f(n) \cdots T^{(k-1)r} f(n) \geq c(k, \delta)$$

for some $c(k, \delta) > 0$ which is independent of N, where $T^r f(n) := f(n-r)$.

It is easy to see that Theorem 4.8 implies Szemerédi's theorem in its original formulation, and it can also be easily used (by using the correspondence (4.2) between functions and sequences) to prove Theorem 4.7 or Theorem 4.5 (in fact it gives a lower bound on (4.3) which depends only on k and the mean $\int_X f \, d\mu$ of f). The converse implication requires an additional averaging argument which is essentially due to Varnavides [48]. We present it here:

PROOF OF THEOREM 4.8 ASSUMING SZEMERÉDI'S THEOREM. First we observe that for any $k \geq 1$ and $\delta > 0$ that there exists an $M = M(\delta)$ such that any subset of $[1, M]$ of density at least δ contains at least one progression of length k. For if this were not the case, then one could find arbitrarily large N and sets $A_M \subset [1, M]$ with $|A_M| \geq \delta M$ which contained no progressions of length k. Taking unions of translates of such sets (with M a rapidly increasing sequence) one can easily find a counterexample to Szemerédi's theorem.

Now we prove Theorem 4.8. It is easy to see that $f(n) \geq \delta/2$ on a set $A \subset \mathbb{Z}/N\mathbb{Z}$ of density at least $\delta/2$. Thus it will suffice to show that

$$\mathbb{E}_{n,r \in \mathbb{Z}/N\mathbb{Z}} 1_{n, n+r, \ldots, n+(k-1)r \in A} \gg_{k,\delta} 1.$$

For N small depending on k, δ this is clear (just from taking the $r = 0$ case) so assume N is large. Let $1 \leq M < N$ be chosen later. It will suffice to show that

$$\mathbb{E}_{n \in \mathbb{Z}/N\mathbb{Z}} \mathbb{E}_{1 \leq r \leq M} 1_{n, n+\lambda r, \ldots, n+(k-1)\lambda r \in A} \gg_{k,\delta} 1$$

for all $\lambda \in \mathbb{Z}/N\mathbb{Z}$, as the claim then follows by averaging in λ. We rewrite this as

$$\mathbb{E}_{n \in \mathbb{Z}/N\mathbb{Z}} \mathbb{E}_{1 \leq m, r \leq M} 1_{n+m, n+m+\lambda r, \ldots, n+m+(k-1)\lambda r \in A} \gg_{k,\delta} 1$$

On the other hand, we have

$$\mathbb{E}_{n \in \mathbb{Z}/N\mathbb{Z}} \mathbb{E}_{1 \leq m \leq M} 1_{n+\lambda m \in A} = |A|/N \geq \delta$$

so we have $\mathbb{E}_{1 \leq m \leq M} 1_{n+\lambda m \in A} \geq \delta/2$ for a set of n of density at least $\delta/2$. For each such n, the set $\{1 \leq m \leq M : n + \lambda m \in A\}$ has density at least $\delta/2$, and so if we choose $M = M(\delta/2)$ we have at least one $1 \leq m, r \leq M$ for which $n + m, n + m + \lambda r, \ldots, n + m + (k-1)\lambda r \in A$, and so

$$\mathbb{E}_{1 \leq m, r \leq M} 1_{n+m, n+m+\lambda r, \ldots, n+m+(k-1)\lambda r \in A} \gg_M 1.$$

Since M depends on k, δ, the claim follows. $\square$

Remark 4.9. One can also deduce Theorem 4.8 directly from Theorem 4.5 by modifying the derivation of Szemerédi's theorem from Theorem 4.5. We sketch the ideas briefly here. One can replace f by a set A in $\mathbb{Z}/N\mathbb{Z}$. One then randomly translates and dilates the function A on $\mathbb{Z}/N\mathbb{Z}$ and then lifts up to $\mathbb{Z}$ to create a random set A in $\mathbb{Z}$. Now one argues as before. See [46] for a detailed argument. See also [4] for further exploration of uniform lower bounds in the Furstenberg recurrence theorem.

5. Some ergodic theory

We will not prove Theorem 4.5 or Theorem 4.7 here; see Bryna Kra's lectures for a detailed treatment of this theory. However we can illustrate some of the key concepts here. For those readers which are more comfortable with finite mathematical structures, a good model of a measure-preserving system to keep in mind here is that of the cyclic shift, where $X = \mathbb{Z}/N\mathbb{Z}$, $\mathcal{B} = 2^X$ is the power set of X (so the atoms are just singleton sets) and $T : n \mapsto n + 1$ is the standard shift. Other finite systems of course exist (though any such system is ultimately equivalent to the disjoint union of finitely many such cyclic shifts).

The basic ergodic theory strategy in proving Theorem 4.7 is to first prove this result for very *structured* types of functions—functions which have a lot of self-correlation between their shifts. As it turns out, this is equivalent to studying very structured *factors* $\mathcal{B}'$ of the σ-algebra $\mathcal{B}$. One then extends the recurrence result from simple factors to more complicated extensions of these factors, continuing in this process (using Zorn's lemma if necessary) until the full σ-algebra is recovered (and so all functions are treated). This is a more complicated version of the topological dynamical situation, in which there was only one type of structured system, namely a minimal system, and the extension from minimal systems to arbitrary systems was trivial (after using Zorn's lemma).

In addition to structured functions, there will also be "anti-structured" or "mixing" functions which can be considered orthogonal to the structured functions. These can be viewed as functions for which there is absolutely no correlation between certain of their shifts. To oversimplify dramatically, one could make the following vague definitions for any $k \geq 2$:

- A function f is *mixing of order $k - 2$* if there is no correlation between the shifts $f, T^n f, \ldots, T^{(k-1)n} f$ for generic n.
- A (possibly vector-valued) function f is *strongly structured of order $k - 2$* if knowledge of $f, T^n f, \ldots, T^{(k-2)n} f$ can be used to predict $T^{(k-1)n} f$ perfectly and "continuously."
- A function f is *structured of order $k - 2$* if it is a component of a strongly structured function of order $k - 2$, or can be approximated to arbitrary accuracy by finite linear combinations of such components.

These definitions can be formalised, for instance using the Gowers–Host–Kra seminorms; see the lectures of Ben Green and Bryna Kra. We will not do so here. However we shall gradually develop some key examples of these concepts in this section. A fundamental observation in the subject is that there is a *structure theorem* that (for any $k \geq 2$) decomposes any function uniquely into a structured component of order $k - 2$ and a mixing component of order $k - 2$; indeed, the structured components end up being precisely those functions which are measurable with respect to a special factor Y_{k-2} of $\mathcal{B}$, known as the *characteristic factor* for k-term recurrence.[3] To prove the Furstenberg recurrence theorem, one first proves

---

[3]We are oversimplifying a lot here, there are some subtleties in precisely how to define this factor; in particular the factor Z_{k-2} constructed by Host and Kra [27] differs slightly from a

recurrence for structured functions of order d for any d (by induction on d), and then shows weakly mixing functions of order $k-2$ are negligible for the purpose of establishing k-term recurrence. Setting $d = k-2$ and applying the structure theorem, one obtains the general case.

These matters will be treated in more detail in Bryna Kra's lectures. Here we shall give only some extremely simple special cases, to build up some intuition. There will be a distinct lack of rigour in this section; for instance, we shall omit certain proofs, and be cavalier about whether a function is bounded or merely square integrable, whether a limit actually exists, etc.

We now consider various classes of functions $f \colon X \to \mathbb{R}$; occasionally we will take f to be complex-valued or vector-valued instead of real-valued. All functions shall be bounded.

The most structured type of functions f are the *invariant* functions, for which $Tf = f$ (up to sets of measure zero, of course). These can be viewed as "(strongly) structured functions of order 0." It is trivial to verify the Furstenberg recurrence theorem for such functions. It is also clear that these (bounded) functions f form a von Neumann algebra,[4] as the space $L^\infty(X)^T$ of bounded invariant functions is closed under uniform limits and algebraic operations. Because of this, we can associate a *factor* Y_0 to these functions, defined as the least σ-algebra with respect to which all functions in $L^\infty(X)^T$ are measurable; because $L^\infty(X)^T$ was a von Neumann algebra, we see that $L^\infty(X)^T$ is in fact *precisely* those functions which are Y_0-measurable. In other words, we take level sets $f^{-1}([a,b])$ of invariant functions and use this to generate the σ-algebra. One can equivalently write Y_0 as the space of essentially invariant sets E, thus TE is equal to E outside of a set of measure zero. For instance, in the finite case Y_0 consists of all sets that are unions of orbits of T; in the cyclic case $X = \mathbb{Z}/N\mathbb{Z}$, $Tx = x + n$, Y_0 consists of all sets that are cosets of the subgroup generated by n (so if n is coprime to N, the only sets in Y_0 are the empty set and the whole set). In the case of the circle shift $X = \mathbb{R}/\mathbb{Z}$, $Tx = x + \alpha$, Y_0 is trivial when α is irrational but contains proper subsets of $\mathbb{R}/\mathbb{Z}$ when α is rational.

Complementary to the invariant functions are the *anti-invariant* functions, which are orthogonal to all invariant functions; these are the "mixing functions of order 0." For instance, given any $g \in L^\infty(X)$, the function $Tg - g$ is an anti-invariant function. In fact, all invariant functions can be approximated to arbitrary accuracy in $L^2(X)$ as linear combinations of such basic anti-invariant functions $Tg - g$. This is because if this were not the case, then by the Hahn–Banach theorem there would exist a noninvariant function f which was orthogonal to all of the $Tg - g$. But then f would be orthogonal to $Tf - f$, which after some manipulation implies that $Tf - f$ has L^2 norm zero and so f is invariant, contradiction. Because of this fact, we see that anti-invariant functions go to zero in the L^2 sense:

(5.1) $\qquad f \perp L^\infty(X)^T \implies \mathbb{E}_{1 \leq r \leq N} T^r f \to_{L^2(X)} 0 \quad \text{as } N \to \infty.$

---

similar factor Y_{k-2} constructed by Ziegler [50] because a slightly different (but closely related) type of averaging is considered, using $k - 1$-dimensional cubes instead of length k progressions. See [30] for a comparison of the two factors.

[4] It seems clear that the theory of von Neumann algebras is somehow lurking in the background of all of this theory, though strangely enough it does not play a prominent role in the current results. An interesting question is to investigate to what extent this theory would survive if $L^\infty(X)$ was replaced by a *noncommutative* von Neumann algebra.

This can be seen by first testing on basic anti-invariant functions $Tg - g$ (in which case one has a telescoping sum), taking linear combinations, and then taking limits. One specific consequence of this is the mixing property

$$\lim_{N \to \infty} \mathbb{E}_{1 \leq r \leq N} \int_X fT^r g \, d\mu = 0 \tag{5.2}$$

whenever *at least one* of f and g is anti-invariant. (Note that there is a symmetry due to the identity $\int_X fT^r g = \int_X gT^{-r}f$.) We will refer to this as the *generalised von Neumann theorem of order 0*.

From Hilbert space theory we know that every function f in $L^2(X)$ uniquely splits as the sum of an invariant function $f_{U^\perp}$ and an anti-invariant f_U function. In fact, since the invariant functions are not only a closed subspace of $L^2(X)$, but are also the measurable functions with respect to a factor Y_0, we can write explicitly $f_{U^\perp} = \mathbb{E}(f \mid Y_0)$ and $f_U = f - \mathbb{E}(f \mid Y_0)$, where the *conditional expectation operator* $f \mapsto \mathbb{E}(f \mid Y_0)$ is simply the orthogonal projection from $L^2(X)$ to the subspace $L^2(Y_0)$ of Y_0-measurable functions.

If f is invariant, then clearly its averages converge back to f:

$$f \in L^\infty(X)^T \implies \mathbb{E}_{1 \leq r \leq N} T^r f \to_{L^2(X)} f \quad \text{as } N \to \infty.$$

Combining this with (5.1) (and taking limits to extend L^∞ to L^2) we obtain the *von Neumann ergodic theorem*

$$f \in L^2(X) \implies \mathbb{E}_{1 \leq r \leq N} T^r f \to_{L^2(X)} \mathbb{E}(f \mid Y_0) \quad \text{as } N \to \infty.$$

This implies in particular that

$$\mathbb{E}_{1 \leq n \leq N} \int_X fT^n f \, d\mu \to \int_X f \mathbb{E}(f \mid Y_0) \, d\mu = \|\mathbb{E}(f \mid Y_0)\|_{L^2(Y_0)}^2$$

which already proves the $k = 2$ case of the Furstenberg recurrence theorem (and gives a precise value for the limit).

Example 5.1. Consider the case of finite systems. Then the invariant functions are those functions which are constant of each of the orbits of T, while the anti-invariant functions are those functions which have mean zero on each of the orbits of T. If $f : X \to \mathbb{R}$ is a general function, then the invariant part $\mathbb{E}(f \mid Y_0)$ is the function which assigns to each orbit of T (i.e., to each atom of Y_0) the average value of f on that orbit, while the anti-invariant part $f - \mathbb{E}(f \mid Y_0)$ is formed by subtracting the mean of each orbit from the original function. It is an instructive exercise to verify all the arguments used to prove the von Neumann ergodic theorem directly in this finite system case.

The factor Y_0 also leads to a useful *ergodic decomposition* of a general measure-preserving system into ergodic ones. A measure preserving system is said to be *ergodic* if Y_0 is trivial, thus every invariant set has measure either zero or one (or equivalently that every invariant function is constant almost everywhere). One can view the space X and the σ-algebra $\mathcal{B}$ as fixed, in which case ergodicity is a property of the shift-invariant probability measure μ. Then it turns out that while a general measure μ is not ergodic, it can always be decomposed (or *disintegrated*) as an integral $\int_Y \mu_y \, d\nu(y)$ of ergodic shift-invariant probability measures μ_x parameterised by some parameter y on another probability space Y. To formalise this decomposition in general requires a certain amount of measure theory, but in the case of a finite system the process is quite simple to describe. Namely, take Y to be the system (X, Y_0, μ), and for each $y \in Y$ let μ_y be the uniform distribution on the T-orbit $\{T^n y : n \in \mathbb{Z}\}$ of y. Then one easily verifies that $\mu = \int_Y \mu_y \, d\nu(y)$, and that each μ_y is an ergodic measure (all invariant sets either have zero measure or full measure). The ergodic decomposition in this case is essentially just the decomposition of X into individual orbits of T, upon each of which T is ergodic. One can easily use the ergodic decomposition to reduce the task of proving Furstenberg's recurrence theorem to the special case in which the system is ergodic; we omit the details. This is somewhat analogous to the reduction in topological dynamics to minimal systems. Unfortunately, whereas in the dynamical case the assumption of minimality was very strong and lead quickly to a proof of the topological recurrence theorem, ergodicity is not by itself a strong enough condition to quickly obtain a direct proof of Furstenberg's recurrence theorem, and further classification and decomposition of the measure-preserving system is needed. As it turns out, one usually cannot usefully disintegrate the measure μ into any smaller invariant measures once one is at an ergodic system; however it is still possible (and useful) to disintegrate the measures into *noninvariant* measures, where the shift map does not act separately on each component, but instead mixes them together using something called a "cocycle." A simple finitary example occurs when considering a finite ergodic system $(X, \mathcal{B}, \mu, T)$ with $\mathcal{B} = 2^X$ which contains a shift-invariant factor $\mathcal{B}' \subset \mathcal{B}$. The ergodicity forces all the atoms in $\mathcal{B}'$ to be the same size, and thus they are all bijective (noncanonically) to a single set Z. This allows one to then parameterise X as $Y \times Z$, where Y is the collection of all the atoms of $\mathcal{B}'$; since the shift T maps one such atom to another, the factor $(X, \mathcal{B}', \mu, T)$ is then equivalent to a system $(Y, 2^Y, \nu, S)$ on Y where ν is the uniform measure on Y, and the original shift can then be described as $T(y, z) := (Sy, \rho_y(z))$ where for each $y \in Y$, the *cocycle* $\rho_y : Z \to Z$ is a permutation on Z. One can view X as an *extension* of Y, by converting each point y to a "vertical fiber" $y \times Z$. We can disintegrate $\mu = \int_Y \mu_y \, d\nu(y)$ where μ_y is uniform measure on $\{y\} \times Z$. These measures are not invariant; instead T will map μ_y to μ_{Sy} for all y. The iterates T^n are then described as $T^n(y, z) = (S^n y, \rho_{y,n}(z))$, where the $\rho_{y,n}$ are defined using the *cocycle equation*

$$\rho_{y, n+m} = \rho_{S^m y, n} \circ \rho_{y, m}.$$

This is a more complicated version of the more familiar equation $T^{n+m} = T^n \circ T^m$, thus cocycles are more complicated versions of shifts (indeed as we just saw, a cocycle is simply the "vertical component" of a shift in a larger product space). The study of cocycles forms an integral part of the higher order recurrence theory but will not be discussed here.

Now let us look at double recurrence (the $k = 3$ case of Theorem 4.7), in which we investigate the limiting behavior of averages such as

$$(5.3) \qquad \lim_{N \to \infty} \mathbb{E}_{1 \leq r \leq N} \int_X f T^r f T^{2r} f \, d\mu.$$

If f is invariant, then again this expression is easy to compute (it is just $\int_X f^3$). One may hope, as in the preceding discussion, that anti-invariant functions are negligible, in the sense that

$$\lim_{N \to \infty} \mathbb{E}_{1 \leq r \leq N} \int_X f T^r g T^{2r} h \, d\mu = 0$$

whenever f, g, h are bounded and at least one of f, g, h is anti-invariant. Unfortunately, this is not the case. For a very simple example, take the small cyclic group $X = \mathbb{Z}/M\mathbb{Z}$ for odd M and let $f = g = h$ be the function which equals $M - 1$ at 0 and -1 elsewhere. Then these functions are all anti-invariant, but the above average can be computed to be $M^2 - 1$; the problem is that periodically (whenever n is a multiple of M) there is a huge "spike" in the value of $\int_X fT^n gT^{2n}h \, d\mu$ which imbalances the average dramatically. Thus periodic functions (ones in which $T^n f = f$ for some $n > 0$) cause a problem. More generally,[5] the *eigenfunctions*, in which $Tf = e^{2\pi i\theta} f$ for some $\theta \in \mathbb{R}/\mathbb{Z}$, will also cause a problem (note that invariant functions correspond to the case $\theta = 0$). Indeed if one sets $h := f$ and $g := \bar{f}^2$, then we see that $T^{2r}h = e^{4\pi ir\theta}h$ and $T^r g = e^{-4\pi ir\theta}g$, and hence[6]

$$\lim_{N \to \infty} \mathbb{E}_{1 \leq r \leq N} \int_X fT^r fT^{2r} f \, d\mu = \int_X |f|^4 \, d\mu \neq 0,$$

despite the fact that such eigenfunctions will necessarily be anti-invariant for $\theta \neq 0$ (as eigenfunctions of the unitary operator T with distinct eigenvalues are necessarily orthogonal).

However, one can simply deal with these problems by devising a suitable factor (larger than Z_0) to contain them. For instance, one can create the factor Y_0 generated by all the periodic functions. This factor can be larger than Z_0 (e.g., in the finite case, Y_0 is in fact everything). The periodic functions form an algebra (they are closed under arithmetic operations) but are not quite a von Neumann algebra because they are not quite closed under limits[7]. Nevertheless, the periodic functions are still *dense* in $L^2(Z_0)$, which turns out to be good enough for most purposes. Even larger than Y_0 is Z_1, the factor generated by all eigenfunctions — this factor is known as the *Kronecker factor*. Now the eigenfunctions are not closed under addition (though they are closed under multiplication), however the space of *quasiperiodic functions* — finite linear combinations of eigenfunctions — is indeed an algebra. The closure of the quasiperiodic functions in L^2 are the *almost periodic functions* — and this is a von Neumann algebra, indeed an L^2 function is almost periodic if it is measurable in Z_1. One can classify all these properties in terms of the orbit $\{T^n f : n \in \mathbb{Z}\}$:

- f is invariant if and only if the orbit $\{T^n f : n \in \mathbb{Z}\}$ is a singleton.
- f is periodic if and only if the orbit $\{T^n f : n \in \mathbb{Z}\}$ is finite.
- f is an eigenfunction if and only if the orbit $\{T^n f : n \in \mathbb{Z}\}$ lives in a one-dimensional complex vector space.

---

[5]A simple application of Fourier analysis or the spectral theorem reveals that every periodic function is a finite linear combination of eigenfunctions, with eigenvalues equal to roots of unity.

[6]This corresponds to the fact that sets of integers such as the *Bohr set* $\{n \in \mathbb{Z} : \|\alpha n\|_{\mathbb{R}/\mathbb{Z}} \leq \varepsilon\}$ have an unexpectedly high number of progressions of length three, due to the identity $\alpha n - 2\alpha(n + r) + \alpha(n + 2r) = 0$, which implies that if two elements of a progression lie in the Bohr set, then the third element has an unexpectedly high probability of doing so also. One should caution that this is not always the case; with the Behrend example in Proposition 1.3, when two elements of a progression lie in the set, then the third element has an unexpectedly *small* probability of lying in the set. Thus certain types of structure can in fact *reduce* the number of progressions present, though Szemerédi or Furstenberg tells us that they cannot destroy these progressions completely. This is another indication that the proof of this theorem has to be somewhat nontrivial (in particular, a naive symmetrisation or variational argument will not work).

[7]There does not seem to be a conventional name for what the uniform or L^2 limit of periodic functions should be called. One possibility is "pro-periodic" or "profinitely periodic" functions.

- f is quasiperiodic if and only if the orbit $\{T^n f : n \in \mathbb{Z}\}$ lives in a finite-dimensional vector space.
- f is almost periodic if and only if the orbit $\{T^n f : n \in \mathbb{Z}\}$ is precompact (its closure is compact).

Functions in these classes will be referred to as "structured functions of order 1" or "linearly structured functions"; the eigenfunctions[8] are "strongly structured functions of order 1." The linear comes from the fact that the action of $T^n f$ behaves "linearly" in n; observe for instance that if f is an eigenfunction with eigenvalue $e^{2\pi i \theta}$ then $T^n f = e^{2\pi i n \theta} f$. Now it turns out that one can get a good handle on the average (5.3) for all f in the linearly structured classes — and more precisely we have a nontrivial lower bound when f is nonnegative and not identically zero. We already saw what happened when f was invariant. If instead f was periodic with some period m, then we get a large positive contribution to (5.3) (specifically, $\int_X f^3 \, d\mu$) when n is a multiple of m, which is already enough for a nontrivial lower bound. For the other cases, one can use a pigeonhole argument to show that almost periodic functions behave very much like periodic functions (hence the name), in the sense that given any ε, we have $\|T^n f - f\|_{L^2(X)} \leq \varepsilon$ for a set of n of positive density. Note that if $T^n f$ is close to f, then (by applying T^n and then the triangle inequality) $T^{2n} f$ is close to f also, which can be used (together with Hölder's inequality and the boundedness of f) to show that $f T^n f T^{2n} f$ is close to f^3. This gives a contribution close to $\int_X f^3 \, d\mu$ for all n in a set of positive density, and one still gets a good lower bound for f. Note that these arguments extend easily to higher averages such as those involving $\int_X f T^n f \cdots T^{(k-1)n} f \, d\mu$. (But problems will emerge with the other half of the argument, as orthogonality to linear structure is not enough to eliminate all problems with triple and higher recurrence.)

There is another proof of recurrence for almost periodic functions which looks more complicated, but ends up being more robust and can extend (with some effort) to higher order cases. We know that the orbit $\{T^n f : n \in \mathbb{Z}\}$ is precompact, which means that for any $\varepsilon > 0$ one can cover this orbit by finitely many balls. This allows us to apply the van der Waerden theorem (or its topological counterpart) and conclude the existence of many progressions $n, n+r, \ldots, n+(k-1)r$ for which $T^n f, T^{n+r} f, \ldots, T^{n+(k-1)r} f$ are all close to each other. This means that $\int_X f T^r f \cdots T^{(k-1)r} f \, d\mu$ is close to $\int_X f^k \, d\mu > 0$, which can be used as before to get a nontrivial lower bound.

Now we say that a function f is "mixing of order 1," or "linearly mixing," if it is orthogonal to all almost periodic functions, or in other words $\mathbb{E}(f \mid Z_1) = 0$. It turns out that a more useful characterisation of this mixing property exists.

Lemma 5.2. *A real-valued function* $f \in L^\infty(X)$ *is mixing of order 1 if and only if the self-correlation functions* $T^n f f$ *are asymptotically mixing of order 0, in the sense that*

(5.4) $$\lim_{N \to \infty} \mathbb{E}_{-N \leq n \leq N} \|\mathbb{E}(T^n f f \mid Z_0)\|_{L^2}^2 = 0.$$

---

[8] An individual quasiperiodic function is usually not strongly structured, in the sense that $f(x)$ does not determine $T^n f(x)$ in a continuous manner; however a quasiperiodic function is always the *component* of a vector-valued function which is strongly structured. For instance, if $X := (\mathbb{R}/\mathbb{Z})^2$ and $T(x_1, x_2) := (x_1 + \alpha_1, x_2 + \alpha_2)$ for rationally independent α_1, α_2, then $f(x_1, x_2) := e^{2\pi i (x_1 + x_2)}$ is quasiperiodic but not strongly structured, however the vector-valued function $(e^{2\pi i (x_1 + x_2)}, e^{2\pi i x_1}, e^{2\pi i x_2})$ *is* strongly structured.

PROOF (Sketch only). Suppose first that f obeys the property (5.4). A Cauchy–Schwarz argument (based on something called the *van der Corput lemma*), which we omit, then shows that

$$\lim_{N\to\infty} \mathbb{E}_{-N\leq n\leq N}\|\mathbb{E}(T^n gf \mid Z_0)\|_{L^2}^2 = 0$$

for any bounded g. If we apply this in the particular case that g is an eigenfunction, we have $\|\mathbb{E}(T^n gf \mid Z_0)\|_{L^2} = \|\mathbb{E}(gf \mid Z_0)\|_{L^2}$ and hence $\mathbb{E}(gf \mid Z_0) = 0$ for all eigenfunctions g. In particular f is orthogonal to all eigenfunctions, hence to all quasiperiodic functions, hence to all almost periodic functions, and is thus mixing of order 1.

Now suppose that (5.4) fails. We rewrite the left-hand side (ignoring issues regarding interchange of limit and integral, which can be justified using the von Neumann ergodic theorem applied to the product space $X \times X$) as

$$\langle f, \lim_{N\to\infty} \mathbb{E}_{-N\leq n\leq N}\mathbb{E}(T^n ff \mid Z_0)T^n f\rangle.$$

Let us introduce the linear operator $S: L^2(X) \to L^2(X)$ by

$$Sf := \lim_{N\to\infty} \mathbb{E}_{-N\leq n\leq N}\mathbb{E}(T^n fg \mid Z_0)T^n f$$

(again, let us ignore the issue regarding whether this limit exists). Thus $\langle f, Sf\rangle \neq 0$. This is a self-adjoint operator (in fact, it is positive definite). Also, being the limit of averages of finite rank operators, it can be shown to be a compact operator. Finally, we have the translation invariance property $T^n S = S T^n$. In particular, this shows that the orbit of Sf lies in the range of S and is thus precompact:

$$\{T^n Sf : n \in \mathbb{Z}\} = \{ST^n f : n \in \mathbb{Z}\} \subset \{Sg : \|g\|_{L^2(X)} \leq \|f\|_{L^2(X)}\}.$$

This shows that Sf is almost periodic. Thus f is not orthogonal to all almost periodic functions, a contradiction. □

By using (5.4) and some applications of Cauchy–Schwarz (more precisely, using the van der Corput lemma) one can show that weakly mixing functions of order 1 are negligible for the purposes of double recurrence; indeed, we have

$$\lim_{N\to\infty} \mathbb{E}_{1\leq r\leq N} \int_X fT^r g T^{2r} h \, d\mu = 0$$

whenever f, g, h are bounded and at least one of f, g, h are mixing of order 1. We can refer to this as the *generalised von Neumann theorem of order* 1. On the other hand, every bounded function f has a unique decomposition $f = \mathbb{E}(f \mid Z_1) + (f - \mathbb{E}(f \mid Z_1))$ as an almost periodic function $\mathbb{E}(f \mid Z_1)$ and a weakly mixing function $f - \mathbb{E}(f|Z_1)$; I like to refer to this as the *Koopman–von Neumann theorem*.[9] Note also that if f is nonnegative with positive mean, then the almost periodic component $\mathbb{E}(f \mid Z_1)$ will be also. Combining this fact with the recurrence already obtained for almost periodic functions, and the negligibility of weakly mixing functions, we obtain recurrence for all functions, i.e., we have established the general $k = 3$ case of Furstenberg's multiple recurrence theorem.

We now give the barest sketch of how things continue onward from here. For $k = 4$ one needs to define notions of almost periodicity and weak mixing of order 2. Of the two, the latter is easier, because we can copy Lemma 5.2, and declare a function

---

[9]Lemma 5.2 is also sometimes known as the Koopman–von Neumann theorem; the two facts are of course closely related.

f to be *weakly mixing of order* 2 if its self-correlations $fT^n f$ are asymptotically weakly mixing of order 1, thus

$$\lim_{N \to \infty} \mathbb{E}_{-N \leq n \leq N} \|\mathbb{E}(fT^n f \mid Z_1)\|_{L^2}^2 = 0.$$

(Many other equivalent definitions are possible.) Repeated application of van der Corput eventually shows that such functions are negligible for the averages

$$\lim_{N \to \infty} \mathbb{E}_{1 \leq r \leq N} \int_X fT^r g T^{2r} h T^{3r} k \, d\mu$$

in the sense that this average vanishes whenever f, g, h, k are bounded and at least one is weakly mixing of order 2. It is not hard to show that there exists a unique factor Z_2 (that extends Z_1) such that the weakly mixing functions of order 2 are precisely those functions f whose conditional expectation $\mathbb{E}(f \mid Z_2)$ vanishes. (In the work of Host and Kra, this factor Z_2 is generated by *nonconventional averages* such as

$$\lim_{N \to \infty} \mathbb{E}_{-N \leq a,b,c \leq N} T^a f T^b f T^c f T^{a+b} f T^{b+c} f T^{a+c} f T^{a+b+c} f;$$

this idea was then adapted for the finite setting in [25] as the notion of a *dual function* to construct a finitary analogue of this factor.) One would then like the almost periodic functions of order 2 to be some dense subclass of $L^2(Z_2)$. This can be done; the trick is to repeat the original definition of almost periodic, but view terms such as "finite-dimensional" or "compact" not in terms of vector spaces over $\mathbb{R}$ (as we have implicitly been doing), but rather[10] as *modules* over the von Neumann algebra $L^\infty(Z_1)$ of bounded almost periodic functions. In particular:

- f is an *eigenfunction of order* 2 (also known as a *quadratic eigenfunction*) if and only if the orbit $\{T^n f : n \in \mathbb{Z}\}$ lives in a one-dimensional module over $L^\infty(Z_1)$.
- f is *quasiperiodic of order* 2 if and only if the orbit $\{T^n f : n \in \mathbb{Z}\}$ lives in a finite-dimensional module over $L^\infty(Z_1)$.
- f is *almost periodic of order* 2 if and only if the orbit $\{T^n f : n \in \mathbb{Z}\}$ can be "approximated to arbitrary accuracy" by subsets of finite-dimensional modules over $L^\infty(Z_1)$. (The precise definition is a little tricky and subtle; see [12].)

A quadratic eigenfunction can equivalently be defined (at least in the ergodic case) as a function f obeying an identity of the form $Tf = gf$, where g is itself a linear eigenfunction, thus $Tg = e^{2\pi i \theta} g$ for some $\theta \in \mathbb{R}/\mathbb{Z}$. The origin of the term "quadratic" can then be observed from an inspection of the phase in the identity

$$T^n f = e^{2\pi i n(n-1)\theta} g^n f.$$

From the closely related identity

$$fT^n(\bar{f}^3) T^{2n}(f^3) T^{3n} \bar{f} = |f|^8$$

---

[10] The combinatorial analogue of this would be to partition the original space X into atoms — in this case, the atoms of Z_1, and somehow work on each atom separately. Of course, things are not this simple because the atoms are usually not shift-invariant and so the shift structure is now more complicated, passing from one atom to the next. The graph theoretic approach, which we will discuss later, also relies heavily on restriction to atoms, but can cope with this with much greater ease because this approach "forgets" all the arithmetic structure and so there is nothing to destroy when passing to an atom.

one also sees that quadratic eigenfunctions are not negligible for the purposes of triple recurrence (indeed they end up being orthogonal to all quadratically mixing functions). Quasiperiodic functions of order 2 are special cases of 2-*step nilsequences*, which will be discussed in Bryna Kra's lectures. They can be viewed as components of *vector-valued* (or matrix-valued) quadratic eigenfunctions, and arise from what are known as *finite rank extensions* of the Kronecker factor Z_1.

At any rate, the almost periodic functions of order 2 now form a dense subclass of $L^2(Z_2)$, and are an algebra, and so one can repeat previous arguments and reduce the proof of the Furstenberg recurrence theorem for $k = 3$ to the task of proving such recurrence for such quadratically almost periodic functions. This turns out to be complicated—in part because this result includes Proposition 1.5 as a special case (the case of quadratic eigenfunctions), and this proposition is itself not entirely trivial (requiring at a bare minimum some form of van der Waerden's theorem). Fortunately, the colouring argument given previously for almost periodic functions—which *does* use van der Waerden's theorem—extends (after nontrivial effort) to this case, and more generally to all orders, thus leading to a proof of the Furstenberg recurrence theorem. See [11, 12, 14], as well as Bryna Kra's lectures.

6. The graph theoretic approach

Now we leave ergodic theory and turn to what (at first glance) appears to be a completely different approach to Szemerédi's theorem, though at a deeper inspection one will find many themes in common. In the ergodic approach, it was the shift operator T which was the primary focus of investigation; the underlying set A of integers merely provided some probability measure for T to leave invariant. We have seen that the dynamical approach focuses almost entirely on the shift operator. In marked contrast, the hypergraph approach discards the shift structure completely; instead, it views the problem of finding an arithmetic progression as that of solving a set of simultaneous relations; these relations initially have some additive structure, but this structure is soon discarded, as these relations are soon modeled abstractly by graphs and hypergraphs. With the forgetting of so much structure it is remarkable that any nontrivial progress can still be made; however there turn out to be deep theorems in (hyper)graph theory, comparable (though not directly equivalent) to the deep recurrence theorems in topological dynamics and ergodic theory, which allow one to proceed even after losing almost all of the arithmetic structure. It is a fascinating question as to what the "true" origin of these deep facts are—it seems to be some very abstract and general dichotomy between randomness and structure—and how they may be united with the ergodic and Fourier-analytic approaches.

To illustrate the power of the graph theoretic approach, let us prove a theorem which looks similar to van der Waerden's theorem though it is slightly different.

Theorem 6.1 (Schur's theorem). *Suppose the positive integers $\mathbb{Z}_+$ are finitely coloured. Then one of the colour classes contains a triple of the form $\{x, y, x + y\}$.*

PROOF. Our task is to find $x, y > 0$ and a colour class $\mathcal{C}$ for which we have the simultaneous relations

$$x \in \mathcal{C}, \quad y \in \mathcal{C}, \quad x + y \in \mathcal{C}.$$

The problem is that these equations (three relations in two unknowns) are coupled together in an unpleasant way. However we can decouple things slightly by making the (somewhat underdetermined) substitution $x = b - a$, $y = c - b$ for some $a < b < c$; our task is then to find such $a < b < c$ and a colour class $\mathcal{C}$ for which we have the simultaneous relations

$$b - a \in \mathcal{C}, \quad c - b \in \mathcal{C}, \quad c - a \in \mathcal{C}.$$

Now we have three relations in three unknowns, which is a bit better for the purposes of finding solutions. Furthermore, the relations are more symmetric in a, b, c, and each relation only involves two of the three unknowns. This is all that we will need to proceed. Indeed, let us now edge-colour the complete graph on the natural numbers by assigning to each edge (a, b) with $b > a$, the colour of $b - a$ in the original colouring (this is known as the *Cayley graph* associated to the original colouring). A solution to the above simultaneous relations is now nothing more than a monochromatic triangle in this graph. But the existence of such a triangle follows immediately from Ramsey's theorem. (Indeed one sees that one can even take a, b, c to be no larger than 6!) □

Note that we only used a very special case of Ramsey's theorem; using the full version of Ramsey's theorem leads to substantial generalisation of Schur's theorem, especially when combined with van der Waerden's theorem, known as *Rado's theorem*; see for instance [21].

Now we see what can similarly be done for progressions of length three in a set A of integers. Actually it will be convenient to localise to a cyclic group $\mathbb{Z}/N\mathbb{Z}$ and prove the following.

Theorem 6.2 (Roth's theorem, cyclic group version). *Let N be a large integer, and let $A \subset \mathbb{Z}/N\mathbb{Z}$ be such that $|A| \geq \delta N$. Then there are at least $c(\delta)N^2$ progressions $x, x + r, x + 2r$ in A for some $c(\delta) > 0$ (we allow r to be zero).*

It is easy to see that this implies the $k = 3$ version of Szemerédi's theorem (and is in fact equivalent to it, thanks to the formulation in Theorem 4.8). Our task is to find many solutions to the system of relations

$$n \in A, \quad n + r \in A, \quad n + 2r \in A.$$

Again this is three equations in two unknowns. We add an unknown by making the underdetermined substitution $n := -x_2 - 2x_3$, $r := x_1 + x_2 + x_3$ and obtain the system

$$\begin{aligned} -x_2 - 2x_3 &\in A \\ x_1 \phantom{{} - x_3} & \\ -x_3 &\in A \\ -2x_1 - x_2 \phantom{{} - x_3} &\in A \end{aligned}$$

This is again three relations in three unknowns, where each relation involves only two of the three variables; our task is to locate $c(\delta).N^3$ solutions. The situation is not quite the same as with Schur's theorem, though; for instance, the three relations are not entirely symmetric. On the other hand, we already know a lot of *degenerate* solutions to this system:

$$\begin{aligned} -x_2 - 2x_3 &\in A \\ x_1 \phantom{{} - x_3} -x_3 &\in A \\ -2x_1 - x_2 \phantom{{} - x_3} &\in A \\ x_1 + x_2 + x_3 &= 0. \end{aligned}$$

Indeed, every element of A generates N such solutions, so we have δN^2 solutions in all. We can rephrase this as a conditional probability bound

(6.1) $\quad \mathbb{P}(-x_2 - 2x_3, x_1 - x_3, 2x_1 + x_2 \in A \mid x_1 + x_2 + x_3 = 0) \geq \delta$

where we think of x, y, z as ranging freely over the cyclic group $\mathbb{Z}/N\mathbb{Z}$, and then conditioned so that $x + y + z = 0$. Our goal seems innocuous, namely to remove this conditional expectation and conclude that

(6.2) $\quad\quad\quad \mathbb{P}(-x_2 - 2x_3, x_1 - x_3, 2x_1 + x_2 \in A) \geq c(\delta)$.

This is less trivial than it first appears. The problem is that the event $x + y + z = 0$ has tiny probability — $1/N$ — and so we only get a tiny lower bound of δ/N if we naively apply Bayes' identity. (This corresponds to the fact that the number of trivial progressions — δN^2 — is negligible compared with the number of progressions that we actually want, which is $c(\delta)N^3$.) However, the point will be that the solution set $\{(x, y, z) : -x_2 - 2x_3, x_1 - x_3, 2x_1 + x_2 \in A\}$, being the intersection of three "second-order" sets $\{(x_1, x_2, x_3) : -x_2 - 2x_3 \in A\}$, $\{(x_1, x_2, x_3) : -x_1 - x_3 \in A\}$, $\{(x_1, x_2, x_3) : 2x_1 + x_2 \in A\}$, is not a completely arbitrary set, and as it turns out it cannot concentrate itself entirely on the "third-order set" $\{(x_1, x_2, x_3) : x_1 + x_2 + x_3 = 0\}$. For instance, observe that given any relation $x_i \sim x_j$ involving just two of the x_1, x_2, x_3, we have

(6.3) $\quad\quad\quad \mathbb{P}(x_i \sim x_j \mid x_1 + x_2 + x_3 = 0) = \mathbb{P}(x_i \sim x_j)$

or given any sets A_1, A_2, we have

$$\mathbb{P}(x_1 \in A_1, x_2 \in A_2 \mid x_1 + x_2 = 0) \leq \min(\mathbb{P}(x_1 \in A_1), \mathbb{P}(x_2 \in A_2))$$
$$\leq \mathbb{P}(x_1 \in A_1, x_2 \in A_2)^{1/2}.$$

So we see that when the structure of the set is sufficiently "low order," one can remove the conditional expectation. Can one do so here? The answer is yes, and it relies on the following abstract result.

Lemma 6.3 (Triangle removal lemma [37]). *Let G be a graph on n vertices that contains fewer than εn^3 triangles for some $0 < \varepsilon < 1$. Then it is possible to delete $o_{\varepsilon \to 0}(n^2)$ edges from G to create a triangle-free graph G'.*

As usual we use $o_{\varepsilon \to 0}(X)$ to denote a quantity which is bounded by $c(\varepsilon)X$ for some function $c(\varepsilon)$ of ε which goes to zero as $\varepsilon \to 0$. Later on we will allow the decay rate to depend on additional parameters, for instance $o_{\varepsilon \to 0; k}(1)$ would be a quantity which decayed to zero as $\varepsilon \to 0$ for each fixed k, but which need not decay uniformly in k. An equivalent formulation of this lemma is:

Lemma 6.4 (Triangle removal lemma, again). *Let G be a graph on n vertices that contains at least δn^2 edge-disjoint triangles for some $0 < \delta < 1$. Then it must in fact contain $c(\delta)n^3$ triangles, where $c(\delta) > 0$ depends only on δ.*

We leave the equivalence of these two formulations to the reader. From the second formulation it is an easy matter to deduce (6.2) from (6.1), by considering the tripartite graph formed by three copies of V (corresponding to x_1, x_2, x_3 respectively), and with the three edge classes between these copies defined by the relations $-x_2 - 2x_3 \in A$, $x_1 - x_3 \in A$, and $2x_1 + x_2 \in A$ respectively; again, we leave this as an exercise for the reader.

There is another way to phrase this lemma in a "several variable measure theory" language that brings it more into line with the ergodic theory approach (and also the Fourier-analytic approach). (See also [31] for a very similar perspective.)

Lemma 6.5 (Triangle removal lemma, several variable version). *Let (X, μ_X), (Y, μ_Y), (Z, μ_Z) be probability spaces, and let $f\colon X \times Y \to [0,1]$, $g\colon Y \times Z \to [0,1]$, and $h\colon Z \times X \to [0,1]$ be measurable functions such that*

$$\Lambda_3(f, g, h) \leq \varepsilon$$

for some $0 < \varepsilon < 1$, where Λ_3 is the trilinear form

$$\Lambda_3(f, g, h) := \int_X \int_Y \int_Z f(x,y) g(y,z) h(z,x) \, d\mu_X(x) \, d\mu_Y(y) \, d\mu_Z(z).$$

Then there exists functions $\tilde{f}\colon X \times Y \to [0,1]$, $\tilde{g}\colon Y \times Z \to [0,1]$, and $\tilde{h}\colon Z \times X \to [0,1]$ which differ from f, g, h in L^1 norm by $o_{\varepsilon \to 0}(1)$, thus

$$\int_X \int_Y |f(x,y) - \tilde{f}(x,y)| \, d\mu_X(x) \, d\mu_Y(y), \quad \int_Y \int_Z |g(y,z) - \tilde{g}(y,z)| \, d\mu_Y(y) \, d\mu_Z(z),$$

$$\int_Z \int_X |h(z,x) - \tilde{h}(z,x)| \, d\mu_Z(z) \, d\mu_X(x) \leq o_{\varepsilon \to 0}(1),$$

and such that $\tilde{f}(x,y) \tilde{g}(y,z) \tilde{h}(z,x)$ vanishes identically (in particular, $\Lambda_3(\tilde{f}, \tilde{g}, \tilde{h}) = 0$).

One can easily deduce Lemma 6.3 from Lemma 6.5 by specialising X, Y, Z to be the finite vertex set V with the uniform probability measure, and let $f = g = h$ be the indicator function of the edge set of the graph G; we omit the details. The converse implication is also true but somewhat tricky (one must discretise the measure spaces X, Y, Z, and split the atoms of such spaces to approximate the probability measures by uniform distributions, and also replace the functions f, g, h by indicator functions); we again omit the details. We will choose to work with the analytic formulation of the triangle removal lemma in these notes because it seems to extend more easily to the hypergraph setting (in which one considers similar expressions in more variables, where now each function can depend on three or more variables).

Lemma 6.5 asserts, roughly speaking, that if a collection of low complexity functions have a small product, then one can "clean" each function slightly *in a low-complexity manner* in order to make the product vanish entirely. Note that the claim would be trivial if one were allowed to modify (say) f in a manner which could depend on all three variables x, y, z. The power of the lemma lies in the fact that the high-complexity expression $\Lambda_3(f, g, h)$ can be manipulated purely in terms of low-complexity operations. This rather deep phenomenon seems to be rather general; in fact there is a similar lemma for any nonnegative combination of functions of various collections of variables (we shall describe one such version a little later below). It is however still not perfectly well understood.

The way one proves Lemma 6.5 is by decomposing f, g, h into "structured" or "low complexity" components, which are easier to clean up, and "error terms," which for one reason or another do not interfere with the cleaning process because they give a negligible contribution to expressions such as $\Lambda_3(f, g, h)$. It turns out that there are two types of error terms which come into play. The first are errors which are "small" in an integral sense, say in L^2 norm, while the second are errors which are (very) small in a weak sense (for instance, they are small when tested against other functions which depend on other sets of variables). The latter will be encoded using a useful norm, the *Gowers $\Box^2$ norm* $\|f\|_{\Box^2(X \times Y)} = \|f\|_{\Box^2}$, defined for measurable bounded $f \colon X \times Y \to \mathbb{R}$ by the formula

$$\|f\|_{\Box^2(X\times Y)}^4$$
$$:= \int_X \int_X \int_Y \int_Y f(x,y)f(x,y')f(x',y)f(x',y') \, d\mu_X(x) \, d\mu_X(x') \, d\mu_Y(y) \, d\mu_Y(y').$$

One easily verifies that the right-hand side is nonnegative. From two applications of the Cauchy–Schwarz inequality one verifies the *Gowers-Cauchy-Schwarz inequality*

(6.4) $$\left| \int_X \int_X \int_Y \int_Y f_{00}(x,y) f_{01}(x,y') f_{10}(x',y) f_{11}(x',y') \right.$$
$$\left. d\mu_X(x) \, d\mu_X(x') \, d\mu_Y(y) \, d\mu_Y(y') \right|$$
$$\leq \|f_{00}\|_{\Box^2} \|f_{01}\|_{\Box^2} \|f_{10}\|_{\Box^2} \|f_{11}\|_{\Box^2}$$

from which one readily verifies that $\Box^2$ obeys the triangle inequality and is thus at least a seminorm. From the Gowers–Cauchy–Schwarz inequality (and bounding the $\Box^2$ norm crudely by the L^∞ norm) one also sees that

(6.5) $$\left| \int_X \int_Y f(x,y) g(y) h(x) \, d\mu_Y(x) \, d\mu_Y(y) \right| \leq \|f\|_{\Box^2}$$

whenever g, h are measurable functions bounded in magnitude by 1; this in particular shows that if $\|f\|_{\Box^2} = 0$ then f is zero almost everywhere. Thus the $\Box^2$ norm is indeed a norm,[11] after the customary convention of identifying two functions that agree almost everywhere. Letting g, h depend on a third variable z in (6.5) and integrating in z, and using symmetry, we thus conclude the *generalised von Neumann inequality*

(6.6) $$|\Lambda_3(f,g,h)| \leq \min(\|f\|_{\Box^2}, \|g\|_{\Box^2}, \|h\|_{\Box^2})$$

whenever $f \colon X \times Y \to [-1,1]$, $g \colon Y \times Z \to [-1,1]$, $h \colon Z \times X \to [-1,1]$ are measurable.

Thus functions with tiny $\Box^2$ norm have a negligible impact on the Λ_3 form; such functions are known as *pseudorandom* or *Gowers uniform*. To exploit this, one would now like to decompose arbitrary functions $f \colon X \times Y \to [0,1]$ into a "structured" component which can be easily analysed and manipulated, plus errors which are small in $\Box^2$ or are otherwise easy to deal with. The first key observation is

Lemma 6.6 (Lack of uniformity implies correlation with structure). *Let $f \colon X \times Y \to [-1,1]$ be such that $\|f\|_{\Box^2} \geq \eta$ for some $\eta > 0$. Then there exists $A \subset X$ and $B \subset Y$ such that*

$$\left| \int_X \int_Y 1_A(x) 1_B(y) f(x,y) \, d\mu_X(x) \, d\mu_Y(y) \right| \geq \eta^4/4.$$

---

[11] One can also identify the $\Box^2$ norm with the Schatten–von Neumann 4-norm of the integral operator with kernel $f(x,y)$; in the important special case when X is a finite set with the uniform distribution, and f is symmetric, then the $\Box^2$ norm is simply the l^4 norm of the eigenvalues of the matrix associated to f. If f is the indicator function of a graph G, the $\Box^2$ norm is a normalised count of the number of 4-cycles in G. See also [31] for some related discussion. However we will not take advantage of these facts as they do not generalise well to hypergraph situations.

PROOF. By definition of the $\Box^2$ norm we have
$$\int_X \int_Y \int_X \int_Y f(x,y)f(x,y')f(x',y)f(x',y')\,d\mu_X(x)\,d\mu_X(x')\,d\mu_Y(y)\,d\mu_Y(y') \geq \eta^4.$$
By the pigeonhole principle and the boundedness of f, we can thus find x', y' such that
$$\left|\int_X \int_Y f(x,y)f(x,y')f(x',y)\,d\mu_X(x)\,d\mu_Y(y)\right| \geq \eta^4.$$
We rewrite this using Fubini's theorem as
$$\left|\int_{-1}^1 \int_{-1}^1 \operatorname{sgn}(s)\operatorname{sgn}(t) \int_X \int_Y 1_{A_s}(x) 1_{B_t(y)} f(x,y)\,d\mu_X(x)\,d\mu_Y(y)\,ds\,dt\right| \geq \eta^4$$
where $A_s := \{x \in X : \operatorname{sgn}(s)f(x,y') \geq |s|\}$ and $B_t := \{y \in Y : \operatorname{sgn}(t)f(x',y) \geq |t|\}$. The claim then follows from another application of the pigeonhole principle. □

To exploit this we borrow some notation from the ergodic theory approach, namely that of σ-algebras and conditional expectation. However, in this simple context we will only need to deal with *finite* σ-algebras. If $\mathcal{B}$ is a finite factor of X (i.e., a finite σ-algebra of measurable sets in X), then $\mathcal{B}$ is essentially just a partition of X into finitely many disjoint atoms $A_1, \ldots, A_M$ (more precisely, $\mathcal{B}$ is the σ-algebra consisting of all finite unions of these atoms). If $f\colon X \to \mathbb{R}$ is measurable, then the conditional expectation $\mathbb{E}(f \mid \mathcal{B})\colon X \to \mathbb{R}$ is the function defined as $\mathbb{E}(f \mid \mathcal{B})(x) := (1/A_i)\int_{A_i} f(x)\,d\mu_X(x)$ whenever x lies in an atom A_i of positive measure. (Conditional expectations are only defined up to sets of measure zero, so we can define $\mathbb{E}(f|\mathcal{B})$ arbitrarily on atoms of measure zero.) We say that a factor has *complexity at most m* if it is generated by at most m sets (and thus it contains at most 2^m atoms). If $\mathcal{B}_X$ is a finite factor of X with atoms $A_1, \ldots, A_M$, and $\mathcal{B}_Y$ is a finite factor of Y with atoms $B_1, \ldots, B_N$, then $\mathcal{B}_X \vee \mathcal{B}_Y$ is a finite factor of $X \times Y$ with atoms $A_i \times B_j$ for $1 \leq i \leq M$ and $1 \leq j \leq N$.

The key relationship between the $\Box^2$ norm and conditional expectation on finite factors is the following.

Lemma 6.7 (Lack of uniformity implies energy increment). *Let $\mathcal{B}_X$, $\mathcal{B}_Y$ be finite factors of X, Y respectively of complexity at most m, and let $f\colon X \times Y \to [0,1]$ be such that*
$$\|f - \mathbb{E}(f|\mathcal{B}_X \vee \mathcal{B}_Y)\|_{\Box^2(X \times Y)} \geq \eta$$
for some $\eta > 0$. Then there exists extensions $\mathcal{B}'_X$, $\mathcal{B}'_Y$ of $\mathcal{B}_X$, $\mathcal{B}_Y$ of complexity at most $m+1$ such that
$$\|\mathbb{E}(f|\mathcal{B}'_X \vee \mathcal{B}'_Y)\|^2_{L^2(X\times Y)} \geq \|\mathbb{E}(f|\mathcal{B}_X \vee \mathcal{B}_Y)\|^2_{L^2(X\times Y)} + \eta^8/16.$$
Here of course $\|F\|^2_{L^2(X\times Y)} := \int_X \int_Y |F(x,y)|^2\,d\mu_X(x)\,d\mu_Y(y)$.

The key point here is that f—which is a "second-order" object, depending on two variables—is correlating with two "first-order" objects $\mathcal{B}'_X$, $\mathcal{B}'_Y$. This ultimately will allow us to approximate the second-order object by a number of first-order objects. It is this kind of reduction—in which a single high-order object is traded in for a large number of lower-order objects—which is the key to proving results such as the triangle removal lemma. The quantity $\|\mathbb{E}(f|\mathcal{B}_X \vee \mathcal{B}_Y)\|^2_{L^2(X \times Y)}$ is known as the *index* of the partition $\mathcal{B}_X \vee \mathcal{B}_Y$ in the graph theory literature; here we shall refer to it as the *energy* of this partition.

PROOF. From Lemma 6.6 we can find measurable $A \subset X$, $B \subset Y$ such that

$$\left| \int_X \int_Y 1_A(x) 1_B(y) \big(f - \mathbb{E}(f \mid \mathcal{B}_X \vee \mathcal{B}_Y)\big) \, d\mu_X(x) \, d\mu_Y(y) \right| \geq \eta^4/4.$$

Let $\mathcal{B}'_X$ be the factor of X generated by $\mathcal{B}_X$ and A, and similarly let $\mathcal{B}'_Y$ be the factor of Y generated by $\mathcal{B}_Y$ and B, then $\mathcal{B}'_X$, $\mathcal{B}'_Y$ have complexity at most $m+1$. Since $1_A(x) 1_B(y)$ is $\mathcal{B}'_X \vee \mathcal{B}'_Y$ measurable, we have

$$\int_X \int_Y 1_A(x) 1_B(y) \big(f - \mathbb{E}(f \mid \mathcal{B}_X \vee \mathcal{B}_Y)\big) \, d\mu_X(x) \, d\mu_Y(y)$$
$$= \int_X \int_Y 1_A(x) 1_B(y) \mathbb{E}\big(f - \mathbb{E}(f \mid \mathcal{B}_X \vee \mathcal{B}_Y) \mid \mathcal{B}'_X \vee \mathcal{B}'_Y\big) \, d\mu_X(x) \, d\mu_Y(y)$$

so by Cauchy–Schwarz

$$\|\mathbb{E}(f - \mathbb{E}(f \mid \mathcal{B}_X \vee \mathcal{B}_Y) \mid \mathcal{B}'_X \vee \mathcal{B}'_Y)\|_{L^2(X \times Y)} \geq \eta^4/4.$$

Now observe that the quantity

$$\mathbb{E}(f - \mathbb{E}(f \mid \mathcal{B}_X \vee \mathcal{B}_Y) \mid \mathcal{B}'_X \vee \mathcal{B}'_Y) = \mathbb{E}(f \mid \mathcal{B}'_X \vee \mathcal{B}'_Y) - \mathbb{E}(f \mid \mathcal{B}_X \vee \mathcal{B}_Y)$$

is orthogonal to $\mathbb{E}(f \mid \mathcal{B}_X \vee \mathcal{B}_Y)$. The claim then follows from Pythagoras' theorem. $\square$

Note that if f is bounded by 1, then the quantity $\|\mathbb{E}(f \mid \mathcal{B}_X \vee \mathcal{B}_Y)\|^2_{L^2(X \times Y)}$ is bounded between 0 and 1. Thus an easy iteration of the above lemma gives

Corollary 6.8 (Koopman–von Neumann decomposition). *Let $\mathcal{B}_X$, $\mathcal{B}_Y$ be finite factors of X, Y respectively of complexity at most m, let $f \colon X \times Y \to [0,1]$ be measurable, and let $\eta > 0$. Then there exists extensions $\mathcal{B}'_X$, $\mathcal{B}'_Y$ of $\mathcal{B}_X$, $\mathcal{B}_Y$ of complexity at most $m + 16/\eta^8$ such that*

$$\|f - \mathbb{E}(f \mid \mathcal{B}'_X \vee \mathcal{B}'_Y)\|_{\square^2(X \times Y)} < \eta.$$

This corollary splits f into a bounded complexity object $\mathbb{E}(f \mid \mathcal{B}'_X \vee \mathcal{B}'_Y)$ and an error which is small in the $\square^2$ norm. In practice, this decomposition is not very useful because the complexity of the structured component $\mathbb{E}(f \mid \mathcal{B}'_X \vee \mathcal{B}'_Y)$ is large compared to the bounds available on the error $f - \mathbb{E}(f \mid \mathcal{B}'_X \vee \mathcal{B}'_Y)$. However one can rectify this by one further iteration of the above decomposition:

Lemma 6.9 (Szemerédi regularity lemma). *Let $f \colon X \times Y \to [0,1]$ be measurable, let $\tau > 0$, and let $F \colon \mathbb{N} \to \mathbb{N}$ be an arbitrary increasing function (possibly depending on τ). Then there exists an integer $M = O_{F,\tau}(1)$ and a decomposition $f = f_1 + f_2 + f_3$ where*
- *(f_1 is structured) We have $f_1 = \mathbb{E}(f \mid \mathcal{B}_X \vee \mathcal{B}_Y)$ for some finite factors $\mathcal{B}_X, \mathcal{B}_Y$ of X, Y respectively of complexity at most M;*
- *(f_2 is small) We have $\|f_2\|_{L^2(X \times Y)} \leq \tau$.*
- *(f_3 is very uniform) We have $\|f_3\|_{\square^2(X \times Y)} \leq 1/F(M)$.*
- *(Positivity) f_1 and $f_1 + f_2$ take values in $[0,1]$.*

This lemma may not immediately resemble the usual Szemerédi regularity lemma for graphs, but it can easily be used to deduce that lemma. See [44]. One can obtain a result similar to this from spectral theory, by viewing f as the kernel of an integral operator and decomposing f using the singular value decomposition of that operator, with f_1, f_2, f_3 corresponding to the high, medium, and low singular values respectively. However it then takes some effort to ensure that f_1 and $f_1 + f_2$ are nonnegative. See [24] for some related discussion. The more "ergodic" approach here, relying on conditional expectation, gives worse quantitative bounds but does easily ensure the positivity property, which is crucial in many applications.

PROOF. Construct recursively a sequence of integers
$$0 = M_0 \leq M_1 \leq M_2 \leq \cdots$$
by setting $M_0 := 0$ and $M_i := M_{i-1} + 16F(M_{i-1})^8$ for $i \geq 1$. Then for each $i \geq 0$, construct recursively factors $\mathcal{B}_X^i$, $\mathcal{B}_Y^i$ of X, Y of complexity at most M_i by setting $\mathcal{B}_X^0$ and $\mathcal{B}_Y^0$ to be the trivial factors of complexity 0, and then applying Corollary 6.8 repeatedly to let $\mathcal{B}_X^i$, $\mathcal{B}_Y^i$ be extensions of $\mathcal{B}_X^{i-1}$, $\mathcal{B}_Y^{i-1}$ such that
$$\|f - \mathbb{E}(f|\mathcal{B}_X^i \vee \mathcal{B}_Y^i)\|_{\square^2(X \times Y)} < 1/F(M_{i-1}).$$
The energies $\|\mathbb{E}(f \mid \mathcal{B}_X^i \vee \mathcal{B}_Y^i)\|_{L^2(X \times Y)}^2$ are monotone increasing in i by Pythagoras' theorem, and are bounded between 0 and 1. Thus by the pigeonhole principle we can find $1 \leq i \leq 1/\tau^2$ for which
$$\|\mathbb{E}(f \mid \mathcal{B}_X^i \vee \mathcal{B}_Y^i)\|_{L^2(X \times Y)}^2 \leq \|\mathbb{E}(f \mid \mathcal{B}_X^{i-1} \vee \mathcal{B}_Y^{i-1})\|_{L^2(X \times Y)}^2 + \tau^2.$$
If one then sets
$$f_1 := \mathbb{E}(f \mid \mathcal{B}_X^{i-1} \vee \mathcal{B}_Y^{i-1}); \qquad f_2 := \mathbb{E}(f \mid \mathcal{B}_X^i \vee \mathcal{B}_Y^i) - \mathbb{E}(f \mid \mathcal{B}_X^{i-1} \vee \mathcal{B}_Y^{i-1});$$
$$f_3 := f - \mathbb{E}(f \mid \mathcal{B}_X^i \vee \mathcal{B}_Y^i); \qquad M := M_{i-1}$$
then we see that the claims are easily verified. $\square$

A slight modification of the above argument allows one to simultaneously regularise several functions at once using the same partition. More precisely, we have

Lemma 6.10 (Simultaneous Szemerédi regularity lemma). *Let $f \colon X \times Y \to [0,1]$, $g \colon Y \times Z \to [0,1]$, $h \colon Z \times X \to [0,1]$ be measurable, let $\tau > 0$, and let $F \colon \mathbb{N} \to \mathbb{N}$ be an arbitrary increasing function (possibly depending on τ). Then there exists an integer $M = O_{F,\tau}(1)$, factors $\mathcal{B}_X$, $\mathcal{B}_Y$, $\mathcal{B}_Z$ of X, Y, Z respectively of complexity at most M and decompositions $f = f_1 + f_2 + f_3$, $g = g_1 + g_2 + g_3$, $h = h_1 + h_2 + h_3$, where*

- *(f_1, g_1, h_1 are structured) We have $f_1 = \mathbb{E}(f \mid \mathcal{B}_X \vee \mathcal{B}_Y)$, $g_1 = \mathbb{E}(g \mid \mathcal{B}_Y \vee \mathcal{B}_Z)$, and $h_1 = \mathbb{E}(f \mid \mathcal{B}_Z \vee \mathcal{B}_X)$.*
- *(f_2, g_2, h_2 are small) We have $\|f_2\|_{L^2(X \times Y)}, \|g_2\|_{L^2(Y \times Z)}, \|h_2\|_{L^2(Z \times X)} \leq \tau$.*
- *(f_3, g_3, h_3 are very uniform) We have $\|f_3\|_{\square^2(X \times Y)}, \|g_3\|_{\square^2(X \times Y)}, \|h_3\|_{\square^2(X \times Y)} \leq 1/F(M)$.*
- *(Positivity) f_1, g_1, h_1 and $f_1 + f_2$, $g_1 + g_2$, $h_1 + h_2$ take values in $[0,1]$.*

We leave the proof of this lemma as an exercise to the reader. With this lemma we can now prove Lemma 6.5. Actually we shall prove a slightly stronger statement, which provides more information about the functions $\tilde{f}$, $\tilde{g}$, $\tilde{h}$ involved.

Lemma 6.11 (Strong triangle removal lemma, several variable version). *Let (X, μ_X), (Y, μ_Y), (Z, μ_Z) be probability spaces, and let $f \colon X \times Y \to [0,1]$, $g \colon Y \times Z \to [0,1]$, and $h \colon Z \times X \to [0,1]$ be measurable functions such that $\Lambda_3(f, g, h) \leq \varepsilon$ for some $0 < \varepsilon < 1$. Then there exists factors $\mathcal{B}_X$, $\mathcal{B}_Y$, $\mathcal{B}_Z$ of X, Y, Z respectively of complexity at most $O_\varepsilon(1)$ and sets $E_{X,Y} \in \mathcal{B}_X \vee \mathcal{B}_Y$,*

$E_{Y,Z} \in \mathcal{B}_Y \vee \mathcal{B}_Z$, $E_{Z,X} \in \mathcal{B}_Z \vee \mathcal{B}_X$ respectively with $1_{E_{X,Y}}(x,y) 1_{E_{Y,Z}}(y,z) 1_{E_{Z,X}}(z,x)$ vanishing identically, such that

$$\int_X \int_Y f(x,y) 1_{E^c_{X,Y}}(x,y) \, d\mu_X(x) \, d\mu_Y(y), \quad \int_Y \int_Z g(y,z) 1_{E^c_{Y,Z}}(y,z) \, d\mu_Y(y) \, d\mu_Z(z),$$

$$\int_Z \int_X h(z,x) 1_{E^c_{Z,X}}(z,x) \, d\mu_Z(z) \, d\mu_X(x) \leq o_{\varepsilon \to 0}(1).$$

Note that Lemma 6.11 immediately implies Lemma 6.5 by setting $\tilde{f} := f 1_{E_{x,y}}$, etc. This strengthened version of the lemma will come in handy in the next section.

PROOF. We apply Lemma 6.10 with $0 < \tau \ll 1$ and F to be chosen later; for now, one should think of τ as being moderately small, but not very small compared to ε, and similarly F will be a moderately growing function. This gives us an integer $M = O_{F,\tau}(1)$, factors $\mathcal{B}_X, \mathcal{B}_Y, \mathcal{B}_Z$ of complexity at most M, and decompositions $f = f_1 + f_2 + f_3$, etc. with the stated properties. In particular

$$\Lambda_3(f_1 + f_2 + f_3, g_1 + g_2 + g_3, h_1 + h_2 + h_3) \leq \varepsilon.$$

The idea shall be to eliminate the uniform errors f_3, g_3, h_3, and then the small errors f_2, g_2, h_2, leaving one with only the structured components f_1, g_1, h_1, which will be easy to deal with directly.

It is easy to eliminate f_3, g_3, h_3. Indeed from repeated application of the generalised von Neumann inequality (6.6) and the $\square^2$ bounds on f_3, g_3, h_3 we have

(6.7) $$\Lambda_3(f_1 + f_2, g_1 + g_2, h_1 + h_2) \leq \varepsilon + O(1/F(M)).$$

We would now like to similarly eliminate f_2, g_2, h_2. A naive application of the L^2 bounds would give an estimate of the form

(6.8) $$\Lambda_3(f_1, g_1, h_1) \leq \varepsilon + O(\tau) + O(1/F(M))$$

but the $O(\tau)$ error turns out to be far too expensive for our purposes. Instead we proceed in a more "local" fashion as follows. Let $E^0_{X,Y} \in \mathcal{B}_X \vee \mathcal{B}_Y$ be the set

$$E^0_{X,Y} := \{(x,y) \in X \times Y : f_1(x,y) \geq \tau^{1/10}; \; \mathbb{E}(f_2(x,y)^2 \mid \mathcal{B}_X \vee \mathcal{B}_Y) \leq \tau\}$$

and define $E^0_{Y,Z} \in \mathcal{B}_Y \vee \mathcal{B}_Z$ and $E^0_{Z,X} \in \mathcal{B}_Z \vee \mathcal{B}_X$ similarly. We first observe that f is small outside of $E^0_{X,Y}$. Indeed we have (by the $\mathcal{B}_X \vee \mathcal{B}_Y$-measurability of $E^0_{X,Y}$)

$$\int_X \int_Y f(x,y) 1_{(E^0_{X,Y})^c}(x,y) \, d\mu_X(x) \, d\mu_Y(y)$$

$$= \int_X \int_Y f_1(x,y) 1_{(E^0_{X,Y})^c}(x,y) \, d\mu_X(x) \, d\mu_Y(y)$$

$$\leq \int_{f_1(x,y) < \tau^{1/10}} f_1(x,y) \, d\mu_X(x) \, d\mu_Y(y) + \int_{\mathbb{E}(f_2(x,y)^2 \mid \mathcal{B}_X \vee \mathcal{B}_Y) > \tau} d\mu(X) \, d\mu(Y)$$

$$\leq \tau^{1/10} + \frac{1}{\tau} \int_X \int_Y \mathbb{E}(f_2(x,y)^2 \mid \mathcal{B}_X \vee \mathcal{B}_Y) \, d\mu(X) \, d\mu(Y)$$

$$= \tau^{1/10} + \frac{1}{\tau} \|f_2\|^2_{L^2(X \times Y)} = o_{\tau \to 0}(1).$$

Let A, B, C be atoms in $\mathcal{B}_X, \mathcal{B}_Y, \mathcal{B}_Z$ respectively such that $A \times B \subset E^0_{X,Y}$, $B \times C \subset E^0_{Y,Z}$, and $C \times A \subset E^0_{Z,X}$, and consider the local quantity

$$\Lambda_3((f_1 + f_2) 1_{A \times B}, (g_1 + g_2) 1_{B \times C}, (h_1 + h_2) 1_{C \times A}).$$

We can estimate this as the sum of a main term
$$\Lambda_3(f_1 1_{A\times B}, g_1 1_{B\times C}, h_1 1_{C\times A})$$
and three error terms
$$O\big(\Lambda_3(|f_2|1_{A\times B}, 1_{B\times C}, 1_{C\times A})\big) + O\big(\Lambda_3(1_{A\times B}, |g_2|1_{B\times C}, 1_{C\times A})\big)$$
$$+ O\big(\Lambda_3(1_{A\times B}, 1_{B\times C}, |h_2|1_{C\times A})\big).$$
By definition of $E^0_{X,Y}$, $E^0_{Y,Z}$, $E^0_{Z,X}$, we have $f_1, g_1, h_1 \geq \tau^{1/10}$ on $A\times B$, $B\times C$, $C \times A$ respectively, and hence the main term is at least
$$\tau^{3/10}\Lambda_3(1_{A\times B}, 1_{B\times C}, 1_{C\times A}).$$
On the other hand, we have by construction
$$\mathbb{E}(f_2(x,y)^2 \mid A\times B) \leq \tau$$
and hence by Cauchy–Schwarz
$$\Lambda_3(|f_2|1_{A\times B}, 1_{B\times C}, 1_{C\times A}) \leq \tau^{1/2}\Lambda_3(1_{A\times B}, 1_{B\times C}, 1_{C\times A}).$$
Similarly for g_2 and h_2. Thus the error terms are $O(\tau^{2/10})$ of the main term. If $\tau \ll 1$ is chosen sufficiently small, we thus have the local estimate
$$\Lambda_3(f_1 1_{A\times B}, g_1 1_{B\times C}, h_1 1_{C\times A})$$
$$= O\Big(\Lambda_3\big((f_1+f_2)1_{A\times B}, (g_1+g_2)1_{B\times C}, (h_1+h_2)1_{C\times A}\big)\Big);$$
summing this over all A, B, C and using (6.7) and the positivity of f_1+f_2, g_1+g_2, h_1+h_2 we conclude that
$$\Lambda_3(f_1 1_{E^0_{X,Y}}, g_1 1_{E^0_{Y,Z}}, h_1 1_{E^0_{Z,X}}) \leq O(\varepsilon) + O\big(1/F(M)\big)$$
(compare this with (6.8)). Since f_1, g_1, h_1 are bounded from below by $\tau^{1/10}$ on these sets, we thus have
$$\Lambda_3(1_{E^0_{X,Y}}, 1_{E^0_{Y,Z}}, 1_{E^0_{Z,X}}) \leq O(\tau^{-3/10}\varepsilon) + O(\tau^{-3/10}/F(M)).$$
Now let $E_{X,Y}$ be the subset of $E^0_{X,Y}$, defined as the union of all products $A\times B \subset E^0_{X,Y}$ of atoms $A \in \mathcal{B}_X$, $B \in \mathcal{B}_Y$ of size at least $\mu_X(A), \mu_Y(B) \geq \tau/2^M$. Since $\mathcal{B}_X$ has complexity at most M, the union of all atoms in $\mathcal{B}_X$ of measure at most $\tau/2^M$ has measure at most τ, and thus we see that
$$\mu_X \times \mu_Y(E^0_{X,Y} \setminus E_{X,Y}) = O(\tau)$$
and hence from preceding computations
$$\int_X \int_Y f(x,y) 1_{E^c_{X,Y}}(x,y) \, d\mu_X(x) \, d\mu_Y(y) = o_{\tau\to 0}(1).$$
We define $E_{Y,Z}$, $E_{Z,X}$ similarly and observe similar bounds. Now suppose that the expression $1_{E_{X,Y}}(x,y) 1_{E_{Y,Z}}(y,z) 1_{E_{Z,X}}(z,x)$ does not vanish identically, then there exist atoms A, B, C of $\mathcal{B}_X$, $\mathcal{B}_Y$, $\mathcal{B}_Z$ with $A\times B \subset E_{X,Y}$, $B\times C \subset E_{Y,Z}$, and $C\times A \subset E_{Z,X}$. In particular
$$\Lambda_3(1_{A\times B}, 1_{B\times C}, 1_{C\times A}) \leq O(\tau^{-3/10}\varepsilon) + O(\tau^{-3/10}/F(M)).$$
On the other hand we have
$$\Lambda_3(1_{A\times B}, 1_{B\times C}, 1_{C\times A}) = \mu_X(A)\mu_Y(B)\mu_Z(C) \geq (\tau/2^M)^3.$$

If we define $F(M) := \lfloor 2^{3M}/\tau^3 \rfloor + 1$, and assume that ε is sufficiently large depending on τ (noting that $M = O_{F,\tau}(1) = O_\tau(1)$), we obtain a contradiction. Thus we see that $1_{E_{X,Y}}(x,y) 1_{E_{Y,Z}}(y,z) 1_{E_{Z,X}}(z,x)$ vanishes identically whenever ε is sufficiently small depending on τ. If we then set τ to be a sufficiently slowly decaying function of ε, the claim follows. □

Observe that the actual decay rate $o_{\varepsilon \to 0}(1)$ obtained by the above proof is very slow (it decays like the reciprocal of the inverse tower-exponential function). It is of interest to obtain better bounds here; it is not known what the exact rate should be, although the Behrend example (Proposition 1.3) does show that the decay cannot be polynomial in nature.

The above arguments extend (with some nontrivial difficulty) to hypergraphs, and to proving Szemerédi's theorem for progressions of length $k > 3$; the $k = 4$ case was handled in [9], [10] (see also [19] for a more recent proof), and the general case in [20, 32–35] (see also [42, 46] for more recent proofs). We sketch the $k = 4$ arguments here (broadly following the ideas from [42, 46]). Finding progressions of length 4 in a set A is equivalent to solving the simultaneous relations

$$\begin{aligned} -x_2 - 2x_3 - 3x_4 &\in A \\ x_1 \phantom{{}+x_2} - x_3 - 2x_4 &\in A \\ 2x_1 + x_2 \phantom{{}-x_3} - x_4 &\in A \\ 3x_1 + 2x_2 + x_3 \phantom{{}-x_4} &\in A. \end{aligned}$$

Because of this, it is not hard to modify the above arguments to deduce the $k = 4$ case of Szemerédi's theorem from the following lemma:

Lemma 6.12 (Strong tetrahedron removal lemma, several variable version). *Let $(X_1, \mu_{X_1}), \ldots, (X_4, \mu_{X_4})$ be probability spaces, and for $ijk = 123, 234, 341, 412$ let $f_{ijk}: X_i \times X_j \times X_k \to [0,1]$ be measurable functions such that*

$$\Lambda_4(f_{123}, f_{234}, f_{341}, f_{412}) \leq \varepsilon$$

for some $0 < \varepsilon < 1$, where Λ_4 is the trilinear form

$$\Lambda_4(f_{123}, f_{234}, f_{341}, f_{412}) \\ := \int_{X_1} \cdots \int_{X_4} \prod_{ijk=123,234,341,412} f_{ijk}(x_i, x_j, x_k) \, d\mu_{X_1}(x_1) \cdots d\mu_{X_4}(x_4).$$

Then for each $ij = 12, 23, 34, 41, 13, 24$ there exists factors $\mathcal{B}_{ij}$ of $X_i \times X_j$ of complexity at most $O_\varepsilon(1)$ and sets $E_{ijk} \in \mathcal{B}_{ij} \vee \mathcal{B}_{ik} \vee \mathcal{B}_{jk}$ for $ijk = 123, 234, 341, 412$ with $\prod_{ijk=123,234,341,412} 1_{E_{ijk}}(x_i, x_j, x_k)$ vanishing identically, such that

$$\int_{X_i} \int_{X_j} \int_{X_k} f_{ijk}(x_i, x_j, x_k) 1_{E_{ijk}^c}(x_i, x_j, x_k) \, d\mu_{X_1}(x_1) \, d\mu_{X_2}(x_2) \, d\mu_{X_3}(x_3) \\ \leq o_{\varepsilon \to 0}(1).$$

One can recast this lemma as a statement concerning 3-uniform hypergraphs; see for instance [42]. We will however not pursue this interpretation here (but see [9, 10, 19, 20, 32–35] for a treatment of this material from a hypergraph perspective).

In the case of the triangle removal lemma, it was the $\Box^2$ norm which controlled the size of Λ_4. Now the role is played by the $\Box^3$ norm, defined for a measurable bounded function $f(x,y,z) \colon X \times Y \times Z \to \mathbb{R}$ of three variables by the formula

$\|f\|_{\square^3(X \times Y \times Z)}^8$

$$:= \int_X \int_X \int_Y \int_Y \int_Z \int_Z f(x,y,z) f(x,y,z') f(x,y',z) f(x,y',z') f(x',y,z)$$
$$\times f(x',y,z') f(x',y',z) f(x',y',z')$$
$$\mathrm{d}\mu_X(x)\, \mathrm{d}\mu_X(x')\, \mathrm{d}\mu_Y(y)\, \mathrm{d}\mu_Y(y')\, \mathrm{d}\mu_Z(z)\, \mathrm{d}\mu_Z(z').$$

By modifying the previous arguments we see that the $\square^3$ norm is indeed a norm (after equating functions that agree almost everywhere) and that we have the generalised von Neumann inequality

$$|\Lambda_4(f,g,h,k)| \leq \min(\|f\|_{\square^3}, \|g\|_{\square^3}, \|h\|_{\square^3}, \|k\|_{\square^3}).$$

The analogue of Lemma 6.6 is

Lemma 6.13 (Lack of uniformity implies correlation with structure). *Let $f \colon X \times Y \times Z \to [-1,1]$ be such that $\|f\|_{\square^3} \geq \eta$ for some $\eta > 0$. Then there exists $A_{X,Y} \subset X \times Y$, $A_{Y,Z} \subset Y \times Z$, and $A_{Z,X} \in Z \times X$*

$$\left| \int_X \int_Y \int_Z 1_{A_{X,Y}}(x,y) 1_{A_{Y,Z}}(y,z) 1_{A_{Z,X}}(z,x) f(x,y,z)\, \mathrm{d}\mu_X(x)\, \mathrm{d}\mu_Y(y)\, \mathrm{d}\mu_Z(z) \right|$$
$$\geq \eta^8/8.$$

This ultimately leads to the following regularity lemma:

Lemma 6.14 (Simultaneous Szemerédi regularity lemma). *For $ijk = 123, 234, 341, 412$, let $f_{ijk} \colon X_i \times X_j \times X_k \to [0,1]$ be measurable, let $\tau > 0$, and let $F \colon \mathbb{N} \to \mathbb{N}$ be an arbitrary increasing function (possibly depending on τ). Then there exists an integer $M = O_{F,\tau}(1)$, factors $\mathcal{B}_{ij}$ of $X_i \times X_j$ of complexity at most M for $ij = 12, 23, 34, 41, 13, 24$ and decompositions $f_{ijk} = f_{ijk,1} + f_{ijk,2} + f_{ijk,3}$ for $ijk = 123, 234, 341, 412$ where*

- *($f_{ijk,1}$ is structured) We have $f_{ijk,1} = \mathbb{E}(f_{ijk}|\mathcal{B}_{ij} \vee \mathcal{B}_{jk} \vee \mathcal{B}_{ik})$.*
- *($f_{ijk,2}$ is small) We have $\|f_{ijk,2}\|_{L^2(X_i \times X_j \times X_k)} \leq \tau$.*
- *($f_{ijk,3}$ is very uniform) We have $\|f_{ijk,3}\|_{\square^3(X_i \times X_j \times X_k)} \leq 1/F(M)$.*
- *(Positivity) $f_{ijk,1}$ and $f_{ijk,1} + f_{ijk,2}$ take values in $[0,1]$.*

One would then like to repeat the proof of Lemma 6.11 by applying this lemma to decompose each function f_{ijk} into three components $f_{ijk,1}, f_{ijk,2}, f_{ijk,3}$, and then somehow eliminate the latter two terms to reduce to the structured component $f_{ijk,1}$. The reason for doing this is that, as $f_{ijk,1}$ is measurable with respect to the bounded complexity factor $\mathcal{B}_{ij} \vee \mathcal{B}_{jk} \vee \mathcal{B}_{ik}$, one can decompose this function (which is a function of three variables x_i, x_j, x_k) as a polynomial combination of functions of just two variables (or more precisely, as a linear combination of functions of the form $f_{ij}(x_i, x_j) f_{jk}(x_j, x_k) f_{ik}(x_i, x_k)$). One can then apply a (slight generalisation of) the triangle removal lemma to handle such functions; more generally, the strategy is to deduce these sort of removal lemmas for functions of k variables, from similar lemmas concerning functions of $k-1$ variables. In executing this strategy, there is little difficulty in disposing of the very uniform components $f_{ijk,3}$, if one takes advantage of the freedom to make the growth function F extremely rapid (one needs to take F to be tower-exponential or faster, to counteract the very weak decay present in the two-variable removal lemmas). To dispose of the small components $f_{ijk,2}$ takes a little more work, however. In the above arguments, one implicitly used the independence of the underlying factors $\mathcal{B}_X, \mathcal{B}_Y, \mathcal{B}_Z$. In the current situation, the factors $\mathcal{B}_{ij}$ are not independent of each other, which makes it difficult to eliminate the $f_{ijk,2}$ factors directly. However, this can be addressed by applying the (two-variable) regularity lemma to simultaneously regularise all the atoms in the factors $\mathcal{B}_{ij}$, making them essentially independent relative to one-variable factors. As one might imagine, making this strategy rigorous is somewhat delicate, and in particular the various large and small parameters (such as τ and F) that appear in the regularity lemmas need to be chosen correctly. See for instance [42] for one such realisation of this type of argument. More recently, an infinitary approach, using a correspondence principle similar in spirit to the Furstenberg correspondence principle, has been employed to give a slightly different proof of the above results, in which the various large and small parameters in the argument have been set to infinity or zero, thus leading to a cleaner (but less elementary) version of the argument; see [46].

7. Relative triangle removal

The triangle removal result proven in the previous section, Lemma 6.3, only has nontrivial content when the underlying graph G is *dense*, or more precisely when it contains more than $o_{\varepsilon \to 0}(n^2)$ edges, since otherwise one could simply delete all the edges in G to remove the triangles. This is related to the fact that Lemma 6.3 only implies the existence of progressions of length three in dense sets of integers, but not in sparse sets. However, it is a remarkable and useful fact that results such as Lemma 6.3, which ostensibly only apply to dense objects, can in fact be extended "for free" to sparse objects, as long as the sparse object has large *relative* density with respect to a sufficiently *pseudorandom* object. This type of "transference principle" from the dense category to the relatively dense category was the decisive new ingredient in the result in [25] that the primes contained arbitrarily long arithmetic progressions. We will not prove that result here, however we present a simplified version of that result which already captures many of the key ideas.

If $n \geq 1$ is an integer and $0 \leq p \leq 1$, let $G(n,p)$ be the standard Erdős–Rényi random graph on n vertices $\{1, \ldots, n\}$, in which each pair of vertices defines an edge in $G(n,p)$ with an identical independent probability of p.

Proposition 7.1 (Relative triangle removal lemma [28, 43]). *Let $n > 1$ and $1/\log n \leq p \leq 1$, let $0 < \varepsilon < 1$, and let $H = G(n,p)$. Then with probability $1 - o_{n \to \infty;\varepsilon}(1)$ the following claim is true: whenever G is a subgraph of H which contains fewer than $\varepsilon p^3 n^3$ triangles, then it is possible to delete $o_{\varepsilon \to 0}(p^2 n^2) + o_{n \to \infty;\varepsilon}(p^2 n^2)$ edges from G to create a new graph G which contains no triangles whatsoever.*

This result in fact extends to much sparser graphs $G(n,p)$, indeed one can take $p = n^{-1/2+\delta}$ for any fixed $0 < \delta < \frac{1}{2}$; see [28]. The argument proceeded by performing a careful generalisation of the usual regularity lemma to the setting of sparse subsets of pseudorandom graphs. As one corollary of their result, one can conclude that if A is a random subset of the positive integers with $\mathbb{P}(n \in A) = n^{-1/2+\delta}$, and with the events $n \in A$ being independent, then almost surely every subset of A of positive density would contain infinitely many progressions of length three. We shall proceed differently, using a "soft" transference argument, inspired by the ergodic theory approach, which follows closely the treatment in [25] (and also [43]). So far, this argument can only handle logarithmic sparsities rather than polynomial, but requires much less randomness on the graph $G(n,p)$; indeed a suitably "pseudorandom" graph would also suffice for this argument. (For the precise definition of the pseudorandomness needed, see [43].)

Let $(X, \mu_X) = (Y, \mu_Y) = (Z, \mu_Z)$ be the vertex set $\{1, \ldots, n\}$ with the uniform distribution. Fix the random graph $H = G(n,p)$, and let $\nu(x,y)$ be the function on $\{1, \ldots, n\} \times \{1, \ldots, n\}$ which equals $1/p$ when (x,y) lies in H and 0 otherwise; we can think of ν as a function on $X \times Y$, $Y \times Z$, or $Z \times X$. Note from Chernoff's inequality that even though ν is not bounded by $O(1)$, with probability $1-o_{n\to\infty}(1)$, ν has average close to 1:

$$\int_X \int_Y \nu(x,y) \, d\mu_X(x) \, d\mu_Y(y) = 1 + o_{n\to\infty}(1).$$

More sophisticated computations of this sort show that many other correlations of ν with itself are close to 1. For instance, one can show that with probability $1 - o_{n\to\infty}(1)$, we have the octahedral correlation estimate

$$(7.1) \quad \int_X \int_X \int_Y \int_Y \int_Z \int_Z \nu(x,y)\nu(x,y')\nu(x',y)\nu(x',y')\nu(y,z)\nu(y,z')$$
$$\times \nu(y',z)\nu(y',z')\nu(z,x)\nu(z,x')\nu(z',x)\nu(z',x')$$
$$d\mu_X(x) \, d\mu_X(x') \, d\mu_Y(y) \, d\mu_Y(y') \, d\mu_Z(z) \, d\mu_Z(z')$$
$$= 1 + o_{n\to\infty}(1).$$

(In [43], this estimate, together with some simpler versions, are referred to as the *linear forms condition* on ν.) To prove Proposition 7.1, it then suffices to prove the following variant of Lemma 6.11:

Lemma 7.2 (Relative strong triangle removal lemma, several variable version). *Let (X, μ_X), (Y, μ_Y), (Z, μ_Z), ν be as above, and let $0 < \varepsilon \leq 1$. With probability $1 - o_{n\to\infty;\varepsilon}(1)$, the following claim is true: whenever $f\colon X \times Y \to [0,1]$, $g\colon Y \times Z \to [0,1]$, and $h\colon Z \times X \to [0,1]$ are measurable functions such that $\Lambda_3(f\nu, g\nu, h\nu) \leq \varepsilon$, then there exists factors $\mathcal{B}_X$, $\mathcal{B}_Y$, $\mathcal{B}_Z$ of X, Y, Z respectively of complexity at most $O_\varepsilon(1)$ and sets $E_{X,Y} \in \mathcal{B}_X \vee \mathcal{B}_Y$, $E_{Y,Z} \in \mathcal{B}_Y \vee \mathcal{B}_Z$, $E_{Z,X} \in \mathcal{B}_Z \vee \mathcal{B}_X$ respectively with $1_{E_{X,Y}}(x,y)1_{E_{Y,Z}}(y,z)1_{E_{Z,X}}(z,x)$ vanishing identically, such that*

$$\int_X \int_Y f(x,y)\nu(x,y) 1_{E^c_{X,Y}}(x,y) \, d\mu_X(x) \, d\mu_Y(y),$$
$$\int_Y \int_Z g(y,z)\nu(y,z) 1_{E^c_{Y,Z}}(y,z) \, d\mu_Y(y) \, d\mu_Z(z),$$
$$\int_Z \int_X h(z,x)\nu(z,x) 1_{E^c_{Z,X}}(z,x) \, d\mu_Z(z) \, d\mu_X(x) \leq o_{\varepsilon\to 0}(1).$$

We leave the deduction of Proposition 7.1 from Lemma 7.2 as an exercise. Note that the only new feature here is the presence of the weight ν, which causes functions such as $f\nu$ to be unbounded. Nevertheless, it turns out to be possible to use arguments similar to those in the preceding section and obtain this result with a little effort from its unweighted counterpart, Lemma 6.11.

The first thing to do is to check that the generalised von Neumann inequality, (6.6), continues to hold in the weighted setting:

Lemma 7.3 (Relative generalised von Neumann inequality [43]). *Let the notation be as above. Then with probability $1 - o_{n\to\infty}(1)$, the following claim is true: whenever $f\colon X \times Y \to \mathbb{R}$, $g\colon Y \times Z \to \mathbb{R}$ and $h\colon Z \times X \to \mathbb{R}$ are bounded in magnitude by $\nu + 1$ (thus for instance $|f(x,y)| \leq \nu(x,y) + 1$ for all $(x,y) \in X \times Y$), then*

$$|\Lambda_3(f, g, h)| \leq 4 \min(\|f\|_{\square^2}, \|g\|_{\square^2}, \|h\|_{\square^2}) + o_{n\to\infty}(1).$$

See also [25] for a closely related computation. We also remark that the estimate (6.5) also continues to hold in this setting because that estimate did not require f to be bounded.

PROOF (Sketch only). By symmetry it suffices to show that

$$\left| \int_X \int_Y \int_Z f(x,y) g(y,z) h(z,x) \, d\mu_X(x) d\mu_Y(y) \, d\mu_Z(z) \right| \leq \|f\|_{\square^2} + o_{n\to\infty}(1).$$

Note that it is easy to verify that $\|\nu + 1\|_{\square^2} = 2 + o_{n\to\infty}(1)$ with high probability, and hence $\|f\|_{\square^2} = O(1)$. We eliminate the h function by Cauchy–Schwarz in the z, x variables and reduce to showing

$$\left| \int_X \int_Y \int_Y \int_Z f(x,y) f(x,y') g(y,z) g(y',z) (\nu(z,x) + 1) \right.$$
$$\left. d\mu_X(x) \, d\mu_Y(y) \, d\mu_Y(y') \, d\mu_Z(z) \right| \leq 8 \|f\|_{\square^2}^2 + o_{n\to\infty}(1)$$

and then eliminate g by a Cauchy–Schwarz in the y, y', z variables and reduce to showing

$$\left| \int_X \int_X \int_Y \int_Y f(x,y) f(x,y') f(x',y) f(x',y') W(x,x',y,y') \right.$$
$$\left. d\mu_X(x) \, d\mu_X(x') \, d\mu_Y(y) \, d\mu_Y(y') \right| \leq 16 \|f\|_{\square^2}^4 + o_{n\to\infty}(1)$$

where

$$W(x, x', y, y') := \int_Z (\nu(y, z) + 1)(\nu(y', z) + 1)(\nu(z, x) + 1)(\nu(z, x') + 1) \, d\mu_Z(z).$$

If $W \equiv 16$ then we would be done by definition of the $\square^2$ norm. So it suffices to show that

$$\left| \int_X \int_X \int_Y \int_Y f(x,y) f(x,y') f(x',y) f(x',y') |W(x,x',y,y') - 16| \right.$$
$$\left. d\mu_X(x) \, d\mu_X(x') \, d\mu_Y(y) \, d\mu_Y(y') \right| \leq o_{n\to\infty}(1).$$

By one last Cauchy–Schwarz this follows from the estimate

$$\left| \int_X \int_X \int_Y \int_Y (\nu(x,y) + 1)(\nu(x,y') + 1)(\nu(x',y) + 1)(\nu(x',y') + 1) \right.$$
$$\left. \times |W(x,x',y,y') - 16|^2 \, d\mu_X(x) \, d\mu_X(x') \, d\mu_Y(y) \, d\mu_Y(y') \right| \leq o_{n\to\infty}(1)$$

which can be easily verified from correlation estimates such as (7.1). $\square$

In light of this lemma, we can continue to neglect errors which are small in $\square^2$ norm as being negligible. The key to establishing Lemma 7.2 now rests with the following decomposition:

Theorem 7.4 (Structure theorem [43]). *Let the notation be as above, let $f\colon X \times Y \to [0,1]$ be a function, and let $\sigma > 0$. Then there exists a decomposition*

$$f\nu = f_1 + f_2 + f_3$$

where f_1 is nonnegative and obeys the uniform upper bound

$$f_1(x,y) \le 1 \quad \text{for all } (x,y) \in X \times Y,$$

f_2 is nonnegative and obeys the smallness bound

(7.2) $$\int_X \int_Y f_2(x,y)\, d\mu_X(x)\, d\mu_Y(y) = o_{n\to\infty;\sigma}(1),$$

and f_3 obeys the uniformity estimate

(7.3) $$\|f_3\|_{\square^2(X\times Y)} = o_{\sigma\to 0}(1).$$

Furthermore $f_1 + f_3$ is also nonnegative.

This theorem should be compared with Lemma 6.9. The key point is that it approximates the function $f\nu$, for which we have no good uniform bounds, for the function f_1, which is bounded by 1. With this theorem (and Lemma 7.3) it is now a simple matter to deduce Lemma 7.2 from Lemma 6.11:

PROOF OF LEMMA 7.2. We may assume that n is sufficiently large depending on ε, as the claim is trivial otherwise. Let $0 < \sigma \le \varepsilon$ be chosen later. We apply Theorem 7.4 to decompose $f\nu = f_1 + f_2 + f_3$, $g\nu = g_1 + g_2 + g_3$, $h\nu = h_1 + h_2 + h_3$, thus

$$\Lambda_3(f_1 + f_2 + f_3, g_1 + g_2 + g_3, h_1 + h_2 + h_3) \le \varepsilon.$$

Since $f_1 + f_3$, $g_1 + g_3$, $h_1 + h_3$, f_2, g_2, h_2 are all nonnegative, we conclude

$$\Lambda_3(f_1 + f_3, g_1 + g_3, h_1 + h_3) \le \varepsilon.$$

Repeated application of Lemma 7.3 and (7.3) (and the hypothesis $\sigma \le \varepsilon$) then gives

$$\Lambda_3(f_1, g_1, h_1) \le o_{\varepsilon\to 0}(1).$$

The functions f_1, g_1, h_1 are bounded, so we may apply Lemma 6.11 and obtain $\mathcal{B}_X, \mathcal{B}_Y, \mathcal{B}_Z$ of X, Y, Z respectively of complexity at most $O_\varepsilon(1)$ and sets $E_{X,Y} \in \mathcal{B}_X \vee \mathcal{B}_Y$, $E_{Y,Z} \in \mathcal{B}_Y \vee \mathcal{B}_Z$, $E_{Z,X} \in \mathcal{B}_Z \vee \mathcal{B}_X$ respectively with $1_{E_{X,Y}}(x,y) \times 1_{E_{Y,Z}}(y,z) 1_{E_{Z,X}}(z,x)$ vanishing identically, such that

$$\int_X \int_Y f_1(x,y) 1_{E^c_{X,Y}}(x,y)\, d\mu_X(x)\, d\mu_Y(y),$$

$$\int_Y \int_Z g_1(y,z) 1_{E^c_{Y,Z}}(y,z)\, d\mu_Y(y)\, d\mu_Z(z),$$

$$\int_Z \int_X h_1(z,x) 1_{E^c_{Z,X}}(z,x)\, d\mu_Z(z)\, d\mu_X(x) \le o_{\varepsilon\to 0}(1).$$

From (7.2) we have similar estimates for f_2, g_2, h_2:

$$\int_X \int_Y f_2(x,y) 1_{E^c_{X,Y}}(x,y)\, d\mu_X(x)\, d\mu_Y(y),$$

$$\int_Y \int_Z g_2(y,z) 1_{E^c_{Y,Z}}(y,z)\, d\mu_Y(y)\, d\mu_Z(z),$$

$$\int_Z \int_X h_2(z,x) 1_{E^c_{Z,X}}(z,x)\, d\mu_Z(z)\, d\mu_X(x) \le o_{n\to\infty;\sigma}(1).$$

Also, from (7.3), (6.5) and the complexity bounds on $\mathcal{B}_X, \mathcal{B}_Y, \mathcal{B}_Z$ we have similar estimates for f_3, g_3, h_3:

$$\int_X \int_Y f_3(x,y) 1_{E^c_{X,Y}}(x,y) \, d\mu_X(x) \, d\mu_Y(y),$$

$$\int_Y \int_Z g_3(y,z) 1_{E^c_{Y,Z}}(y,z) \, d\mu_Y(y) \, d\mu_Z(z),$$

$$\int_Z \int_X h_3(z,x) 1_{E^c_{Z,X}}(z,x) \, d\mu_Z(z) \, d\mu_X(x) \leq o_{\sigma \to 0; \varepsilon}(1).$$

If we choose σ sufficiently small depending on ε, we thus have

$$\int_X \int_Y f(x,y) 1_{E^c_{X,Y}}(x,y) \, d\mu_X(x) \, d\mu_Y(y),$$

$$\int_Y \int_Z g(y,z) 1_{E^c_{Y,Z}}(y,z) \, d\mu_Y(y) \, d\mu_Z(z),$$

$$\int_Z \int_X h(z,x) 1_{E^c_{Z,X}}(z,x) \, d\mu_Z(z) \, d\mu_X(x) \leq o_{\varepsilon \to 0}(1) + o_{n \to \infty; \varepsilon}(1)$$

and the claim follows. $\square$

Notice how the complexity estimates on $\mathcal{B}_X, \mathcal{B}_Y, \mathcal{B}_Z$ were essential in allowing one to transfer the unweighted triangle removal lemma, Lemma 6.11, to the weighted setting, Lemma 7.2.

It remains to prove the structure theorem, Theorem 7.4. A full proof (in much greater generality) of this theorem can be found in [43], while a closely related theorem appears in [25]. We give only a brief summary of the argument here. Broadly speaking, we follow the energy increment strategy as used to prove Corollary 6.8. However, we cannot use Lemma 6.6 as it only applies for functions f which are bounded. We must therefore redefine the notion of "structure," replacing the notion of a tensor product $1_A(x) 1_B(y)$ with the notion of a *dual function* $\mathcal{D}f(x,y)$ of a function $f \colon X \times Y \to \mathbb{R}$, defined as

$$\mathcal{D}f(x,y) := \int_X \int_Y f(x,y') f(x',y) f(x',y') \, d\mu_X(x') \, d\mu_Y(y').$$

Observe that we have the identity

$$\int_X \int_Y f(x,y) \mathcal{D}f(x,y) \, d\mu_X(x) \, d\mu_Y(y) = \|f\|^4_{\square^2(X \times Y)}.$$

Thus if a function f has large $\square^2$ norm then it correlates with its own dual function. This fact will be used as a substitute for Lemma 6.6. One key property of dual functions are that they can be bounded even when f is unbounded; in particular, with high probability we have $\mathcal{D}(\nu + 1)$ bounded pointwise by $O(1)$, and hence $\mathcal{D}f$ will also be bounded for any f bounded pointwise in magnitude by $\nu + 1$. Each of these dual functions can define finite factors $\mathcal{B}_{\mathcal{D}f, \varepsilon}$ for any resolution $\varepsilon > 0$ by partitioning the range of $\mathcal{D}f$ into intervals of length ε and letting $\mathcal{B}_{\mathcal{D}f, \varepsilon}$ be the factor generated by the inverse image of these intervals. (For technical reasons it is convenient to randomly shift this partition in order to negate certain boundary effects — which ultimately lead to the small error f_2 appearing in Theorem 7.4 — but let us gloss over this minor detail here.) Define a *dual factor of complexity M and resolution* ε to be a factor of the form $\mathcal{B} = \mathcal{B}_{\mathcal{D}f_1, \varepsilon} \vee \cdots \vee \mathcal{B}_{\mathcal{D}f_M, \varepsilon}$ where $f_1, \ldots, f_M$ are bounded in magnitude by $\nu + 1$. These factors are the counterparts of the factors $\mathcal{B}_X \vee \mathcal{B}_Y$ studied in the previous section. A crucial feature of these factors is (with high probability) that the random weight function ν is uniformly distributed with respect to all of these factors; more precisely, with probability $1 - o_{n \to \infty; \varepsilon, M}(1)$ we have $\mathbb{E}(\nu|\mathcal{B}) = 1 + o_{n \to \infty; M, \varepsilon}(1)$ outside of an exceptional set $\Omega = \Omega_\mathcal{B}$ with $\int_X \int_Y 1_\Omega(x, y)(\nu(x, y) + 1) \, d\mu_X \, d\mu_Y = o_{n \to \infty; M, \varepsilon}(1)$ for all dual factors of complexity M. This fact is somewhat nontrivial to prove; one needs to invoke the Weierstrass approximation theorem to approximate the indicator function of atoms in $\mathcal{B}$ by polynomial combinations of the dual functions $\mathcal{D}f$ (with the approximation being uniform outside of a small exceptional set Ω), and then using tools such as the Gowers-Cauchy–Schwarz inequality one can control the inner product of ν with such polynomials. See [25, 43] for details.

Once one has these dual factors with respect to which ν is (essentially) uniformly distributed, one can then develop a counterpart of Lemma 6.7, which roughly speaking asserts that if f is a function bounded in magnitude by ν, and $\mathcal{B}$ is a dual factor of some complexity M and resolution ε for which $\|f - \mathbb{E}(f \mid \mathcal{B})\|_{\square^2} \geq \eta$, then with high probability one can find an extension $\mathcal{B}'$ of $\mathcal{B}$ which is a dual factor of complexity $M+1$ and resolution ε, for which the energy $\|\mathbb{E}(f|\mathcal{B}')\|_{L^2}^2$ has increased from $\|\mathbb{E}(f \mid \mathcal{B})\|_{L^2}^2$ by some factor $c(\eta) - o_{\varepsilon \to 0}(1) - o_{n \to \infty; M, \varepsilon}(1)$ for some $c(\eta) > 0$. This is essentially proven by the same Pythagoras theorem argument used to establish Lemma 6.7, though one has to take some care because f, being bounded by ν, does not enjoy good L^2 bounds (though the conditional expectations $\mathbb{E}(f \mid \mathcal{B})$, $\mathbb{E}(f \mid \mathcal{B}')$ enjoy uniform bounds outside of a small exceptional set). One can then iterate this as in the proof of Corollary 6.8 to obtain Theorem 7.4 (with some additional $o_{n \to \infty; \varepsilon}(1)$ errors arising from exceptional sets etc. that can be placed in the small error f_2). See [25, 43] for details.

8. Szemerédi's original proof

We now discuss some of the ideas behind Szemerédi's original proof [41] of his theorem. This is a remarkably subtle combinatorial argument, and there is no chance that we can describe the full argument here, but we can at least begin to motivate part of the argument. Rather than plunge directly into the full setup of the argument, we will begin with some naive first attempts at the problem, which do not fully work, but which indicate the steps that need to be taken to obtain a full proof.

The task is, given $k \geq 3$, to show that any subset A of integers whose upper density $\delta = \delta[A] := \limsup_{N \to \infty} |A \cap [-N, N]|/(2N + 1)$ is positive contains at least one progression of length k. The first idea dates back to the original argument of Roth [36] for the $k = 3$ case, which is to try to induct downwards on the upper density of the set (this is known as the *density increment method*). If δ is extremely large, say $\delta > 1 - 1/2k$, then the result is easy, because even a randomly chosen progression will have a good chance of being entirely contained in A. Now one assumes inductively that A has some given upper density $\delta > 0$, and that the theorem has already been proven for higher values of δ. It is not hard to show that the set of δ for which Szemerédi's theorem holds must be open, so if we can verify in this "maximal bad density" case[12] that progressions of length k exist, then we are done.

---

[12]This trick is vaguely reminiscent of the reduction to minimal topological dynamical systems, or to ergodic measure-preserving systems. Unfortunately these tricks seem to be mutually exclu-

Suppose for contradiction that the set A of this critical density δ did not have any progressions of length k, even though all sets of higher density did have progressions. What this means is that A cannot contain within it arbitrarily large progressions on which A has higher density. In other words, we cannot find a sequence of progressions $P_1, P_2, \ldots$ in $\mathbb{Z}$ with length tending to infinity for which $\limsup_{n \to \infty} |A \cap P_n|/|P_n| > \delta$, since if this were the case it would not be difficult to piece together out of the $A \cap P_n$ a set with slightly higher upper density than A, but which still had no progressions, contradicting the hypothesis on δ. Thus we must have $\limsup_{n \to \infty} |A \cap P_n|/|P_n| \leq \delta$ whenever $|P_n| \to \infty$. In other words, we have the upper bound

(8.1) $$|A \cap P| \leq \bigl(\delta + o_{|P| \to \infty; A}(1)\bigr)|P|$$

for all progressions P. [Incidentally, if we knew Szemerédi's theorem in the first place, one would deduce immediately that the only such sets A are those sets with density $\delta = 0$ or density $\delta = 1$, but of course we cannot use Szemerédi's theorem to prove itself in such a circular manner!]

Thus on a long progression P, the density of A in P cannot significantly exceed δ. It is still possible for the density of A to be significantly *less* than δ on such progressions — but this cannot happen too often, as this would (in conjunction with the upper bound) eventually cause A itself to have density less than δ. This idea can be easily quantified, and leads to the statement that given any length N, the set

$$\bigl\{ n \in \mathbb{Z} : |A \cap [n, n+N]| = \bigl(\delta + o_{N \to \infty; A}(1)\bigr) N \bigr\}$$

has upper density $1 - o_{N \to \infty; A}(1)$. Thus "most" progressions of length N have density $\delta + o_{N \to \infty; A}(1)$.

This then leads to the next idea, which is to partition the integers into *blocks* $[nN, (n+1)N)$ — progressions of length N, in which n is a multiple of N. Call such a block *saturated* if it has the expected density $\delta + o_{N \to \infty; A}(1)$, thus most blocks (in an upper density sense) are saturated. Suppose temporarily that we could in fact assume that *all* blocks are saturated. Then we could conclude the argument as follows. We can colour the nth block $[nN, (n+1)N)$ in one of 2^N colours depending on how A is situated inside that block; more precisely, we can color the block $[nN, (n+1)N)$ by the set $\{0 \leq i < N : nN + i \in A\}$. Actually we only need $2^N - 1$ colours because the block, being saturated, cannot be completely devoid of elements of A. We have thus coloured all the integers into finitely many colours, and hence by van der Waerden's theorem there is a monochromatic progression of blocks of length k. These blocks have A contained in them in identical fashions, and the blocks are not completely devoid of elements of A, so it is not hard to see that the progression of blocks induces a progression of elements of A of the same length, and we are done.

Unfortunately, life is not so simple, and we have the unsaturated blocks to deal with. While the (lower) density of these exceptional blocks is somewhat small in an absolute sense — it is $o_{N \to \infty; A}(1)$ — it is not very small when compared against the number of colours, $2^N - 1$ (or against the reciprocal of this number, to be precise). Van der Waerden's theorem is nowhere near robust enough to handle such a severe influx of "uncoloured" elements. (It can however deal with a rather

---

sive; if one takes sequences of maximal density then it becomes difficult to convert the argument into a dynamical setting.

easy degenerate case in which the density of saturated blocks unexpectedly happens to be incredibly close to 1, say at least $1 - c(N)$ for some explicit but extremely small $c(N) > 0$ whose exact value depends on the constants arising from van der Waerden's theorem.) Here we encounter a recurring problem in this field: we are always dealing with quantities which are small, but not small enough. One is always seeking ways to somehow iteratively improve the smallness, or at least convert the smallness to another type of smallness which is more robust, in order to get around this basic issue.

Let's try something else for now. Suppose we can locate k large blocks of integers, say $[0, N), [N, 2N), \ldots, [(k-1)N, kN)$, which are all saturated. (This is not hard since the upper density of saturated blocks easily exceeds $1 - 1/2k$ when N is large enough.) Let's try to find progressions of length k in A with one element in each block. Suppose we have somehow (presumably by some sort of an inductive hypothesis) managed to already find many progressions of length $k-1$ in A with one element in each of the first $k-1$ of these blocks. We can extend each of these progressions by one element, which will most likely lie in the final block $[(k-1)N, kN)$. (Some of them will not. However observe that A has to be more or less uniformly distributed on any saturated block, because on any subinterval of proportional size, A has to have density not much larger δ, and thus on subtraction it must have density not much less than δ either. Because of this it is very plausible that a significant fraction of the progressions of $k-1$ located from the induction step will have kth element in the final block as claimed.) Let B denote the set of all such additional elements of these progressions in $[(k-1)N, kN)$. If we had a lot of progressions of length $k-1$, it is plausible to expect (by simple counting heuristics) that B should have some positive density in $[(k-1)N, kN)$ (indeed, one expects the density to be comparable to δ^{k-1}). If B intersects A, then we are done.

Unfortunately, B and A are both rather sparse sets inside $[(k-1)N, kN)$—one has density about δ^{k-1} (assuming some appropriate induction hypothesis), and the other has density about δ. These are too sparse to force an intersection unconditionally. However, we do know that A obeys some good uniform distribution bounds on progressions—its density is always bounded from above, and often bounded from below. This would be useful if B was somehow made out of progressions (or even better, if the *complement* of B was made out of progressions, since upper bounds on the density of A in the complement of B translate to lower bounds on the density of A in B), but we do not have such good structural control on B and it could well be just a generic sparse subset of $[(k-1)N, kN)$, and we are stuck. Indeed, there is nothing right now that stops B from simply being some subset of the complement of A, and no matter how structured or uniformly distributed A is, we cannot prevent such an event from happening.

Szemerédi's ingenious solution to this problem is to *extend* this sequence of k blocks in an additional direction, which gives B (and more importantly, the complement of B) enough of an "arithmetic progression" structure that one can eventually get lower bounds on the density of A in B.

To get a preliminary idea of how this idea works, suppose that we have a moderately long progression of saturated blocks $P_1, \ldots, P_L$, thus we have $P_i = [a + ir, a + ir + N)$ for some $a \in \mathbb{Z}$ and $r \geq N$, and

(8.2) $$|A \cap P_i| = \bigl(\delta + o_{N \to \infty; A}(1)\bigr)N \quad \text{for all } 1 \leq i \leq L.$$

Here L is a moderately large number, though it will be smaller than the length N of each block: $1 \leq L \leq N$. (Given that the set of saturated blocks has upper density $1 - o_{N \to \infty; A}(1)$, it would be unreasonable to hope to obtain a progression of saturated blocks of length comparable to N or more.) Let us define $A_i \subset [0, N)$ to be the set $A \cap P_i$, translated backwards by $a + ir$.

Now let $B \subset [0, N)$ be a set of some size αN. Then heuristically we expect $A_i \cap B$ to have size $\approx \delta \alpha N$. Now, as discussed before, any *individual* A_i need not have any intersection with B. However, once one considers the sequence $A_1, \ldots, A_L$ there is a kind of "mixing" phenomenon that forces at least one of the A_i to have at least the right number of elements inside B:

Lemma 8.1 (Single lower mixing). *Let $P_1, \ldots, P_L$ be a progression of saturated blocks, with attendant sets $A_1, \ldots, A_L \subset [0, N)$ and let $B \subset [0, N)$ be a set of cardinality αN. Then there exists $1 \leq i \leq L$ such that*

$$|A_i \cap B| \geq (\alpha \delta - o_{L \to \infty; A}(1))N.$$

PROOF. By summing (8.2) for $1 \leq i \leq L$ we have

$$\left| A \cap \bigcup_{i=1}^{L} P_i \right| = (\delta + o_{N \to \infty; A}(1))NL.$$

On the other hand, the set $\bigcup_{i=1}^{L}(P_i \setminus (B + a + ir))$ can be viewed as the union of $(1 - \alpha)N$ arithmetic progressions of length L. Applying (8.1) on each such progression and taking unions, we obtain

$$\left| A \cap \bigcup_{i=1}^{L}(P_i \setminus (B + a + ir)) \right| \leq (\delta + o_{L \to \infty; A}(1))(1 - \alpha)NL.$$

Subtracting the latter estimate from the former, we obtain

$$\left| A \cap \bigcup_{i=1}^{L}(B + a + ir) \right| \geq (\delta \alpha - o_{L \to \infty; A}(1) - o_{N \to \infty; A}(1))NL.$$

Since $L \leq N$, the latter error term can be absorbed into the former. The claim then follows from the pigeonhole principle, noting that $A \cap (B + a + ir)$ is just a translate of $A_i \cap B$. □

We can amplify this result substantially. Firstly, we may work with multiple sets $B_1, \ldots, B_m$ instead of a single set B.

Lemma 8.2 (Multiple lower mixing). *Let $P_1, \ldots, P_L$ be a progression of saturated blocks, with attendant sets $A_1, \ldots, A_L \subset [0, N)$ and let $B_1, \ldots, B_m \subset [0, N)$ be sets of cardinality $\alpha_1 N, \ldots, \alpha_m N$ respectively. Then there exists $1 \leq i \leq L$ such that*

$$|A_i \cap B_j| \geq (\alpha_j \delta - o_{L \to \infty; A, m}(1))N \quad \text{for all } 1 \leq j \leq m.$$

PROOF. Suppose that this claim failed. Then for each $1 \leq i \leq L$ there exists a j for which

$$|A_i \cap B_j| < (\alpha_j \delta - o_{L \to \infty; A, m}(1))N.$$

This is an m-colouring of $\{1, \ldots, L\}$. By van der Waerden's theorem, $\{1, \ldots, L\}$ must then contain a monochromatic progression of length $\omega_{L \to \infty; m}(1)$, where $\omega_{L \to \infty; m}(1) = 1/o_{L \to \infty; m}(1)$ denotes a quantity which goes to infinity as $L \to \infty$

for any fixed m. But then this contradicts Lemma 8.1 if the $o(\cdot)$ constants are chosen properly. □

Corollary 8.3 (Multiple mixing). *Let $P_1, \ldots, P_L$ be a progression of saturated blocks, with attendant sets $A_1, \ldots, A_L \subset [0, N)$ and let $B_1, \ldots, B_m \subset [0, N)$ be sets of cardinality $\alpha_1 N, \ldots, \alpha_m N$ respectively. Then there exists $1 \leq i \leq L$ such that*
$$|A_i \cap B_j| = (\alpha_j \delta + o_{L \to \infty; A, m}(1)) N \quad \text{for all } 1 \leq j \leq m.$$

PROOF. Apply the preceding lemma, but with m replaced by $2m$ and with $B_{j+m} := [0, N) \setminus B_j$ for $1 \leq j \leq m$. □

This type of result is useful when m is small compared with L. Since L is in turn small compared to N, this means that we can only hope to exploit this mixing property when the number m of sets that we wish to be uniformly distributed with respect to A is small compared with the size N of the block. At first glance, this will severely limit the usefulness of this mixing property; however, we can use the Szemerédi regularity lemma to get around this problem (the key point being that the complexity of the partition created by the regularity lemma — which will be m — does not depend on the number of underlying vertices, which is essentially N):

Proposition 8.4 (Graph mixing). *Let $P_1, \ldots, P_L$ be a progression of saturated blocks, with attendant sets $A_1, \ldots, A_L \subset [0, N)$ and let $G_1, \ldots, G_m \subset [0, N) \times [0, N)$ be bipartite graphs connecting two copies of $[0, N)$. Then there exists $1 \leq i \leq L$ such that*
$$\sum_{b \in [0, N)} \big| |\{a \in A_i : (a, b) \in G_j\}| - \delta|\{a \in [0, N) : (a, b) \in G_j\}| \big| = o_{L \to \infty; A, m}(N^2)$$
for all $1 \leq j \leq m$.

This is a remarkably strong assertion that the set A_i becomes uniformly distributed with density δ on the interval $[0, N)$ for many values of i. Note that the error term is completely uniform in the graphs $G_1, \ldots, G_m$ (although it does depend of course on the number m of graphs involved) and also is independent of N (after normalising out the natural $1/N^2$ factor).

PROOF (Sketch). By van der Waerden's theorem as before we can reduce to the case $m = 1$. Pick an $\varepsilon > 0$ and apply the Szemerédi regularity lemma to G to obtain an ε-regular approximation to G_1 induced by a partition of complexity $O_\varepsilon(1)$. Apply Corollary 8.3 to estimate the contribution of the approximation to obtain a net error of $o_{L \to \infty; A, \varepsilon}(N^2) + o_{\varepsilon \to 0}(N^2)$. The claim then follows by choosing ε to be a sufficiently slowly decaying function of L. (One could also proceed here using a weaker regularity lemma such as Corollary 6.8.) □

Let us now informally discuss how one can exploit such strong mixing properties to extend progressions of length $k - 1$ to progressions of length k. (Actually, for technical inductive reasons we will also need to extend progressions of length $i - 1$ to progressions of length i for $1 \leq i \leq k$; we shall return to this point later.) Suppose we have a sequence of k-tuples $(P_{1,i}, P_{2,i}, \ldots, P_{k,i})$ of saturated blocks for $1 \leq i \leq L$, where each k-tuple is in progression, and furthermore the final blocks $P_{k,1}, \ldots, P_{k,L}$ of each k-tuple are also in progression. We can then define sets $A_{j,i} \subset [0,N)$ for $1 \leq j \leq k$ and $1 \leq i \leq L$ as before by intersecting A with $P_{j,i}$ and then translating back to $[0,N)$. We also make the assumption that A "looks the same" in the nonfinal blocks $P_{1,i}, \ldots, P_{k-1,i}$, in the sense that for any $1 \leq j \leq k-1$, the sets $A_{j,i}$ are in fact independent of i. Suppose also that in each k-tuple $(P_{1,i}, P_{2,i}, \ldots, P_{k,i})$, we have found "many" ($\gg \delta^{k-1}N^2$, in fact) progressions of length k, with the jth element of the progression in $P_{j,i}$, and with the first $k-1$ elements in A. Note that in fact once a single k-tuple, say $(P_{1,1}, P_{2,1}, \ldots, P_{k,1})$ has this property, then all k-tuples do, since this property depends only on the distribution of A in the nonfinal blocks $P_{1,i}, \ldots, P_{k-1,i}$ and we are assuming that this distribution is independent of i. Later we shall address the rather important question of *how* one could construct such a strange sequence of k-tuples; for now, let us simply assume that such a sequence exists. This sequence shows that A has many progressions of length $k-1$. We now show that some of these progressions of length $k-1$ can be extended to progressions of length k in A; this is a model of the key inductive step in Szemerédi's argument.

Consider the sets $A_{1,i}, \ldots, A_{k-1,i}, A_{k,i}$ in $[0,N)$, which describe the distribution of A in the k-tuple $(P_{1,i}, \ldots, P_{k,i})$. The first $k-1$ of these sets are independent of i, while the final set $A_{k,i}$ varies in i; however, because the blocks $P_{k,1}, \ldots, P_{k,L}$ are in progression, the final set $A_{k,i}$ obeys the strong mixing properties described earlier. By hypothesis, we have many progressions of length k in $[0,N)$, with the jth element of such progressions lying in $A_{j,i}$ for $1 \leq j \leq k-1$. The kth elements of such progressions can be collected into a subset of $[0,N)$ which we shall call B; we can then get a reasonable lower bound on the density of B in $[0,N)$ (roughly speaking, we have $|B| \gg \delta^{k-1}N$). The objective is to get B to intersect $A_{k,i}$ for at least one i, as this will generate a progression of length k in A. But this happens for at least one i if L is large enough (depending on δ, but not on N), thanks to Lemma 8.1. (Note that we did not use the strongest mixing properties available; we will utilise those later.) Indeed the intersection of B with $A_{k,i}$ will be rather large, and by arguing slightly more carefully one can then show that the ith k-tuple $(A_{1,i}, \ldots, A_{k,i})$ will contain quite a large number of progressions of length k ($\gg \delta^k N^2$, in fact).

To summarise, by using the mixing properties, we can convert a long sequence of k-tuples of blocks, each of which contain many progressions of length $k-1$ in A, into a single k-tuple of blocks, which contains many progressions of length k in A, provided that we have the following two additional properties:

- The distribution of A in the $k-1$ nonfinal blocks of the k-tuples is fixed as one moves along the sequence.
- The final block of the k-tuples are in progression as one moves along the sequence.

This looks like a promising induction-type step. However it cannot by itself be iterated to generate progressions of length k unconditionally for two reasons. Firstly, there is the minor objection that we will need a generalisation of the above statement in which progressions of length $k-1$ and k in A are replaced by progressions of length $i-1$ and i in A for various $1 \leq i \leq k$. This is not hard to address. The more important objection is that we will need a way of generating not only individual k-tuples of blocks that contain progressions of length (say) $k-1$ in A, but entire *sequences* of such k-tuples which obey additional structural properties.

The key to obtaining this type of superstructure atop a k-tuple of blocks in [41] is by passing to a "coarser" level, and viewing each block as a single element of $\mathbb{Z}$; the saturated blocks (as well as a subset of the saturated blocks which are known as the "perfect" blocks) then become subsets of $\mathbb{Z}$. These sets in turn have upper densities, and one can also define notions of saturated blocks of these sets, which are thus "blocks of blocks." The point is that the task of finding sequences of k-tuples of blocks simplifies, on moving to this coarser scale, to the task of finding sequences of k-term progressions, which is easier and in fact will follow once one has a suitable k-tuple of saturated blocks at this coarse scale.

The details are very technical, but let us just mention some brief highlights here. Write $A_0 = A$. One picks a large number N_0 for which there are lots of saturated blocks of length N_0 (the upper density of such blocks should be $1 - o_{N_0 \to \infty; A}(1)$). We subdivide the integers into blocks of length N_0, and identify the set of such blocks again with $\mathbb{Z}$, creating a "coarse scale" view of the set A_0. (Objects in the coarse scale will be subscripted by 1, while objects in the fine scale subscripted by 0.) The saturated blocks then form a subset S_1 of $\mathbb{Z}$ of upper density close to 1. Each element of S_1 corresponds to a saturated block, with respect to which A_0 is distributed in one of 2^{N_0} ways. This can be viewed as a colouring of S_1 into 2^{N_0} colours. One of the colour classes must be somewhat prevalent (in particular, occuring with positive upper density); we designate this as the "perfect" colour, and let $A_1 \subset S_1$ be the associated colour class. (The precise definition of "prevalent" is slightly technical - it is sort of an upper density "relative" to S_1 — and we omit it here.) A_1 has some upper density δ_1; it is possible (after some notational trickery) to run a density increment argument for A_1 and reduce to the case where A_1 obeys an analogue of the bound (8.1). In particular we can pick a large number N_1 (much larger than N_0) and construct many saturated blocks of A_1 of length N_1. The definition of "saturated" is a little technical; we require that these blocks not only contain A_1 to approximately the right density (i.e. $\delta_1 + o_{N_1 \to \infty; A_1}(1)$), but also contains S_1 to approximately the right density ($1 - o_{N_0 \to \infty; A}(1)$, if N_1 is large enough). This can be done by tinkering with the notion of upper density appropriately, as mentioned briefly before; we omit the details.

Now suppose one has a k-tuple $P_1, \ldots, P_k$ of saturated blocks of A_1, and suppose that one can find many k-term progressions with the ith term in P_i for $1 \leq i \leq k$, and also in A_1 for $1 \leq i \leq k-1$. Specifically, let us suppose that for almost all (e.g., with density $1 - o_{N_0 \to \infty; A}(1)$) of the integers n in the middle third of the final block P_i, that there are many k-term progressions ending in n with the first $k-1$ terms in $P_1 \cap A_1, P_2 \cap A_1, \ldots, P_{k-1} \cap A_1$ respectively. Most of these integers n are going to also lie in S_1 (since S_1 fills almost all of P_k), and so there should be no difficulty obtaining an arithmetic progression of such n of some moderate length L_0 (which can be a slowly growing function of N_0), thus each element of this progression is the final element of a k-term progression which is mostly in A_1. Now recall that each integer in this coarse representation corresponds to a block of length N_0 in the original fine-scale representation. Thus this arithmetic progression can be identified with a sequence of L_0 k-tuples of such blocks, where the final block in each k-tuple is in arithmetic progression, and all the other blocks have the "perfect" colour. This is essentially the very structure we need in order to run our inductive step and convert the progressions with $k - 1$ elements in A_0, to progressions with k elements in A_0.

To summarize, by coarsening the scale it is possible to convert k-tuples of blocks to sequences of k-tuples of blocks (and more generally to a type of "homogeneous, well-arranged" family of k-tuples, as defined in [41]). These sequences can then be traded in via the mixing properties to upgrade short progressions in a set A to longer progressions. By alternating these two arguments in a moderately sophisticated induction argument (passing from fine scales to coarse scales approximately 2^k times), one can start with progressions with 0 elements in one of the A sets and eventually upgrade to progressions with k elements in the original set A. There are some technical issues at intermediate stages of the argument, when descending a scale in a case when only the first i elements of a progression are guaranteed to have the perfect colour, when it becomes important that the remaining elements are unsaturated. To achieve this, the graph mixing properties in Proposition 8.4 become essential; the progressions are reinterpreted as edges connecting the elements of one block to another. We omit the details.

References

1. I. Assani, *Pointwise convergence of ergodic averages along cubes*, preprint.
2. F. A. Behrend, *On sets of integers which contain no three terms in arithmetic progression*, Proc. Nat. Acad. Sci. U.S.A. **32** (1946), 331–332.
3. V. Bergelson, Host B., and B. Kra, *Multiple recurrence and nilsequences*, Invent. Math. **160** (2005), no. 2, 261–303.
4. V. Bergelson, B. Host, R. McCutcheon, and F. Parreau, *Aspects of uniformity in recurrence*, Colloq. Math. **85** (2000), 549–576.
5. V. Bergelson and A. Leibman, *Polynomial extensions of van der Waerden's and Szemerédi's theorems*, J. Amer. Math. Soc. **9** (1996), no. 3, 725–753.
6. V. Bergelson and I. Ruzsa, *Squarefree numbers, IP sets and ergodic theory*, Paul Erdős and his Mathematics, I (Budapest, 1999) (G. Halász, L. Lovász, M. Simonovits, and V. T. Sós, eds.), Bolyai Soc. Math. Stud., vol. 11, János Bolyai Math. Soc., Budapest, 2002, pp. 147–160.
7. J. Bourgain, *On triples in arithmetic progression*, Geom. Funct. Anal. **9** (1999), no. 5, 968–984.
8. P. Erdős and P. Turán, *On some sequences of integers*, J. London Math. Soc. **11** (1936), 261–264.
9. P. Frankl and V. Rödl, *The uniformity lemma for hypergraphs*, Graphs Combin. **8** (1992), no. 4, 309–312.
10. _____, *Extremal problems on set systems*, Random Structures Algorithms **20** (2002), no. 2, 131–164.
11. H. Furstenberg, *Ergodic behavior of diagonal measures and a theorem of Szemerédi on arithmetic progressions*, J. Analyse Math. **31** (1977), 204–256.
12. _____, *Recurrence in ergodic theory and combinatorial number theory*, M. B. Porter Lectures, Princeton Univ. Press, Princeton, NJ, 1981.
13. H. Furstenberg and Y. Katznelson, *An ergodic Szemerédi theorem for commuting transformations*, J. Analyse Math. **34** (1978), 275–291.
14. H. Furstenberg, Y. Katznelson, and D. Ornstein, *The ergodic theoretical proof of Szemerédi's theorem*, Bull. Amer. Math. Soc. (N.S.) **7** (1982), 527–552.
15. H. Furstenberg and B. Weiss, *Topological dynamics and combinatorial number theory*, J. Analyse Math. **34** (1978), 61–85.
16. _____, *A mean ergodic theorem for $(1/N)\sum_{n=1}^{N} f(T^n x)g(T^{n^2} x)$*, Convergence in Ergodic Theory and Probability (Columbus, OH, 1993), Ohio State Univ. Math. Res. Inst. Publ., vol. 5, de Gruyter, Berlin, 1996, pp. 193–227.
17. W. T. Gowers, *A new proof of Szemerédi's theorem for arithmetic progressions of length four*, Geom. Funct. Anal. **8** (1998), no. 3, 529–551.
18. _____, *A new proof of Szemerédi's theorem*, Geom. Funct. Anal. **11** (2001), no. 3, 465–588.
19. _____, *Quasirandomness, counting and regularity for 3-uniform hypergraphs*, Combin. Probab. Comput. **15** (2006), no. 1-2, 143–184.
20. _____, *Hypergraph regularity and the multidimensional Szemerédi theorem*, preprint.

21. R. Graham, B. Rothschild, and J. H. Spencer, *Ramsey theory*, Wiley-Intersci. Ser. Discrete Math., John Wiley & Sons, New York, 1980.
22. B. J. Green, *A Szemerédi-type regularity lemma in abelian groups, with applications*, Geom. Funct. Anal. **15** (2005), no. 2, 340–376.
23. _____, *Finite field models in additive combinatorics*, Surveys in Combinatorics 2005 (Durham, 2005) (B. S. Webb, ed.), London Math. Soc. Lecture Note Ser., vol. 327, Cambridge Univ. Press, Cambridge, 2005, pp. 1–27.
24. B. J. Green and S. Konyagin, *On the Littlewood problem modulo a prime*, preprint.
25. B. J. Green and T. Tao, *The primes contain arbitrarily long arithmetic progressions*, preprint.
26. A. W. Hales and R. I. Jewett, *Regularity and positional games*, Trans. Amer. Math. Soc. **106** (1963), 222–229.
27. B. Host and B. Kra, *Nonconventional ergodic averages and nilmanifolds*, Ann. of Math. (2) **161** (2005), no. 1, 397–488.
28. Y. Kohayakawa, T. Luczsak, and V. Rödl, *Arithmetic progressions of length three in subsets of a random set*, Acta Arith. **75** (1996), no. 2, 133–163.
29. B. Kra, *The Green–Tao theorem on arithmetic progressions in the primes: an ergodic point of view*, Bull. Amer. Math. Soc. (N.S.) **43** (2006), no. 1, 3–23.
30. A. Leibman, *Host–Kra and Ziegler factors, and convergence of multiple averages*, Handbook of Dynamical Systems (B. Hasselblatt and A. Katok, eds.), Vol. 1B, Elsevier, Amsterdam, 2006, pp. 745–841.
31. L. Lovász and B. Szegedy, *Szemerédi's lemma for the analyst*, preprint.
32. B. Nagle, V. Rödl, and M. Schacht, *The counting lemma for regular k-uniform hypergraphs*, Random Structures Algorithms **28** (2006), no. 2, 113–179.
33. V. Rödl and M. Schacht, *Regular partitions of hypergraphs*, preprint.
34. V. Rödl and J. Skokan, *Regularity lemma for k-uniform hypergraphs*, Random Structures Algorithms **25** (2004), no. 1, 1–42.
35. _____, *Applications of the regularity lemma for uniform hypergraphs*, Random Structures Algorithms **28** (2006), no. 2, 180–194.
36. K. F. Roth, *On certain sets of integers*, J. London Math. Soc. **28** (1953), 104–109.
37. I. Ruzsa and E. Szemerédi, *Triple systems with no six points carrying three triangles*, Combinatorics, Vol. II (Keszthely, 1976) (A. Hajnal and V. T. Sós, eds.), Colloq. Math. Soc. János. Bolyai, vol. 18, North-Holland, Amsterdam–New York, 1978, pp. 939–945.
38. I. Schur, *Über die Kongruenz $x^m + y^m = z^m \pmod{p}$*, Jber. Deutsch. Math.-Verein. **25** (1916), 114–116.
39. S. Shelah, *Primitive recursive bounds for van der Waerden numbers*, J. Amer. Math. Soc. **1** (1988), no. 3, 683–697.
40. E. Szemerédi, *On sets of integers containing no four elements in arithmetic progression*, Acta Math. Acad. Sci. Hungar. **20** (1969), 89–104.
41. _____, *On sets of integers containing no k elements in arithmetic progression*, Acta Arith. **27** (1975), 199–245.
42. T. Tao, *A variant of the hypergraph removal lemma*, J. Combin. Theory Ser. A **113** (2006), no. 7, 1257–1280.
43. _____, *The Gaussian primes contain arbitrarily shaped constellations*, J. Anal. Math. **99** (2006), 109–176.
44. _____, *Szemerédi's regularity lemma revisited*, Contrib. Discrete Math. **1** (2006), no. 1, 8–28.
45. _____, *A quantitative ergodic theory proof of Szemerédi's theorem*, preprint.
46. _____, *A correspondence principle between (hyper)graph theory and probability theory, and the (hyper)graph removal lemma*, preprint.
47. B. L. van der Waerden, *Beweis einer Baudetschen Vermutung*, Nieuw Arch. Wisk. **15** (1927), 212–216.
48. P. Varnavides, *On certain sets of positive density*, J. London Math. Soc. **34** (1959), 358–360.
49. T. Ziegler, *A non-conventional ergodic theorem for a nilsystem*, Ergodic Theory Dynam. Systems **25** (2005), no. 4, 1357–1370.
50. _____, *Universal characteristic factors and Furstenberg averages*, J. Amer. Math. Soc. **20** (2007), no. 1, 53–97.

Department of Mathematics, University of California at Los Angeles, Los Angeles, CA 90095, USA
E-mail address: tao@math.ucla.edu

Cardinality Questions About Sumsets

Imre Z. Ruzsa

1. Introduction

This paper will tell you about questions of the following kind. Suppose we know the cardinality of a (finite) set and we know also the number of sums of pairs. What can we say about the number of differences, or of sums of triples?

First we explain the most important tool, Plünnecke's inequality, then two further inequalities independent of it. These will be applied to study the connection between $|A|$, $|A+B|$ and $|A+2B|$, with particular emphasis on the case $B=A$.

Only new results will be proved. However, examples illustrating the sharpness of some results will be quoted even if they had been published before.

We end this introduction by mentioning some basic ideas.

1.1. Direct product. Assume $A_1, A_2, \ldots, A_k$ are subsets of a group G with cardinalitities of sumsets

$$|A_{i_1} + A_{i_2} + \cdots + A_{i_m}| = N(i_1, \ldots, i_m).$$

Let $A_1', \ldots$ be another collection of sets in another group G' with corresponding values $N'(\ldots)$. If we form the direct products

$$B_i = A_i \times A_i' = \{(a,b) : a \in A, b \in B\} \subset G \times G',$$

then we have

$$|B_{i_1} + B_{i_2} + \cdots + B_{i_m}| = N(i_1, \ldots, i_m) N'(i_1, \ldots, i_m).$$

This explains the multiplicative nature of many of the results — when a quantity is estimated in terms of others, this is mostly in the form of a product of powers. This method can often be used to build large examples starting from a single one.

2000 *Mathematics Subject Classification.* 11B75, 11P70.

Supported by Hungarian National Foundation for Scientific Research (OTKA), Grants No. T 43623, T 42750, K 61908.

This is the final form of the paper.

1.2. Projection. If we start from sets of integers, the above construction gives us sets of integral vectors. This is, however, not an essential difference. If we have sets $A_i \subset \mathbb{Z}^k$ and a *finite* number of sum-cardinalities are prescribed, then we can construct sets of integers that behave the same way. Indeed, the linear map

$$(x_1, \ldots, x_k) \to x_1 + mx_2 + \cdots + m^{k-1}x_k$$

will not add any new coincidence between sums if m is large enough.

This observation will be used without any further mentioning. If we construct a set in $\mathbb{Z}^k$ with certain properties, we shall tacitly realize that a set of integers can also be constructed if necessary; a set in several dimensions often exhibits the structure more clearly.

1.3. Torsion. The above consideration shows that from our point of view the structure of $\mathbb{Z}^k$ is not richer than that of $\mathbb{Z}$. We can add that no torsionfree group produces anything new either. Indeed, let G be a torsionfree group and take a finite subset (the union of all finite sets which we want to add). This generates a subgroup G'; and, as a finitely generated torsionfree group, G' is isomorphic to $\mathbb{Z}^k$ for some k.

2. Plünnecke's method

Plünnecke [6] developed a graph-theoretic method to estimate the density of sumsets $A + B$, where A has a positive density and B is a basis. I published a simplified version of his proof [8,9]. Other accounts (of my version) were published by Malouf [4] and Nathanson [5].

Plünnecke observed that the cardinality properties of the sets A, $A+B$, $A+2B$, $\ldots$, are well reflected by the following directed graph. We take $h+1$ copies of the group where these sets are situated, and build a graph on these sets as vertices by connecting an $x \in A + jB$ to an $y \in A + (j+1)B$ if $y = x + b$ with some $b \in B$. We call this graph the *addition graph*. These graphs have certain properties which follow from the commutativity of addition, and hence Plünnecke called them *commutative*; we shall retain this terminology.

We consider directed graphs $G = (V, E)$, where V is the set of vertices and E is that of the edges. If there is an edge from x to y, then we also write $x \to y$. A graph is *semicommutative*, if for every collection $(x; y; z_1, z_2, \ldots, z_k)$ of distinct vertices such that $x \to y$ and $y \to z_i$ there are distinct vertices $y_1, \ldots, y_k$ such that $x \to y_i$ and $y_i \to z_i$. G is *commutative*, if both G and the graph $\hat{G}$ obtained by reversing the direction of every edge of G are semicommutative.

Our graphs will be of a special kind we call *bridging*. By an *h-bridging* graph we mean a graph with a fixed partition of the set of vertices

$$V = V_0 \cup V_1 \cup \cdots \cup V_h$$

into $h+1$ disjoint sets (islands) such that every edge goes from some V_{i-1} into V_i.

For $X, Y \subset V$, we define the *image* of X in Y as

$$\mathrm{im}(X, Y) = \{y \in Y : \text{there is a directed path from some } x \in X \text{ to } y\}.$$

The *magnification ratio* is defined by

$$D(X, Y) = \min\left\{ \frac{|\mathrm{im}(Z, Y)|}{|Z|} : Z \subset X, Z \neq \varnothing \right\}.$$

For a bridging graph we write
$$D_i(G) = D(V_0, V_i).$$
Now Plünnecke's main result can be stated as follows.

Theorem 2.1 (Plünnecke [6]). *In a commutative bridging graph $D_i^{1/i}$ is decreasing.*

That is, for $i < h$ we have $D_h \leq D_i^{h/i}$. An obvious (and typically the only available) upper estimate for D_i is $|V_i|/|V_0|$. This yields the following corollary (in fact, an equivalent assertion).

Theorem 2.2. *Let $i < h$ be integers, G a commutative bridging graph on the islands $V_0, \ldots, V_h$. Write $|V_0| = m$, $|V_i| = s$. There is an $X \subset V_0$, $X \neq \emptyset$ such that*
$$|\operatorname{im}(X, V_h)| \leq (s/m)^{h/i}|X|.$$

An application of the above theorem to the addition graph yields the following result.

Theorem 2.3. *Let $i < h$ be integers, A, B sets in a commutative group and write $|A| = m$, $|A + iB| = \alpha m$. There is an $X \subset A$, $X \neq \emptyset$ such that*
$$|X + hB| \leq \alpha^{h/i}|X|.$$

Since $|X + hB| \geq |hB|$ and $|X| \leq m$, we get the following immediate consequence.

Corollary 2.4. *Let $i < h$ be integers, A, B sets in a commutative group and write $|A| = m$, $|A + iB| = \alpha m$. We have*
$$|hB| \leq \alpha^{h/i} m.$$

In the torsionfree case, using $|X + hB| \geq |X| + |hB| - 1$ instead, we obtain the following result, which is stronger for α near to 1 (and gives the correct order of magnitude).

Corollary 2.5. *Let $i < h$ be integers, A, B sets in a torsionfree commutative group and write $|A| = m$, $|A + iB| = \alpha m$. We have*
$$|hB| \leq (\alpha^{h/i} - 1)m + 1.$$

An application to different summands is less straightforward, however, I proved [8] the following.

Theorem 2.6. *Let $A, B_1, \ldots, B_k$ be sets and write $|A| = m$, $|A + B_i| = \alpha_i m$. There is an $X \subset A$, $X \neq \emptyset$ such that*
$$|X + B_1 + \cdots + B_h| = \alpha_1 \alpha_2 \ldots \alpha_h |X|.$$

Besides the complete addition graph we used above, a more general graph may be useful. Given three sets A, B, C we build on them the *restricted addition graph* as follows. The islands will be $V_0 = A$, $V_1 = (A+B)\setminus C$, $V_j = (A+jB)\setminus(C+(j-1)B)$ for $j > 1$. (We can omit this distinction by defining $0B = \{0\}$.) Again, there is an edge from an $x \in V_j$ to a $y \in V_{j+1}$ if $y = x + b$ with some $b \in B$. The case $C = \emptyset$ returns the complete addition graph. An important case is $C = A$, where in each stage we get the "new sums."

Lemma 2.7. *The restricted addition graph is commutative.*

PROOF. Consider a typical path of length 2, $x \to y \to z$ with $x \in V_{j-1}$, $y \in V_j$, $z \in V_{j+1}$. This means $y = x + b$, $z = y + b'$ with $b, b' \in B$. We claim that $x \to x + b' \to x + b' + b = z$ is also a path in our graph. To see this we only need to check $x + b' \in V_j$, that is, $x + b' \in A + jB$ and $x + b' \notin C + (j-1)B$. The first follows from $x \in V_{j-1}$, and the negation of the second would imply $z = x + b' + b \in C + jB$, which would contradict $z \in V_{j+1}$.

We apply this substitution to a collection $x \to y \to z_i$ to find distict y_i with $x \to y_i \to z_i$, and to a collection $x_i \to y \to z$ to find $x_i \to y_i \to z$; this is what we need to establish commutativity. □

By applying Plünnecke's theorem 2.1 to this graph we obtain the following.

Theorem 2.8. *Let $i < h$ be integers, A, B, C sets in a commutative group and write $|A| = m$, $|(A + iB) \setminus (C + (i-1)B)| = \alpha m$. There is an $X \subset A$, $X \neq \emptyset$ such that*
$$|(X + hB) \setminus (C + (h-1)B)| \leq \alpha^{h/i} |X|.$$

We will later see that the applicability of this result may depend on the size of the set X. We devote the next section to modifying Plünnecke's inequality in a way that gives us some control over the size of X.

3. Plünnecke's inequality with a large subset

We show an extension of Theorem 2.2 with a bound on the size of the selected subset.

Theorem 3.1. *Let $i < h$ be integers, G a commutative bridging graph on the islands $V_0, \ldots, V_h$. Write $|V_0| = m$, $|V_i| = s$, $\gamma = h/i$. Let an integer k be given, $1 \leq k \leq m$. There is an $X \subset V_0$, $|X| \geq k$ such that*

$$(3.1) \quad |\mathrm{im}(X, V_h)| \leq \left(\frac{s}{m}\right)^\gamma + \left(\frac{s}{m-1}\right)^\gamma + \ldots$$
$$+ \left(\frac{s}{m-k+1}\right)^\gamma + (|X| - k)\left(\frac{s}{m-k+1}\right)^\gamma.$$

PROOF. We use induction on k. The case $k = 1$ is Theorem 2.2.

Assume we know it for k; we prove it for $k + 1$. The assumption gives us a set X, $|X| \geq k$ with a bound on $|\mathrm{im}(X, V_h)|$ as given by (3.1). We want to find a set X' with $|X'| \geq k + 1$ and

$$(3.2) \quad |\mathrm{im}(X', V_h)| \leq \left(\frac{s}{m}\right)^\gamma + \left(\frac{s}{m-1}\right)^\gamma + \ldots$$
$$+ \left(\frac{s}{m-k}\right)^\gamma + (|X'| - k - 1)\left(\frac{s}{m-k}\right)^\gamma.$$

If $|X| \geq k+1$, we can put $X' = X$. If $|X| = k$, we apply Theorem 2.2 to the graph obtained from G by omitting the vertices in X. This yields a set $Y \subset V_0 \setminus X$ such that
$$|\mathrm{im}(Y, V_h)| \leq \left(\frac{s}{m-k}\right)^\gamma |Y|$$
and we put $X' = X \cup Y$. □

The following variant will be more comfortable for calculations.

Theorem 3.2. *Let $i < h$ be integers, G a commutative bridging graph on the islands $V_0, \ldots, V_h$. Write $|V_0| = m$, $|V_i| = s$, $\gamma = h/i$. Let a real number t be given, $0 \leq t < m$. There is an $X \subset V_0$, $|X| > t$ such that*

$$(3.3) \qquad |\mathrm{im}(X, V_h)| \leq \frac{s^\gamma}{\gamma}\left(\frac{1}{(m-t)^{\gamma-1}} - \frac{1}{m^{\gamma-1}}\right) + (|X| - t)\left(\frac{s}{m-t}\right)^\gamma.$$

PROOF. We apply Theorem 3.1 with $k = [t] + 1$. The right side of (3.3) can be written as $s^\gamma \int_0^{|X|} f(x)\,dx$, where $f(x) = (m-x)^{-\gamma}$ for $0 \leq x \leq t$, and $f(x) = (m-t)^{-\gamma}$ for $t < x \leq |X|$. Since f is increasing, the integral is $\geq f(0) + f(1) + \cdots + f(|X|-1)$. This exceeds the right side of (3.1) by a termwise comparison. $\square$

We state the consequences of this result for the complete and restricted addition graphs.

Theorem 3.3. *Let $i < h$ be integers, A, B sets in a commutative group and write $|A| = m$, $|A + iB| = s$, $\gamma = h/i$. Let a real number t be given, $0 \leq t < m$. There is an $X \subset A$, $|X| > t$ such that*

$$|X + hB| \leq \frac{s^\gamma}{\gamma}\left(\frac{1}{(m-t)^{\gamma-1}} - \frac{1}{m^{\gamma-1}}\right) + (|X| - t)\left(\frac{s}{m-t}\right)^\gamma.$$

Theorem 3.4. *Let $i < h$ be integers, A, B, C sets in a commutative group and write $|A| = m$, $|(A + iB) \setminus (C + (i-1)B)| = s$, $\gamma = h/i$. Let a real number t be given, $0 \leq t < m$. There is an $X \subset A$, $|X| > t$ such that*

$$|(X + hB) \setminus (C + (h-1)B)| \leq \frac{s^\gamma}{\gamma}\left(\frac{1}{(m-t)^{\gamma-1}} - \frac{1}{m^{\gamma-1}}\right) + (|X| - t)\left(\frac{s}{m-t}\right)^\gamma.$$

We state separately the case $i = 1$, $h = 2$ which will be applied in the sequel.

Corollary 3.5. *Let A, B sets in a commutative group and write $|A| = m$, $|A + iB| = s$. Let a real number t be given, $0 \leq t < m$. There is an $X \subset A$, $|X| > t$ such that*

$$|X + 2B| \leq \frac{s^2}{(m-t)^2}\left(|X| - \frac{t(t+m)}{2m}\right).$$

Corollary 3.6. *Let A, B, C be sets in a commutative group and write $|A| = m$, $|(A + B) \setminus C| = s$. Let a real number t be given, $0 \leq t < m$. There is an $X \subset A$, $|X| > t$ such that*

$$|(X + 2B) \setminus (C + B)| \leq \frac{s^2}{(m-t)^2}\left(|X| - \frac{t(t+m)}{2m}\right).$$

4. Sums and differences

With Plünnecke's method one can get various inequalities for cardinalities of sumsets, but it stops to work when differences are also involved (we shall give reasons why).

As far as I know the first inequality connecting sums and differences is due to Freĭman and Pigaev [1]. They prove that

$$|A + A|^{3/4} \leq |A - A| \leq |A + A|^{4/3}.$$

I proved [7] the following.

Theorem 4.1. *Let A, Y, Z be finite sets in a (not necessarily commutative) group. We have*

(4.1) $$|A||Y - Z| \leq |A - Y||A - Z|.$$

Substituting $Y = Z = -A$ we obtain the following inequality: if $|A| = m$, $|2A| \leq \alpha m$, then $|A - A| \leq \alpha^2 m$. The exponent 2 is best possible (though an improvement to something like $\alpha^2/\log \alpha$ is conceivable). This can be seen by considering the lattice points inside a d-dimensional simplex

$$\{(x_1, \ldots, x_d) \in \mathbb{Z}^d : x_i \geq 0, \sum x_i \leq n\},$$

where $2^d \approx \alpha$. This example is analyzed in detail by Hennecart, Robert and Yudin [3]; they attribute the underlaying idea to Freĭman and Pigaev's abovementioned paper [1].

One difference from the Plünnecke inequalities is the noncommutative nature of the above result. From Corollary 2.4 we obtain a similar implication: if $|A| = m$, $|A - A| \leq \alpha m$, then $|2A| \leq \alpha^2 m$. This fails in noncommutative groups. Take a free group with generators a, b and put

$$A = \{ia + b : 1 \leq i \leq m\}.$$

Then both difference sets $A - A$ and $-A + A$ have $2m - 1$ elements, while $|2A| = m^2$.

Another difference is the following. To go to differences would require the case "$h = -1$" of Theorem 2.3, which might be expected to sound as follows:

"if $|A| = n$, $|A + B| = \alpha n$, then there is a nonempty $X \subset A$ such that $|X - B| \leq \alpha' |X|$, with α' depending only on α."

This is, however, false; we have the following result [11].

Theorem 4.2. *For infinitely many n there are sets of integers A, B such that $|A| = m$, $|A + B| \leq 3m$ and*

$$|X - B| \geq (c \log m)|X|$$

for every $X \subset A$, where c is a positive absolute constant.

It is possible to improve the logarithmic factor, and the best value is around $\exp \sqrt{\log m}$ (we are still uncertain about the power of $\log \log$ in the exponent). More details will be given in a paper by Katalin Gyarmati, François Hennecart and me [2].

Inequality (4.1) together with Plünnecke's can be used to deduce the following one, which is sufficient for most of the applications [10, Lemma 3.3].

Theorem 4.3. *Let A, B be finite sets in a commutative group and write $|A| = m$, $|A + B| = \alpha m$. For arbitrary nonnegative integers k, l we have*

$$|kB - lB| \leq \alpha^{k+l} m.$$

The sum-sum analogue of (4.1) can be deduced from Plünnecke's inequality:

(4.2) $$|A||Y + Z| \leq |A + Y||A + Z|.$$

Indeed, applying Theorem 2.6 we get a set $X \subset A$ such that

$$|X + Y + Z| \leq |X| \frac{|A + Y|}{|A|} \frac{|A + Z|}{|A|},$$

and to obtain (4.2) we just have to use $|X + Y + Z| \geq |Y + Z|$ and $|X| \leq |A|$.

5. Double and triple sums

We present an inequality which sometimes nicely complements Plünnecke's.

Theorem 5.1. *Let X, Y, Z be finite sets in a commutative group. We have*

(5.1) $$|X+Y+Z|^2 \leq |X+Y||Y+Z||X+Z|.$$

This inequality may be extended to the noncommutative case as follows.

Theorem 5.2. *Let X, Y, Z be finite sets in a not necessarily commutative group. We have*

(5.2) $$|X+Y+Z|^2 \leq |X+Y||Y+Z| \max_{y \in Y} |X+y+Z|.$$

PROOF. We use induction on $|Y|$. For $|Y|=1$ (5.2) reduces to the obvious inequality
$$|X+y+Z| \leq |X||Y|.$$

Assume now we know (5.2) for smaller sets. Fix y as the element of Y which maximizes $|X+y+Z|$. Write $|X+y+Z|=m$, $Y \setminus \{y\} = Y'$, $|(X+Y+Z) \setminus (X+Y'+Z)| = a$, $|(X+Y) \setminus (X+Y')| = b$, $|(Y+Z) \setminus (Y'+Z)| = c$. With these notations (5.2) can be rewritten as

(5.3) $$(|X+Y'+Z|+a)^2 \leq m(|X+Y'|+b)(|Y'+Z|+c).$$

We shall obtain (5.3) as the sum of the following three inequalities:

(5.4) $$|X+Y'+Z|^2 \leq m|X+Y'||Y'+Z|,$$

(5.5) $$2a|X+Y'+Z| \leq m(c|X+Y'|+b|Y'+Z|),$$

(5.6) $$a^2 \leq mbc.$$

Of these inequalities (5.4) follows from the induction hypothesis.

Clearly every element of $(X+Y+Z) \setminus (X+Y'+Z)$ is of the form $x+y+z$ with $x \in X$, $z \in Z$, hence $a \leq m$. We can map this set into the Cartesian product of $(X+Y) \setminus (X+Y')$ and $(Y+Z) \setminus (Y'+Z)$ by mapping a typical element $x+y+z$ into the pair $(x+y, y+z)$. This pair determines $x+y+z$ uniquely and clearly $x+y \notin X+Y'$ as otherwise we would have $x+y+z \in X+Y'+Z$; similarly $y+z \notin Y'+Z$. This mapping shows $a \leq bc$. The product of these inequalities gives (5.6).

By multiplying inequalities (5.4) and (5.6) and taking the square root we obtain
$$a|X+Y'+Z| \leq m\sqrt{bc}|X+Y'||Y'+Z|;$$

(5.5) now follows from the arithmetic-geometric mean inequality. □

We show by an example that the maximum cannot be omitted and cannot be replaced by an average, even in the case of identical sets. Take a free group with generators a, b and put
$$X = Y = Z = \{a, 2a, \ldots, na, b\}.$$

We have $|X| = n+1$, $|2X| = 4n$ and $|3X| > n^2$ since all the elements $ia+b+ja$, $1 \leq i, j \leq n$ are distinct. From the $n+1$ sets $X+y+X$, $y \in X$ only one is of size n^2, namely the one with $y=b$, all the others have $O(n)$ elements.

A particular case of Theorem 5.1 is the inequality
$$|3A|^{1/3} \leq |2A|^{1/2}.$$

As a possible generalization of this, I asked whether $|kA|^{1/k}$ is always decreasing. Vsevolod Lev observed (personal communication) that this is indeed the case, and this follows immediately from Corollary 2.4, by applying it for a one-element set A.

Problem 5.3. Can Theorem 5.1 be generalized to more than three sets? The first case would be

$$|X+Y+Z+U|^3 \leq |Y+Z+U||X+Z+U||X+Y+U||X+Y+Z|.$$

Problem 5.4 (A noncommutative Plünnecke?). Theorems 5.1 and 5.2 suggest a way to find noncommutative analogues of inequalities that for commutative groups were proved by Plünnecke's method. I formulate the simplest possible of them. Let A, B be finite sets in a noncommutative group, and define α by

$$\max_{b \in B} |A+b+B| = \alpha |A|.$$

Must there exist a nonempty $X \subset A$ such that

$$|X+2B| \leq \alpha'|X|$$

with an α' depending only on α?

We finish this section by a meditation on the sizes of $2A$ and $3A$.

Write $|A| = m$, $|2A| = n$. Corollary 2.4 implies $|3A| \leq n^3/m^2$, and Theorem 5.1 implies $|3A| \leq n^{3/2}$. The first is better for $n \leq m^{4/3}$, the second for larger values. The two together describe the maximal possible value of $|3A|$ up to a constant.

Theorem 5.5. *Let m, n be positive integers satisfying $m \leq n \leq m^2$. There is a set A of integers such that $|A| \asymp m$, $|2A| \asymp n$ and*

$$|3A| \asymp \min(n^3/m^2, n^{3/2}).$$

PROOF. We construct A in $\mathbb{Z}^3$. Take two integers k, l such that $k \leq l \leq k^3$ and put

$$A_1 = \{(x,y,z) : 0 \leq x,y,z < k\},$$
$$A_2 = \{(x,0,0), (0,x,0), (0,0,x) : 0 \leq x < l\},$$

and $A = A_1 \cup A_2$.

We have $m = |A| = k^3 + 3(l-k) \asymp k^3$, so the proper choice is $k \sim m^{1/3}$. Further $2A = 2A_1 \cup (A_1 + A_2) \cup 2A_2$. The cardinality of the parts is of order k^3, $k^2 l$ and l^2, respectively. The first is always smaller than the second, hence

$$n = |2A| \asymp \max(k^2 l, l^2);$$

the threshold of behaviour is at $l = k^2$. Hence the proper choice of l is

$$l \sim \min(\sqrt{n}, n/m^{2/3})$$

and the claim follows from the fact that $|3A| \geq |3A_2| \geq l^3$. □

6. $A+B$ and $A+2B$

In this section we consider the following problem. Let $|A| = m$, $|A+B| = \alpha m$. How large can $|A+2B|$ be? In the case $B = A$ the answer was given at the end of the last section. A similar bound can be found by Plünnecke's method if A and B are about the same size. Without any assumption on B, however, the situation changes.

An application of Theorem 5.1 immediately yields

(6.1) $$|A+2B| \leq |A+B|\sqrt{|2B|};$$

this inequality was already proved differently in [11, Theorem 7.2]. To estimate $|2B|$ we can use Corollary 2.4 to obtain $|2B| \leq \alpha^2 m$; combined with (6.1) we get

$$|A+2B| \leq \alpha^3 m^{3/2}.$$

In [11, Theorem 7.1] examples are given (for every rational α and infinitely many m) such that

(6.2) $$|A+2B| \geq \left(\frac{\alpha-1}{4}\right)^2 m^{3/2}.$$

These results describe the order of magnitude for fixed $\alpha > 1$ unless α is near to 1.

We now explore what happens for small values of a. In the extremal case $\alpha = 1$ clearly also $A+2B = m$. The transition is somewhat less clear.

Theorem 6.1. *Let A, B be finite sets in a commutative group G, $|A| = m$, $|A+B| = \alpha m$, $1 < \alpha \leq 2$. We have*

(6.3) $$|A+2B| \leq \alpha m + \tfrac{3}{2}(\alpha-1)m\sqrt{|2B|},$$

consequently

(6.4) $$|A+2B| \leq \alpha m + 3(\alpha-1)m^{3/2};$$

if G is torsionfree, then

(6.5) $$|A+2B| \leq \alpha m + 3(\alpha-1)^{3/2}m^{3/2};$$

PROOF. We apply Corollary 3.6 with the choice $C = A+b$, where b is an arbitrary element of B. The s in the hypothesis will be

$$s = |(A+B) \setminus (A+b)| = (\alpha-1)m,$$

and we obtain (for every $0 \leq t < m$) the existence of an $X \subset A$, $|X| > t$ such that

$$|(X+2B) \setminus (C+B)| \leq \frac{s^2}{(m-t)^2}\left(|X| - \frac{t(t+m)}{2m}\right).$$

Since $|C+B| = |A+B| = \alpha m$, this implies

$$|(X+2B)| \leq \alpha m + \frac{s^2}{(m-t)^2}\left(|X| - \frac{t(t+m)}{2m}\right).$$

For $A \setminus X$ we use an obvious estimate:

$$|(A \setminus X) + 2B| \leq |A \setminus X||2B| = (m-|X|)|2B|,$$

and sum the last two inequalities to get

(6.6) $$|(A+2B)| \leq \alpha m + \left(\frac{s^2}{(m-t)^2} - |2B|\right)|X| + m|2B| - \frac{s^2}{(m-t)^2}\frac{t(t+m)}{2m}.$$

We choose t so that the coefficient of $|X|$ vanishes, that is,

(6.7) $$\frac{s^2}{(m-t)^2} = |2B|.$$

Such a t exists in the interval $(0, m)$ as long as $|2B| \geq s^2/m^2 = (\alpha-1)^2$, which certainly holds under our assumption $\alpha \leq 2$. (We do not really need this restriction;

however, for $\alpha > 2$ this estimate is weaker than (6.1), due to the factor $\frac{3}{2}$.) With this choice (6.6) becomes

$$(6.8) \quad |(A+2B)| \leq \alpha m + |2B|\left(m - \frac{t(t+m)}{2m}\right) = \alpha m + |2B|\frac{(m-t)(2m+t)}{2m}.$$

We estimate $2m+t$ by $3m$, and we express $m-t$ by (6.7):

$$m - t = \frac{s}{\sqrt{|2B|}} = \frac{(\alpha-1)m}{\sqrt{|2B|}}.$$

After these substitutions (6.8) becomes (6.3).

To deduce (6.4) we use Corollary 2.4 and $\alpha \leq 2$.

To deduce (6.5) we use Corollary 2.5: in a torsionfree group

$$|2B| \leq 1 + (\alpha^2 - 1)m \leq 4(\alpha - 1)m,$$

since $\alpha = |A+B|/m \geq 1 + 1/m$, and put this into (6.3). $\square$

We remark that the summand αm in these estimates can actually be the main term, as α may be as small as $1 + O(1/m)$. In the general estimate (6.4) the threshold is $1 + O(m^{-1/2})$, in the torsionfree estimate (6.5) it is $1 + O(m^{-1/3})$.

Still there is a gap between the exponent 2 of $\alpha - 1$ in the example (6.2) and $\frac{3}{2}$ in the upper estimate (6.5). We now show by an example that the exponent 1 of $\alpha - 1$ for general groups in (6.4) is exact. Take a group G which has two k-element subgroups H_1, H_2 such that $H_1 \cap H_2 = \{0\}$. Write $H = H_1 + H_2$ and let

$$A = H \cup \{a_1, \ldots, a_t\}, \quad B = H_1 \cup H_2,$$

where $a_1, \ldots, a_t$ lie in different nonzero cosets of H. Observe that $2B = H$. We have

$$m = |A| = k^2 + t,$$
$$\alpha m = |A+B| = k^2 + t(2k-1),$$
$$\alpha - 1 = \frac{2t(k-1)}{m},$$
$$|A+2B| = (t+1)k^2 = \alpha m + (\alpha-1)(k-1)m/2.$$

Since $k - 1 \sim \sqrt{m}$ as long as $t = o(k^2)$ (and in the interesting case $t = O(k)$), the only difference from the upper estimate (6.4) is a factor of 6.

Acknowledgement. I am grateful to a referee for several corrections.

References

1. Freĭman. G. A and V. P. Pigarev, *The relation between the invariants R and T*, Number-Theoretic Studies in the Markov Spectrum and in the Structural Theory of Set Addition, Kalinin. Gos. Univ., Moscow, 1973, pp. 172–174 (Russian).
2. K. Gyarmati, F. Hennecart, and I. Z. Ruzsa, *Sums and differences of finite sets*, Funct. Approx. Comment. Math., to appear.
3. F. Hennecart, G. Robert, and A. Yudin, *On the number of sums and differences*, Structure Theory of Set Addition, Astérisque, vol. 258, Soc. Math. France, Paris, 1999, pp. 173–178.
4. J. L. Malouf, *On a theorem of Plünnecke concerning the sum of a basis and a set of positive density*, J. Number Theory **54** (1955), no. 1, 12–22.
5. M. B. Nathanson: *Inverse problems and the geometry of sumsets*, Grad. Texts in Math., vol. 165, Springer, New York, 1996.
6. H. Plünnecke, *Eine zahlentheoretische Anwendung der Graphtheorie*, J. Reine Angew. Math. **243** (1970), 171–183.

7. I. Z. Ruzsa, *On the cardinality of $A+A$ and $A-A$*, Combinatorics, Vol. II (Keszthely, 1976) (A. Hajnal and V. T. Sós, eds.), Colloq. Math. Soc. János Bolyai, vol. 18, North-Holland, Amsterdam–New York, 1978, pp. 933–938.
8. _____, *An application of graph theory to additive number theory*, Sci. Ser. A Math. Sci. (N.S.) **3** (1989), 97–109.
9. _____, *Addentum to: An application of graph theory to additive number theory*, Sci. Ser. A Math. Sci. (N.S.) **4** (1990/91), 93–94.
10. _____, *Arithmetical progressions and the number of sums*, Period. Math. Hungar. **25** (1992), no. 1, 105–111.
11. _____, *Sums of finite sets*, Number Theory (New York, 1991–1995), Springer, New York, 1996, pp. 281–293.

ALFRÉD RÉNYI INSTITUTE OF MATHEMATICS, HUNGARIAN ACADEMY OF SCIENCES, POB 127, 1364 BUDAPEST, HUNGARY

E-mail address: `ruzsa@renyi.hu`

Open Problems in Additive Combinatorics

Ernest S. Croot III and Vsevolod F. Lev

ABSTRACT. A brief historical introduction to the subject of additive combinatorics and a list of challenging open problems, most of which are contributed by the leading experts in the area, are presented.

In this paper we collect assorted problems in additive combinatorics, including those which we qualify as classical, those contributed by our friends and colleagues, and those raised by the present authors. The paper is organized accordingly: after a historical survey (Section 1) we pass to the classical problems (Section 2), then proceed with the contributed problems (Sections 3–6), and conclude with the original problems (Section 7). Our problem collection is somewhat eclectic and by no means pretends to be complete; the number of problems can be easily doubled or tripled. We tried to include primarily those problems we came across in our research, or at least lying close to the area of our research interests.

1. Additive combinatorics: a brief historical overview

As the name suggests, additive combinatorics deals with combinatorial properties of algebraic objects, typically abelian groups, rings, or fields. That is, one is interested in those combinatorial properties of the set of elements of an algebraic structure, where the corresponding algebraic operation plays a crucial role. This subject is filled with many wondrous and deep theorems; the earliest of them is, perhaps, the basic Cauchy–Davenport theorem, proved in 1813 by Cauchy [17] and independently rediscovered in 1935 by Davenport [25, 26]. This theorem says that if p is a prime, $\mathbb{F}_p$ denotes the finite field with p elements (notation used throughout the rest of the paper), and the subsets $A, B \subseteq \mathbb{F}_p$ are nonempty, then the *sumset* $A + B := \{a + b : a \in A, b \in B\}$ has at least $\min\{p, |A| + |B| - 1\}$ elements. The analogue of this theorem for the set $\mathbb{Z}$ of integers is the almost immediate assertion (left as a simple exercise to the interested reader) that $|A+B| \geq |A|+|B|-1$ holds for any finite nonempty subsets $A, B \subseteq \mathbb{Z}$.

The $\mathbb{F}_p$-version of the problem is considerably more difficult, and all presently known proofs of the Cauchy–Davenport theorem incorporate a nontrivial idea, such as the transform method (sometimes called the "intersection-union trick"), the polynomial method (as in [1]), or Fourier analysis (see [96]). The situation

2000 *Mathematics Subject Classification.* Primary 11B75; Secondary 11P99.
This is the final form of the paper.

becomes even more complicated when one considers subsets of a general abelian group. An extension of the Cauchy-Davenport theorem onto this case was provided by Kneser, this celebrated result [59, 60] asserts that if A and B are finite, nonempty subsets of an abelian group with $|A + B| < |A| + |B| - 1$, then $A + B$ is a union of cosets of a nonzero subgroup. Further refinement of Kneser's theorem was given by Kemperman in [57].

Over a century passed between Cauchy's paper [17] and the next major result in the subject, proved by Schur [87] in the early 1900s. Schur's theorem states that for every fixed integer $r > 0$ and every r-coloring of the set $\mathbb{N}$ of natural number, there is a monochromatic triple $(x, y, z) \in \mathbb{N} \times \mathbb{N} \times \mathbb{N}$ with $x + y = z$. This theorem, followed by van der Waerden's theorem [99] and its generalization due to Rado [77], eventually developed into the whole area of arithmetic Ramsey theory. In this context we mention an important extension of van der Waerden's theorem by Hales and Jewett [52], and a different proof of the Hales–Jewett theorem by Shelah [88], leading to primitive recursive bounds for the van der Waerden numbers $W(r, k)$ (defined to be the least integer N such that every r-coloring of $[1, N]$ possesses a monochromatic k-term arithmetic progression).

Schur's theorem would follow immediately if for any set $A \subseteq \mathbb{N}$ of positive upper density there were triples $(a_1, a_2, a_3) \in A \times A \times A$ with $a_1 + a_2 = a_3$. The simple example of the set of all *odd* naturals shows that this is not the case, for the equality $x + y = z$ cannot hold with $x, y, z \in \mathbb{N}$ all odd. The situation changes drastically if we are looking for triples $(a_1, a_2, a_3) \in A \times A \times A$, satisfying $a_1 + a_2 = 2a_3$; in other words, for arithmetic progressions of length 3, contained in A. In this case no obvious counterexample can be constructed; perhaps, this is what led Erdős and Turán [34] to conjecture that for every $\varepsilon \in (0, 1]$ and N sufficiently large, any subset of $[1, N]$ with at least εN elements contains a three-term arithmetic progression. (Indeed, Erdős and Turán conjectured that for any integer $k \geq 3$ and N large enough, a subset of $[1, N]$ with at least εN elements necessarily contains a k-term arithmetic progression.) Roth [80] gave an ingenious proof of this conjecture using Fourier analysis, opening the flood gates to applying Fourier methods in additive combinatorial problems.[1]

At first sight, the above conjecture of Erdős and Turán may appear rather weak, for the following reason. Suppose that N is a large positive integer, and take a random integer subset $A \subseteq [1, N]$ with about εN elements, where $\varepsilon \in (0, 1]$. How many three-term arithmetic progressions would we expect A to contain? The number of pairs $(x, y) \in A \times A$ such that $x < y$ are of the same parity is about $\varepsilon^2 N^2 / 4$, and $(x + y)/2 \in A$ holds for about $\varepsilon^3 N^2 / 4$ pairs; that is, A contains about $\varepsilon^3 N^2 / 4$ arithmetic progressions. This exceeds 1 if $\varepsilon^3 N^3 > 4N$, and hence one can naively expect that a subset of $[1, N]$ with at least $CN^{1/3}$ elements (where C is a sufficiently large absolute constant) is guaranteed to contain a three-term arithmetic progression. Perhaps this simple heuristic is what motivated Erdős and Turán [34] to ask whether for some fixed $\varepsilon > 0$ and all sufficiently large N, any subset of $[1, N]$ with at least $N^{1-\varepsilon}$ elements contains a three-term arithmetic progression.

Behrend [7] showed that the answer to the Erdős and Turán question is negative, by constructing for some absolute constant $c > 0$ and any sufficiently large

---

[1]Though Fourier analysis (exponential sums) was used by Hardy, Ramanujan, Littlewood, and others, to deal with Waring's and similar problems, Roth addressed sets of arbitrary structure with density constraints, an altogether different type of problem.

integer N a subset of $[1, N]$, free of three-term arithmetic progressions, with at least $N \exp(-c\sqrt{\log N})$ elements. Showing that the heuristic above is false, this result exhibits sets in which the number of arithmetic progressions differs substantially from what one expects from a random set of the same density. This profoundly alters the way we think about arithmetic progressions.

We now return to the line of research which stems from the Cauchy–Davenport theorem; specifically, to sumset estimates. In the early 1930s, Schnirelmann showed [85] that the set of primes forms an asymptotic basis of $\mathbb{N}$ of finite order; in other words, there is an integer n such that any sufficiently large integer can be represented as a sum of at most n primes. As a technical tool, he introduced the notion of a *lower density* of a set A of nonnegative integers (often called now "the Schnirelmann density" — not to be confused with the lower *asymptotic* density), which he defined by

$$d(A) := \inf\{|A \cap [1, N]|/N : N \in \mathbb{N}\}.$$

A simple yet important lemma from Schnirelmann's paper states that if A and B are sets of nonnegative integers with $0 \in A \cap B$, then $d(A+B) \geq d(A)+d(B)-d(A)d(B)$. A famous conjecture, proposed jointly by Schnirelmann himself and Landau, is that the last inequality can be replaced with the much stronger statement: namely, either $A + B$ contains all positive integers, or $d(A + B) \geq d(A) + d(B)$. Having attracted much attention (including that of such distinguished mathematicians as Besicovich, Brauer, Khinchin, Landau, and Schur, who have established some partial results), this conjecture was eventually solved by Mann in 1942, see [70]. This activity has spanned much interest; it is enough to mention that Davenport rediscovered Cauchy's result as an $\mathbb{F}_p$-analog of the Landau–Schnirelmann conjecture, and that the theorem of Kneser, mentioned above, has appeared as an auxiliary result in his proof of the analog of Mann's theorem for the asymptotic density.

In the middle 1950s Freĭman initiated a systematic study of sumsets of finite integer sets and more generally, of finite subsets of torsion-free abelian groups. In particular, Freĭman introduced the basic notion of local isomorphism, and as a culmination of his research proved in 1964 (see [37]) the result which is now often referred to as "Freĭman's theorem." To discuss this fundamental theorem, we start with a few simple observations.

If $P \subseteq \mathbb{Z}$ is a (finite) arithmetic progression, then the sumset $P + P$ is an arithmetic progression, too, and $|P + P| = 2|P| - 1$ holds; conversely, it is easy to show that if $P \subseteq \mathbb{Z}$ is a finite set with $|P + P| = 2|P| - 1$, then P is an arithmetic progression. Slightly more sophisticated are sets of the form

$$\{a_0 + x_1 d_1 + x_2 d_2 : 0 \leq x_1 < X_1,\, 0 \leq x_2 < X_2\},$$

where a_0 and $d_1, d_2, X_1, X_2 > 0$ are fixed integers. Just like arithmetic progressions, these sets have "small doubling": it is not difficult to see (though this is not completely obvious as $|P| \neq X_1 X_2$ in general) that if P is a set of the above indicated form, then $|P + P| \leq 4|P|$ holds. To further generalize this construction, consider the sets

$$\{a_0 + x_1 d_1 + \cdots + x_r d_r : 0 \leq x_i < X_i;\ i \in [1, r]\},$$

where a_0 and $r, d_1, \ldots, d_r, X_1, \ldots, X_r > 0$ are fixed integers. A set of integers, representable in this form, is called a generalized arithmetic progression of rank (or dimension) r and volume $X_1 \cdots X_r$. Again, it is not difficult to see that if P is a generalized arithmetic progression of rank r, then $|2P| \leq 2^r|P|$. Consequently, if A is a finite set of integers, contained in a generalized arithmetic progression P of rank r so that $|A| \geq \alpha|P|$ with $\alpha \in (0,1]$, then

$$|2A| \leq |2P| \leq 2^r|P| \leq (\alpha^{-1}2^r)|A|.$$

This shows that dense subsets of generalized arithmetic progressions have small doubling, and Freĭman's theorem says that they are the *only* finite integer sets with the small doubling. More precisely, Freĭman's theorem (in its now-standard form due to Ruzsa) says that for every $c \geq 2$ there exist $C > 0$ and $r \in \mathbb{N}$ such that if A is a finite set of integers with $|A + A| < c|A|$, then A is contained in a generalized arithmetic progression of rank at most r and volume at most $C|A|$ (so that the density of A in this generalized arithmetic progression is at least C^{-1}). Ruzsa [83] gave Freĭman's theorem a final shape and a new elegant proof, and Chang [18] greatly refined the dependence of C and r on c. Green and Ruzsa [50] extended Freĭman's theorem to arbitrary abelian groups.

The 1960s and early 1970s saw several new developments in the subject, two of which are the Hales–Jewett theorem [52] and the Szemerédi proof of the general Erdős–Turán conjecture [93]. The Hales–Jewett theorem, which arguably is the most versatile result of Ramsey theory, is often stated in terms of *combinatorial lines*. For integers $N, d \geq 1$, a combinatorial line in the cube $[1, N]^d$ is a subset of the cube of the form $L = \{x + jv : j = 0, \ldots, N - 1\}$ with some $x \in [1, N]^d$ and $v \in \{0, 1\}^d$. Clearly, if $x = (x_1, \ldots, x_d)$ and $v = (v_1, \ldots, v_d)$, then for L to be contained in $[1, N]^d$ it is necessary and sufficient that for each $i \in [1, d]$ we have either $v_i = 0$ (in which case all points of L agree in the ith coordinate), or $x_i = 1$. For instance, a typical example of a combinatorial line for $N = 5$ and $d = 8$ is the set

$$\{(1, 3, 5, 3, 1, 1, 1, 4),$$
$$(2, 3, 5, 3, 2, 2, 1, 4),$$
$$(3, 3, 5, 3, 3, 3, 1, 4),$$
$$(4, 3, 5, 3, 4, 4, 1, 4),$$
$$(5, 3, 5, 3, 5, 5, 1, 4)\}.$$

The Hales–Jewett theorem says that for any integers $r, N \geq 1$ there exists $d_0(r, N)$ such that if $d > d_0(r, N)$ is an integer, then every r-coloring of $[1, N]^d$ possesses a monochromatic combinatorial line. Alternatively, the Hales–Jewett theorem can be stated in terms of words over a finite alphabet.

Note, that van der Waerden's theorem is a simple consequence of the Hales–Jewett theorem. To see this, fix an integer $N \geq 2$ and, writing all integers in $[0, N^d - 1]$ in the base-N form, associate them with the points of the cube $[0, N - 1]^d$. It is immediate then that a combinatorial line in $[0, N - 1]^d$ corresponds to a progression in $[0, N^d - 1]$ of length N.

The density version of the Hales–Jewett theorem, due to Furstenberg and Katznelson [43], asserts that for any fixed real $\varepsilon \in (0, 1]$ and integer $N \geq 1$ there exists $d_0(\varepsilon, N)$ so that if $d \geq d_0(\varepsilon, N)$ is an integer, then any subset of the cube $[1, N]^d$ of density at least ε contains a combinatorial line. Unfortunately, the Furstenberg–Katznelson proof provides no bound for $d_0(\varepsilon, N)$ in terms of ε and N.

Szemerédi's proof of the aforementioned Erdős–Turán conjecture ("for any fixed integer $k \geq 3$, every subset of $\mathbb{N}$ of positive lower density contains a k-term arithmetic progression"), besides establishing a wonderful result, brought with it a powerful new tool, the Szemerédi regularity lemma, which has greatly affected graph theory, combinatorics in general, and additive combinatorics in particular. Hypergraph versions of the regularity lemma, studied by Kohayakawa, Nagle, Rödl, Schact, and Skokan [79], Gowers [46], and Tao [97], have just recently led to a new proof of Szemerédi's theorem.

In the late 1970s Furstenberg [41] gave a new remarkable ergodic-theoretic proof of Szemerédi's theorem. The proof was later generalized in various directions, leading to the multidimensional Szemerédi theorems of Furstenberg and Katznelson [42], and to the polynomial Szemerédi theorem of Bergelson and Leibman [8]. Here we confine ourselves to stating the following corollary of the latter theorem: if k is a positive integer, $f_1, \ldots, f_k$ are polynomials with rational coefficients such that $f_1(0) = \cdots = f_k(0)$, and $S \subseteq \mathbb{N}$ is a set of positive upper density, then there are infinitely many pairs $(m, n) \in \mathbb{Z} \times \mathbb{Z}$ with

$$n + f_1(m) \in S, \ldots, n + f_k(m) \in S.$$

No combinatorial proof of the results just mentioned is presently known.

Besides the growth of ergodic-theoretic methods, the 1980s and 1990s have seen the further expansion of Fourier methods and the emergence of sum-product inequalities. Three famous results from this period, the proofs of which use Fourier analysis, are the theorems of Szemerédi [94], Heath-Brown [53], and Bourgain [12] on three-term arithmetic progressions. Szemerédi and Heath-Brown made the initial breakthrough showing that if $N \in \mathbb{N}$ is large enough and $A \subseteq [1, N]$ satisfies $|A| > N/\log^c N$ (for a certain absolute constant $c > 0$), then A contains a three-term arithmetic progression. Bourgain showed that $A \subseteq [1, N]$ contains a three-term progression whenever $|A| > CN\sqrt{\log \log N / \log N}$, which is considerably stronger than the results of Szemerédi and Heath-Brown's (as their value of c is much smaller than $\frac{1}{2}$). These results gave one the hope to settle the case $k = 3$ of a famous problem of Erdős and Turán, presented below as Problem 2.3; for, it is easy to show that if the assumption of Bourgain's theorem can be relaxed to $|A| > CN/\log^c N$ with some $c > 1$, then any subsets of $\mathbb{N}$ whose sum of reciprocals diverges contains a three-term arithmetic progression.

Given a set A of elements of a ring, let $A \cdot A := \{a_1 a_2 : a_1, a_2 \in A\}$. The history of sum-product inequalities began with the following conjecture of Erdős and Szemerédi (cf. Problem 2.4): for every $\varepsilon > 0$ there exists $c > 0$ such that if $A \subseteq \mathbb{N}$ is finite, then

$$\max\{|A + A|, |A \cdot A|\} > c|A|^{2-\varepsilon}.$$

This conjecture remains the central unsolved problem in the subject, though a lot of progress has been made on it. Erdős and Szemerédi [33] themselves proved that for some $\delta > 0$ there exists $c > 0$ such that

$$\max\{|A + A|, |A \cdot A|\} > c|A|^{1+\delta}$$

holds for every finite subset $A \subseteq \mathbb{N}$. Nathanson [73] showed that one can take $\delta = 1/31$, and Ford [35] later improved this to $\delta = 1/15$. Perhaps the most elegant result in this direction is due to Elekes [29], who used the Szemerédi–Trotter theorem [95] on point-line incidences to prove that

$$|A + A||A \cdot A| \geq c|A|^{5/2};$$

this improves Ford's result, leading to $\delta = 1/4$. The most recent and strongest result is due to Solymosi [89], who showed that any value, smaller than 3/11, can be taken for δ.

From the year 2000 to the present, there has been a tremendous explosion of new and deep results in additive combinatorics. There are too many of them to list in this short summary; instead of attempting this, we just focus on the three most famous: Gowers' new proof of Szemerédi's theorem [45], the Bourgain–Katz–Tao [16] and Bourgain–Glibichuk–Konyagin [15] sum-product estimates in finite fields, and the Green–Tao proof [51] that the set of primes contains arbitrarily long arithmetic progressions.

Gowers' proof of Szemerédi's theorem was a phenomenal breakthrough, partly because it substantially improved the dependence between the density of a subset $A \subseteq [1, N]$ and the largest integer k such that A is guaranteed to contain a k-term arithmetic progression. As Gowers has shown, for every $k \geq 3$ there exists $c > 0$ such that if N is sufficiently large and $A \subseteq [1, N]$ satisfies $|A| \geq N(\log \log N)^{-c}$, then A contains a k-term arithmetic progression. Another reason for Gower's argument to be of great importance is that it introduced into the subject new tools and ideas, the use of which has expanded well beyond additive combinatorics. Just to give one example, within the field of combinatorics, but not specifically additive combinatorics, Conlon [22] has used some of Gowers's methods to produce a new upper bound on the diagonal Ramsey number $r(n, n)$ (which is the smallest r such that if the edges of a complete graph on r vertices are 2-colored, then there is a monochromatic clique of n vertices); his bound is

$$r(n+1, n+1) \leq \binom{2n}{n} \exp\left(-c\left(\frac{\log n}{\log \log n}\right)^2 \log \log \log n\right),$$

where $c > 0$ is an absolute constant. Two of the tools in Gowers's new proof of Szemerédi's theorem, which have been used in many other fields of mathematics and even computer science, are the concept of the Gowers uniformity norm and a strong version of an important theorem of Balog and Szemerédi [6], sometimes called now the Balog–Szemerédi–Gowers theorem. A slightly relaxed form of Gowers' version of the Balog–Szemerédi theorem is as follows: if $\gamma \in (0, 1)$ and A is a finite subset of an abelian group such that the equation $x_1 + x_2 = x_3 + x_4$ has at least $\gamma |A|^3$ solutions in the elements of A, then there is a subset $A_0 \subseteq A$ with $|A_0| \geq \gamma^c |A|$ and $|A_0 + A_0| \leq \gamma^{-d} |A_0|$, where c and d are positive absolute constants.

The Bourgain–Katz–Tao and Bourgain–Glibichuk–Konyagin estimates expanded the Erdős–Szemerédi sum-product problem onto the finite field setting. They show that for every $\varepsilon > 0$ there exists $\delta > 0$ such that if p is a sufficiently large prime and $A \subseteq \mathbb{F}_p$ satisfies $|A| \leq p^{1-\varepsilon}$, then

$$\max\{|A + A|, |A \cdot A|\} \geq |A|^{1+\delta}.$$

The proof, using among other ideas the Balog–Szemerédi–Gowers theorem, is itself a centerpiece in many new, remarkable results in the area. For example, Bourgain [13] has used sum-product estimates to bound the size of certain exponential sums, involving sparse polynomials of high degree. Such bounds were thought to be out of reach of any method that currently exists, including those coming from arithmetic geometry and analytic number theory.

Our survey would be incomplete without mentioning the crowning achievement of the last years, which required the combination of ideas from different areas, as well as the invention of new ideas; namely, resolving by Green and Tao the old conjecture that the set of prime numbers contains arbitrarily long arithmetic progressions. In their proof, Green and Tao introduced the concept of quasi-randomness and used the results of Goldston, Pintz, and Yilidrim [44] to show that functions like

$$f(n) := \frac{1}{\log^2 n} \left(\sum_{d|n: d < N^\theta} \mu(d) \log(n/d) \right)^2,$$

restricted to certain arithmetic progressions, are "quasirandom to a high degree." It is easily seen, on the other hand, that if g denotes the indicator function of the set of primes, then f majorizes g, at least in the range $(N^\theta, N]$; that is, if n is in this range, then $f(n) \geq g(n)$. Furthermore, one can show that $g(n)$ "eats up a positive proportion of the mass of $f(n)$"; more precisely,

$$\sum_{n \leq N} f(n) < c \sum_{n \leq N} g(n)$$

holds for some positive constant c (depending on θ). Green and Tao showed that these properties (quasi-randomness, majorization, and "eating positive proportion of the mass") imply that the primes contain arithmetic progressions of any prescribed length.

Just recently, Tao and Ziegler [98] have proved the even more general result that the primes contain arbitrarily long "polynomial progressions"; specifically, if $f_1, \ldots, f_k$ are any polynomials with integer coefficients such that $f_1(0) = \cdots = f_k(0)$, then there exist infinitely many integers $m, n \in \mathbb{N}$ such that $m, m+f_1(n), m+f_2(n), \ldots, m + f_k(n)$ are all simultaneously prime.

The story of additive combinatorics is far from over. It continues to thrive to a large extent due to the many excellent problems that researchers have brought to the subject; some of these problems are listed below.

2. Classical problems

2.1. Dense progression-free integer sets. For a large positive integer N, what is the largest size of a subset of the interval $[1, N]$, free of three-term arithmetic progressions? The present records are due to Behrend (who constructed in [7] a progression-free set $A \subseteq [1, N]$, satisfying $|A| > N \exp(-c\sqrt{\log N})$ with an absolute constant $c > 0$) and Bourgain (who proved in [12] that if $A \subseteq [1, N]$ is progression-free, then $|A| < CN\sqrt{\log \log N / \log N}$ with an absolute constant C). Narrow the gap between these estimates.

2.2. Dense progression-free sets in abelian groups. It is natural to ask how large can be a progression-free subset of an abelian group, other than the group of integers. For example, consider the additive group of the finite field $\mathbb{F}_q$ with $q = 3^r$ elements, where r is a positive integer. What is the largest size of a subset of this group, free of three-term arithmetic progressions? How does this quantity behave as r grows? (The finite geometry interpretation of this problem stems from the observation that $x, y, z \in \mathbb{F}_q$ form an arithmetic progression if and only if they lie on a line.)

It follows from a result of Meshulam [71] (see also [10]) that if $A \subseteq \mathbb{F}_q$ is progression-free, then $|A| \leq 2 \cdot 3^r/r$. On the other hand, it is easy to construct a progression-free set $A \subseteq \mathbb{F}_q$ such that $|A| = 2^r$: just fix arbitrarily a basis

$\{e_1, \ldots, e_r\}$ of $\mathbb{F}_q$ over $\mathbb{F}_3$ and let $A := \{\varepsilon_1 e_1 + \cdots + \varepsilon_r e_r : \varepsilon_1, \ldots, \varepsilon_r \in \{0,1\}\}$. The best known lower bound is due to Edel [28], who has constructed progression-free sets in $\mathbb{F}_q$ of size $(2.217\ldots)^r$ by finding a particular example in rather large dimension and then taking a product of several copies of it.

It would be of much interest to improve Meshulam's estimate, to show that any progression-free subset of $\mathbb{F}_q$ has size $o(3^r/r)$, and/or to determine whether there is an absolute constant $c < 3$ such that any progression-free subset of $\mathbb{F}_q$ has size, smaller than c^r.

2.3. Arithmetic progressions in sets with diverging reciprocals (Erdős–Turán).
Suppose that $A \subseteq \mathbb{N}$ has the property that the sum of the reciprocal of the elements of A diverges: $\sum_{a \in A} 1/a = \infty$. Must A contain k-term arithmetic progressions for all $k \geq 3$? Even the case $k = 3$ is open.

2.4. Sum-product estimate for integers (Erdős–Szemerédi).
Prove (or disprove) that for every $\varepsilon > 0$ there exists $c > 0$ (depending on ε) such that if A is a finite set of integers, then
$$\max\{|A+A|, |A \cdot A|\} > c|A|^{2-\varepsilon}.$$

See Section 1 for comments on this problem.

2.5. Quantitative Hales–Jewett theorem.
Obtain reasonable bounds for the Hales–Jewett theorem and for the density version of it.

See Section 1 for the discussion on the Hales–Jewett theorem.

2.6. van der Waerden's numbers.
For $k \in \mathbb{N}$, the van der Waerden number $W(k) = W(2,k)$ is defined to be the least positive integer N such that for any 2-coloring of $[1,N]$ there is a monochromatic k-term arithmetic progression with the elements in $[1,N]$. What is the order of growth of $W(k)$? Say, is it true that $W(k) \leq 2^{k^2}$?

Berlekamp [9] proved that if p is a prime, then
$$W(p+1) \geq p2^p,$$
and from the work of Gowers [45] we know that
$$W(k) \leq 2^{2^{2^{2^{2^{2^{k+9}}}}}}.$$

2.7. Just bases in $\mathbb{N}$ (Erdős–Turán).
A set $B \subseteq \mathbb{N}$ is called a *basis of order* 2 if any positive integer is representable as a sum of two elements of B. Does there exist a basis B of order 2 such that the number-of-representations function $\nu_B(n) := |\{(b_1, b_2) \in B \times B : b_1 + b_2 = n\}|$ is uniformly bounded by a constant, independent on n? Erdős conjectured that the answer is negative.

In connection with this problem, Nešetřil and Serra ask in [74] whether it is true that for any finite partition of a basis of order 2, at least one of the partite sets has unbounded number-of-representations function. A positive answer would imply Erdős' conjecture.

We refer the reader to [81] for exciting partial results towards the solution of this problem and its finite analogues.

2.8. Difference sets in quadratic residues. Given a prime $p \equiv 1 \pmod{4}$, how large can a set $A \subseteq \mathbb{F}_p$ be given that the difference between any two elements of A is a quadratic residue modulo p? In other words, what is the clique number of the Paley graph over $\mathbb{F}_p$?

The existence of a set, possessing the property in question and of size $(0.5 + o(1))\log_2 p$, is established in [21]. In [48] the lower bound $c \log p \log \log \log p$ is proved for infinitely many primes p. Finding a reasonable upper bound is an old problem on which nothing is known beyond the estimate $|A| < \sqrt{p}$. A simple elementary proof is as follows. Suppose that $|A| > \sqrt{p}$. Then for any $x \in \mathbb{F}_p$ there exist $a_1, b_1, a_2, b_2 \in A$ such that $a_1 x + b_1 = a_2 x + b_2$ and $a_1 \neq a_2$. Consequently, $x = (b_1 - b_2)/(a_2 - a_1)$ and since *any* $x \in \mathbb{F}_p$ has a representation of this form, the set of all nonzero elements of $A - A$ is not contained in a multiplicative subgroup of $\mathbb{F}_p$.

This is probably a very hard problem: just observe that the estimate $|A| < p^{\varepsilon}$ would imply that the least quadratic nonresidue modulo p is smaller than p^{ε}.

One can consider the sumset $A + A$ instead of the set of differences, or ask the question for the multiplicative subgroups of $\mathbb{F}_p$, other than the subgroup of quadratic residues, in the following spirit: given a proper subgroup H of the multiplicative group of $\mathbb{F}_q$, what is the largest size of a subset $A \subseteq \mathbb{F}_p$ with $A - A \subseteq H$? One can also ask whether there exists $A \subseteq \mathbb{F}_p$ such that $A + A = H$, or such that the symmetric difference of $A + A$ and H is small.

2.9. Large sum-free subsets of integer sets. What is the largest constant c with the property that any finite, sufficiently large set A of integers contains a sum-free subset of size at least $c|A|$?

Recall, that a subset A of an additively written group is called *sum-free*, if $A \cap (A + A) = \varnothing$; that is, the equation $x + y = z$ has no solutions in the elements of A. (Thus, Schur's theorem, discussed in the Introduction, says that the set of positive integers cannot be partitioned into finitely many sum-free subsets.) Alon and Kleitman showed in [3], as a slight improvement of a result of Erdős from [30], that any finite set A of nonzero integers contains a sum-free subset with at least $\lceil (|A| + 1)/3 \rceil$ elements. Bourgain [11], using an elaborate Fourier analysis technique, improved this further to $\lceil (|A| + 2)/3 \rceil$. There is no indication that the factor $\frac{1}{3}$ is best possible here; however, it is shown in [3] that it cannot be replaced by a number, larger than $12/29$, and J. Malouf has recently improved this to $\frac{2}{5}$ in her Ph.D. thesis.

3. Contributed problems. I: sets, free of particular structures

3.1. Dense subsets of $\mathbb{F}_p$ with few arithmetic progressions (contributed by B. Green). For a prime p, what is the least number of three-term arithmetic progressions that a subset $A \subseteq \mathbb{F}_p$ with $|A| = (p-1)/2$ can have? What happens if $|A| = \delta p$, where $\delta < 0.5$?

It follows from a result by Varnavides [100] that this number is at least cp^2 with some $c = c(\delta) > 0$, and Croot [23] has recently shown that it is in fact $cp^2(1 + o(1))$ as $p \to \infty$. It seems to be a difficult problem to determine the order of magnitude of the constant $c(\delta)$ as $\delta \to 0$.

3.2. Uniform sets in $\mathbb{Z}/N\mathbb{Z}$ with few four-term arithmetic progressions (contributed by I. Ruzsa).
Is it true that for any fixed $k \geq 1$ and sufficiently small $c > 0$, there exist integers $N_1, N_2, \ldots$ and sets $A_1 \subseteq \mathbb{Z}/N_1\mathbb{Z}, A_2 \subseteq \mathbb{Z}/N_2\mathbb{Z}, \ldots$ such that for all $i \geq 1$

(i) $|A_i| \geq cN_i$;
(ii) A_i is α_i-uniform, where $\lim_{i \to \infty} \alpha_i = 0$;
(iii) A_i has at most $c^k N_i^2$ arithmetic progressions of length 4?

Recall that $A \subseteq \mathbb{Z}/N\mathbb{Z}$ is said to be α-uniform if, letting $\hat{A}(u) = \sum_{a \in A} e^{2\pi i a u / N}$, one has
$$\sum_{\substack{u \in \mathbb{Z}/N\mathbb{Z} \\ u \neq 0}} |\hat{A}(u)|^4 \leq \alpha N^4.$$

3.3. Large product-free sets in finite groups (contributed by V. Sós).
How large can a product-free subset of a finite group be?

A subset A of a group is called *product-free* if the equation $xy = z$ has no solutions in the elements of A (cf. Problem 2.9). Babai and Sós proved in [5] that any finite group G contains a product-free subset with at least $c|G|^{4/7}$ elements, where c is a positive absolute constant; Kedlaya improved this to $c|G|^{11/14}$ in [56]. As shown by Alon and Kleitman [3], any finite *abelian* group G contains a product-free subset of size at least $2|G|/7$, and this is the best possible bound. In the nonabelian case, already the alternating groups A_n are of interest; Green conjectures that the largest size of a product-free subset of A_n is $o(|A_n|)$ (as $n \to \infty$).

Recently, Gowers [47] has shown that any product-free subset of the group $G = \mathrm{PSL}_2(p)$ has fewer than $c|G|^{8/9}$ elements, for a suitable absolute constant C. (The group $\mathrm{PSL}_2(p)$, short for *projective special linear group*, is the quotient group $\mathrm{SL}_2(p)/\{I, -I\}$; here $\mathrm{SL}_2(p)$ is the multiplicative group of 2×2 matrices over $\mathbb{F}_p$ with unit determinant, and I is the 2×2 identity matrix over $\mathbb{F}_p$.)

3.4. Sets, free of solutions of a linear equation (contributed by Y. Stanchescu).
Fix an integer $t \geq 1$ and suppose that $A \subseteq [1, N]$ has the property that none of the t^2 equations $mx + ny = (m+n)z$ with $1 \leq m, n \leq t$ has a nontrivial solution in the variables $x, y, z \in A$. How large can A be under this assumption?

A small modification (cf. [92]) of Behrend's construction yields a set $A \subseteq [1, N]$ with $|A| = N \exp(-C(t)\sqrt{\log N})$, possessing the property under consideration. On the other hand, one evidently has $|A| \leq r_3(N)$, where $r_3(N)$ is the largest size of a subset of $[1, N]$, free of three-term arithmetic progressions.

3.5. Sequences, locally free of arithmetic progressions (contributed by G. Freĭman).
Fix an integer $s \geq 3$ and suppose that $A = \{a_1, a_2, \ldots\}$ is a strictly increasing sequence of nonnegative integers, such that no segment of this sequence of the form $(a_{i+1}, a_{i+2}, \ldots, a_{i+s})$ for $i = 0, 1, \ldots$ contains a three-term arithmetic progression. How large can the density of A be under this assumption?

The contributor observes that for $s = 4$ one can take
$$A = \{0, 1, 3, 4, 6, 7, 9, 10, \ldots\}$$
(the set of all nonnegative integers, congruent to 0 or 1 modulo 3) which has density $\frac{2}{3}$; similarly, for $s = 8$ one can take
$$A = \{0, 1, 3, 4, 9, 10, 12, 13, 18, 19, 21, \ldots\}$$

(the set of all nonnegative integers, congruent to $0, 1, 3$, or 4 modulo 9) which has density $\frac{4}{9}$.

Konyagin indicates that if $n(s)$ denotes the smallest positive integer N such that there exists an s-element subset of $[1, N]$, free of three-term arithmetic progressions, then the upper asymptotic density of A does not exceed $s/n(s)$; on the other hand, there exists A with the property in question and with the lower asymptotic density at least $s/(2n(s))$.

3.6. van der Waerden related numbers (contributed by R. Graham).
Define $W^*(k)$ to be the size of the smallest set A of integers such that any 2-coloring of A has a monochromatic k-term arithmetic progression; thus, $W^*(k) \leq W(k)$ (the "classical" van der Waerden number). Is $W(k) - W^*(k)$ unbounded as $k \to \infty$? Is it true that
$$\lim_{k \to \infty} \frac{W^*(k)}{W(k)} = 1?$$

The contributor offers \$100 for the answer to the first question. He also remarks that $W^*(3) = W(3) = 9$ and $W^*(4) \leq 27$, while $W(4) = 35$.

3.7. The plane analogue of Problem 2.3 (contributed by R. Graham).
Suppose that a set $A \subseteq \mathbb{Z} \times \mathbb{Z}$ has the property that $\sum_{(x,y) \in A} 1/(x^2 + y^2) = \infty$. Must A contain the four vertices of a square, i.e., four points of the form (x, y), $(x + d, y)$, $(x, y + d)$, and $(x + d, y + d)$ with $x, y \in \mathbb{Z}$, $d \in \mathbb{N}$?

The contributor conjectures that the answer is positive and offers \$1000 for the proof (or disproof) of this conjecture. More generally, he conjectures that any set A with the above property contains a $k \times k$ square grid, for any integer $k \geq 2$.

3.8. The number of monochromatic solutions (contributed by R. Graham).
Let $\mathbf{E}$ be a set of homogeneous linear equations which is partition regular; that is, $\mathbf{E}$ has a nontrivial monochromatic solution for any r-coloring of $\mathbb{Z}$. For positive integers N and r, what is the minimum number $f_{\mathbf{E}}(N, r)$ of monochromatic solutions to $\mathbf{E}$ which can occur for an r-coloring of $[1, N]$?

It follows from general results of Frankl, Graham, and Rödl [36] that a positive fraction (depending only on $\mathbf{E}$ and r) of all solutions are monochromatic. However, it seems to be difficult to determine exactly the best possible constant.

It is known that if $r = 2$ and $\mathbf{E}$ consists of a single equation $x + y = z$, then $f_{\mathbf{E}}(N, r) = N^2(1 + o(1))/22$ (Robertson–Zeilberger [78], Schoen [86]). On the other hand, when $r = 2$ and $\mathbf{E}$ consists of the equation $x + y = 2z$ (corresponding to three-term arithmetic progressions), then we only know that
$$\frac{189}{4096} N^2(1 + o(1)) < f_{\mathbf{E}}(N, 2) < \frac{117}{2192} N^2(1 + o(1));$$
here the lower bound is due to Parrilo, Robertson, and Saracino (preprint), and the upper bound is due to these authors and independently to Butler, Costello, and Graham (unpublished). Note that a random 2-coloring of $[1, N]$ would have $N^2(1 + o(1))/16$ monochromatic three-term arithmetic progressions, and that
$$\frac{189}{4096} = \frac{1}{21.671957\ldots} < \frac{117}{2192} = \frac{1}{18.73504\ldots} < \frac{1}{16}.$$

There is some evidence that the upper bound is actually the truth here.

Let, again, **E** be the single equation $x+y=2z$. Alon reports that he can prove the existence of an absolute constant c such that

$$f_{\mathbf{E}}(N,r) \leq r^{-c\log r} N^2$$

holds for all $r, N \in \mathbb{N}$; thus, for large r the number of monochromatic triples can be *much* smaller than $(1+o(1))N^2/(4r^2)$, obtained for the random coloring.

3.9. Partition regularity of the Pythagorean equation (contributed by R. Graham). Is the equation

$$x^2 + y^2 = z^2$$

partition regular? This is an old problem of Erdős and Graham [31] for which we have little evidence either way. It is perhaps the simplest question involving partition regularity of homogeneous nonlinear equations. The contributor offers $250 for resolving this problem.

3.10. A Pisier-type problem (contributed by J. Nešetřil). Does there exist a positive real number ε and a set A of positive integers with the following properties:

  (i) for every finite partition of A, one of the classes contains a monochromatic three-term arithmetic progression;
  (ii) for every finite subset $B \subseteq A$ there exists $C \subseteq B$ with $|C| > \varepsilon |B|$ such that C does not contain a three-term arithmetic progression.

It is easy to see that if A satisfies (ii), then A does not contain arithmetic progressions that are "too long" (as a density ε subset of a sufficiently long arithmetic progression contains a three-term arithmetic progression by Szemerédi's theorem). Also, Spencer and Nešetřil, Rödl gave examples of sets, satisfying (i), which do not contain four-term arithmetic progressions.

Notice that the Hales–Jewett theorem cannot be used (at least, in the standard way) to prove the existence of A satisfying (i) and (ii); for, if one uses Hales–Jewett to prove that a certain set A satisfies (i), then the Furstenberg–Katznelson density version of Hales–Jewett shows that A does not satisfy (ii).

Erdős, Nešetřil, and Rödl call statements, involving properties (i) and (ii) (for example for longer arithmetic progressions, but also for other structures), *Pisier-type theorems*; see [32, 58].

4. Contributed problems. II: sumsets

4.1. Sequences with locally small sumsets (contributed by V. Sós).
Let A be a strictly increasing infinite sequence of integers, and denote by A_n the set of the n smallest elements of A. What can be said about the structure of A given that

$$|A_n + A_n| < Cn$$

holds for any positive integer n and an absolute constant C?

4.2. Covering vector spaces with subset sums (contributed by G. Martin).
A consequence of the Cauchy–Davenport theorem (see the Introduction) is that given any prime p and any multiset A of $p-1$ nonzero elements of $\mathbb{F}_p$, any element of $\mathbb{F}_p$ is representable as a subset sum of A. What is the natural generalization of this assertion onto finite-dimensional vector spaces over $\mathbb{F}_p$? That is, what are natural conditions that guarantee that the set of subset sums is the whole vector space (even when the multiset under consideration is not too large)?

Let A be a subset of the r-dimensional vector space V over $\mathbb{F}_p$, and let $\Sigma(A)$ denote the set of all subset sums of A. To avoid the situation where $\Sigma(A)$ is trapped in or heavily concentrated on subspaces, it is natural to bring into consideration the quantities

$$\sigma_j(A) = \max_{\substack{W \le V \\ \dim(W)=j}} |A \cap W|; \quad j \in [1, r].$$

Given the numbers $\alpha_1, \ldots, \alpha_{r-1} > 0$, find a "reasonable" estimate (as a function of $\alpha_1, \ldots, \alpha_{r-1}$) for the size of the largest set A such that $\Sigma(A) \ne V$ and

$$\sigma_1(A) \le \alpha_1, \sigma_2(A) \le \alpha_2, \ldots, \sigma_{r-1}(A) \le \alpha_{r-1}.$$

In particular, if $\sigma_j(A) < jp$ for all $j = 1, \ldots, r-1$, does $\Sigma(A) \ne V$ imply $|A| \le rp - 2$?

4.3. Sumsets of progression-free sets (contributed by G. Freĭman).
Given that n is a positive integer and $A \subseteq \mathbb{Z}$ is an n-element set, free of three-term arithmetic progressions, how small can $|2A|$ be?

Freĭman [38] proved that $|2A|/n$ tends to infinity, and Ruzsa [82] proved that this quotient is at least $0.5(n/r_3(n))^{1/4}$, where $r_3(N)$ denotes the largest size of a subset of $[1, N]$, free of three-term arithmetic progressions. On the other hand, it is immediate that if N is so chosen that $r_3(N) = n$ (which is possible for any given n), then there is an n-element set A, free of three-term arithmetic progressions and such that $|2A|/n < 2N/n = 2N/r_3(N)$. Thus, for instance, Behrend's construction yields a set A with $|2A|/n = O(e^{c\sqrt{\log n}})$, where c is an absolute constant.

4.4. Sumsets of no-three-points-on-a-line sets (contributed by G. Freĭman).
Given that n is a positive integer and $A \subseteq \mathbb{Z} \times \mathbb{Z}$ is an n-element set, no three points of which are collinear, how small can $|2A|$ be?

As above, denote by $r_3(N)$ the largest size of a subset of $[1, N]$, free of three-term arithmetic progressions. Using results of [82] (see previous problem), Stanchescu showed in [92] that $|2A| \ge 0.5n(n/r_3(n))^{1/4}$, and on the other hand, that there are arbitrarily large $n \in \mathbb{N}$ and corresponding n-element sets $A \subseteq \mathbb{Z} \times \mathbb{Z}$, free of collinear triples, such that $|2A| \le n\exp(C\sqrt{\ln n})$ (with an absolute constant C).

4.5. Freĭman's theorem for distinct set summands (contributed by T. Tao).
Is it true that for any $K > 1$ there exists $C > 0$ with the following property: if A and B are finite, nonempty integer sets, satisfying $|A + B| < K|A|$ and $|B| \le |A|$, then there is a generalized arithmetic progression P of rank at most C and a set $X \subseteq \mathbb{Z}$ so that $B \subseteq P$, $A \subseteq X + P$, and $|X + P| < C|A|$?

The case $|A| = |B|$ is, essentially, Freĭman's theorem (in conjunction with the "covering lemma" of Ruzsa, implicit in [82]), and the case where $|A|$ and $|B|$ are of the same order of magnitude follows easily.

As the contributor indicates, using Plünnecke's inequalities (cf. [82]) one can establish a weaker assertion, with the requirement that P is an arithmetic progression relaxed to a hypothesis of the sort $|P + P| \leq c(K, \varepsilon)|A|^{\varepsilon}|P|$.

4.6. Doubling the squares (contributed by B. Green and T. Tao). How small can $|2A|$ be for an n-element subset A of the set of squares of integers?

The contributors indicate that this problem is implicit in a paper of Chang [19] on Rudin's problem ("are the squares a $\Lambda(p)$-set?"), and that a result from [19] implies that $|2A| \geq cn(\ln n)^{1/12}$ with an absolute constant $c > 0$ (see comments on Problem 6.5).

4.7. Small doubling in binary spaces (contributed by I. Ruzsa). Is it true that for any $K > 1$, $r \in \mathbb{N}$, and any subset $A \subseteq \mathbb{F}_2^r$, satisfying $|A + A| \leq K|A|$, there exists a linear subspace $V \subseteq \mathbb{F}_2^r$ such that $|V| < K^c|A|$ and $|A \cap V| \geq K^{-c}|A|$, with some absolute constant $c > 0$?

The contributor has shown that the problem can be equivalently restated as follows: is it true that any function $f \colon \mathbb{F}_2^r \to \mathbb{F}_2^{\infty}$ can be written as a sum of a linear function and a function, whose image has size, polynomial in the cardinality of the set
$$\{f(x+y) + f(x) + f(y) : x, y \in \mathbb{F}_2^r\}?$$

4.8. Growth of higher sumsets (contributed by T. Tao). For a finite set $A \subseteq \mathbb{Z}$ and real $K > 0$, how fast can $|nA|$ grow (as $n \to \infty$) given that $|2A| < K|A|$? Estimate the quantity
$$f(n, K) := \sup\{|nA|/|A| : A \subseteq \mathbb{Z} \text{ is finite and } |2A| < K|A|\}.$$
Plünnecke–Ruzsa inequalities [82] imply that $f(n, K) \leq K^n$.

4.9. Balog–Szemerédi theorem for distinct set summands (contributed by T. Tao). Let m and n be positive integers with $m \geq n$, and let $K, \delta > 0$. Suppose that A and B are finite sets of integers with $|A| = m$ and $|B| = n$, and that $G \subseteq A \times B$ satisfies $|G| \geq \delta mn$. Does
$$|\{a + b : a \in A, \ b \in B, \ (a, b) \in G\}| < Km$$
imply anything about the structure of A and B? In particular, does it imply that there are $A' \subseteq A$ and $B' \subseteq B$ with $|A'| \geq cm$, $|B'| \geq cn$ such that $|A' + B'| \leq Cm$, where c and C depend only on δ and K?

The case $m = n$ is the Balog–Szemerédi's theorem. For $\lambda > 1$ and $n \leq m \leq \lambda n$ the assertion is easy to derive with the constants c and C, depending on λ (in addition to the dependence on δ and K).

4.10. Sumset and difference set (contributed by G. Freĭman). For a finite set $A \subseteq \mathbb{Z}$ and a subset $G \subseteq A \times A$, set
$$A \stackrel{G}{+} A := \{a' + a'' : (a', a'') \in G\}, \quad A \stackrel{G}{-} A := \{a' - a'' : (a', a'') \in G\}.$$
Estimate $|A \stackrel{G}{-} A|$ from above in terms of $|A \stackrel{G}{+} A|$ and $|A|$ and describe those sets A with the largest possible value of $|A \stackrel{G}{-} A|$.

The contributor indicates that, writing $n := |A|$, we have

(i) If $|A \stackrel{G}{+} A| = 1$, then $|A \stackrel{G}{-} A| \leq n$; moreover, if equality is attained, then A is symmetrical;

(ii) if $|A \stackrel{G}{+} A| = 2$, then $|A \stackrel{G}{-} A| \leq 2n - 1$; moreover, if equality is attained, then A is an arithmetic progression.

For the (much more delicate) complete solution in the case $|A \stackrel{G}{+} A| = 3$, see [40].

4.11. Recognizing sumsets algorithmically (contributed by A. Granville). Given a finite subset of an abelian group, can one give an efficient algorithm to determine whether it is of the form $A + A$, where A is yet another subset of the group?

A strong necessary condition for a subset S of an abelian group to be of the indicated form is that the difference set $S - S$ contains many "popular differences"; more precisely, if $K = \lfloor \sqrt{2|S|} \rfloor$ then there is a subset $D \subseteq S - S$ with $|D| \geq \sqrt{K|S|}$ such that any element of D has at least K representations as a difference of two elements of S. To see that this condition is necessary, assuming that $S = A + A$ let $D := A - A$. From $|S| \leq \binom{|A|}{2} + |A|$ we derive that $|A| > \sqrt{2|S|} - 1$, whence $K \leq |A|$. Next, the well-known "Ruzsa triangle inequality" (see [84]) gives $|A - A|^2 \geq |A+A||A|$, implying $|D| \geq \sqrt{K|S|}$. Finally, for any $a', a'' \in A$ the number of representations of $a' - a''$ as a difference of two elements of S is $|(S-a') \cap (S-a'')|$, which is at least $|A|$ as both $S - a'$ and $S - a''$ contain A.

A simple algorithm, exponential in the size of the subset under investigation, stems from the observation that if $S = A+A$, then for any $a \in A$ we have $2(A+a) = S + 2a$ and $A + a \subseteq S$. Consequently, to check whether S is a sumset one can run through all subsets of S one-by-one, for every subset computing its sumset. The set S is a sumset if and only if there exists $B \subseteq S$ such that $2B = S + g_0$, where $g_0 \in \{2g : g \in G\}$. No better algorithm is known even when the underlying group is torsion-free, cyclic, or when it is the additive group of a finite field.

4.12. Large sets in $\mathbb{F}_p$ which are not sumsets (contributed by B. Green). For a prime p, what is the largest size of a subset of $\mathbb{F}_p$, which is not of the form $A + A$ (with $A \subseteq \mathbb{F}_p$)?

Improving the contributor's original estimate

$$p - p^{2/3+\varepsilon} < \max_{\substack{S \subseteq \mathbb{F}_p \\ S \neq A+A}} |S| < p - \tfrac{1}{9} \log p$$

(for any fixed $\varepsilon > 0$ and p large enough), Alon proves in [2] that

$$p - C \frac{p^{2/3}}{(\log p)^{1/3}} < \max_{\substack{S \subseteq \mathbb{F}_p \\ S \neq A+A}} |S| < p - c \frac{p^{1/2}}{(\log p)^{1/2}}$$

with some absolute constants $c, C > 0$ for all sufficiently large p; as indicated in [2], the upper bound is likely to be close to the truth.

5. Contributed problems. III: combinatorial and finite geometry

5.1. Small Besicovich sets in finite geometries (contributed by T. Tao). Let $\mathbb{F}$ be a finite field, and suppose that $A \subseteq \mathbb{F} \times \mathbb{F} \times \mathbb{F}$ is a Besicovich set; i.e., A contains a line in every direction. It is known from the work of Wolff that $|A| \geq |\mathbb{F}|^{5/2}$; prove that in fact $|A| \geq |\mathbb{F}|^{5/2+\varepsilon}$ holds for some $\varepsilon > 0$.

5.2. Small sets, determining all possible directions (contributed by A. Granville). Given a finite field $\mathbb{F}$ and an integer $r \geq 1$, find the smallest size of a subset $E \subseteq \mathbb{F}^r$ which determines all directions in $\mathbb{F}^r$. That is, determine the smallest size of a subset $A \subseteq \mathbb{F}^r$ with the property that for any $d \in \mathbb{F}^r$ there exist $a_1, a_2 \in A$ such that $a_1 - a_2$ is a scalar multiple of d.

Let $q := |\mathbb{F}|$. It is immediate that if $A \subseteq \mathbb{F}^r$ determines all $(q^r - 1)/(q - 1)$ directions in $\mathbb{F}^r$, then $|A| \geq \sqrt{2}\, q^{(r-1)/2}$, and Konyagin indicates that this estimate can be matched up to the constant factor, as follows. Writing $\mathbb{F}^r = \mathbb{F}^{r-1} \oplus \mathbb{F}$, find a subset $D \subseteq \mathbb{F}^{r-1}$ with $|D| < Cq^{(r-1)/2}$, where C is an absolute constant, so that any element of $\mathbb{F}^{r-1}$ can be represented as a difference of two elements of D. (The existence of such a subset follows from a general result, proved in [64], and also is not difficult to establish directly.) Now the set $D \oplus \{0, 1\}$ determines all directions in $\mathbb{F}^r$.

5.3. Szemerédi-Trotter in $\mathbb{F}_p \times \mathbb{F}_p$ (contributed by T. Tao). Find an analogue for the Szemerédi–Trotter theorem [95] for $\mathbb{F}_p \times \mathbb{F}_p$. More precisely, determine whether for any prime p and any system of n points and l lines in $\mathbb{F}_p \times \mathbb{F}_p$, assuming that the values of n and l are in some "reasonable" range, the number I of point-line incidences satisfies

$$I \ll (nl)^{2/3} + n + l$$

(with an absolute implicit constant). In particular, if both n and l are about $\log p$, is it true that $I = O((nl)^{2/3})$?

If n and l are both large, then the estimate in question may fail: say, for $n = p^2$ we have $I = pl$, which is not bounded by $(nl)^{2/3} + n + l$ if $l/p \to \infty$. For $n = l = p$ a paper by Bourgain, Katz, and the contributor [16] shows that the trivial bound $(nl)^{3/2}$ can be improved to $(nl)^{3/2 - \epsilon}$ for some explicit, but very small $\epsilon > 0$.

5.4. Sets in $\mathbb{F}_p^2$ with many equidistant pairs (contributed by T. Tao). For a prime p, how many pairs of points at distance 1 apart can there be in a p-element subset of $\mathbb{F}_p \times \mathbb{F}_p$? That is, how large can be the set

$$\{((x_1, y_1), (x_2, y_2)) \in A \times A : (x_1 - x_2)^2 + (y_1 - y_2)^2 = 1\}$$

for a p-element subset $A \subseteq \mathbb{F}_p \times \mathbb{F}_p$?

The set $A = \{(x, 0) : x = 0, \ldots, p-1\}$ has just $2p$ pairs of points at distance 1. A simple upper bound is $p^{3/2}$, which can be established as follows. For $u \in \mathbb{F}_p \times \mathbb{F}_p$, let A_u denote the set of all those $a \in A$ which are at the distance 1 from u. Then the intersection of any two distinct sets A_u contains at most two elements, and the union of these sets for all $u \in A$ has at most $|A| = p$ elements. With a little effort (hint: show first that $|A| \geq |A_{u_1} \cup \cdots \cup A_{u_n}| \geq |A_{u_1}| + \cdots + |A_{u_n}| - O(n^2)$ holds for any pairwise distinct $u_1, \ldots, u_n \in A$, then choose $n \approx \sqrt{p}$ and average over all n-tuples $(u_1, \ldots, u_n)$), it can be deduced that $\sum_{u \in A} |A_u| \leq p^{3/2}$. It remains to notice that the sum on the left-hand side is the number of pairs in question.

Iosevich and Rudnev proved in [55] some results on this and related problems when $|A|$ is much larger than p.

5.5. A Szemerédi–Trotter type problem (contributed by J. Bourgain). Find a lower bound for the size of a set $A \subseteq \mathbb{R}^3$, given that there is a system of n^2 lines, no n of which are co-planar, and such that every line contains n points from A. Is it true that $|A| \geq n^{3-\varepsilon}$ for any $\varepsilon > 0$ and all sufficiently large n?

5.6. Joints in $\mathbb{R}^3$ (contributed by T. Tao). Given a system of lines in $\mathbb{R}^3$, define a *joint* as a point where three non-coplanar lines from our system meet. For a positive integer n, what is the largest possible number of joints in a system of n lines?

The contributor remarks that there are configurations with as many as $cn^{3/2}$ joints (with an absolute constant c), and the trivial upper bound is $\binom{n}{2}$. The same question can be asked with $\mathbb{R}$ replaced by a finite field.

5.7. Structure Szemerédi–Trotter (contributed by T. Tao). Given n lines and n points in $\mathbb{R}^2$, the number of point-line incidences by the Szemerédi–Trotter theorem is $O(n^{4/3})$. Suppose that the number of incidences is, indeed, of this order; what can then be said about the structure of our configuration of points and lines?

5.8. Sets in $\mathbb{Z}^r$ with small difference set (contributed by Y. Stanchescu). Let $r \in \mathbb{N}$, and suppose that $A \subseteq \mathbb{Z}^r$ is a finite set, not contained in a hyperplane of dimension smaller than r. Determine the smallest possible value of $|A - A|$ as a function of $|A|$ and r.

Freĭman, Heppes, and Uhrin proved in [39] that $|A-A| \geq (r+1)|A| - \frac{1}{2}r(r+1)$ holds for every $r \in \mathbb{N}$, and this inequality is best possible for $r \in \{1, 2\}$. The contributor showed in [90] that for $r = 3$ the best possible estimate is $|A - A| \geq 4.5|A| - 9$, and conjectured in [91] that for $r \geq 4$ one has

$$|A - A| \geq \left(2r - 2 + \frac{1}{r-1}\right)|A| - C_r,$$

with a constant C_r, depending on r. As shown in [91], the last inequality, if true, is best possible.

6. Contributed problems. IV: miscellany

6.1. Nonvanishing transversals (contributed by N. Alon, cf. [4]). Is it true that for any $n \in \mathbb{N}$ and any collection of finite sets $A_1, \ldots, A_n \subseteq \mathbb{Z}$ with $\min\{|A_1|, \ldots, |A_n|\} \geq n + 1$ one can select the elements $a_1 \in A_1, \ldots, a_n \in A_n$ so that $\sum_{i \in I} a_i \neq 0$ for every nonempty subset $I \subseteq [1, n]$?

If true, this is best possible: there are "many" collections $A_1, \ldots, A_n$ with $|A_1| = \cdots = |A_n| = n$ which do not admit such a choice of $a_1, \ldots, a_n$. On the other hand, it is shown in [4] that for any $\varepsilon > 0$ there is $C > 0$ such that the answer is positive, provided that $\min\{|A_1|, \ldots, |A_n|\} \geq n + 1$ is replaced with the stronger assumption $\min\{|A_1|, \ldots, |A_n|\} \geq Cn^{1+\varepsilon}$.

6.2. Sumsets of a multiplicative subgroup (contributed by J. Bourgain). Given $\delta \in (0, 1)$, what is the smallest integer $k \geq 1$ such that for any prime p and any subgroup $H \leq \mathbb{F}_p^\times$ with $|H| > p^\delta$ one has $kH(:= H + \cdots + H) = \mathbb{F}_p$?

It is known [15] that one can take $\log k > \delta^{-C}$ with a sufficiently large absolute constant C, and Glibichuk and Konyagin have recently shown (work in progress) that $k > C4^{1/\delta}$ suffices.

6.3. Exponential sums over multiplicative subgroups (contributed by J. Bourgain).
Let p be a prime. How large $H \leq \mathbb{F}_p^\times$ must be in order for

$$\left| \sum_{x \in H} e^{2\pi i a x / p} \right| = o(|H|)$$

to hold for all $a \in \mathbb{F}_p^\times$?

6.4. Sets in $\mathbb{Z}/q\mathbb{Z}$ with few sums and products (contributed by M.-C. Chang).
Is it true that for any $\varepsilon > 0$ there exists $\delta > 0$ with the following property: if $A \subseteq \mathbb{Z}/q\mathbb{Z}$ (with a sufficiently large integer q) satisfies $\max\{|A+A|,|A \cdot A|\} < q^\varepsilon |A|$, then either $|A| > q^{1-\delta}$, or there exists $d \mid q$, $d > 1$ such that the canonical image of A in $\mathbb{Z}/d\mathbb{Z}$ has at most q^δ elements?

6.5. A quadratic Diophantine equation (contributed by M.-C. Chang).
What is the largest possible number of solutions of the equation

$$x_1^2 + x_2^2 = x_3^2 + x_4^2$$

where the variables $x_1, \ldots, x_4$ attain values from an integer set A with prescribed size?

In [19] it is shown that this number of solutions is $O(|A|^3/(\ln |A|)^{1/12})$, which readily implies that for any finite set S of squares one has $|S+S| \gg |S|(\ln |S|)^{1/12}$ (cf. Problem 4.6). It is also conjectured in [19] that for any fixed $\varepsilon > 0$ the number of solutions is $O(|A|^{2+\varepsilon})$ (with the implicit constant depending on ε).

For a thorough discussion on this and related problems see [20].

6.6. A mixed sumset problem (contributed by I. Łaba).
Given an integer $n \geq 1$, how small can $|A+\alpha A|$ be for an n-element set $A \subseteq \mathbb{R}$ and transcendental α?

Konyagin and Łaba showed in [61] that $|A + \alpha A| \geq cn \log n/(\log \log n)$ with an absolute constant $c > 0$. On the other hand, an example due to Green (also presented in [61]) shows that $|A + \alpha A| \ll ne^{c\sqrt{\log n}}$ is possible.

6.7. Hypergraph regularity (contributed by T. Tao).
Is there a hypergraph regularity lemma for subsets of pseudorandom sparse hypergraphs of large density? If so, it would give a new proof that there are arbitrarily long arithmetic progressions among the primes, which may possibly extend to a more general situation. The following analogue for graphs is known: If $|A| = |B| = N$, $G_0 \subseteq A \times B$ is "sparsely $c(\varepsilon, \delta)$-quasirandom," $G \subseteq G_0$, $|G| > \delta|G_0|$, then there exist equitable partitions

$$A = A_1 \cup \cdots \cup A_s, \quad B = B_1 \cup \cdots \cup B_t,$$

where $s, t < C(\varepsilon, \delta)$, such that for $(1-\varepsilon)st$ of the pairs (i, j) the restriction of G to $A_i \times B_j$ is ε-regular relative to G_0.

The contributor remarks that, despite being a generalization of the already rather difficult hypergraph regularity lemmas, if done correctly the proof of such a result may be *easier* than that of the existing regularity lemmas. This is because the induction used to prove such lemmas may be cleaner.

6.8. Product sets in $\mathrm{SL}_2(\mathbb{F}_p)$ (contributed by A. Venkatesh). Let p be a prime, and suppose that $A \subseteq \mathrm{SL}_2(\mathbb{F}_p)$ satisfies $|A| \sim p^{5/2}$. Does it follow that, writing $A \cdot A := \{a'a'' : a', a'' \in A\}$, one has

$$|A \cdot A| > p^{5/2+\delta}$$

for some fixed $\delta > 0$ and all sufficiently large p?

Helfgott [54] showed, among other things, that if $|A| < p^{3-\delta}$ with $\delta > 0$, and A is not contained in any proper subgroup of $\mathrm{SL}_2(\mathbb{F}_p)$, then $|A \cdot A \cdot A| > c|A|^{1+\varepsilon}$, where $c > 0$ and $\varepsilon > 0$ depend only on δ and $A \cdot A \cdot A$ is defined in the natural way. He has also shown that there is an absolute constant C such that if A is a set of generators of $\mathrm{SL}_2(\mathbb{F}_p)$, then every element of $\mathrm{SL}_2(\mathbb{F}_p)$ is a product of at most $O\big((\log p)^C\big)$ elements from $A \cup A^{-1}$.

7. Original problems

7.1. Polynomials with large image modulo a prime (the first named author; inspired by, and very similar to a problem of J. Bourgain). Given $\varepsilon \in (0, 1/2]$, classify all those polynomials $P \in \mathbb{Z}[x, y]$ for which there exists $\delta = \delta(\varepsilon) > 0$ with the following property: for all primes $p > p_0(\varepsilon)$, if $A \subseteq \mathbb{F}_p$ satisfies $|A| \leq p^\varepsilon$, then

$$|\{P(a', a'') : a', a'' \in A\}| \geq |A|^{1+\delta}.$$

Bourgain [14] has shown that $P(x, y) = x(x+y)$ has the above property, and it is implicit in the work of Pudlak [76] that the property holds for $P(x, y) = x^2 + y$. (The first-named author has a different, though perhaps related, proof of this result of Pudlak).

7.2. Arithmetic progressions in sumsets of dense sets (cf. [24]). Let $\ell(A)$ denote the maximum length of an arithmetic progression, contained in the set $A \subseteq \mathbb{Z}$. Given a real $\theta \in [0, 1)$ and an integer $N \geq 1$, estimate

$$\min\{\ell(A + A) : A \subseteq [1, N], |A| \geq N^{1-\theta}\}.$$

In [24] it is shown that this minimum is at least $2/\theta + O(1)$, and also that it is less than $\exp(C\theta^{-2/3-o(1)})$ as $N \to \infty$ and θ is fixed (with an absolute constant C).

7.3. Arithmetic progressions in large subsets of thin sumsets (J. Solymosi and the first named author). Is it true that for every $\varepsilon \in (0, 1]$ there exists $\delta > 0$ with the following property: if A is a finite set of integers with $n := |A|$ sufficiently large and $|A + A| \leq n^{1+\delta}$, then any subset $S \subseteq A + A$ satisfying

$$\sum_{s \in S} |\{(a', a'') \in A \times A : a' + a'' = s\}| \geq \varepsilon n^2$$

contains a three-term arithmetic progression?

7.4. Inverse problem for square-like sets (C. Elsholtz and the first named author). Given an integer $N \geq 1$, classify all sets $A \subseteq [1, N]$ such that $|A| > N^{1/3+\varepsilon}$, and A occupies at most $2p/3$ residue classes modulo p for every prime $p < \sqrt{N}$. Must any such A essentially be contained in the set of values of a quadratic polynomial? By "essentially" we mean that all, but $N^{o(1)}$ elements of A, lie in such a set.

Both the above properties can be weakened, and still the problem would be interesting and difficult; for example, the $2p/3$ can be replaced with $(1-\delta_1)p$, and the $1/3 + \varepsilon$ can be replaced with N^{δ_2} (though this would mean there are more possibilities than just quadratic polynomials to consider).

Naively, one may think that such sets A cannot exist, upon applying the following heuristic: if A occupies at most $2/3$ of the residue classes mod p for k different primes p, then one would expect that A has size at most $(2/3)^k N$, which can be made smaller than 1 by choosing $k > c \log N$, for a certain $c > 0$. However, this simple heuristic does not give accurate predictions, as can be seen by considering the set of all squares in $[1, N]$.

7.5. Arithmetic progressions in nonabelian groups (the first named author).
For a group G define $r_3(G)$ to be the largest size of a subset of G, containing no three-term arithmetic progressions. (In this context, a three-term arithmetic progression is a triple of the form (a, ad, ad^2) with $a, d \in G$, $d \neq 1$.) For finite abelian groups G known upper and lower bounds for $r_3(G)$ are appallingly far apart. Can one exhibit an infinite family of finite *nonabelian* groups G and give lower and upper bounds for $r_3(G)$ which are within a constant factor? Does there exist an infinite family of nonabelian groups G for which $r_3(G) > |G|/\log^K |G|$, with an absolute constant K?

Gowers considers in [47] several related problems, and in particular the following one. Fix $\theta \in (0, 1]$. Do there exist infinitely many primes p such that if $G = \mathrm{PSL}_2(p)$ (see Problem 3.3 for the definition) and $A, B, C \subseteq G$ satisfy $\min\{|A|, |B|, |C|\} > \theta |G|$, then $A \times B \times C$ contains a triple $(a, da, d^2 a)$ with $a, d \in G$? Note that the similar property fails for groups with a "large" cyclic subgroup, at least for small θ; for example, if N is a positive integer, $A = B = (0, N/4) \subseteq \mathbb{Z}/N\mathbb{Z}$, and $C = (N/2, 3N/4) \subseteq \mathbb{Z}/N\mathbb{Z}$, then $A \times B \times C$ does not contain any triple of the form $(a, a+d, a+2d)$ with $a, d \in \mathbb{Z}/N\mathbb{Z}$.

Letting $b = da$ and $c = d^2 a$ one sees that Gowers's question is equivalent to asking whether there is a solution to $c = ba^{-1}b$ with $(a, b, c) \in A \times B \times C$. Replacing A with the set of its inverses, one can further restate the question to seek triples $(a, b, c) \in A \times B \times C$ with $c = bab$.

It is worth noting that writing arithmetic progressions as $(a, da, d^2 a)$ (as in Gower's paper) is equivalent to writing them as (a, ad, ad^2); indeed, letting $\delta = a^{-1}da$ we find that $(a, da, d^2 a) = (a, a\delta, a\delta^2)$.

Two further questions in the spirit of Gowers's problem are the following. Given $K > 0$, do there exist infinitely many groups G such that for any subset $A \subseteq G$ with $|A| \geq |G|/\log^K |G|$ there are $a, b, c \in A$, satisfying $c = bab$? Do there exist infinitely many groups G and subsets $A \subseteq G$ with $|A| > |G|/\log^K |G|$ for which there are no such $a, b, c \in A$? This second question is equivalent to the question above of whether $r_3(G) > |G|/\log^K |G|$.

7.6. Arithmetic progressions and the Fourier transform (the first named author).
If the L^1-norm of the Fourier transform of a large subset of $\mathbb{F}_p$ is small, must the set contain a three-term arithmetic progression? More precisely, is it true that for any fixed $C, D > 0$, if p is a sufficiently large prime, then any set $A \subseteq \mathbb{F}_p$ with

$$|A| > p/\log^C p, \quad \sum_{z \in \mathbb{F}_p} |\hat{A}(z)| < p \log^D p$$

contains a three-term arithmetic progression? (Here $\hat{A}$ is defined as in the Problem 3.2; that is, $\hat{A}(z) = \sum_{a \in A} e^{2\pi i a z/p}$.)

7.7. The Fourier spectrum of functions restricted to subsets (the first named author). Given a prime p, for a function $f \colon \mathbb{F}_p \to \mathbb{R}$ set $\hat{f}(z) := \sum_{u \in \mathbb{F}_p} f(u) e^{2\pi i u z/p}$. Fix $\varepsilon \in (0, 1]$ and $A, B > 0$. Is it true for all sufficiently large primes p that if $f, g \colon \mathbb{F}_p \to [0, 1]$ satisfy $\hat{f}(0) = \hat{g}(0) > p/(\log p)^A$ and

$$\max\{|\hat{f}(z) - \hat{g}(z)| : z \in \mathbb{F}_p^\times\} < p \exp(-\sqrt{\log p}),$$

then for every function $h \colon \mathbb{F}_p \to [0, 1]$ with $h(u) \le f(u)$ ($u \in \mathbb{F}_p$) and $\hat{h}(0) \ge \varepsilon \hat{f}(0)$ there is a function $h_0 \colon \mathbb{F}_p \to [0, 1]$ such that $h_0(u) \le g(u)$ ($u \in \mathbb{F}_p$) and

$$\max\{|\hat{h}_0(z) - \hat{h}(z)| : z \in \mathbb{F}_p^\times\} < p/(\log p)^B?$$

Roughly, what we are asking is as follows: assuming that the Fourier spectrums of $f, g \colon \mathbb{F}_p \to [0, 1]$ are very close, must each function $h \colon \mathbb{F}_p \to [0, 1]$, majorized by f (with a positive fraction of the mass of f), have a partner function $h_0 \colon \mathbb{F}_p \to [0, 1]$, majorized by g, such that the Fourier spectrums of h and h_0 are close? In this problem $\exp(-\sqrt{\log p})$ can be replaced with any function, decaying to 0 faster than any power of $\log p$.

7.8. Covering subsets of $\mathbb{F}_p$ by arithmetic progressions (cf. [65]). For an integer $n \ge 2$ and prime p, let $l_n(p)$ denote the smallest integer l such that any n-element subset of $\mathbb{F}_p$ is contained in an arithmetic progression of length l. It is conjectured in [65] that if n is fixed and $p \to \infty$, then

$$l_n(p) = 2n^{-1/(n-1)} p^{1-1/(n-1)} \big(1 + o(1)\big);$$

prove (or disprove) this conjecture.

If $n = 2$ the assertion is immediate, for $n = 3$ it is established in [65], for $n \ge 4$ it is shown in [65] that

$$p^{1-1/(n-1)} \big(1 + o(1)\big) < l_n(p) < 2n^{-1/(n-1)} p^{1-1/(n-1)}.$$

7.9. Arithmetic and geometric progressions in $\mathbb{F}_p$ (the second named author). For a prime p, an element $\lambda \in \mathbb{F}_p$, and a subset $A \subseteq \mathbb{F}_p$, set $\lambda * A = \{\lambda a : a \in A\}$. Does there exist $\varepsilon > 0$ with the property that for any sufficiently large prime p there is $\lambda \in \mathbb{F}_p$ such that every subset $A \subseteq \mathbb{F}_p$ with $|A| < p/2$ satisfies

$$|A \cup (A + 1) \cup (\lambda * A)| > (1 + \varepsilon)|A|?$$

A positive answer would lead to a simple construction of good expanders. (For the construction to be effective, though, one has to specify λ effectively.)

There are reasons to believe that $\lambda = O(1)$ does *not* work.

7.10. Large sum-free sets in ternary spaces (cf. [69]). We say that the sum-free subset A of an abelian group G is *induced* if there is a nonzero subgroup $H < G$ such that A is the full inverse image of a sum-free subset of the quotient group G/H under the canonical homomorphism $G \to G/H$. (See Problem 2.9 for the definition of a sum-free subset.)

For an integer $r \ge 1$, how large can a sum-free subset $A \subseteq \mathbb{F}_3^r$ be given that A is not contained in an induced sum-free subset? In [69] examples of such subsets with $|A| = (3^{r-1} + 1)/2$ are constructed, and it is conjectured that if $A \subseteq \mathbb{F}_3^r$ is sum-free and satisfies $|A| > (3^{r-1}+1)/2$, then A is contained in an induced sum-free subset; prove (or disprove) this conjecture.

As shown in [69], the conjecture holds true for $r \leq 4$ at least.

We mention that for all finite abelian groups G, the largest size of a sum-free subset of G is known; see [49]. In contrast, "primitive" sum-free subsets (those not contained in induced sum-free subsets) remain mostly unexplored, with the exception of the elementary abelian 2-groups. Observe, that a sum-free subset A is induced if and only if it is periodic; that is, A is a union of cosets of a nonzero subgroup.

7.11. General properties of the sum spectrum (cf. [67]). Given two finite integer sets A and B, write
$$\nu_{A,B}(n) := |\{(a,b) \in A \times B : a+b = n\}|; \quad n \in \mathbb{Z}.$$
The spectrum of ν defines a partition of the integer $|A||B|$ which can be visualized using a Ferrers diagram; that is, an arrangement of $|A||B|$ square boxes in bottom-aligned columns such that the height of the leftmost column is the largest value attained by ν, the height of next column is the second largest value of ν, and so on. It is not difficult to show that if r_k denotes the height of the kth column of the diagram (that is, the kth largest value attained by ν), then

(*) $$r_k^2 \leq r_k + r_{k+1} + r_{k+2} + \cdots .$$

for any $k \geq 1$. What are the general properties shared by the functions ν for all finite sets $A, B \subseteq \mathbb{Z}$, other than that reflected by this inequality?

Notice that for any $t \in \mathbb{N}$, the length of the tth row of the above described diagram (counting the rows from the bottom) is $N_t := |\{n : \nu(n) \geq t\}|$. From a well-known result of Pollard [75] it follows that $N_1 + \cdots + N_t \geq t(|A|+|B|-t)$ for any $t \leq \min\{|A|,|B|\}$, and this can be derived also as a corollary of (*).

7.12. Scherk's theorem for restricted addition (cf. [68]). Is there an analog of Scherk's theorem for the restricted sumset
$$A \dotplus B := \{a+b : a \in A,\, b \in B,\, a \neq b\}?$$
It is conjectured in [68] that for any finite subsets A and B of an abelian group, satisfying $A \cap (-B) = \{0\}$, one has
$$|A \dotplus B| \geq |A| + |B| - 3;$$
prove (or disprove) this conjecture.

Solving a problem by Moser, Scherk proved in [72] that if A and B are finite subsets of an abelian group such that $A \cap (-B) = \{0\}$, then $|A+B| \geq |A|+|B|-1$. (The condition $A \cap (-B) = \{0\}$ means that there is a unique representation of the sort $0 = a+b$ with $a \in A$ and $b \in B$; specifically, that with $a = b = 0$.) The estimate of Scherk's theorem is best possible: equality is attained, for instance, if A and B are arithmetic progressions with the same difference, the order of which is at least $|A|+|B|-1$.

The conjecture reduces to the special case $B \subseteq A$ by considering the sets $A^* = A \cup B$ and $B^* = A \cap B$. We have verified computationally this case (and hence the general conjecture) for all cyclic groups of order up to 25, and in the case $B = A$ for cyclic groups of order up to 36. The conjecture holds true also for torsion-free abelian groups, for cyclic groups of prime order, and for elementary abelian 2-groups.

7.13. Sumsets, restricted by an injective mapping (cf. [66]). Let p be a prime. For $A, B \subseteq \mathbb{F}_p$ and $\tau \colon A \to B$ set $A \stackrel{\tau}{+} B := \{a + b : a \in A,\, b \in B,\, b \neq \tau(a)\}$. Is it true that for any prime p, any nonempty subsets $A, B \subseteq \mathbb{F}_p$ with $|A| + |B| < p$, and any *injective* mapping $\tau \colon A \to B$, one has
$$|A \stackrel{\tau}{+} B| \geq |A| + |B| - 3?$$

If true, this would extend a result of Dias da Silva and Hamidoune [27], establishing a well-known conjecture of Erdős and Heilbronn [31, p. 95]. For discussion and some partial results see [66] where it is shown, in particular, that $|A \stackrel{\tau}{+} B| \geq |A| + |B| - 2\sqrt{\min\{|A|, |B|\}} - 1$ for any (not necessarily injective) mapping τ.

7.14. Popular differences (the second named author). Let A be a finite nonempty subset of an abelian group G, and write $D := A - A$. Given that any $d \in D$ has at least $|A|/2$ representations of the form $d = a' - a''$ with $a', a'' \in A$, is it necessarily true that D is either a subgroup, or a union of three cosets?

If any $d \in D$ has *strictly more* than $|A|/2$ representations, then D is a subgroup: indeed, by the pigeonhole principle for any $d_1, d_2 \in D$ there exists a pair of representations $d_1 = a_1' - a_1''$, $d_2 = a_2' - a_2''$ such that $a_1'' = a_2''$, and it follows that $d_1 - d_2 = a_1' - a_2' \in D$.

If any $d \in D$ is only guaranteed to have *at least* $|A|/2$ representations, then the argument above doesn't work, and in fact, the conclusion is not true either. To see this, consider the set $A := H \cup (g + H)$, where $H < G$ is a finite subgroup and $g \in G$ is chosen so that the order of g in the quotient group G/H is at least 4. Then $D = (-g + H) \cup H \cup (g + H)$ is not a subgroup; at the same time, it is easily seen that any $d \in D$ has at least $|H| = |A|/2$ representations of the form $d = a' - a''$. The question is whether this example is essentially unique.

7.15. The maximal length of an integer set (cf. [62]). Is it true that for $n \geq 7$, any n-element set of integers is isomorphic (in Freĭman's sense) to a subset of $[0, 2^{n-2}]$?

For any integer $n \geq 2$ the set $\{0, 1, 2, 4, \ldots, 2^{n-2}\}$ is "linear" (has Freĭman's dimension 1) and not contained in an arithmetic progression with difference larger than 1, hence it is not isomorphic to a set of integers of length smaller than 2^{n-2}. It is conjectured in [62] that this is the extremal case; that is, in any class of isomorphic n-element sets there is a set of length at most 2^{n-2}.

Note, that all Sidon sets with the same number of elements are isomorphic to each other, and it is well-known that for N large enough the interval $[0, N]$ contains a Sidon set of cardinality about $\sqrt{N}$. Thus, any n-element Sidon set is isomorphic to a subset of $\left[0, n^2(1 + o(1))\right]$. For $n \leq 6$, however, this $n^2(1 + o(1))$ turns out to be larger than 2^{n-2}: more precisely, $[0, 2^{n-2}]$ does not contain an n-element Sidon set. This explains the restriction $n \geq 7$ above.

7.16. Weighted distances on the unit circle (cf. [63]). Suppose that we are given $r \geq 1$ complex numbers $z_1, \ldots, z_r$ with $|z_1| = \cdots = |z_r| = 1$, to which correspond real weights $p_1, \ldots, p_r \geq 0$, normalized by the condition $p_1 + \cdots + p_r = r$. We want to find yet another complex number z with $|z| = 1$, which should be as far as possible from all numberss z_j in the sense that the product $\prod_{j=1}^{r} |z - z_j|^{p_j}$

is to be maximized. Prove (or disprove) that for any system of numberss z_j and weights p_j as above, there exists z such that

$$\prod_{j=1}^{r} |z - z_j|^{p_j} \geq 2.$$

The constant 2 in the right-hand side is easily seen to be best possible. The assertion is established in [63] in a variety of special cases: in particular if all weights p_j equal each other, and also if z_j are equally spaced on the unit circle. It can be re-stated as an assertion about the maximum possible value of a polynomial on the unit circle.

7.17. Hamiltonicity of addition Cayley graphs (the second named author). Is it true that any connected addition Cayley graph, induced on a finite cyclic group by its 4-element subset, is Hamiltonian?

Recall, that the addition Cayley graph, induced on a finite abelian group G by its subset $S \subseteq G$, is the graph with the vertex set G and the edge set $\{(g_1, g_2) \in G \times G : g_1 + g_2 \in S\}$. It is easy to see that for this graph to be connected it is necessary and sufficient that S is not contained in a coset of a proper subgroup of G, save, perhaps, for the nonzero coset of a subgroup of index 2. Computations suggest that if G is cyclic, $|S| \geq 4$, and the graph under consideration is connected, then it is Hamiltonian. On the other hand, there exist non-Hamiltonian (though 2-connected) addition Cayley graphs on finite cyclic groups, generated by 3-element subsets.

Acknowledgement. Some of the problems presented in this paper originate from the list, compiled by the present authors as a follow-up to the Workshop on Recent Trends in Additive Combinatorics, organized in 2004 by the American Institute of Mathematics (http://www.aimath.org). More problems were communicated to us during the Workshop on Additive Combinatorics, organized in 2006 by Centre de recherches mathématiques at the Université de Montréal. We are grateful to these institutions for bringing together a large number of distinguished mathematicians, which ultimately allowed us to write this paper.

References

1. N. Alon, *Combinatorial Nullstellensatz*, Combin. Probab. Comput. **8** (1999), no. 1-2, 7–29.
2. _____, *Large sets in finite fields and sumsets*, submitted.
3. N. Alon and D. J. Kleitman, *Sum-free subsets*, A Tribute to Paul Erdős, Cambridge Univ. Press, Cambridge, 1990, pp. 13–26.
4. N. Alon and I. Ruzsa, *Non-averaging subsets and non-vanishing transversals*, J. Combin. Theory Ser. A **86** (1999), no. 1, 1–13.
5. L. Babai and V. Sós, *Sidon sets in groups and induced subgraphs of Cayley graphs*, European J. Combin. **6** (1985), no. 2, 101–114.
6. A. Balog and E. Szemerédi, *A statistical theorem of set addition*, Combinatorica **14** (1994), no. 3, 263–268.
7. F. A. Behrend, *On sets of integers which contain no three terms in arithmetic progression*, Proc. Nat. Acad. Sci. U.S.A. **32** (1946), 331–332.
8. V. Bergelson and A. Leibman, *Polynomial extensions of van der Waerden's and Szemerédi's theorems*, J. Amer. Math. Soc. **9** (1996), no. 3, 725–753.
9. E. A. Berlekamp, *Construction for partitions which avoid long arithmetic progressions*, Canad. Math. Bull. **11** (1968), 409–414.
10. J. Bierbrauer and Y. Edel, *Bounds on affine caps*, J. Combin. Des. **10** (2002), no. 2, 111–115.

11. J. Bourgain, *Estimates related to sumfree subsets of sets of integers*, Israel J. Math. **97** (1997), 71–92.
12. _____, *On triples in arithmetic progression*, Geom. Funct. Anal. **9** (1999), no. 5, 968–984.
13. _____, *Mordell's exponential sum estimate revisited*, J. Amer. Math. Soc. **18** (2005), no. 2, 477–499.
14. _____, *More on the sum-product phenomenon in prime fields and its applications*, Int. J. Number Theory **1** (2005), no. 1, 1–32.
15. J. Bourgain, A. A. Glibichuk, and S. V. Konyagin, *Estimates for the number of sums and products and for exponential sums in fields of prime order*, J. London Math. Soc. (2) **73** (2006), no. 2, 380–398.
16. J. Bourgain, N. Katz, and T. Tao, *A sum-product estimate in finite fields and their applications*, Geom. Funct. Anal. **14** (2004), no. 1, 27–57.
17. A. Cauchy, *Recherches sur les nombres*, J. École Polytech. **9** (1813), 99–116.
18. M.-C. Chang, *A polynomial bound in Freĭman's theorem*, Duke Math. J. **113** (2002), 399–419.
19. _____, *On problems of Erdős and Rudin*, J. Funct. Anal. **207** (2004), no. 2, 444–460.
20. J. Cilleruelo and A. Granville, *Lattice points on circles, squares in arithmetic progressions and sumsets of squares*, submitted.
21. S. D. Cohen, *Clique numbers of Paley graphs*, Quaestiones Math. **11** (1988), no. 2, 225–231.
22. D. Conlon, *A new upper bound for diagonal Ramsey numbers*, submitted.
23. E. Croot, *The minimal number of three-term arithmetic progressions modulo a prime converges to a limit*, Canad. Math. Bull., to appear.
24. E. Croot, I. Ruzsa, and T. Schoen, *Long arithmetic progressions in sparse sumsets*, Integers, to appear.
25. H. Davenport, *On the addition of residue classes*, J. London Math. Soc. **10** (1935), 30–32.
26. _____, *A historical note*, J. London Math. Soc. **22** (1947), 100–101.
27. J. A. Dias da Silva and Y. O. Hamidoune, *Cyclic spaces for Grassmann derivatives and additive theory*, Bull. London Math. Soc. **26** (1994), no. 2, 140–146.
28. Y. Edel, *Extensions of generalized product caps*, Des. Codes Cryptogr. **31** (2004), no. 1, 5–14.
29. G. Elekes, *On the number of sums and products*, Acta Arith. **81** (1997), 365–367.
30. P Erdős, *Extremal problems in number theory*, Proc. Sympos. Pure Math., vol. 8, Amer. Math. Soc., Providence, R.I., 1965, pp. 181–189.
31. P. Erdős and R. L. Graham, *Old and new problems and results in combinatorial number theory*, Monogr. Enseign. Math., vol. 28, L'Enseignement Mathématique, Geneva, 1980.
32. P. Erdős, J. Nešetřil, and V. Rödl, *On Pisier type problems and results (combinatorial applications to number theory)*, Mathematics of Ramsey Theory, Algorithms Combin., vol. 5, Springer, Berlin, 1990, pp. 214–231.
33. P. Erdős and E. Szemerédi, *On sums and products of integers*, Studies in Pure Mathematics, Birkhaüser, Basel, 1983, pp. 213–218.
34. P. Erdős and P. Turán, *On some sequences of integers*, J. London Math. Soc. **11** (1936), 261–264.
35. K. Ford, *Sums and products from a finite set of real numbers*, Ramanujan J. **2** (1998), no. 1-2, 59–66.
36. P. Frankl, R. L. Graham, and V. Rödl, *Quantitative theorems for regular systems of equations*, J. Combin. Theory Ser. A **47** (1988), no. 2, 246–261.
37. G. A. Freĭman, *On the addition of finite sets*, Dokl. Akad. Nauk SSSR **158** (1964), 1038–1041 (Russian).
38. _____, *Elements of a structural theory of set addition*, Kazan. Gosudarstv. Ped. Inst; Elabuž. Gosudarstv. Ped. Inst., Kazan, 1966 (Russian).
39. G. A. Freĭman, A. Heppes, and B. Uhrin, *A lower estimation for the cardinality of finite difference sets in R^n*, Number Theory, Vol. I (Budapest, 1987), Colloq. Math. Soc. János Bolyai, vol. 51, North-Holland, Amsterdam, 1990, pp. 125–139.
40. G. A. Freĭman and Y. V. Stanchescu, *On a Kakeya type problem*, submitted.
41. H. Furstenberg, *Ergodic behavior of diagonal measures and a theorem of Szemerédi on arithmetic progressions*, J. Analyse Math. **31** (1977), 204–256.
42. H. Furstenberg and Y. Katznelson, *An ergodic Szemerédi theorem for commuting transformations*, J. Analyse Math. **34** (1978), 275–291.

43. _____, *A density version of the Hales–Jewett theorem*, J. Analyse Math. **57** (1991), 64–119.
44. D. Goldston, J. Pintz, and C. Y. Yildirim, *Primes in tuples.* I, Ann. of Math. (2), to appear.
45. W. T. Gowers, *A new proof of Szemerédi's theorem*, Geom. Funct. Anal. **11** (2001), no. 3, 465–588.
46. _____, *Hypergraph regularity and the multidimensional Szemerédi theorem*, preprint.
47. _____, *Quasirandom groups*, preprint.
48. S. W. Graham and C. J. Ringrose, *Lower bounds for least quadratic nonresidues*, Analytic Number Theory (Allterton Park, IL, 1989), Progr. Math., vol. 85, Birkhäuser, Boston, 1990, pp. 269–309.
49. B. Green and I. Z. Ruzsa, *Sum-free sets in abelian groups*, Israel J. Math. **147** (2005), 157–189.
50. _____, *Freĭman's theorem in an arbitrary abelian group*, J. London Math. Soc. (2), to appear.
51. B. Green and T. Tao, *The primes contain arbitrarily long arithmetic progressions*, Ann. of Math. (2), to appear.
52. A. W. Hales and R. I. Jewett, *Regularity and positional games*, Trans. Amer. Math. Soc. **106** (1963), 222–229.
53. D. R. Heath-Brown, *Integer sets containing no arithmetic progressions*, J. London Math. Soc. (2) **35** (1987), no. 3, 385–394.
54. H. Helfgott, *Growth and generation in* $SL_2(\mathbb{Z}/p\mathbb{Z})$, submitted.
55. A. Iosevich and M. Rudnev, *Erdős distance problem in vector spaces over finite fields*, Trans. Amer. Math. Soc., to appear.
56. K. Kedlaya, *Large product-free subsets of finite groups*, J. Combin. Theory Ser. A **77** (1997), no. 2, 339–343.
57. J. H. B. Kemperman, *On small sumsets in an abelian group*, Acta Math. **103** (1960), 63–88.
58. M. Klazar, J. Kratochvíl, M. Loebl, J. Matoušek, R. Thomas, and P. Valtr (eds.), *Topics in discrete mathematics*, Algorithms Combin., vol. 26, Springer, Berlin, 2006.
59. M. Kneser, *Abschätzung der asymptotischen Dichte von Summenmengen*, Math. Z. **58** (1953), 459–484.
60. _____, *Ein Satz über abelsche Gruppen mit Anwendungen auf die Geometrie der Zahlen*, Math. Z. **61** (1955), 429–434.
61. S. V. Konyagin and I. Laba, *Distance sets of well-distributed planar sets for polygonal norms*, Israel J. Math. **152** (2006), 157–179.
62. S. V. Konyagin and V. F. Lev, *Combinatorics and linear algebra of Freĭman's isomorphism*, Mathematika **47** (2000), 39–51.
63. _____, *On the maximum value of polynomials with given degree and number of roots*, ChebyshevskiĭSb. **3** (2003), no. 2(4), 165–170.
64. G. Kozma and A. Lev, *Bases and decomposition numbers of finite groups*, Arch. Math. (Basel) **58** (1992), no. 5, 417–424.
65. V. F. Lev, *Simultaneous approximations and covering by arithmetic progressions in* $\mathbb{F}_p$, J. Combin. Theory Ser. A **92** (2000), no. 2, 103–118.
66. _____, *Restricted set addition in groups. II: A generalization of the Erdős–Heilbronn conjecture*, Electron. J. Combin. **7** (2000), Research Paper 4.
67. _____, *Reconstructing integer sets from their representation functions*, Electron. J. Combin. **11** (2004), no. 1, Research Paper 78.
68. _____, *Restricted set addition in abelian groups: results and conjectures*, J. Théor. Nombres Bordeaux **17** (2005), no. 1, 181–193.
69. _____, *Large sum-free sets in ternary spaces*, J. Combin. Theory Ser. A **111** (2005), no. 2, 337–346.
70. H. B. Mann, *A proof of the fundamental theorem on the density of sets of positive integers*, Ann. of Math. (2) **43** (1942), 523–527.
71. R. Meshulam, *On subsets of finite abelian groups with no 3-term arithmetic progressions*, J. Combin. Theory Ser. A **71** (1995), no. 1, 168–172.
72. L. Moser and P. Scherk, *Advanced problems and solutions: Solutions: 4466*, Amer. Math. Monthly **62** (1955), no. 1, 46–47.
73. M. B. Nathanson: *Inverse problems and the geometry of sumsets*, Grad. Texts in Math., vol. 165, Springer, New York, 1996.

74. J. Nešetřil and O. Serra, *The Erdős–Turán property for a class of bases*, Acta Arith. **115** (2004), no. 3, 245–254.
75. J. M. Pollard, *A generalization of the theorem of Cauchy and Davenport*, J. London Math. Soc. (2) **8** (1974), 460–462.
76. P. Pudlák, *On explicit Ramsey graphs and estimates of the number of sums and products*, Topics in Discrete Mathematics (M. Klazar, J. Kratochvíl, M. Loebl, J. Matoušek, R. Thomas, and P. Valtr, eds.), Algorithms Combin., vol. 26, Springer, Berlin, 2006, pp. 169–175.
77. R. Rado, *Studien zur Kombinatorik*, Math. Z. **36** (1933), no. 1, 424–470.
78. A. Robertson and D. Zeilberger, *A 2-coloring of $[1,n]$ can have $(1/22)n^2 + O(n)$ monochromatic Schur triples, but not less!*, Electron. J. Combin. **5** (1998), Research Paper 19.
79. V. Rödl, B. Nagle, J. Skokan, M. Schacht, and Y. Kohayakawa, *The hypergraph regularity method and its applications*, Proc. Nat. Acad. Sci. U.S.A. **102** (2005), no. 23, 8109–8113.
80. K. F. Roth, *On certain sets of integers*, J. London Math. Soc. **28** (1953), 104–109.
81. I. Z. Ruzsa, *A just basis*, Monatsh. Math. **109** (1990), no. 2, 145–151.
82. _____, *Arithmetical progressions and the number of sums*, Period. Math. Hungar. **25** (1992), no. 1, 105–111.
83. _____, *Generalized arithmetical progressions and sumsets*, Acta Math. Acad. Sci. Hungar. **65** (1994), no. 4, 379–388.
84. _____, *Sums of finite sets*, Number Theory (New York, 1991–1995), Springer, New York, 1996, pp. 281–293.
85. L. Schnirelmann, *Über additive Eigenschaften der Zahlen*, Math. Ann. **107** (1933), 649–690.
86. T. Schoen, *The number of monochromatic Schur triples*, European J. Combin. **20** (1999), no. 8, 855–866.
87. I. Schur, *Über die Kongruenz $x^m + y^m = z^m$ (mod p)*, Jber. Deutsch. Math.-Verein. **25** (1916), 114–116.
88. S. Shelah, *Primitive recursive bounds for van der Waerden numbers*, J. Amer. Math. Soc. **1** (1988), no. 3, 683–697.
89. J. Solymosi, *On the number of sums and products*, Bull. London Math. Soc. **37** (2005), no. 4, 491–494.
90. Y. V. Stanchescu, *On finite difference sets*, Acta Math. Acad. Sci. Hungar. **79** (1998), no. 1-2, 123–138.
91. _____, *An upper bound for d-dimensional difference sets*, Combinatorica **21** (2001), no. 4, 591–595.
92. _____, *Planar sets containing no three collinear points and nonaveraging sets of integers*, Discrete Math. **256** (2002), no. 1-2, 387–395.
93. E. Szemerédi, *On sets of integers containing no k elements in arithmetic progression*, Acta Arith. **27** (1975), 199–245.
94. _____, *Integer sets containing no arithmetic progressions*, Acta Math. Acad. Sci. Hungar. **56** (1990), no. 1-2, 155–158.
95. E. Szemerédi and W. Trotter, *Extremal problems in discrete geometry*, Combinatorica **3** (1983), no. 3-4, 381–392.
96. T. Tao, *An uncertainty principle for cyclic groups of prime order*, Math. Res. Lett. **12** (2005), no. 1, 121–127.
97. _____, *A variant of the hypergraph removal lemma*, J. Combin. Theory Ser. A, to appear.
98. T. Tao and T. Ziegler, *The primes contain arbitrarily long polynomial progressions*, submitted.
99. B. L. van der Waerden, *Beweis einer Baudetschen Vermutung*, Nieuw Arch. Wisk. **15** (1927), 212–216.
100. P. Varnavides, *On certain sets of positive density*, J. London Math. Soc. **34** (1959), 358–360.

DEPARTMENT OF MATHEMATICS, GEORGIA INSTITUTE OF TECHNOLOGY, ATLANTA, GA 30332, USA
E-mail address: `ecroot@math.gatech.edu`

DEPARTMENT OF MATHEMATICS, THE UNIVERSITY OF HAIFA AT ORANIM, TIVON 36006, ISRAEL
E-mail address: `seva@math.haifa.ac.il`

Some Problems Related to Sum-Product Theorems

Mei-Chu Chang

The study of sum-product phenomena and product phenomena is an emerging research direction in combinatorial number theory that has already produced several striking results. Many related problems are not yet fully understood, or are far from being resolved. In what follows we propose several questions where progress can be expected and should lead to advances in this general area.

For two finite subsets A and B of a ring, the *sum set* and the *product set* of A, B are
$$A + B := \{a + b : a \in A, b \in B\},$$
and
$$A \cdot B := \{ab : a \in A, b \in B\}.$$

1. Product theorems in matrix spaces

In a recent remarkable result, H. Helfgott [17] proved that if $A \subset \mathrm{SL}_2(\mathbb{F}_p)$, $\mathbb{F}_p$ being the prime field with p elements, then either A is contained in a proper subgroup of $\mathrm{SL}_2(\mathbb{F}_p)$ or $|A \cdot A \cdot A| > |A|^{1+\varepsilon}$, where $\varepsilon > 0$ is an absolute constant. This result has already caused a number of striking developments — in particular, in the theory of expanders [4] and the spectral theory of Hecke operators. (See [5, 14].) Underlying this nonabelian "product theorem" is the "sum-product theorem" in $\mathbb{F}_p$, stating that if $A \subset \mathbb{F}_p$ and $1 < |A| < p^{1-\varepsilon}$, then $|A + A| + |A \cdot A| > |A|^{1+\delta}$, for some $\delta = \delta(\varepsilon) > 0$. (See [6, 7].) Understanding how Helfgott's result generalizes to higher dimensions is a natural question. Thus

Problem 1. Is there a generalization of Helfgott's theorem to subsets of $\mathrm{SL}_3(\mathbb{F}_p)$?

The proof of Helfgott's theorem is based on an explicit examination of $\mathrm{SL}_2(\mathbb{F}_p)$ using product operations on a general subset. A similar procedure for $\mathrm{SL}_3(\mathbb{F}_p)$ does not yet seem to be understood. However, one could imagine that the validity of Kazhdan's property may be relevant to this situation (see [19]). Instead of considering the characteristic p case, one might start with a characteristic 0 case such as $\mathrm{SL}_d(\mathbb{C})$ or $\mathrm{SL}_d(\mathbb{R})$. The issue is closely related to the theory of "growth" and "random walks" in finitely generated subgroups, centered around the polynomial versus exponential dichotomy: Tits' alternative for linear groups G over a field of

---

2000 *Mathematics Subject Classification.* Primary 11B75; Secondary 11D04.

This is the final form of the paper.

©2007 American Mathematical Society

characteristic 0, states that either G contains a free group on two generators, or G is virtually solvable (i.e., contains a solvable subgroup of finite index). A solvable group that does not have any nilpotent subgroup of finite index, is of exponential growth. On the other hand, nilpotent groups are of polynomial growth, so that a solvable group that has a nilpotent subgroup of finite index, is of polynomial growth (see [16]). Our problem is however, different, since we are not iterating the product operation but rather only performing a few of them. In a recent paper, we showed the following

Theorem 1 ([12]). *For all $\varepsilon > 0$, there is $\delta > 0$ such that if $A \subset SL_3(\mathbb{Z})$ is a finite set, then either $|A \cdot A \cdot A| > |A|^{1+\delta}$ or A intersects a coset of a nilpotent subgroup of $SL_3(\mathbb{Z})$ in a set of size at least $|A|^{1-\varepsilon}$.*

Several examples show that as a general statement the previous one is basically the best possible, up to the dependence of ε and δ.

In $SL_2(\mathbb{Z})$ or more generally, $SL_2(\mathbb{C})$, we have the following stronger statement that may be proved using Helfgott's method.

Theorem 2 ([12]). *There is $\delta > 0$ such that if $A \subset SL_2(\mathbb{C})$ is a finite set not contained in a virtually abelian subgroup (that is, a group with an abelian subgroup of finite index), then $|A \cdot A \cdot A| > c|A|^{1+\delta}$.*

In addition to certain subgroups introduced by Helfgott, the proof of Theorem 1 uses, in an essential way, the finiteness results of Evertse–Schlickewei–Schmidt [13] on solutions of linear equations

$$x_1 + x_2 + \cdots + x_r = 1$$

with variables x_i taken in a multiplicative subgroup of $\mathbb{C}$ of bounded rank (derived from quantitative versions of the subspace theorem). The Evertse–Schlickewei–Schmidt result turns out to be a powerful ingredient in our approach — one for which we have yet to find a characteristic-p substitute.

Problem 2. Find a different proof of Theorem 1 not using the subspace theorem and that hopefully will permit progress on Problem 1.

In fact, in this respect we may even put forward the following

Problem 3. Prove the analogue of Theorem 1 for subsets of $SL_3(\mathbb{R})$.

On the other hand, one can reasonably expect, with the same technique as in [12] and additional work, to settle

Problem 4. Generalize Theorem 1 to subsets of $SL_d(\mathbb{Z})$ for d arbitrary.

Returning to Theorem 2 and observing that the free group F_2 is a subgroup of $SL_2(\mathbb{Z})$ (which in fact has the former as a subgroup of finite index), we obtain the following

Theorem 3 ([12]). *There exists $\delta > 0$ such that if $A \subset F_2$ is a finite subset not contained in a cyclic subgroup, then $|A \cdot A \cdot A| > |A|^{1+\delta}$.*

Problem 5. (i) Find a direct combinatorial proof of Theorem 3 (or a stronger result).
(ii) What may be said about the value of $\delta > 0$?

Related to (ii), Razborov [20] recently proved that if A is a finite subset of a free group with at least two noncommuting elements then

$$|A \cdot A \cdot A| \geq \frac{|A|^2}{(\log |A|)^{O(1)}}.$$

We conclude this section with the following vaguely formulated question expressing a "nilpotent subgroup versus product theorem" type dichotomy beyond linear groups.

Problem 6. To what extent does Theorem 1 generalize to general groups?

The obvious groups to explore in this context are the Grigorchuk counterexamples [15] to the polynomial versus exponential dichotomy. These groups are not linear and have growth that is between polynomial and exponential.

2. Explicit functions on finite fields with "large range"

A. Wigderson raised the question of finding *explicit* (algebraic) functions $f \colon \mathbb{F}_p \times \mathbb{F}_p \to \mathbb{F}_p$ with the property that whenever $A, B \subset \mathbb{F}_p$ and $|A|, |B| > \sqrt{p}$, we have that $|f(A, B)| > p^{1/2+\varepsilon}$ for some absolute constant $\varepsilon > 0$. A similar problem may be posed replacing the exponent $\frac{1}{2}$ by any $0 < \alpha < 1$. The first such example, $f(x, y) = x(x + y)$, was obtained by Bourgain. The proof relies on a version of the Szemerédi–Trotter theorem in the $\mathbb{F}_p$-plane (proved in [7]), itself derived from the sum-product theorem in $\mathbb{F}_p$. It is reasonable to expect that many other 2-variable polynomials have the "expanding property" described above. The proper question would be

Problem 7. What may be said about a polynomial $f(x, y) \in \mathbb{F}_p[x, y]$ that does not have the expanding property?

Results along this line were obtained in characteristic 0, for instance in the work of Elekes and Ruzsa. Problem 7 is closely related to

Problem 8. Prove a Szemerédi–Trotter type theorem in characteristic p for systems of algebraic curves.

This problem is largely unresolved. It will require different methods from the characteristic 0 case.

The following expander problem was also posed by Wigderson: It is in a similar vein but now restricted to known structures in the finite field $\mathbb{F}_{2^n}$ in characteristic 2.

For a set S and a field K, we denote the vector space generated by S over K by $\langle S \rangle_K$.

Problem 9. Let J be some constant. For $1 \leq j \leq J$, find an explicit system of linear maps $\phi_j \colon \mathbb{F}_{2^n} \to \mathbb{F}_{2^n}$ with the property that whenever E is a linear subspace of $\mathbb{F}_{2^n}$ and $\dim_{\mathbb{F}_2} E > n/2$, then

$$\dim \Big\langle \bigcup_{1 \leq j \leq J} \phi_j(E) \Big\rangle_{\mathbb{F}_2} > (\tfrac{1}{2} + \varepsilon)n,$$

where $\varepsilon > 0$ is a fixed constant.

There are other variants of this question of a similar flavor. Of relevance to Problem 9 is our result, which we formulate in nontechnical terms:

Theorem 4 ([10]). *There is a function $c(K)$, with $c(K) \to \infty$ as $K \to \infty$, such that for any sufficiently large prime p, if θ is not a root of any polynomial in $\mathbb{F}_p[x]$ of degree at most K and coefficients bounded by K (as integers), then*
$$|A + \theta A| > c(K)|A|$$
for any subset $A \subset \mathbb{F}_p$ for which $p^{1/10} < |A| < p^{9/10}$.

The proof of Theorem 4 uses Freĭman's theorem. We do not know much more about the behavior of $c(K)$, except for an upper bound of the form
$$\log c(K) \lesssim \sqrt{\log K}.$$

In particular one can show that the inequality in Problem 9 does not hold even for simple maps such as $\phi_j(y) = xy$ with $x \in \mathbb{F}_{2^n}$. (See [18].) Also Theorem 4 should be compared with the recent result of Konyagin and Łaba in characteristic 0.

Theorem 5 ([18]). *If $\theta \in \mathbb{R}$ is transcendental, then*
$$|A + \theta A| > c(\log N)N$$
for any finite subset $A \subset \mathbb{R}$ with $|A| = N$.

Here the optimal result is not known and the "true" expansion factor, $c(N)$, lies somewhere between $\log N$ and $2^{\sqrt{\log N}}$.

Let V be an infinite dimensional vector space (say over a field of characteristic 0). Freĭman's lemma states that if $A \subset V$ is a finite set and $|A + A| < K|A|$, then $\dim\langle A \rangle \leq K$. Freĭman's theorem also gives a bound on the size of the smallest parallelogram in $\langle A \rangle$ containing A, as a function of K. In [18] the authors only use the more elementary Freĭman's lemma instead of Freĭman's theorem. One might hope to strengthen Freĭman's lemma, in the spirit of the "polynomial Freĭman conjecture" proposed by B. Green and I. Ruzsa, in that one might hope to find a large subset A' of A for which $\dim\langle A' \rangle$ is much smaller. For instance one might ask

Problem 10. Can one find a subexponential (perhaps polynomial) function $f(K)$ such that if $|A + A| < K|A|$, there is $A' \subset A$ satisfying $|A'| > |A|/f(K)$ and $\dim\langle A' \rangle < \log f(K)$?

Any result of this type would lead to an improvement of Theorem 5.

3. The Erdős–Szemerédi sum-product problem

Recall the conjecture due to Erdős–Szemerédi about the size of sum and product sets of finite subsets A of $\mathbb{Z}$ or $\mathbb{R}$.

Conjecture. $|A + A| + |A \cdot A| \gg |A|^{2-\varepsilon}$ for all $\varepsilon > 0$.

To date, the strongest general result is due to Solymosi [21]
$$|A + A| + |A \cdot A| \gg |A|^{14/11-\varepsilon}.$$

Over the recent year, a number of estimates have been obtained by the author ([8, 9]), establishing the conjecture under the additional assumption that either $|A + A|$ or $|A \cdot A|$ is small. The methods we used are different from the geometric approach of [21] and borrow especially from the theory of algebraic number fields, such as prime factorization and consequences of the subspace theorem. The present state of affairs suggests that the solution to the Erdős–Szemerédi conjecture, if true, will be deep. We also believe that the methods involved in the problem so far have not been fully exploited. In [1], the following was shown

Theorem 6. *There is a function $\delta(\varepsilon) \to 0$, as $\varepsilon \to 0$ such that if $A \subset \mathbb{Z}$ is a finite set satisfying $|A \cdot A| < |A|^{1+\varepsilon}$, then $|A + A| > |A|^{2-\delta}$.*

The corresponding statement for subsets A of $\mathbb{R}$ is not even known. In [11], a similar conclusion was reached under the stronger assumption $|A \cdot A| < K|A|$ (using the Evertse–Schlickewei–Schmidt theorem mentioned earlier) and in very recent work [2], the result in [1] was generalized to sets A of algebraic numbers of bounded degree. Perhaps one can eliminate the dependence on the degree. Thus we ask

Problem 11. Prove Theorem 6 for finite sets of algebraic numbers without dependence on the degree.

As is clear from this report, despite lots of progress in combinatorial number theory, there is still a wealth of natural problems remaining open in both commutative or noncommutative settings. It is likely that such results would also imply analogous results for subsets of $\mathbb{R}$, using a transference argument in the spirit of [9] or [13].

References

1. J. Bourgain and M.-C. Chang, *On the size of k-fold sum and product sets of integers*, J. Amer. Math. Soc. **17** (2003), no. 2, 473–497.
2. _____, *Sum-product theorems in algebraic number fields*, in preparation.
3. J. Bourgain and A. Gamburd, *New results on expanders*, C. R. Math. Acad. Sci. Paris **342** (2006), no. 10, 717–721.
4. _____, *Uniform expansion bounds for Cayley graphs of* $\mathrm{SL}^2(\mathbb{F}_p)$, Ann. of Math. (2), to appear.
5. _____, *On the spectral gap for finitely generated subgroups of* $\mathrm{SU}(2)$, Invent. Math., to appear.
6. J. Bourgain, A. A. Glibichuk, and S. V. Konyagin, *Estimates for the number of sums and products and for exponential sums in fields of prime order*, J. London Math. Soc. (2) **73** (2006), 380–398.
7. J. Bourgain, N. Katz, and T. Tao, *A sum-product estimate in finite fields and their applications*, Geom. Funct. Anal. **14** (2004), no. 1, 27–57.
8. M.-C. Chang, *Erdős–Szemerédi problem on sum set and product set*, Ann. of Math. (2) **157** (2003), no. 3, 939–957.
9. _____, *Factorization in generalized arithmetic progressions and applications to the Erdős–Szemerédi sum-product problems*, Geom. Funct. Anal. **13** (2003), no. 2, 720–736.
10. _____, *On sum-product representations in* $\mathbb{Z}_q$, J. Eur. Math. Soc. (JEMS) **8** (2006), no. 3, 435–463.
11. _____, *Sum and product of different sets*, Contrib. Discrete Math. **1** (2006), no. 1, 47–56.
12. _____, *Product theorems in* SL_2 *and* SL_3, J. Inst. Math. Jussieu, to appear.
13. J.-H. Evertse, H. P. Schlickewei, and W. M. Schmidt, *Linear equations in variables which lie in a multiplicative group*, Ann. of Math. (2) **155** (2002), no. 3, 807–836.
14. A. Gamburd, D. Jakobson, and P. Sarnak, *Spectra of elements in the group ring of* $\mathrm{SU}(2)$, J. Eur. Math. Soc. (JEMS) **1** (1999), no. 1, 51–85.
15. R. I. Grigorchuk, *On growth in group theory*, Proceedings of the International Congress of Mathematicians. Vol. I (Kyoto, 1990), Math. Soc. Japan, Tokyo, 1991, pp. 325–338.
16. M. Gromov, *Groups of polynomial growth and expanding maps*, Inst. Hautes Études Sci. Publ. Math. **53** (1981), 53–73.
17. H. Helfgott, *Growth and generation in* $\mathrm{SL}_2(\mathbb{Z}/\mathbb{Z}_p)$, Ann. of Math. (2), to appear.
18. S. Konyagin and I. Laba, *Distance sets of well-distributed planar sets for polygonal norms*, Israel J. Math. **152** (2006), 157–179.
19. A. Lubotzky, *Discrete groups, expanding graphs and invariant measures*, Progr. Math., vol. 125, Birkhäuser, Basel, 1994.
20. A. Razborov, *A product theorem in free groups*, preprint.

21. J. Solymosi, *On the number of sums and products*, Bull. London Math. Soc. **37** (2005), no. 4, 491–494.

Department of Mathematics, University of California, Riverside, CA 92521, USA
E-mail address: mcc@math.ucr.edu

Lattice Points on Circles, Squares in Arithmetic Progressions and Sumsets of Squares

Javier Cilleruelo and Andrew Granville

ABSTRACT. We discuss the relationship between various additive problems concerning squares.

1. Squares in arithmetic progression

Let $\sigma(k)$ denote the maximum of the number of squares in $a+b, \ldots, a+kb$ as we vary over positive integers a and b. Erdős conjectured that $\sigma(k) = o(k)$ which Szemerédi [30] elegantly proved as follows: If there are more than δk squares amongst the integers $a+b, \ldots, a+kb$ (where k is sufficiently large) then there exists four indices $1 \leq i_1 < i_2 < i_3 < i_4 \leq k$ in arithmetic progression such that each $a + i_j b$ is a square, by Szemerédi's theorem. But then the $a + i_j b$ are four squares in arithmetic progression, contradicting a result of Fermat. This result can be extended to any given field L which is a finite extension of the rational numbers: From Faltings' theorem we know that there are only finitely many six term arithmetic progressions of squares in L, so from Szemerédi's theorem we again deduce that there are $o_L(k)$ squares of elements of L in any k term arithmetic progression of numbers in L. (Xavier Xarles [31] recently proved that there are never six squares in arithmetic progression in $\mathbb{Z}[\sqrt{d}]$ for any d.)

In his seminal paper *Trigonometric series with gaps* [27] Rudin stated the following conjecture:

Conjecture 1. $\sigma(k) = O(k^{1/2})$.

It may be that the most squares appear in the arithmetic progression $-1 + 24i$, $1 \leq i \leq k$ once $k \geq 8$ yielding that $\sigma(k) = \sqrt{\frac{8}{3}k} + O(1)$. Conjecture 1 evidently implies the following slightly weaker version:

Conjecture 2. *For any $\varepsilon > 0$ we have $\sigma(k) = O(k^{1/2+\varepsilon})$.*

Bombieri, Granville and Pintz [4] proved that $\sigma(k) = O(k^{2/3+o(1)})$, and recently Bombieri and Zannier [5] have proved that $\sigma(k) = O(k^{3/5+o(1)})$.

2000 *Mathematics Subject Classification.* 11N36.

The idea for this paper emerged while the first author was visiting the Centre de recherches mathématiques of Montreal and he is extremely grateful for their hospitality.

This is the final form of the paper.

©2007 American Mathematical Society

2. Rudin's approach

Let $e(\theta) := e^{2i\pi\theta}$ throughout. The following well-known conjecture was discussed by Rudin (see the end of Section 4.6 in [27]):

Conjecture 3. *For any $2 \leq p < 4$ there exists a constant C_p such that, for any trigonometric polynomial $f(\theta) = \sum_k a_k e(k^2\theta)$ we have*

(2.1) $$\|f\|_p \leq C_p \|f\|_2.$$

Here, as usual, we define $\|f\|_p^p := \int_0^1 |f(t)|^p \, dt$ for a trigonometric polynomial f. Conjecture 3 says that the set of squares is a $\Lambda(p)$-set for any $2 \leq p < 4$, where E is a $\Lambda(p)$-set if there exists a constant C_p such that (2.1) holds for any f of the form $f(\theta) = \sum_{n_k \in E} a_k e(n_k \theta)$ (a so-called E-polynomial). By Hölder's inequality we have, for $r < s < t$,

(2.2) $$\|f\|_s^{s(t-r)} \leq \|f\|_r^{r(t-s)} \|f\|_t^{t(s-r)};$$

taking $r = 2$ we see that if E is a $\Lambda(t)$-set then it is a $\Lambda(s)$-set for all $s \leq t$.

Let $r(n)$ denote the number of representations of n as the sum of two squares (of positive integers). Taking $f(\theta) = \sum_{1 \leq k \leq x} e(k^2\theta)$, we deduce that $\|f\|_2^2 = x$, whereas $\|f\|_4^4 = \sum_n \#\{1 \leq k, l \leq x : n = k^2 + l^2\}^2 \geq \sum_{n \leq x^2} r(n)^2 \asymp x^2 \log x$; so we see that (2.1) does not hold in general for $p = 4$.

Conjecture 3 has not been proved for any $p > 2$, though Rudin [27] has proved the following theorem.

Theorem 1. *If E is a $\Lambda(p)$-set then any arithmetic progression of N terms contains $\ll N^{2/p}$ elements of E. In particular, if Conjecture 3 holds for p then $\sigma(k) = O(k^{2/p})$.*

PROOF. We use Fejér's kernel $\kappa_N(\theta) := \sum_{|j| \leq N} (1 - |j|/N) \, e(j\theta)$. Note that $\|\kappa_N\|_1 = 1$ and $\|\kappa_N\|_2^2 = \sum_{|j| \leq N}(1 - |j|/N)^2 \ll N$ so, by (2.2) with $r = 1 < s = q < t = 2$ we have $\|\kappa_N\|_q^q \ll 1^{2-q} N^{q-1}$ so that $\|\kappa_N\|_q \ll N^{1/p}$ where $1/q + 1/p = 1$.

Suppose that $n_1, n_2 \ldots, n_\sigma$ are the elements of E which lie in the arithmetic progression $a + ib, 1 \leq i \leq N$. If $n_l = a + ib$ for some i, $1 \leq i \leq N$ then $n_l = a + mb + jb$ where $m = [(N+1)/2]$ and $|j| \leq N/2$; and so $1 - |j|/N \geq \frac{1}{2}$. Therefore, for $g(\theta) := \sum_{1 \leq l \leq \sigma} e(n_l \theta)$, we have

$$\int_0^1 g(-\theta) e\big((a+bm)\theta\big) \kappa_N(b\theta) \, d\theta \geq \frac{\sigma}{2}.$$

On the other hand, we have $\|g\|_p \leq C_p \|g\|_2 \ll \sqrt{\sigma}$ since E is a $\Lambda(p)$-set and g is an E-polynomial. Therefore, by Hölder's inequality,

(2.3) $$\left| \int_0^1 g(-\theta) e\big((a+bm)\theta\big) \kappa_N(b\theta) \, d\theta \right| \leq \|g\|_p \|\kappa_N\|_q \ll \sqrt{\sigma} N^{1/p}$$

and the result follows by combining the last two displayed equations. $\square$

It is known that Conjecture 3 is true for polynomials $f(\theta) = \sum_{k \leq N} e(k^2\theta)$ and Antonio Córdoba [18] proved that Conjecture 3 also holds for polynomials $f(\theta) = \sum_{k \leq N} a_k e(k^2\theta)$ when the coefficients a_k are positive real numbers and non-increasing.

3. Sumsets of squares

For a given finite set of integers E let $f_E(\theta) = \sum_{k \in E} e(k\theta)$. Mei-Chu Chang [11] conjectured that for any $\varepsilon > 0$ we have

$$\|f_E\|_4 \ll_\varepsilon \|f_E\|_2^{1+\varepsilon}$$

for any finite set of squares E. As $\|f_E\|_4^4 = \sum_n r_{E+E}^2(n)$ where $r_{E+E}(n)$ is the number of representations of n as a sum of two elements of E, her conjecture is equivalent to:

Conjecture 4 (Mei-Chu Chang). *For any $\varepsilon > 0$ we have that*

(3.1) $$\|f_E\|_4^4 = \sum_n r_{E+E}^2(n) \ll |E|^{2+\varepsilon} = \|f_E\|_2^{4+2\varepsilon}$$

for any finite set E of squares.

We saw previously that $\sum_n r_{E+E}^2(n) \gg |E|^2 \log |E|$ in the special case $E = \{1^2, \ldots, k^2\}$, so Conjecture 4 is sharp, in the sense one cannot entirely remove the ε.

Trivially we have

$$\|f_E\|_4^4 = \sum_n r_{E+E}^2(n) \leq \max_n r_{E+E}(n) \sum_n r_{E+E}(n) \leq |E| \cdot |E|^2 = |E|^3$$

for any set E; it is surprisingly difficult to improve this estimate when E is a set of squares. The best such result is due to Mei-Chu Chang [11] who proved that

$$\sum_n r_{E+E}^2(n) \ll |E|^3 / \log^{1/12} |E|$$

for any set E of squares. Assuming a major conjecture of arithmetic geometry we can improve Chang's result, in a proof reminiscent of that in [4].

Theorem 2. *Assume the Bombieri–Lang conjecture. Then, for any set E of squares, and any set of integers A, we have*

$$\sum_n r_{E+A}^2(n) \ll |A|^2 |E|^{2/3} + |A| \, |E|.$$

In particular

$$\sum_n r_{E+E}^2(n) \ll |E|^{8/3}.$$

PROOF. One consequence of [10] is that there exists an integer κ, such that if the Bombieri–Lang conjecture is true then for *any* polynomial $f(x) \in \mathbb{Z}[x]$ of degree six which does not have repeated roots, there are no more than κ rational numbers m for which $f(m)$ is a square. For any given set of four elements $a_1, \ldots, a_4 \in A$ consider all integers n for which there exist $b_1, \ldots, b_4 \in E$ with $n = a_1 + b_1^2 = \cdots = a_4 + b_4^2$. Evidently $f(b_1) = (b_2 b_3 b_4)^2$ where $f(x) = (x^2 + a_1 - a_2)(x^2 + a_1 - a_3)(x^2 + a_1 - a_4)$, and so there cannot be more than κ

such integers n. Therefore,

$$\sum_n \binom{r_{E+A}(n)}{4}$$
$$= \sum_n \#\{a_1,\ldots,a_4 \in A : \exists b_1^2,\ldots,b_4^2 \in E, \text{ with } n = a_i + b_i^2, i = 1,\ldots,4\}$$
$$= \sum_{a_1,\ldots,a_4 \in A} \#\{n : \exists b_1^2,\ldots,b_4^2 \in E, \text{ with } n = a_i + b_i^2, i = 1,\ldots,4\} \leq \kappa\binom{|A|}{4}.$$

We have $\sum_n r_{E+A}(n) = |E||A|$; and so $\sum_n r_{E+A}(n)^4 \ll \sum_n \binom{r_{E+A}(n)}{4} + \sum_n r_{E+A}(n) \ll |A|^4 + |A||E|$. Therefore the result follows from Holder's inequality since

$$\sum_n r_{E+A}^2(n) \leq \left(\sum_n r_{E+A}(n)\right)^{2/3} \left(\sum_n r_{E+A}^4(n)\right)^{1/3} \ll |A|^2 |E|^{2/3} + |A||E|. \quad \square$$

The final form of Theorem 2 was inspired by email correspondance with Jozsef Solymosi, who also suggested the following:

Theorem 3. *Assume the Bombieri–Lang conjecture. There exists a constant $c > 0$ such that for any finite set E of squares we have $r_{E+E}(n) \leq c\sqrt{|E+E|}$ for all integers n.*

PROOF. Suppose, on the contrary, that $r_{E+E}(n) > \sqrt{\kappa|E+E|}$ for some integer n, where κ is the constant in the Bombieri–Lang conjecture. Let $A := \{a : a+a' = n, a, a' \in E\}$, so that

$$\max_m r_{A+A}(m) \geq \frac{1}{|A+A|}\sum_m r_{A+A}(m) = \frac{|A|^2}{|A+A|} = \frac{r_{E+E}(n)^2}{|A+A|} \geq \frac{\kappa|E+E|}{|E+E|} = \kappa.$$

Now for each a for which $m - a \in A \subset E$, we have $m - a, n - a, n - m + a \in E$, and therefore an integral point on the curve $y^2 = (x^2 - m)(x^2 - n)(x^2 + n - m)$, contradicting the Bombieri–Lang conjecture. $\quad\square$

By the Cauchy–Schwarz inequality we have

$$|E|^2|A|^2 = \left(\sum r_{E+A}(n)\right)^2 \leq |E+A|\sum r_{E+A}^2(n) \ll |E+A|(|A|^2|E|^{2/3} + |A||E|),$$

so we can deduce from Theorem 2 that

$$|E+A| \gg |A| + |E|\min\{|A|, |E|^{1/3}\},$$

still under the assumption of the Bombieri–Lang conjecture. In particular,

$$|E+A| \gg |E||A|$$

when $|A| \ll |E|^{1/3}$ which is sharp, and also

(3.2) $$|E+E| \gg |E|^{4/3}.$$

This leads us to the following conjecture.

Conjecture 5 (Ruzsa). *For all $\varepsilon > 0$ we have*

$$|E+E| \gg |E|^{2-\varepsilon},$$

for every finite set, E, of squares of integers.

Theorem 4. *Conjecture 4 implies Conjecture 5 with the same ε.*

PROOF. By the Cauchy–Schwarz inequality we have

$$|E|^4 = \left(\sum_n r_{E+E}(n)\right)^2 \leq |E+E| \cdot \sum_n r_{E+E}^2(n)$$

and the result follows. □

Theorem 5. *Conjecture 5 implies Conjecture 2 with $\varepsilon_{\mathrm{cj}\,2} = \varepsilon_{\mathrm{cj}\,5}/(4 - 2\varepsilon_{\mathrm{cj}\,5})$.*

PROOF. If E is a set of squares which is a subset of an arithmetic progression P of length k then $E + E \subset P + P$. From Conjecture 5 we deduce that

$$|E|^{2-\varepsilon} \ll |E+E| \leq |P+P| = 2k - 1$$

and the result follows. □

In particular, applying this result to (3.2) shows that the Bombieri–Lang conjecture implies $\sigma(k) \ll k^{3/4}$. However one can easily do better than this: Suppose that there are $\sigma_{r,s}$ squares amongst $a + ib$, $1 \leq i \leq k$ which are $\equiv r \pmod{s}$; that is the squares amongst $a + rb + jsb$, $0 \leq j \leq [k/s]$. This gives rise to $\binom{\sigma_{r,s}}{6}$ rational points on the set of curves $y^2 = x \prod_{j=1}^{5}(x + n_j)$ for $0 \leq n_1 < n_2 < \cdots < n_5 \leq [k/s]$. Summing over all $r \pmod{s}$ and all $s > \sigma/10$ we get

$$\frac{k^5}{\sigma^5} \gg \binom{10k/\sigma}{5} \gg \sum_{s > \sigma/10} \sum_{r \,(\mathrm{mod}\, s)} \binom{\sigma_{r,s}}{6} \gg \sum_{s > \sigma/10} s\binom{[\sigma/s]}{6} \gg \sigma^2$$

and we obtain $\sigma(k) \ll k^{5/7}$. This is all irrelevant since stronger upper bounds were proved unconditionally in [4, 5].

An *affine cube* of dimension d in $\mathbb{Z}$ is a set of integers $\{b_0 + \sum_{i \in I} b_i : I \subset \{1, \ldots, d\}\}$ for non-zero integers $b_0, \ldots, b_d$. In [28], Solymosi states

Conjecture 6 (Solymosi). *There exists an integer $d > 0$ such that there is no affine cube of dimension d of distinct squares.*

This was asked earlier as a question by Brown, Erdős and Freedman in [9], and Hegyvári and Sárközy [21] have proved that an affine cube of squares, all $\leq n$, has dimension $\leq 48(\log n)^{1/3}$.

This conjecture follows from the Bombieri–Lang conjecture for if there were an affine cube of dimension d then for any $x^2 \in \{b_0 + \sum_{i \in I} b_i : I \subset \{3, \ldots, d\}\}$ we have that $x^2 + b_1$, $x^2 + b_2$, $x^2 + b_1 + b_2$ are also squares, in which case there are $\geq 2^{d-2}$ integers x for which $f(x) = (x^2 + b_1)(x^2 + b_2)(x^2 + b_1 + b_2)$ is also square; and so $2^{d-2} \leq \kappa$, as in the proof of Theorem 2.

In [28], Solymosi gives a beautiful proof that for any set of real numbers A, if $|A + A| \ll_d |A|^{1 + 1/(2^{d-1} - 1)}$ then A contains many affine cubes of dimension d. Therefore we deduce a weak version of Ruzsa's conjecture from Solymosi's conjecture:

Theorem 6. *Conjecture 6 implies that there exists $\delta > 0$ for which $|E + E| \gg |E|^{1+\delta}$, for any set E of squares.*

The Erdős–Szemerédi conjecture states that for any set of integers A we have

$$|A + A| + |A \diamond A| \gg_\varepsilon |A|^{2-\varepsilon}.$$

In fact they gave a stronger version, reminiscent of the Balog–Szemerédi–Gowers theorem:

Conjecture 7 (Erdős–Szemerédi). *If A is a finite set of integers and $G \subset A \times A$ with $|G| \gg |A|^{1+\varepsilon/2}$ then*

(3.3) $\qquad |\{a+b : (a,b) \in G\}| + |\{ab : (a,b) \in G\}| \gg_\varepsilon |G|^{1-\varepsilon}.$

Mei-Chu Chang [11] proved that a little more than Conjecture 7 implies Conjecture 4:

Theorem 7. *If (3.3) holds whenever $|G| \geq \frac{1}{2}|A|$ then Conjecture 4 holds.*

PROOF. Let B be a set of k non-negative integers and $E = \{b^2 : b \in B\}$. Define $G_M := \{(a_+, a_-) : \exists b, b' \in B \text{ with } a_+ = b + b', a_- = b - b', \text{ and } b^2 - b'^2 \in M\}$ where $M \subset E - E$; and so $A_M := \{a_+, a_- : (a_+, a_-) \in G_M\} \subset (B + B) \cup (B - B)$. Therefore $|A_M| \leq 2|G_M|$.

Since $\{a + a' : (a, a') \in G\}$ and $\{a - a' : (a, a') \in G\}$ are subsets of $\{2b : b \in B\}$, they have $\leq k$ elements; and $\{aa' : (a, a') \in G\} \subset M$. Therefore (3.3) implies that $|G_M|^{1-\varepsilon} \ll_\varepsilon |M| + k$. Since, trivially, $|G_M| \leq k^2$ we have $\sum_{m \in M} r_{E-E}(m) = |G_M| \ll k^{2\varepsilon}(|M| + k)$.

Now let M be the set of integers m for which $r_{E-E}(m) \geq k^{3\varepsilon}$, so that $\sum_{m \in M} r_{E-E}(m) \geq k^{3\varepsilon}|M|$ and hence $\sum_{m \in M} r_{E-E}(m) \ll k^{1+2\varepsilon}$ by combining the last two equations. Therefore, as $r_{E-E}(m) \leq k$,

$$\|f_E\|_4^4 = \sum_m r_{E-E}(m)^2 \leq \sum_{m \in E-E} k^{6\varepsilon} + k \sum_{m \in M} r_{E-E}(m) \ll k^{2+6\varepsilon}. \qquad \square$$

She also proves a further, and stronger, result along similar lines:

Conjecture 8 (Mei-Chu Chang). *If A is a finite set of integers and $G \subset A \times A$ then*

(3.4) $\quad |\{a+b : (a,b) \in G\}| \cdot |\{a-b : (a,b) \in G\}| \cdot |\{ab : (a,b) \in G\}| \gg_\varepsilon |G|^{2-\varepsilon}.$

Theorem 8. *Conjecture 8 holds if and only if Conjecture 4 holds.*

PROOF. Assume Conjecture 8 and define B, A and G_M as in the proof of Theorem 7, so that $\left(\sum_{m \in M} r_{E-E}(m)\right)^2 = |G_M|^2 \ll k^{2+2\varepsilon}|M|$. We partition $E - E$ into the sets $M_j := \{m : 2^{j-1} \leq r_{E-E}(m) < 2^j\}$ for $j = 1, 2, \ldots, J := [\log(2k)/\log 2]$; then $(2^{j-1}|M_j|)^2 \leq \left(\sum_{m \in M_j} r_{E-E}(m)\right)^2 \ll k^{2+2\varepsilon}|M_j|$ so that $\sum_{m \in M_j} r_{E-E}(m)^2 \leq 2^{2j}|M_j| \ll k^{2+2\varepsilon}$. Hence

$$\|f_E\|_4^4 = \sum_m r_{E-E}(m)^2 < \sum_j \sum_{m \in M_j} r_{E-E}(m)^2 \ll Jk^{2+2\varepsilon} \ll k^{2+3\varepsilon},$$

as desired.

Now assume Conjecture 4 and let $G_n := \{(a,b) \in G : ab = n\}$. Then $|G|^2 = (\sum_n |G_n|)^2 \leq |\{ab : (a,b) \in G\}| \cdot \sum_n |G_n|^2$, while

$$\sum_n |G_n|^2 = \int_0^1 \left|\sum_{(a,b) \in G} e(4abt)\right|^2 dt = \int_0^1 \left|\sum_{(a,b) \in G} e\bigl((a+b)^2 t\bigr)e\bigl(-(a-b)^2 t\bigr)\right|^2 dt$$

which, letting $E_\pm := \{r^2 : r = a \pm b, (a,b) \in G\}$, is

$$\leq \int_0^1 \left| \sum_{r^2 \in E_+} e(r^2 t) \sum_{s^2 \in E_-} e(-s^2 t) \right|^2 dt \leq \|f_{E_+}\|^2 \|f_{E_-}\|^2$$

by the Cauchy–Schwarz inequality. Since $\|f_{E_\pm}\|^2 \ll |\{a \pm b : (a,b) \in G\}| \cdot |G|^{2\varepsilon}$ by Conjecture 4, our result follows by combining the above information. □

4. Solutions of a quadratic congruence in short intervals

We begin with a connection between additive combinatorics and the Chinese remainder theorem. Suppose that $n = rs$ with $(r,s) = 1$; and that for given sets of residues $\Omega(r) \subset \mathbb{Z}/r\mathbb{Z}$ and $\Omega(s) \subset \mathbb{Z}/s\mathbb{Z}$ we have $\Omega(n) \subset \mathbb{Z}/n\mathbb{Z}$ in the sense that $m \in \Omega(n)$ if and only if there exists $u \in \Omega(r)$ and $v \in \Omega(s)$ such that $m \equiv u \pmod{r}$ and $m \equiv v \pmod{s}$. When $(r, n/r) = 1$ consider the map which embeds $\mathbb{Z}/r\mathbb{Z} \to \mathbb{Z}/n\mathbb{Z}$ by taking $u \pmod{r}$ and replacing it by $U \pmod{n}$ for which $U \equiv u \pmod{r}$ and $U \equiv 0 \pmod{n/r}$; we denote the image of $\Omega(r)$ by $\Omega(r,n)$ under this map. The key remark, which follows immediately from the definitions, is that

$$\Omega(n) = \Omega(r,n) + \Omega(s,n).$$

Thus if $n = p_1^{e_1} \cdots p_k^{e_k}$ where the primes p_i are distinct then

$$\Omega(n) = \Omega(p_1^{e_1}, n) + \Omega(p_2^{e_2}, n) + \cdots + \Omega(p_k^{e_k}, n).$$

Particularly interesting is where $\Omega_f(n)$ is the set of solutions $m \pmod{n}$ to $f(m) \equiv 0 \pmod{n}$, for given $f(x) \in \mathbb{Z}[x]$. Can there be many elements of $\Omega_f(n)$ in a short interval when f has degree two? A priori this seems unlikely since the elements of the $\Omega(r,n)$ are so well spread out, that is the distance between any pair of elements is $\geq n/r$ because they are all divisible by n/r.

The next theorem involves the distribution of the elements of $\Omega(n)$ in the simplest non-trivial case, in which each $\Omega(p_j^{e_j})$ has just two elements, namely $\{0,1\}$, so that $\Omega(n)$ is the set of solutions of $x(x-1) \equiv 0 \pmod{n}$ (see also [3]).

Theorem 9. *Let $\Omega(n)$ be the set of solutions of $x(x-1) \equiv 0 \pmod{n}$. Then*

(1) *$\Omega(n)$ has an element in the interval $(1, n/k + 1)$.*
(2) *For any $\varepsilon > 0$ there exists $n = p_1 \cdots p_k$ such that $\Omega(n) \cap (1, (\frac{1}{k} - \varepsilon)n] = \emptyset$.*
(3) *For any $\varepsilon > 0$ there exists $n = p_1 \cdots p_k$ such that if $x \in \Omega(n)$ then $|x| < \varepsilon n$.*

PROOF. Let $\Omega(p_j^{e_j}, n) = \{0, x_j\}$ where $x_j \equiv 1 \pmod{p_j^{e_j}}$ and $x_j \equiv 0 \pmod{p_i^{e_i}}$ for any $i \neq j$. Then $\Omega(n) = \{0, x_1\} + \cdots + \{0, x_k\}$. Let $s_0 = n$ and s_r be the least positive residue of $x_1 + \cdots + x_r \pmod{n}$ for $r = 1, \ldots, k$ so that $s_k = 1$. By the pigeonhole principle, there exists $0 \leq l < m \leq k$ such that s_l and s_m lie in the same interval $(jn/k, (j+1)n/k]$, and so $|s_l - s_m| < n/k$ with $s_m - s_l \equiv x_{l+1} + \cdots + x_m \pmod{n} \in \Omega(n)$. If $s_m - s_l > 1$ then we are done. Now $s_k - s_0 \equiv 1 \pmod{n}$ but is not $= 1$, so $s_m - s_l \neq 0, 1$. Thus we must consider when $s_m - s_l < 0$. In this case $x_1 + \cdots + x_l + x_{m+1} + \cdots + x_k \pmod{n} \in \Omega(n)$ and is $\equiv s_k - (s_m - s_l) \equiv 1 - (s_m - s_l)$, and the result follows.

To prove (2) let us take $k-1$ primes $p_1, \ldots, p_{k-1} > k$, and integers $a_j = [p_j/k]$ for $j = 1, \ldots, k-1$. Let $P = p_1 \cdots p_{k-1}$ and determine $r \pmod{P}$ by the Chinese remainder theorem satisfying $ra_j(P/p_j) \equiv 1 \pmod{p_j}$ for $j = 1, \ldots, k-1$. Now let p_k be a prime $\equiv r \pmod{P}$, and let a_k the least positive integer satisfying $a_k P \equiv 1 \pmod{p_k}$. Let $n = p_1 \cdots p_k$ so that $x_i = a_i n/p_i$ for $i = 1, \ldots, k$. Now

$n/k \geq x_i > n/k - n/p_i > 0$ for $i = 1, \ldots, k-1$ and since $x_1 + \cdots + x_k \equiv 1 \pmod{n}$ with $1 \leq x_k < n$ we deduce that $x_1 + \cdots + x_k = n+1$ and therefore $1 + n/k \leq x_k < 1 + n/k + n\sum_{i=1}^{k-1} 1/p_i$. Now elements of $\Omega(n)$ are of the form $\sum_{i \in I} x_i$ and we have $|\sum_{i \in I} x_i - n|I|/k| \leq 1 + 2n\sum_{i=1}^{k-1} 1/p_i$, and this is $< \varepsilon n$ provided $p_i > 2k/\varepsilon$ for each p_i. Finally, since the cases $I = \emptyset$ and $I = \{1, \ldots, k\}$ correspond to the cases $x = 0$ and $x = 1$ respectively, we have that any other element is greater than $(1/k - \varepsilon)n$.

To prove (3) we mimic the proof of (2) but now choosing non-zero integers a_j satisfying $|a_j/p_j| < \varepsilon/(2k)$ for $j = 1, \ldots, k-1$. This implies that $|a_k/p_k| < \varepsilon/2$ and then $|\sum_{i \in I} x_i| < \varepsilon n$. □

In the other direction, we give a lower bound for the length of intervals containing k elements of $\Omega(n)$.

Theorem 10. *Let integer $d \geq 2$ be given, and suppose that for each prime power q we are given a set of residues $\Omega(q) \subset (\mathbb{Z}/q\mathbb{Z})$ which contains no more than d elements. Let $\Omega(n)$ be determined for all integers n using the Chinese remainder theorem, as described at the start of this section. Then, for any $k \geq d$, there are no more than k integers $x \in \Omega(n)$ in any interval of length $n^{\alpha_d(k)}$, where $\alpha_d(k) = (1 - \varepsilon_d(k))/d > 0$ with $0 < \varepsilon_d(k) = (d-1)/k + O(d^2/k^2)$.*

PROOF. Let $x_1, \ldots, x_{k+1}$ elements of $\Omega(n)$ such that $x_1 < \cdots < x_{k+1} < x_1 + n^{\alpha_d(k)}$. Let q a prime power dividing n. Each x_i belongs to one of the d classes (mod q) in $\Omega(q)$. Write $r_1, \ldots, r_d$ to denote the number of these x_i belonging to each class. Then $\prod_{1 \leq i < j \leq k+1}(x_j - x_i)$ is a multiple of $q^{\sum_{i=1}^d \binom{r_i}{2}}$. The minimum of $\sum_{i=1}^d \binom{r_i}{2}$ under the restriction $\sum_i r_i = k+1$ is $d\binom{r}{2} + rs$ where r, s are determined by $k+1 = rd + s$, $0 \leq s < d$. Finally
$$n^{\alpha_d(k)\binom{k+1}{2}} > \prod_{1 \leq i < j \leq k+1}(x_i - x_j) > n^{d\binom{r}{2}+rs}$$
and we get a contradiction, by taking $\alpha_d(k) = \left(d\binom{r}{2} + rs\right)/\binom{k+1}{2}$. □

The next theorem is an easy consequence of the proof above.

Theorem 11. *If $x_1 < \cdots < x_k$ are solutions to the equation $x_i^2 \equiv a \pmod{b}$, then $x_k - x_1 > b^{1/2 - 1/2l}$, where l is the largest odd integer $\leq k$.*

FIRST PROOF. For any maximal prime power q dividing b, (a, q) must be a square so we can write $x_i = y_i \prod_q (a, q)^{1/2}$ with $y_i^2 \equiv a' \pmod{q'}$ where $q' = q/(a,q)$ and $(a', q') = 1$. Let $\Omega(q')$ be the solutions of $y^2 \equiv a' \pmod{q'}$. Now, since $(a', q') = 1$ we have that $|\Omega(q')| \leq 2$ and we can apply Theorem 10 to obtain that
$$x_k - x_1 = (y_k - y_1) \prod_q (a,q)^{1/2}$$
$$\geq \left(\prod_q q/(a,q)\right)^{\alpha_2(k-1)} \prod_q (a,q)^{1/2} \geq \left(\prod_q q\right)^{\alpha_2(k-1)}.$$

Now, notice that $\alpha_2(k-1) = \frac{1}{2} - 1/(2l)$ where l is the largest odd number $\leq k$. □

SECOND PROOF. Write $x_j^2 = a + r_j b$ where $r_1 = 1 < r_2 < \cdots < r_k$ (if necessary, by replacing a in the hypothesis by $x_1^2 - b$). Consider the k-by-k Vandermonde matrix V with (i,j)th entry x_j^{i-1}. The row with $i = 1 + 2I$ has jth entry $(a+r_jb)^I$; by subtracting suitable multiples of the rows $1 + 2l$, $l < I$, we obtain a matrix V_1 with the same determinant where the $(2I+1, j)$ entry is now $(r_jb)^I$. Similarly the row with $i = 2I + 2$ has jth entry $x_j(a+r_jb)^I$; by subtracting suitable multiples of the rows $2 + 2l$, $l < I$, we obtain a matrix V_2 with the same determinant where the $(2I+2, j)$ entry is now $x_j(r_jb)^I$. Finally we arrive at a matrix W by dividing out b^I from rows $2I+1$ and $2I+2$ for all I. Then the determinant of V, which is $\prod_{1 \leq i < j \leq k}(x_j - x_i)$, equals $b^{[(k-1)^2/4]}$ times the determinant of W, which is also an integer, and the result follows. □

The advantage of this new proof is that if we can get non-trivial lower bounds on the determinant of W then we can improve Theorem 11. We note that W has $(2I+1, j)$ entry r_j^I, and $(2I+2, j)$ entry $x_j r_j^I$.

Remark. Taking $k = l$ to be the smallest odd integer $\geq \log b/\log 4$, then we can split our interval into two pieces to deduce from Theorem 11 a weak version of Conjecture 9: There are no more than $\log(4b)/\log 2$ solutions x to the equation $x^2 \equiv a \pmod{b}$ in any interval of length $b^{1/2}$. From this it follows that the number of solutions x to the equation $x^2 \equiv a \pmod{b}$ in any interval of length L is

$$\ll 1 + \frac{\log L}{\log(1 + b^{1/2}/L)}.$$

This result, with "$\frac{1}{2}$" replaced by "$1/d$," was proved for the roots of any degree d polynomial mod b by Konyagin and Steger in [23].

A slight improvement on the theorem above would have interesting consequences.

Conjecture 9. *There exists a constant N such that there are no more than N solutions $0 < x_1 < x_2 < \cdots < x_N < x_1 + b^{1/2}$ to the equation $x_i^2 \equiv a \pmod{b}$, for any given a and b.*

Theorem 12. *Conjecture 9 implies Conjecture 1.*

PROOF. Suppose that there are $l \gg k^{1/2}$ squares amongst $a + b, a + 2b, \ldots, a + kb$, which we will denote $x_1^2 < x_2^2 < \cdots < x_l^2$. By Conjecture 9 we have $x_l - x_1 \geq [(l-1)/N]b^{1/2}$, whereas $(k-1)b \geq (x_l + x_1)(x_l - x_1) \geq (x_l - x_1)^2$. Therefore $[(l-1)/N]^2 \leq (k-1)$ which implies that $l \leq N(1 + \sqrt{k-1})$. □

Conjecture 9 would follow easily from Theorem 10 if we could get the exponent $\frac{1}{2}$ for some k, instead of $\frac{1}{2} - \varepsilon_2(k)$. Conjecture 9 can be strengthened and generalized as follows:

Conjecture 10. *Let integer $d \geq 1$ be given, and suppose that for each prime power q we are given a set of residues $\Omega(q) \subset (\mathbb{Z}/q\mathbb{Z})$ which contains no more than d elements. $\Omega(b)$ is determined for all integers b using the Chinese remainder theorem, as described at the start of this section. Then, for any $\varepsilon > 0$ there exists a constant $N(d, \varepsilon)$ such that for any integer b there are no more than $N(d, \varepsilon)$ integers n, $0 \leq n < b^{1-\varepsilon}$ with $n \in \Omega(b)$.*

In Theorem 10 we proved such a result with the exponent "$1 - \varepsilon$" replaced by "$1/d - \varepsilon$." We strongly believe Conjecture 10 with '$1 - \varepsilon$' replaced by "$1/d$," analogous to Conjecture 9. In a 1995 email to the second author, Bjorn Poonen asked Conjecture 10 with "$1-\varepsilon$" replaced by "$\frac{1}{2}$" for $d=4$; his interest lies in the fact that this would imply the uniform boundedness conjecture for rational prepreriodic points of quadratic polynomials (see [25]).

Conjecture 10 does not cover the case $\Omega_f(b) = \{m \pmod{b} : f(m) \equiv 0 \pmod{n}\}$ for all monic polynomials f of degree d since, for example, the polynomial $(x-a)^d \equiv 0 \pmod{p^k}$ has got $p^{k-\lceil k/d \rceil}$ solutions $\pmod{p^k}$, rather than d. One may avoid this difficulty by restricting attention to squarefree moduli (as in a conjecture posed by Croot [19]); or, to be less restrictive, note that if $f(x)$ has more than d solutions $\pmod{p^k}$ then f must have a repeated root mod p, so that p divides the discriminant of f:

Conjecture 11. *Fix integer $d \geq 2$. For any $\varepsilon > 0$ there exists a constant $N(d, \varepsilon)$ such that for any monic $f(x) \in \mathbb{Z}[x]$ there are no more than $N(d, \varepsilon)$ integers n, $0 \leq n < b^{1-\varepsilon}$, with $f(n) \equiv 0 \pmod{b}$ for any integer b such that if p^2 divides b then p does not divide the discriminant of f.*

5. Lattice points on circles

Conjecture 12. *There exists $\delta > 0$ and integer $m > 0$ such that if $a_i^2 + b_i^2 = n$ with $a_i, b_i > 0$ and $a_i^2 \equiv a_1^2 \pmod{q}$ for $i = 1, \ldots, m$ then $q = O(n^{1-\delta})$.*

Theorem 13. *Conjecture 12 implies Conjecture 1.*

PROOF. Suppose that $x_1^2 < \cdots < x_r^2$ are distinct squares belonging to the arithmetic progression $a+b, a+2b, \ldots, a+kb$ with $(a,b) = 1$, where $r > \sqrt{8lk}$, with l sufficiently large $> m$. We may assume that $(a, b) = 1$ and that b is even. There are r^2 sums $x_i^2 + x_j^2$ each of which takes one of the values $2a+2b, 2a+3b, \ldots, 2a+2kb$, and so one of these values, say n, is taken $\geq r^2/(2k-1) > 4l$ times. Therefore we can write $n = r_j^2 + s_j^2$ for $j = 1, 2, \ldots, 4l$ for distinct pairs (r_j, s_j), and let $v_j = r_j + is_j$. Note that $n \equiv 2 \pmod{8}$. Let $\Pi = \prod_{1 \leq i < j \leq 4l}(v_j - v_i) \neq 0$. We will prove that $|\Pi| \geq b^{4\binom{l}{2}}(n/2)^{\binom{2l}{2}}$, by considering the powers of the prime divisors of b and n which divide Π. Note that $(n/2, b) = 1$.

Suppose $p^e \| b$ where p is a prime, and select $w \pmod{p^e}$ so that $w^2 \equiv a \pmod{p^e}$. Note that each $r_j, s_j \equiv w$ or $-w \pmod{p^e}$: We partition the v_j into four subsets J_1, J_2, J_3, J_4 depending on the value of $(r_j \pmod{p^e}, s_j \pmod{p^e})$. Note then that p^e divides $v_j - v_i$ if v_i, v_j belong to the same subset, and so p^e to the power $\sum_i \binom{|J_i|}{2} > 4\binom{l}{2}$ divides Π.

Now let p be an odd prime with $p^e \| n$. If $p \equiv 3 \pmod{4}$ then $p^{e/2}$ must divide each r_j and s_j so that then $p^{(e/2)\binom{4l}{2}}$ divides Π. If $p \equiv 1 \pmod{4}$ let us suppose π is a prime in $\mathbb{Z}[i]$ dividing p; then $\pi^{e_j}\bar{\pi}^{e-e_j}$ divides v_j for some $0 \leq e_j \leq e$. If $e_i \leq e_j$ we deduce that $\pi^{e_i}\bar{\pi}^{e-e_j}$ divides $v_j - v_i$. We now partition the values of j into sets $J_0, \ldots, J_e$ depending on the value of e_j. The power of π dividing Π is thus $\sum_{i=0}^{e} i \sum_{g=i+1}^{e} |J_i||J_g| + \sum_{i=0}^{e}(e-i)\binom{|J_i|}{2}$, and the power of $\bar{\pi}$ dividing Π is thus $\sum_{i=0}^{g}(e-g)\sum_{i=0}^{g-1}|J_i||J_g| + \sum_{i=0}^{e}(e-i)\binom{|J_i|}{2}$. It is easy to show that $\sum_{0 \leq i < g \leq e}(i+e-g)m_i m_g + e\sum_{i=0}^{e}\binom{m_i}{2}$, under the conditions that $\sum_i m_i$ is fixed and each $m_i \geq 0$, is minimized when $m_0 = m_e$, $m_1 = \cdots = m_{e-1} = 0$. Therefore the power of π plus the power of $\bar{\pi}$ dividing Π is $\geq 2e\binom{2l}{2}$.

Finally $|r_j - r_i|, |s_j - s_i| \leq (x_r^2 - x_1^2)/(x_r + x_1) \leq (k-1)b/(2\sqrt{a+b})$, and so $|v_j - v_i|^2 \leq (k-1)^2 b^2/(2(a+b))$, giving that $|\Pi| \leq \left(k^2 b^2/(2(a+b))\right)^{(1/2)\binom{4l}{2}}$.

Putting these all together with the fact that $n > 2(a+b)$, we obtain $2^{2l-1}(a+b)^{3l-1} \leq k^{4l-1}b^{3l}$. However this implies that $n \leq 2k(a+b) \leq 2^{1/2}k^{5/2}b^{1+1/(3l-1)} \ll k^{5/2}n^{(1+1/(3l-1))(1-\delta)} \ll k^{5/2}n^{1-\delta/2}$, for l sufficiently large; and therefore $a+b < n \ll k^{O(1)}$.

Let $u_1, \ldots u_d$ be the distinct integers in $[1, b/2]$ for which each $u_j^2 \equiv a \pmod{b}$, so that $d \asymp 2^{\omega(b)}$, by the Chinese remainder theorem, where $\omega(b)$ denotes the number of prime factors of b. The number of $x_i \equiv u_j \pmod{b/2}$ is $\leq 1 + ((a+kb)^{1/2} - a^{1/2})/(b/2) \leq 1 + 2(k/b)^{1/2}$; and thus $r \ll 2^{\omega(b)} + k^{1/2}2^{\omega(b)}/b^{1/2}$. This is $\ll k^{1/2}$ provided $\omega(b) \ll \log k$, which happens when $b \ll k^{O(\log \log k)}$ by the prime number theorem. The result follows. $\square$

Below are two flowcharts (both parts of Figure 1) which describe the relationships between the conjectures above.

Conjecture 13. *For any $\alpha < \frac{1}{2}$, there exists a constant C_α such that for any N we have*
$$\#\{(a,b), a^2 + b^2 = n, N \leq |b| < N + n^\alpha\} \leq C_\alpha.$$

A special case of interest is where $N = 0$:

(5.1) $$\#\{(a,b), a^2 + b^2 = n, |b| < n^\alpha\} \leq C_\alpha.$$

Heath-Brown pointed out that one has to be careful in making an analogous conjecture in higher dimension as the following example shows: Select integer r which has many representations as the sum of two squares; for example, if r is the product of k distinct primes that are $\equiv 1 \pmod 4$ then r has 2^k such representations. Now let N be an arbitrarily large integer and consider the set of representations of $n = N^2 + r$ as the sum of three squares. Evidently we have $\geq 2^k$ such representations in an interval whose size, which depends only on k, is independent of n. However, one can get around this kind of example in formulating the analogy to Conjecture 13 in 3-dimensions, since all of these solutions live in a fixed hyperplane. Thus we may be able to get a uniform bound on the number of such lattice points in a small box, no more than three of which belong to the same hyperplane.

It is simple to prove (5.1) for any $\alpha \leq \frac{1}{4}$ (and Conjecture 13 for $\alpha \leq \frac{1}{4}$ with $N \ll n^{1/2-\alpha}$), but we cannot prove (5.1) for any $\alpha > \frac{1}{4}$. Conjecture 13 and the special case (5.1) are equivalent to the following conjectures respectively:

Conjecture 14. *The number of lattice points $\{(x,y) \in \mathbb{Z}^2 : x^2 + y^2 = R^2\}$ in an arc of length $R^{1-\varepsilon}$ is bounded uniformly in R.*

Conjecture 15. *The number of lattice points $\{(x,y) \in \mathbb{Z}^2 : x^2 + y^2 = R^2\}$ in an arc of length $R^{1-\varepsilon}$ around the diagonal is bounded uniformly in R.*

Conjectures 13 and 14 are just a rephrasing of one another, and obviously imply (5.1) and Conjecture 15. In the other direction, if we have points $\alpha_j := x_j + iy_j$ on $x^2 + y^2 = R^2$ in an arc of length $R^{1-\varepsilon}$ then we have points $\alpha_j \overline{\alpha_0} = a_j + ib_j$ satisfying $a_j^2 + b_j^2 = R^2$ with $|b_j| \ll R^{1-\varepsilon}$ contradicting (5.1), and we have points $(1+i)\alpha_j\overline{\alpha_0}$ on $x^2 + y^2 = 2R^2$ in an arc of length $\ll R^{1-\varepsilon}$ around the diagonal, contradicting Conjecture 15.

The following result is proved in [15]:

Theorem 14. *There are no more than k lattice points $\{(x,y) \in \mathbb{Z}^2 : x^2 + y^2 = R^2\}$ in an arc of length $R^{1/2-1/(4[k/2]+2)}$.*

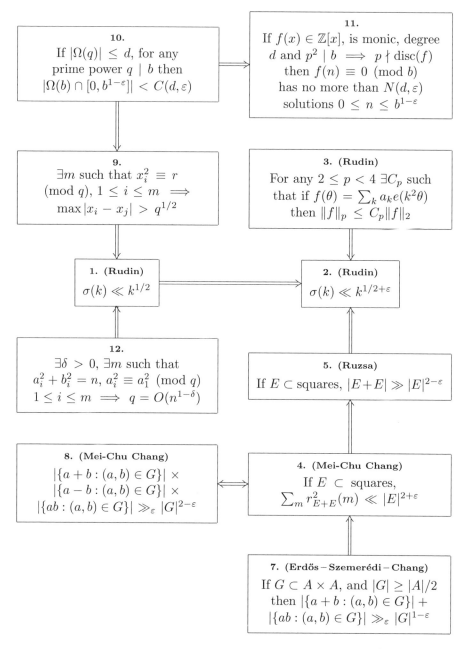

FIGURE 1. *Dependencies.* $\varepsilon_2 = \varepsilon_5/(4 - 2\varepsilon_5)$, $\varepsilon_2 = 2/p - \frac{1}{2}$, $\varepsilon_4 = 3\varepsilon_7/2$, $\varepsilon_4 = 3\varepsilon_8/4$, $\varepsilon_5 = \varepsilon_4$, $\varepsilon_8 = 4\varepsilon_4$.

PROOF. We may assume that $R^2 = \prod_{p \equiv 1 \pmod 4} p^e$, as the result for general R^2 is easily deduced from this case. Let $\mathbf{p}\overline{\mathbf{p}}$ be the Gaussian factorization of p. Then each lattice point ν_i, $1 \le i \le k+1$ can be identified with a divisor of R^2 of the form $\nu_i = \prod_{\mathbf{p}} \mathbf{p}^{e_i} \overline{\mathbf{p}}^{e-e_i}$. Therefore $\nu_i - \nu_j$ is divisible by $\mathbf{p}^{\min\{e_i, e_j\}} \overline{\mathbf{p}}^{\min\{e-e_i, e-e_j\}}$, so that $|\nu_i - \nu_j|^2$ is divisible by $p^{e-|e_i - e_j|}$. Hence, since $\sum_{1 \le i < j \le k+1} |e_i - e_j| \le$

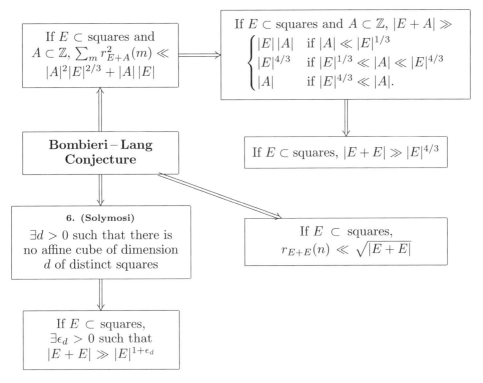

FIGURE 1. (suite)

$e[(k+1)/2](k - [(k+1)/2])$, we have

$$\prod_{1 \leq i < j \leq k+1} |\nu_i - \nu_j|^2 \geq \prod_p p^{\sum_{1 \leq i < j \leq k+1} e - |e_i - e_j|} \geq \left(\prod_p p^e\right)^{\binom{k+1}{2} - [(k+1)/2](k - [(k+1)/2])}$$

and the result follows. □

It seems difficult to decide whether the exponent $1/2 - 1/(4[k/2] + 2)$ is sharp for each k in Theorem 14. We know that it is sharp for $k = 1, 2, 3$ but we don't know what happens for larger k. More precisely:

(1) Obviously an arc of length $\sqrt{2}$ contains no more than one lattice point; whereas the lattice points $(n, n+1)$, $(n+1, n)$ lie on an arc of length $\sqrt{2} + o(1)$.

(2) It was shown in [12] that an arc of length $(16R)^{1/3}$ contains no more than two lattice points. On the other hand the lattice points $(4n^3 - 1, 2n^2 + 2n)$, $(4n^3, 2n^2 + 1)$, $(4n^3 + 1, 2n^2 - 2n)$ lie on an arc of length $(16R_n)^{1/3} + o(1)$.

(3) It was shown in [17] that an arc of length $(40 + (40/3)\sqrt{10})^{1/3} R^{1/3}$, with $R > \sqrt{65}$, contains no more than three lattice points, whereas there exists an infinite family of circles $x^2 + y^2 = R_n^2$ containing four lattice points on an arc of length $(40 + (40/3)\sqrt{10})^{1/3} R_n^{1/3} + o(1)$. Other than in the examples arising from this family, an arc of length $(40 + 20\sqrt{5})^{1/3} R^{1/3}$ contains no more than three lattice points, whereas the four lattice points $(x_0 - 2G_{n-2}, y_0 - 2G_{n+1})$, $(x_0 + G_{n-3}, y_0 + G_n)$, $(x_0 + G_{n-2}, y_0 + G_{n+1})$, $(x_0 - G_{n-1}, y_0 - G_{n+2})$, where $x_0 := \frac{1}{2}F_{3n+2}$, $y_0 = \frac{1}{2}F_{3n-1}$,

$G_m = (-1)^m F_m$ and F_m is the mth Fibonacci number, lie on the circle $x^2 + y^2 = \frac{5}{2} F_{2n-2} F_{2n} F_{2n+2} = R_n^2$ on an arc of length $(40 + 20\sqrt{5})^{1/3} R_n^{1/3} + o(1)$.

(4) Theorem 14 is the best result known for all $k \geq 4$. In particular it implies that an arc of length $R^{2/5}$ contains at most 4 lattice points, and we do not know whether the exponent $\frac{2}{5}$ can be improved: Are there infinitely many circles $x^2 + y^2 = R_n^2$ with four lattice points on an arc of length $\ll R_n^{2/5}$?

6. Incomplete trigonometric sums of squares

The L_4 norm of a trigonometric polynomial has an interesting number theory interpretation. For $f(\theta) = \sum_{n_k \in E} a_k e(n_k \theta)$ we can write

$$\|f\|_4^4 = \int_0^1 \left| \sum_k a_k e(n_k \theta) \right|^4 d\theta = \int_0^1 \left| \sum_m \left(\sum_{n_k + n_j = m} a_k a_j \right) e(m\theta) \right|^2 d\theta$$

$$= \sum_m \left| \sum_{n_k + n_j = m} a_k a_j \right|^2 \leq \sum_m r_{E+E}(m) \sum_{n_k + n_j = m} |a_k|^2 |a_j|^2$$

$$\leq \left(\sum_k |a_k|^2 \right)^2 \max_m r_{E+E}(m)$$

using the Cauchy–Schwarz inequality to obtain the first inequality, so that

$$(6.1) \qquad \|f\|_4 \leq \|f\|_2 \left(\sum_k \max_m r_{E+E}(m) \right)^{1/4}.$$

If E is the set of squares then $r_{E+E}(m) \leq \tau(m) \ll m^\varepsilon$; so, by (6.1), we have

$$\|f\|_4 \ll N^\varepsilon \|f\|_2$$

for any E-polynomial f where $E = \{1^2, \ldots, N^2\}$. Bourgain [6] conjectured the more refined:

Conjecture 16. *There exists a constant δ such that for any E-polynomial f where $E = \{1^2, \ldots, N^2\}$, we have*

$$\|f\|_4 \ll \|f\|_2 (\log N)^\delta.$$

Note that δ must be $\geq \frac{1}{4}$; since we saw, in the second section, that $\|f\|_4 \sim C(\log N)^{1/4} \|f\|_2$ for $f(\theta) = \sum_{1 \leq k \leq N} e(k^2 \theta)$.

The corresponding conjecture when $f(\theta) = \sum_{k \in E} e(k^2 \theta)$ and $E \subset \{1^2, \ldots, N^2\}$ is the following.

Conjecture 17. *There exists $C > 0$ such that if $E \subset \{1^2, \ldots, N^2\}$ then $\sum_m r_{E+E}^2(m) \ll |E|^2 (\log N)^C$.*

This conjecture implies that $|E + E| \gg N^2/(\log N)^C$, using the Cauchy–Schwarz inequality, a strengthening of Ruzsa's conjecture (Conjecture 5). Actually we can prove that the last two conjectures are equivalent:

Theorem 15. *Conjectures 16 and 17 are equivalent.*

PROOF. Conjecture 17 is a special case of Conjecture 16, so we must prove that Conjecture 16 follows from Conjecture 17. We may divide the coefficients of f by $\|f\|_2$ to ensure that $\|f\|_2 = (\sum_k |a_k|^2)^{1/2} = 1$, and therefore every $|a_k| \leq 1$.

Define $E_0 = \{k, |a_k| \leq N^{-1}\}$ and $E_j = \{k, 2^{j-1}/N < |a_k| \leq 2^j/N\}$ for all $j \geq 1$. Since $f = \sum_{j \geq 0} f_j$ (where each f_j is the appropriate E_j-polynomial), we have $\|f\|_4 \leq \sum_{j \geq 0} \|f_j\|_4$ by the triangle inequality. By Conjecture 17 we have

$$\|f_j\|_4^4 = \sum_n \left| \sum_{\substack{k^2+j^2=n \\ k,j \in E_j}} a_k a_j \right|^2 \leq (2^j/N)^4 \sum_n r_{E_j+E_j}^2(n) \ll (\log N)^C (2^j/N)^4 |E_j|^2.$$

Now $\sum_{k \in E_j} |a_k|^2 \asymp |E_j|(2^{2j}/N^2)$ for all $j \geq 1$, and $|E_0|/N^2, \sum_{k \in E_0} |a_k|^2 \ll 1/N$, which imply that $\sum_{j \geq 0} |E_j|(2^{2j}/N^2) \asymp 1$. Since $|E_j| = 0$ for $j > \lceil \log_2 N \rceil$, we deduce that

$$\frac{1}{(\log N)^{C/4}} \sum_{j \geq 0} \|f_j\|_4 \ll \sum_{j \geq 0} \frac{2^j |E_j|^{1/2}}{N} \ll \left(\sum_{j=0}^{\lceil \log_2 N \rceil} 1 \sum_{j \geq 1} \frac{2^{2j} |E_j|}{N^2} \right)^{1/2} \ll (\log N)^{1/2}.$$

Therefore Conjecture 16 follows with $\delta = C/4 + \frac{1}{2}$. □

Also we prove the following related result which slightly improves on Theorem 2 of [15].

Theorem 16. If $E = \{k^2 : N \leq k \leq N + \Delta\}$ with $\Delta \leq N$ and $f(\theta) = \sum_{r \in E} e(r\theta)$, so that $\|f\|_2^2 \sim \Delta$, then

$$\|f\|_4^4 \asymp \Delta^2 + \Delta^3 \cdot \frac{\log N}{N}.$$

In particular, $\|f\|_4 \ll \|f\|_2$ if and only if $\Delta \ll (\log N)/N$.

PROOF. Note that $\|f\|_2^2 = |E|$ and

$$\|f\|_4 = \sum_n r_{E+E}(n)^2 = 2|E|^2 - |E| + 2 \sum_n \left(\binom{r_{E+E}(n)}{2} - \left[\frac{r_{E+E}(n)}{2} \right] \right);$$

and that the sum counts twice the number of representations $k_1^2 + k_2^2 = k_3^2 + k_4^2$ with $N \leq k_1, k_2, k_3, k_4 \leq N+\Delta$ and $\{k_1, k_2\} \neq \{k_3, k_4\}$. Let $a+ib = \gcd(k_1+ik_2, k_3+ik_4)$ and so $k_1 + ik_2 = (a+ib)(x-iy)$ with $k_3 + ik_4 = (a+ib)(x+iy)u$ for some integers a, b, x, y where $u = 1, -1, i$ or $-i$ is a unit. Therefore $k_1 = ax+by$, $k_2 = bx-ay$, and the four values of u lead to the four possibilities $\{k_3, k_4\} = \{\pm(bx+ay), \pm(ax-by)\}$. All four cases work much the same so just consider $k_3 = bx + ay$, $k_4 = ax - by$. Then $N \leq ax = (k_1+k_4)/2$, $bx = (k_3+k_2)/2 \leq N+\Delta$ and $|by| = |k_1-k_4|/2$, $|ay| = |k_3-k_2|/2 \leq \Delta/2$. Multiplying through a, b, x, y by -1 if necessary, we may assume $a > 0$. Therefore $1 + \Delta/N \geq b/a \geq (1+\Delta/N)^{-1}$ so that

$$b = a + O(a\Delta/N), \quad N/a \leq x \leq N/a + \Delta/a, \quad |y| \leq \Delta/2a.$$

We may assume that $a < \Delta$ else $y = 0$ in which case $\{k_1, k_2\} \neq \{k_3, k_4\}$. Therefore, for a given a the number of possibilities for b, x and y is $\ll (a\Delta/N)(\Delta/a)^2 = \Delta^3/aN$. Summing up over all a, $1 \leq a \leq \Delta$, gives that $\|f\|_4 \ll \Delta^3 (\log \Delta)/N$.

On the other hand if integers a, b, x, y satisfy

$$a \in [7N/\Delta, \Delta/2], \quad b \in [a(1-\Delta/7N), a], \quad ax \in [N+\Delta/2, N+2\Delta/3], \quad ay \in [1, \Delta/3],$$

then $N \leq k_1 = ax + by < k_2 = bx - ay$, $k_3 = bx + ay < k_4 = ax - by \leq N + \Delta$ for $\Delta \leq N/3$, and so $\|f\|_4 \gg \Delta^2 + \Delta^3 (\log(\Delta^2/N))/N$. □

Conjecture 18. *The exists η such that for any E-polynomial f with $E = \{N^2, \ldots, (N + N/(\log N)^\eta)^2\}$, we have*

$$\|f\|_4 \ll \|f\|_2.$$

Conjecture 18 probably holds with $\eta = 1$. If $E = \cup_{i=1}^r E_i$ then we can write any E-polynomial f as $f = \sum_{i=1}^r f_i$, and by the triangle inequality we have $|f|^4 \leq \sum_{i=1}^r |f_i|^4$. Therefore Conjecture 18 implies Bourgain's Conjecture 16 with $\delta = \eta/2$. In [15] the following weaker conjecture was posed.

Conjecture 19. *For any $\alpha < 1$, for any trigonometric polynomial f with frequencies in the set $\{N^2, \ldots, (N + N^\alpha)^2\}$, we have*

$$\|f\|_4 \ll_\alpha \|f\|_2.$$

Conjecture 19 is trivial for $\alpha \leq \frac{1}{2}$, yet is completely open for any $\alpha > \frac{1}{2}$. From (6.1) we immediately deduce:

Theorem 17. *Conjecture 13 implies Conjecture 19.*

The next Conjectures 20 and 21 correspond to Conjectures 18 and 19, respectively, in the particular case $f(\theta) = \sum_{k^2 \in E} e(k^2\theta)$ and are also open.

Conjecture 20. *There exists $\delta > 0$ such that if $E \subset \{k^2, N \leq k \leq N + N/\log^\delta N\}$ then $\sum_m r_{E+E}^2(m) \ll |E|^2$.*

Conjecture 21. *If $E \subset \{k^2, N \leq k \leq N + N^{1-\varepsilon}\}$ then $\sum_m r_{E+E}^2(m) \ll |E|^2$.*

On the next page we give a flowchart (Figure 2) describing the relationships between the conjectures in this middle part of the paper.

7. Sidon sets of squares

A set of integers A is called a Sidon set if we have $\{a, b\} = \{c, d\}$ whenever $a + b = c + d$ with $a, b, c, d \in A$. More generally A is a $B_2[g]$-set if there are $\leq g$ solutions to $n = a + b$ with $a, b \in A$, for all integers n (so that a Sidon set is a $B_2[1]$-set). The set of squares is not a Sidon set, nor a $B_2[g]$-set for any g; however it is close enough to have inspired Rudin in his seminal article [27], as well as several sections of this paper.

One question is to find the largest Sidon set $A \subset \{1^2, \ldots, N^2\}$. Evidently $A = \{(N - [\sqrt{N}] + k)^2, k = 0, \ldots, [\sqrt{N}] - 1\}$ is a Sidon set of size $[\sqrt{N}]$. Alon and Erdős [1] used the probabilistic method to obtain a Sidon set $A \subset \{1^2, \ldots, N^2\}$ with $|A| \gg_\varepsilon N^{2/3-\varepsilon}$ (and Lefmann and Thiele [24] improved this to $|A| \gg_\varepsilon N^{2/3}$).

We "measure" the size of infinite Sidon sets $\{a_k\}$ by giving an upper bound for a_k. Erdős and Rényi [20] proved that there exists an infinite $B_2[g]$-set $\{a_k\}$ with $a_k \ll k^{2+2/g+o(1)}$, for any g. In [13], the first author showed that one may take all the a_k to be squares; and in [14] he showed that there exists an infinite $B_2[g]$-set $\{a_k\}$ with $a_k \ll k^{2+1/g}(\log k)^{1/g+o(1)}$. Here we adapt this latter approach to the set of squares.

Theorem 18. *For any positive integer g there exists an infinite $B_2[g]$ sequence of squares $\{a_k\}$ such that*

$$a_k \ll k^{2+1/g}(\log k)^{O_g(1)}.$$

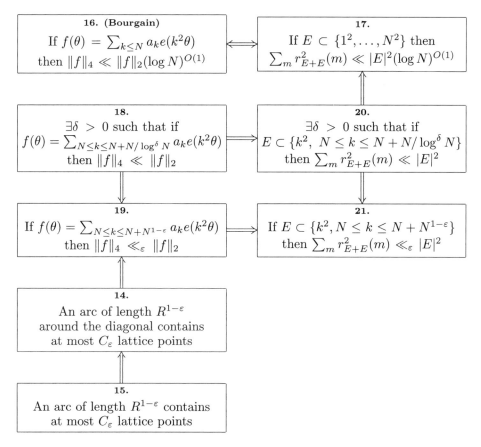

FIGURE 2.

PROOF. Let $X_1, X_2, \ldots$ be an infinite sequence of independent random variables, each of which take values 0 or 1, where

$$p_b := \mathbf{P}(X_b = 1) = \frac{1}{b^{1/(2g+1)}\big(\log(2+b)\big)^{\beta_g}}.$$

where $\beta_g > 1$ is a number we will choose later. For each selection of random variables we construct a set of integers $\mathcal{B} = \{b \geq 1 : X_b = 1\} = \{b_1 < b_2 < \cdots\}$. By the central limit theorem we have $\mathcal{B}(x) \sim cx^{1-1/(2g+1)}/(\log x)^{\beta_g}$ with probability 1 or, equivalently, $b_k \sim c'(k(\log k)^{\beta_g})^{1+1/(2g)}$.

We will remove from our sequence of integers $\mathcal{B}$ any integer b_0 such that there exists n for which there are $g + 1$ distinct representations of n as the sum of two squares of elements of $\mathcal{B}$, in which b_0 is the very largest element of $\mathcal{B}$ involved. Let $\mathcal{D} \subset \mathcal{B}$ denote the set of such integers b_0. Then the set $\{c^2 : c \in \mathcal{B} \setminus \mathcal{D}\}$ is the desired $B_2[g]$ sequence of squares.

Now, if $b_0 \in \mathcal{D}$ then, by definition, there exits $b'_0, b_1, b'_1, \ldots b_g, b'_g \in \mathcal{B}$ with $b'_g \leq b_g < \cdots < b_1 < b_0$, for which

$$n = b_0^2 + b'^2_0 = b_1^2 + b'^2_1 = \cdots = b_g^2 + b'^2_g.$$

Define $R(n) = \{(b,b'),\, b \geq b',\, b^2+b'^2 = n\}$, and $r(n) = |R(n)|$. Then the probability that $b_0 \in \mathcal{D}$ because of this particular value of n is

$$\mathbf{E}\left(X_{b_0}X_{b'_0} \sum_{\substack{(b_1,b'_1),\ldots,(b_g,b'_g) \in R(b_0^2+b'^2_0) \\ b_g < \cdots < b_1 < b_0}} X_{b_1}X_{b'_1}\cdots X_{b_g}X_{b'_g}\right).$$

The b_j, b'_j are all distinct except in the special case that $n = 2b_g^2$ with $b_g = b'_g$. Thus, other than in this special case, $\mathbf{E}(X_{b_0}X_{b'_0}X_{b_1}X_{b'_1}\cdots X_{b_g}X_{b'_g}) = \prod_{j=0}^{g} p_{b_j}p_{b'_j} \leq (p_{b_0}p_{b'_0})^{g+1}$, since $p_{b_j}p_{b'_j} \leq p_{b_0}p_{b'_0}$ for all j. This gives a contribution above of $\leq (p_{b_0}p_{b'_0})^{g+1}\binom{r(n)-1}{g}$. The terms with $n = 2b_g^2$ similarly contribute $\leq (p_{b_0}p_{b'_0})^{g+1/2}\binom{r(n)-2}{g-1} \leq p_{b_0}^{2g+1}r(n)^{g-1} \ll r(n)^{g-1}/b'_0$. Therefore

$$\mathbf{E}\big(\mathcal{D}(x) - \mathcal{D}(x/2)\big) \ll \sum_{\substack{b'_0 \leq b_0 \\ x/2 < b_0 \leq x}} (p_{b_0}p_{b'_0})^{g+1}r(b_0^2+b'^2_0)^g + \sum_{\substack{b'_0 < b < b_0 \leq x \\ b'^2_0+b_0^2=2b^2}} \frac{1}{b'_0}r(2b^2)^{q-1}.$$

For the second sum note that $r(m) \ll m^{o(1)}$ and that for any n (and in particular for $n = b_0'^2$) we have $\#\{(y,z), n = 2z^2 - y^2, y, z \leq x\} \ll (nx)^{o(1)}$, and so its total contribution is $\ll x^{o(1)}\sum_{b'_0 \leq x} 1/b_0 = x^{o(1)}$.

For the first term we apply Hölder's inequality with $p = 2 - 1/(g+1)$ and $q = 2 + 1/g$ to obtain

$$\leq \left(\sum_{\substack{b'_0 \leq b_0 \\ x/2 < b_0 \leq x}} (p_{b_0}p_{b'_0})^{2g+1}\right)^{(g+1)/(2g+1)} \left(\sum_{b'_0 \leq b_0 \leq x} r^{2g+1}(b_0^2+b'^2_0)\right)^{g/(2g+1)}.$$

As $\beta_g > 1$, we have

$$\sum_{\substack{b'_0 \leq b_0 \\ x/2 < b_0 \leq x}} (p_{b_0}p_{b'_0})^{2g+1} \ll \sum_{x/2 < b_0 \leq x} \frac{1}{b_0(\log b_0)^{\beta_g(2g+1)}} \sum_{b'_0 \leq b_0} \frac{1}{b'_0(\log b'_0)^{\beta_g(2g+1)}}$$

$$\ll \frac{1}{(\log x)^{\beta_g(2g+1)}},$$

and

$$\sum_{b'_0 \leq b_0 \leq x} r^{2g+1}(b_0^2+b'^2_0) \leq \sum_{n \leq 2x^2} r^{2g+2}(n) \ll x^2(\log x)^{2^{2g+1}-1},$$

so that

$$\mathbf{E}\big(\mathcal{D}(x) - \mathcal{D}(x/2)\big) \ll x^{2g/(2g+1)}(\log x)^{e_g} \quad \text{where } e_g := g\left(\frac{2^{2g+1}-1}{2g+1}\right) - \beta_g(g+1).$$

Markov inequality's tells us that $\mathbf{P}\big(\mathcal{D}(2^j) \geq j^2 \mathbf{E}(\mathcal{D}(2^j) - \mathcal{D}(2^{j-1}))\big) \leq 1/j^2$ so that $\sum_{j \geq 1} \mathbf{P}\big(\mathcal{D}(2^j) - \mathcal{D}(2^{j-1}) \gg j^{2+e_g}(2^j)^{2g/(2g+1)})\big) < \infty$. The Borel–Cantelli lemma then implies that

$$\mathcal{D}(2^j) - \mathcal{D}(2^{j-1}) \ll j^{2+e_g}(2^j)^{2g/(2g+1)} = o(\mathcal{B}(2^j) - \mathcal{B}(2^{j-1}))$$

with probability 1, provided $\beta_g > (2^{2g+1}-1)/(2g+1) + 2/g$. Thus there exists a $B_2[g]$-sequence of the form $\mathcal{A} := \{b^2 : b \in \mathcal{B} \setminus \mathcal{D}\} = \{a_k\}$, where $a_k \ll k^{2+1/g}(\log k)^{\beta_g(2+1/g)}$. □

Corollary 1. *There exists an infinite Sidon sequence of squares $\{a_k\}$ with $a_k \ll k^3 (\log k)^{12}$.*

PROOF. Take $g = 1$ and $\beta = 4$ in the proof above. A more careful analysis would allow to put $10 + o(1)$ instead of 12. □

8. Generalized arithmetic progressions of squares

A generalized arithmetic progression (GAP) is a set of numbers of the form $\{x_0 + \sum_{i=1}^{d} j_i x_i : 0 \le j_i \le J_i - 1\}$ for some integers $J_1, J_2, \ldots, J_d$ and each $x_i \ne 0$. We have seen that many questions in this article are closely related to GAPs of squares of integers and, at the beginning we noted that Fermat proved there are no arithmetic progressions of squares of length 4, and so we may assume each $J_d \le 3$. We also saw Solymosi's Conjecture 6 which claims that there are no GAPs of squares with each $J_i = 2$ and d sufficiently large. This leaves us with only a few cases left to examine:

We begin by examining arithmetic progressions of length 3 of squares: If x^2, y^2, z^2 are in arithmetic progression then they satisfy the Diophantine equation $x^2 + z^2 = 2y^2$. All integer solutions to this equation can be parameterized as

$$x = r(t^2 - 2t - 1), \; y = r(t^2 + 1), \; z = r(t^2 + 2t - 1), \quad \text{where } t \in \mathbb{Q} \text{ and } r \in \mathbb{Z}.$$

Therefore the common difference Δ of this arithmetic progression is given by $\Delta = z^2 - y^2 = 4r^2(t^3 - t)$. Integers which are a square multiple of numbers of the form $t^3 - t$, $t \in \mathbb{Q}$ are known as *congruent numbers* and have a rich, beautiful history in arithmetic geometry (see Koblitz's delightful book [22]). They appear naturally when we study right-angled triangles whose sides are rationals because these triangles can be parameterized as $s(t^2 - 1)$, $2st$, $s(t^2 + 1)$ with $s, t \in \mathbb{Q}$, and so have area $s^2(t^3 - t)$ (there is a direct correspondence here since we may take the right-angled triangle to have sides $x + z$, $z - x$, $2y$ which has area $z^2 - x^2 = 2\Delta$). It is a highly non-trivial problem to classify the congruent numbers; indeed this is one of the basic questions of modern arithmetic geometry, see [22].

So can we have a 2-by-3 GAP? This would require having two different ways to obtain the same congruent number. The theory of elliptic curves tells us exactly how to do this: We begin with the *elliptic curve*

(8.1) $$E_\Delta : \Delta Y^2 = X^3 - X$$

and the 3-term arithmetic progressions of rational squares are in 1-to-1 correspondence with the rational points $(t, 1/2r)$ on (8.1). Now the rational points on an elliptic curve form an abelian group and so if $P = (t, 1/2r)$ is a rational point on E_Δ then there are rational points $2P, 3P, \ldots$. This is all explained in detail in [22]. All we need is to note that $2P = (T, 1/2R)$ where

$$T = \frac{(t^2 + 1)^2}{4(t^3 - t)} = \frac{y^2}{\Delta} \quad \text{and} \quad R = \frac{8r(t^3 - t)^2}{(t^2 + 1)(t^2 + 2t - 1)(t^2 - 2t - 1)} = \frac{\Delta^2}{2xyz}.$$

So we have infinitely many 2-by-3 GAPs of squares where the common difference of the 3-term arithmetic progressions is Δ, for any congruent number Δ.

How about 3-by-3 GAPs of squares? Let us suppose that the common difference in one direction is Δ; having a 3-by-3 GAP is then equivalent to having y_1^2, y_2^2, y_3^2 in arithmetic progression. But note that $y_i^2 = \Delta T_i = \Delta x(2P_i)$ (where $x(Q)$ denotes the x-coordinate of Q on a given elliptic curve). Therefore 3-by-3 GAPs of squares are in 1-to-1 correspondence with the sets of congruent numbers and triples of rational points, $(\Delta; P_1, P_2, P_3) : P_1, P_2, P_3 \in E_\Delta(\mathbb{Q})$ for which the x-coordinates $x(2P_1), x(2P_2), x(2P_3)$ are in arithmetic progression (other than the triples $-1, 0, 1$ which do not correspond to squares of interest).

In [8] it is proved that if there is such an arithmetic progression of rational points then the *rank* of E_Δ must be at least 2; that is there are at least two points of infinite order in the group of points that are independent.

Bremner became interested in the same issue from a seemingly quite different motivation:

A 3-by-3 *magic square* is a 3-by-3 array of numbers where each row, column and diagonal has the same sum. Solving the linear equations that arise it may be parameterized as

$$\begin{pmatrix} u+v & u-v-\Delta & u+\Delta \\ u-v+\Delta & u & u+v-\Delta \\ u-\Delta & u+v+\Delta & u-v \end{pmatrix}.$$

The entries of the magic square form the 3-by-3 GAP $\{(u - v - \Delta) + j_1 v + j_2 \Delta : 0 \le j_1, j_2 \le 2\}$. Hence the question of finding a non-trival 3-by-3 magic square with entries from a given set E is *equivalent* to the question of finding a non-trival 3-by-3 GAP with entries from a given set E; in particular when E is the set of squares. (This connection is beautifully explained in [26].)

We believe that the existence of non-trivial 3-by-3 GAPs of squares, and equivalently of non-trivial 3-by-3 magic squares of squares, remain open.

9. The abc-conjecture

In [4] it was shown that the large sieve implies that if there are $\gg \sqrt{k} \log k$ squares amongst $a+b, a+2b, \ldots, a+kb$ then $b \ge e^{\sqrt{k}}$. We wish to obtain an upper bound on b also. We shall do so assuming one of the most important conjectures of arithmetic geometry:

Conjecture 22 (The abc-conjecture). *If $a + b = c$ where a, b and c are coprime positive integers then $r(abc) \gg c^{1-o(1)}$ where $r(abc)$ is the product of the distinct primes dividing abc.*

Unconditional results on the abc-conjecture are still far from this objective, giving only that $r(abc) \gg (\log c)^{3-o(1)}$, for some $A > 0$ (see [29]). Nonetheless, by considering the strongest feasible version of certain results on linear forms of logarithms, Baker [2] made a conjecture which implies the stronger

(9.1) $$r(abc) \gg c/\exp((\log c)^\tau),$$

with $\tau = \frac{1}{2} + o(1)$.

Lemma 1. *Suppose that $A + t_j B$ is a square for $j = 1, 2, 3, 4, 5$, where A, B and the t_j are integers and $(A, B) = 1$. Let $T = \max_j |t_j|$. Then (9.1) implies that $A + B \ll \exp(O(T^{9\tau/(1-\tau)}))$. Moreover if $B \gg A^{5/6-\varepsilon}$ then we may improve this to $B \ll \exp(O(T^{6\tau/(1-\tau)}))$.*

PROOF. There is always a partial fraction decomposition

$$\frac{1}{\prod_{j=1}^{5}(x+t_j)} = \sum_{j=1}^{5} \frac{e_j}{x+t_j} \quad \text{where } e_j = \frac{1}{\prod_{i=1, i \ne j}^{5}(t_i - t_j)},$$

so that $\sum_j e_j t_j^l = 0$ for $0 \leq l \leq 3$. Let L be the smallest integer such that each $E_j := L e_j$ is an integer. Define the polynomials

$$h(x) := \prod_{\substack{1 \leq j \leq 5 \\ E_j > 0}} (x + t_j)^{E_j} \text{ and } g(x) := \prod_{\substack{1 \leq j \leq 5 \\ E_j < 0}} (x + t_j)^{-E_j}, \text{ with } f(x) := h(x) - g(x).$$

If $h(x)$ has degree D then the coefficient of x^{D-i} in $f(x)$ is a polynomial in the $\sum_j e_j t_j^l$ with $0 \leq l \leq i$, so we deduce that $f(x)$ has degree $D - 4$. Now let $a = B^D h(A/B)$, $b = B^D g(A/B)$, $c = B^4 \cdot B^{D-4} f(A/B)$ and then $a' = a/(a,b)$, $b' = b/(a,b)$, $c' = c/(a,b)$. Thus $r(a'b'c') \leq r(\prod_{j=1}^5 (A + t_j B))|B||c'/B^4| \leq \prod_{j=1}^5 (A + t_j B)^{1/2} |c'|/B^3$. Now $\prod_{j=1}^5 (A + t_j B) \ll B^{6-2\varepsilon}$ provided $T = B^{o(1)}$ and $A \ll B^{6/5-\varepsilon}$, in which case $r(a'b'c') \ll |c'|/B^\varepsilon$. Then, by (9.1), we have $(\log c)^\tau \gg \log B$. Now $c = a + b \ll (A + TB)^D$ so that $\log c \ll D \log B$; we deduce that $B \ll \exp(O(D^{\tau/(1-\tau)}))$. Finally note that $D \ll \max_l |E_l| \leq \prod_{1 \leq i < j \leq 5, i,j \neq l} |t_i - t_j| \ll T^6$, and the second result follows.

In case that $A \gg B^{6/5-\varepsilon}$ we may replace t_j by $1/t_j$ in our construction of polynomials given above. In that case we get new exponents $e_j^* = e_j t_j^3 \prod_{i=1}^5 t_i$ and therefore $|E_j^*| \leq |t_j|^3 E_j$. We now have integers $a^* = \kappa A^d h^*(B/A)$, $b^* = \kappa A^d g^*(B/A)$, $c^* = \kappa A^4 \cdot A^{d-4} f^*(B/A)$ where $\kappa := \prod_j t_j^{|E_j^*|}$ and d is the degree of h^*. Thus we have that either $A \ll T^{O(1)}$ or $A \ll \exp(O(d^{\tau/(1-\tau)}))$ and $d \ll T^9$. □

We can apply this directly: If there are $\gg \sqrt{k}$ squares amongst $a + b, a + 2b, \ldots, a + kb$ then there must be $i_1 < \cdots < i_5$ with $i_5 < i_1 + O(\sqrt{k})$ such that each $a + i_j b$ is a square. Thus by Lemma 1 with $A = a + i_i b$, $B = b$, $t_j = i_j - i_1$, assuming (9.1) with Baker's $\tau = \frac{1}{2} + o(1)$, we obtain $a + b \ll \exp(k^{9/2 + o(1)})$. Therefore we may, in future, restrict our attention to the case $k^{1/2} \ll \log(a + b) \ll k^{9/2 + o(1)}$.

Acknowledgements. Many thanks to Bjorn Poonen for his permission to discuss his unpublished work at the end of Section 4, and for the ideas shared by Jozsef Solymosi which helped us improve and generalize several results at the start of Section 3. Thanks are also due to Norbert Hegyvári, Sergei Konyagin and Umberto Zannier for their comments and especially to Antonio Cordoba for his careful reading of this paper.

References

1. N. Alon and P. Erdős, *An application of graph theory to additive number theory*, European J. Combin. **6** (1985), no. 3, 201–203.
2. A. Baker, *Logarithmic forms and the abc-conjecture*, Number Theory (Eger, 1996), de Gruyter, Berlin, 1998, pp. 37–44.
3. D. Berend, *On the roots of certain sequences of congruences*, Acta Arith. **67** (1994), no. 1, 97–104.
4. E. Bombieri, A. Granville, and J. Pintz, *Squares in arithmetic progressions*, Duke Math. J. **66** (1992), no. 3, 165–204.
5. E. Bombieri and U. Zannier, *A note on squares in arithmetic progressions. II*, Atti Accad. Naz. Lincei Cl. Sci. Fis. Mat. Natur. Rend. Lincei (9) Mat. Appl. **13** (2002), no. 2, 69–75.
6. J. Bourgain, *On $\Lambda(p)$ subsets of squares*, Israel J. Math. **67** (1989), no. 3, 291–311.
7. A. Bremner, *On squares of squares*, Acta Arith. **88** (1999), no. 3, 289–297.
8. A. Bremner, J. H. Silverman, and N. Tzanakis, *Integral points in arithmetic progression on $y^2 = x(x^2 - n^2)$*, J. Number Theory **80** (2000), no. 2, 187–208.

9. T. C. Brown, P. Erdős, and A. R. Freedman, *Quasi-progressions and descending waves*, J. Combin. Theory Ser. A **53** (1990), no. 1, 81–95.
10. L. Caporaso, J. Harris, and B. Mazur, *Uniformity of rational points*, J. Amer. Math. Soc. **10** (1997), no. 1, 1–35.
11. M.-C. Chang, *On problems of Erdős and Rudin*, J. Funct. Anal. **207** (2004), no. 2, 444–460.
12. J. Cilleruelo, *Arcs containing no three lattice points*, Acta Arith. **59** (1991), no. 1, 87–90.
13. _____, $B_2[g]$ *sequences whose terms are squares*, Acta Math. Acad. Sci. Hungar. **67** (1995), no. 1-2, 79–83.
14. _____, *A variant of the probabilistic method applied to sequences of integers*, in preparation.
15. J. Cilleruelo and A. Córdoba, *Trigonometric polynomials and lattice points*, Proc. Amer. Math. Soc. **115** (1992), no. 4, 899–905.
16. _____, $B_2[\infty]$-*sequences of square numbers*, Acta Arith. **61** (1992), no. 3, 265–270.
17. J. Cilleruelo and A. Granville, *Close lattice points*, in preparation.
18. A. Córdoba, *Translation invariant operators*, Fourier Analysis (Madrid, 1980), Asoc. Mat. Española, vol. 1, Asoc. Mat. Española, Madrid, 1980, pp. 117–176.
19. E. S. Croot III, *Research problems in arithmetic combinatorics*, in preparation.
20. P. Erdős and A. Rényi, *Additive properties of random sequences of positive integers*, Acta Arith. **6** (1960), 83–110.
21. N. Hegyvári and A. Sárközy, *On Hilbert cubes in certain sets*, Ramanujan J. **3** (1999), 303–314.
22. N. Koblitz, *Introduction to elliptic curves and modular forms*, 2nd ed., Grad. Texts in Math., vol. 97, Spinger, New York, 1993.
23. S. V. Konyagin and T. Steger, *Polynomial congruences*, Mat. Zametki **55** (1994), no. 6, 73–79 (Russian); English transl., Math. Notes **55** (1994), no. 5-6, 596–600.
24. H. Lefmann and T. Thiele, *Point sets with distinct distances*, Combinatorica **15** (1995), no. 3, 379–408.
25. B. Poonen, *The classification of rational preperiodic points of quadratic polynomials over Q: a refined conjecture*, Math. Z. **228** (1998), no. 1, 11–29.
26. J. P. Robertson, *Magic squares of squares*, Math. Mag. **69** (1996), no. 4, 289–293.
27. W. Rudin, *Trigonometric series with gaps*, J. Math. Mech. **9** (1960), no. 2, 203–227.
28. J. Solymosi, *Elementary additive combinatorics*, in this volume.
29. C. L. Stewart and K. Yu, *On the abc-conjecture. II*, Duke Math. J. **108** (2001), no. 1, 169–181.
30. E. Szemerédi, *The number of squares in an arithmetic progression*, Studia Sci. Math. Hungar. **9** (1974), no. 3-4, 417.
31. X. Xarles, *Squares in arithmetic progressions over number fields*, preprint.

DEPARTAMENTO DE MATEMÁTICAS, UNIVERSIDAD AUTÓNOMA DE MADRID, 28049 MADRID, SPAIN
  E-mail address: `franciscojavier.cilleruelo@uam.es`

DÉPARTEMENT DE MATHÉMATIQUES ET DE STATISTIQUE, UNIVERSITÉ DE MONTRÉAL, C.P. 6128, SUCC. CENTRE-VILLE, MONTRÉAL, QC H3C 3J7, CANADA
  E-mail address: `andrew@dms.umontreal.ca`

Problems in Additive Number Theory. I

Melvyn B. Nathanson

ABSTRACT. Talk at the *Atelier en combinatoire additive* (Workshop on Additive Combinatorics) at the Centre de recherches mathématiques at the Université de Montréal on April 8, 2006.

Definition. A *problem* is a problem I cannot solve, not necessarily an unsolved problem.

1. Sums and differences

We begin with a fundamental mystery about sums and differences of integers. For any set A of integers, we define the *sumset*
$$2A = A + A = \{a + a' : a, a' \in A\}$$
and the *difference set*
$$A - A = \{a - a' : a, a' \in A\}.$$
In this section we consider finite sets of integers, and the relative sizes of their sumsets and difference sets. If A is a finite set of integers and $x, y \in \mathbb{Z}$, then the *translation* of A by x is the set $x + A = \{x + a : a \in A\}$ and the *dilation* of A by y is $y * A = \{ya : a \in A\}$. We have
$$(x + A) + (x + A) = 2x + 2A$$
and
$$(x + A) - (x + A) = A - A.$$
Similarly,
$$y * A + y * A = y * (A + A)$$
and
$$y * A - y * A = y * (A - A).$$
It follows that
$$|(x + y * A) + (x + y * A)| = |2A|$$

2000 *Mathematics Subject Classification.* 11B05, 11B13, 11B34.

Key words and phrases. Additive number theory, sumscts, difference sets, representation functions.

Supported in part by grants from the NSA Mathematical Sciences Program and the PSC-CUNY Research Award Program.

This is the final form of the paper.

and
$$|(x + y * A) - (x + y * A)| = |A - A|$$
so the cardinalities of the sum and difference sets of a finite set of integers are invariant under affine transformations of the set.

The set A is *symmetric* with respect to the integer z if $A = z - A$ or, equivalently, $a \in A$ if and only if $z - a \in A$. For example, the set $\{4, 6, 7, 9\}$ is symmetric with respect to $z = 13$. If A is symmetric, then
$$A + A = A + (z - A) = z + (A - A)$$
and so $|A + A| = |A - A|$.

If $A = \{a, b, c\}$ with $a < b < c$, then
$$2a < a + b < 2b < b + c < 2c$$
and
$$a + b < a + c < b + c.$$
Similarly,
$$a - c < a - b < 0 < b - a < c - a$$
and
$$a - c < b - c < 0 < c - b < c - a.$$
If $a + c \neq 2b$, then
$$|A + A| = 6 < 7 = |A - A|.$$
Considering the set
$$A = \{0, 2, 3, 4, 7\}$$
we obtain
$$A + A = [0, 14] \setminus \{1, 12, 13\}, \quad A - A = [-7, 7] \setminus \{-6, 6\}$$
and so
$$|A + A| = 12 < 13 = |A + A|.$$
This is the typical situation. Since
$$2 + 7 = 7 + 2$$
but
$$2 - 7 \neq 7 - 2$$
it is natural to expect that in any finite set of integers there are always at least as many differences as sums. There had been a conjecture, often ascribed incorrectly to John Conway[1] that asserted that $|A + A| \leq |A - A|$ for every finite set A of integers. This conjecture is false, and a counterexample is the set

(1) $$A = \{0, 2, 3, 4, 7, 11, 12, 14\},$$

for which
$$A + A = [0, 28] \setminus \{1, 20, 27\}, \quad A - A = [-14, 14] \setminus \{\pm 6, \pm 13\}$$
and
$$|A + A| = 26 > 25 = |A - A|.$$

---

[1]The confusion may be due to the fact that the first published paper on the conjecture, by John Marica [5], is entitled "On a conjecture of Conway." I asked Conway about this at the Logic Conference in Memory of Stanley Tennenbaum at the CUNY Graduate Center on April 7, 2006. He said that he had actually found a counterexample to the conjecture, and that this is recorded in unpublished notes of Croft [2].

On the other hand, the conjecture really *should* be true, and suggests the first somewhat philosophical problem.

Problem 1. Why do there exist finite sets A of integers such that $|A + A| > |A - A|$?

Given the existence of such aberrant sets, we can ask for the smallest one. The set A in (1) satisfies $|A| = 8$.

Problem 2. What is
$$\min\{|A| : A \subseteq \mathbb{Z} \text{ and } |A + A| > |A - A|\}?$$

Sets A with the property that $|A + A| > |A - A|$ should have structure. By structure I do not mean that A contains arithmetic progressions or generalized arithmetic progressions or even subsets of some polynomial type. There is a significant part of combinatorial and additive number theory, sometimes called *additive combinatorics*, one of the main goals of which is to look for arithmetic progressions inside sets of integers, and to prove that certain sets can be approximated by generalized arithmetic progressions. The results are beautiful, deep, and difficult, but it is hard to ignore the fact that arithmetic progressions are fundamentally boring, and dense or even relatively dense sets of integers must contain vastly more interesting structures that we have not yet imagined.

The astronomers are trying to understand the large-scale structure of the universe. If they found an arithmetic progression or a generalized arithmetic progression of galaxies, they would be ecstatic, but it would also be obvious to them that this fascinating and unexpected curiosity is only a small part of the universe, and they would keep looking for other structures. Since the complexity of sets of integers is comparable to that of the universe, we should also keep looking.

Problem 3. What is the structure of finite sets satisfying $|A + A| > |A - A|$?

If A is a finite set of integers and m is a sufficiently large positive integer (for example, $m > 4\max(\{|a| : a \in A\})$, then the set

$$A_t = \left\{\sum_{i=0}^{t-1} a_i m^i : a_i \in A \text{ for } i = 0, 1, \ldots, t-1\right\}$$

has the property that
$$|A_t + A_t| = |A + A|^t$$
and
$$|A_t - A_t| = |A - A|^t.$$

It follows that if $|A + A| > |A - A|$, then $|A_t + A_t| > |A_t - A_t|$ and, moreover,
$$\lim_{t \to \infty} \frac{|A_t + A_t|}{|A_t - A_t|} = \lim_{t \to \infty} \left(\frac{|A + A|}{|A - A|}\right)^t = \infty.$$

The sequence of sets $\{A_t\}_{t=1}^{\infty}$ is a standard parametrized family of sets with more sums than differences.

Problem 4. Are there other parametrized families of sets satisfying $|A + A| > |A - A|$?

Even though there exist sets A that have more sums than differences, such sets should be rare, and it must be true with the right way of counting that the vast majority of sets satisfy $|A - A| > |A + A|$.

Problem 5. Let $f(n)$ denote the number sets $A \subseteq [0, n-1]$ such that $|A-A| < |A+A|$, and let $f(n,k)$ denote the number of such sets $A \subseteq [0, n-1]$ with $|A| = k$ and $|A - A| < |A + A|$. Compute

$$\lim_{n \to \infty} \frac{f(n)}{2^n}$$

and

$$\lim_{n \to \infty} \frac{f(n,k)}{\binom{n}{k}}.$$

The functions $f(n)$ and $f(n,k)$ are not necessarily the best functions to count finite sets of nonnegative integers with respect to sums and differences.

Problem 6. Prove that $|A - A| > |A + A|$ for almost all sets A with respect to other appropriate counting functions.

2. Binary linear forms

The problem of sums and differences is a special case of a more general problem about binary linear forms

$$f(x, y) = ux + vy$$

where u and v are nonzero integers. For every finite set A of integers, let

$$f(A) = \{f(a, a') : a, a' \in A\}.$$

For example, the sets associated to the binary linear forms

$$s(x, y) = x + y$$

and

$$d(x, y) = x - y$$

are the sumset $s(A) = A + A$ and the difference set $d(A) = A - A$. We are interested in the cardinality of the sets $f(A)$.

Every binary linear form is associated to a unique normalized binary linear form $f(x, y) = ux + vy$ such that

$$u \geq |v| \geq 1 \quad \text{and} \quad (u, v) = 1.$$

We construct this normalized form as follows. Let $f_0(x, y) = u_0 x + v_0 y$. If $d = (u_0, v_0) > 1$, let $f_1(x, y) = (u_0/d)x + (v_0/d)y = u_1 x + v_1 y$. If $|u_1| < |v_1|$, let $f_2(x, y) = v_1 x + u_1 y = u_2 x + v_2 y$. If $u_2 < 0$, let $f_3(x, y) = -u_2 x - v_2 y = u_3 x + v_3 y$. The binary linear form $f_3(x, y)$ is normalized, and

$$|f_0(A)| = |f_1(A)| = |f_2(A)| = |f_3(A)|$$

for every finite set A.

The natural question is: If $f(x, y)$ and $g(x, y)$ are two distinct normalized binary linear forms, do there exist finite sets A and B of integers such that $|f(A)| > |g(A)|$ and $|f(B)| < |g(B)|$, and, if so, is there an algorithm to construct A and B?

Brooke Orosz gave constructive solutions to this problem in some important cases. For example, she proved the following: Let $u > v \geq 1$ and $(u, v) = 1$, and consider the normalized binary linear forms

$$f(x, y) = ux + vy$$

and

$$g(x, y) = ux - vy.$$

For $u \geq 3$, the sets

$$A = \{0, u^2 - v^2, u^2, u^2 + uv\}$$

and

$$B = \{0, u^2 - uv, u^2 - v^2, u^2\}$$

satisfy the inequality

$$|f(A)| = 14 > 13 = |g(A)|$$

and

$$|f(B)| = 13 < 14 = |g(B)|.$$

For $u = 2$, we have $f(x, y) = 2x + y$ and $g(x, y) = 2x - y$. The sets

$$A = \{0, 3, 4, 6\}$$

and

$$B = \{0, 4, 6, 7\}$$

satisfy the inequality

$$|f(A)| = 13 > 12 = |g(A)|$$

and

$$|f(B)| = 13 < 14 = |g(B)|.$$

The problem of pairs of binary linear forms has been completely solved by Nathanson, O'Bryant, Orosz, Ruzsa, and Silva [14].

Theorem. *Let $f(x, y)$ and $g(x, y)$ be distinct normalized binary linear forms. There exist finite sets A, B, C with $|C| \geq 2$ such that*

$$|f(A)| > |g(A)|, \quad |f(B)| < |g(B)|, \quad |f(C)| = |g(C)|.$$

Problem 7. Let $f(x, y)$ and $g(x, y)$ be distinct normalized binary linear forms. Determine if $|f(A)| > |g(A)|$ for most or for almost all finite sets of integers.

These results should be extended to linear forms in three or more variables.

Problem 8. Let $f(x_1, \ldots, x_n) = u_1 x_1 + \cdots + u_n x_n$ and $g(x_1, \ldots, x_n) = v_1 x_1 + \cdots + v_n x_n$ be linear forms with integer coefficients. Does there exist a finite set A of integers such that $|f(A)| > |g(A)|$?

3. Polynomials over finite sets of integers and congruence classes

An integer-valued function is a function $f(x_1, x_2, \ldots, x_n)$ such that if $x_1, x_2, \ldots, x_n \in \mathbb{Z}$, then $f(x_1, x_2, \ldots, x_n) \in \mathbb{Z}$. The binomial polynomial

$$\binom{x}{k} = \frac{x(x-1)(x-2)\cdots(x-k+1)}{k!}$$

is integer-valued, and every integer-valued polynomial is a linear combination with integer coefficients of the polynomials $\binom{x}{k}$. For any set $A \subseteq \mathbb{Z}$, we define

$$f(A) = \{f(a_1, a_2, \ldots, a_n) : a_i \in A \text{ for } i = 1, 2, \ldots, n\} \subseteq \mathbb{Z}.$$

Problem 9. Let $f(x_1, \ldots, x_n)$ and $g(x_1, \ldots, x_n)$ be integer-valued polynomials. Determine if there exist finite sets A, B, C of positive integers with $|C| \geq 2$ such that

$$|f(A)| > |g(A)|, \quad |f(B)| < |g(B)|, \quad |f(C)| = |g(C)|.$$

There is a strong form of Problem 9.

Problem 10. Let $f(x_1, \ldots, x_n)$ and $g(x_1, \ldots, x_n)$ be integer-valued polynomials. Does there exist a sequence $\{A_i\}_{i=1}^{\infty}$ of finite sets of integers such that

$$\lim_{i \to \infty} \frac{|f(A_i)|}{|g(A_i)|} = \infty?$$

There is also the analogous modular problem. For a polynomial $f(x_1, x_2, \ldots, x_n)$ with integer coefficients and for a set $A \subseteq \mathbb{Z}/m\mathbb{Z}$, we define

$$f(A) = \{f(a_1, a_2, \ldots, a_n) : a_i \in A \text{ for } i = 1, 2, \ldots, n\} \subseteq \mathbb{Z}/m\mathbb{Z}.$$

Problem 11. Let $f(x_1, \ldots, x_n)$ and $g(x_1, \ldots, x_n)$ be polynomials with integer coefficients and let $m \geq 2$. Do there exist sets $A, B, C \subseteq \mathbb{Z}/m\mathbb{Z}$ with $|C| > 1$ such that

$$|f(A)| > |g(A)|, \quad |f(B)| < |g(B)|, \quad |f(C)| = |g(C)|.$$

Problem 12. Let $f(x_1, \ldots, x_n)$ and $g(x_1, \ldots, x_n)$ be polynomials with integer coefficients. Let $M(f, g)$ denote the set of all integers $m \geq 2$ such that there exists a finite set A of congruence classes modulo m such that $|f(A)| > |g(A)|$. Compute $M(f, g)$.

Note that if there exists a finite set A of integers with $|f(A)| > |g(A)|$, then $M(f, g)$ contains all sufficiently large integers.

4. Representation functions of asymptotic bases

A central topic in additive number theory is the study of bases for the integers and for arbitrary abelian groups and semigroups, written additively. The set A is called an *additive basis of order h* for the set X if every element of X can be written as the sum of exactly h not necessarily distinct elements of A. The set A is called an *asymptotic basis of order h* for the set X if all but at most finitely many elements of X can be written as the sum of h not necessarily distinct elements of A. The classical bases in additive number theory for the set $\mathbf{N}_0$ of nonnegative integers are the squares (Lagrange's theorem), the cubes (Wieferich's theorem), the k-th powers (Waring's problem and Hilbert's theorem), the polygonal numbers (Cauchy's theorem), and the primes (Shnirel'man's theorem for sufficently large integers). These classical results in additive number theory are in Nathanson [8,9].

The *representation function* for a set A is the function $r_{A,h}(n)$ that counts the number of representations of n as the sum of h elements of A. More precisely, $r_{A,h}(n)$ is the number of h-tuples $(a_1, a_2, \ldots, a_h) \in A^h$ such that

$$a_1 + a_2 + \cdots + a_h = n$$

and

$$a_1 \leq a_2 \leq \cdots \leq a_h.$$

The set A is an asymptotic basis of order h if $r_{A,h}(n) \geq 1$ for all but finitely many elements of X. A fundamental unsolved problem in additive number theory is the *classification problem for representation functions*.

Problem 13. Let $f \colon \mathbb{N}_0 \to \mathbb{N}_0$ be a function and let $h \geq 2$. Find necessary and sufficient conditions on f in order that there exists a set A in $\mathbb{N}_0$ such that $r_{A,h}(n) = f(n)$ for all $n \in \mathbb{N}_0$.

A special case is the classification problem for representation functions for asymptotic bases for the nonnegative integers.

Problem 14. Let $\mathcal{F}_0(\mathbb{N}_0)$ denote the set of all functions $f \colon \mathbb{N}_0 \to \mathbb{N}_0$ such that $f(n) = 0$ for only finitely many nonnegative integers n. Let $h \geq 2$. For what functions $f \in \mathcal{F}_0(\mathbb{N}_0)$ does there exist a set $A \subseteq \mathbb{N}_0$ such that $r_{A,h}(n) = f(n)$ for all $n \in \mathbb{N}_0$?

Nathanson [10] introduced these problems, and recently began to study the representation functions of asymptotic bases for the set $\mathbb{Z}$ of integers. He proved [12] that if $f \colon \mathbb{Z} \to \mathbb{N}_0 \cup \{\infty\}$ is *any* function such that $f(n) = 0$ for only finitely many integers n, then there exists a set A in $\mathbb{Z}$ such that $r_{A,h}(n) = f(n)$ for all integers n. Moreover, arbitrarily sparse infinite sets A can be constructed with the given representation function f. The important new problem is to determine the maximum density of a set A of integers with given representation function f. For any set A of integers, we define the *counting function*

$$A(x) = \sum_{\substack{a \in A \\ |a| \leq x}} 1.$$

Problem 15. Let $f \colon \mathbb{Z} \to \mathbb{N}_0 \cup \{\infty\}$ be any function such that $f(n) = 0$ for only finitely many integers n. Let $\mathcal{R}(f)$ denote the set of all sets $A \subseteq \mathbb{Z}$ such that

$$r_{A,2}(n) = f(n) \quad \text{for all } n \in \mathbb{Z}.$$

Compute
$$\sup\{\alpha : A \in \mathcal{R}(f) \text{ and } A(x) \gg x^\alpha \text{ for all } x \geq x_0\}.$$

Nathanson [11] proved that for any function f there exists a set $A \in \mathcal{R}(f)$ with

$$A(x) \gg x^{1/3}.$$

Cilleruelo and Nathanson [1] recently improved this to

$$A(x) \gg x^{\sqrt{2}-1+\varepsilon}$$

for any $\varepsilon > 0$.

A related problem is the *inverse problem for representation functions*. Associated to a function $f \colon \mathbb{Z} \to \mathbb{N}_0 \cup \{\infty\}$ can be infinitely many sets A of integers such that $A \in \mathcal{R}(f)$. On the other hand, the semigroup $\mathbb{N}_0$ of nonnegative integers is more rigid than the group $\mathbb{Z}$ of integers. Given $f\colon \mathbb{N}_0 \to \mathbb{N}_0$, there may be a unique set $A \subseteq \mathbb{N}_0$ such that $r_{A,h}(n) = f(n)$ for all $n \in \mathbb{N}_0$.

Problem 16. Let $f\colon \mathbb{N}_0 \to \mathbb{N}_0$ be the representation function of a set of integers. Determine all sets $A \subseteq \mathbb{N}_0$ such that $r_{A,h}(n) = f(n)$ for all sufficiently large integers n.

The problem was first studied by Nathanson [7], and subsequently by Lev [4] and others. There is an excellent survey of additive representation functions by Sárközy and Sós [15].

5. Addendum, May, 2007

In the year since this paper was presented in Montreal, there has been striking progress on some of the problems concerning sets with more sums than differences (MSTD sets). Hegarty [3] proved that the minimum size of an MSTD set is 8, and that the set $A - \{0, 2, 3, 4, 7, 11, 12, 14\}$ is, up to affine transformations, the unique MSTD set of cardinality 8. Hegarty [3] and Nathanson [13] constructed explicit infinite families of MSTD sets. Martin and O'Bryant [6] proved that a positive proportion of the subsets of $\{0, 1, 2, \ldots, n\}$ are MSTD sets.

References

1. J. Cilleruelo and M. B. Nathanson, *Dense sets of integers with prescribed representation fucntions*, preprint.
2. H. T. Croft, *Research problems*, Cambridge, 1967, Problem 7, Section 6, p. 24, mimeographed notes.
3. P. Hegarty, *Some explicit constructions of sets with more sums than differences*, available at arXiv:math/0611582.
4. V. F. Lev, *Reconstructing integer sets from their representation functions*, Electron. J. Combin. **11** (2004), no. 1, Research Paper 78.
5. J. Marica, *On a conjecture of Conway*, Canad. Math. Bull. **12** (1969), 233–234.
6. G. Martin and K. O'Bryant, *Many sets have more sums than differences*, in this volume.
7. M. B. Nathanson, *Representation functions of sequences in additive number theory*, Proc. Amer. Math. Soc. **72** (1978), no. 1, 16–20.
8. _____, *Additive number theory: The classical bases*, Grad. Texts in Math., vol. 164, Springer, New York, 1996.
9. _____, *Elementary methods in number theory*, Grad. Texts in Math., vol. 195, Springer, New York, 2000.
10. _____, *Unique representation bases*, Acta Arith. **108** (2003), no. 1, 1–8.
11. _____, *The inverse problem for representation functions of additive bases*, Number Theory (New York, 2003), Springer, New York, 2004, pp. 253–262.
12. _____, *Every function is the representation function of an additive basis for the integers*, Port. Math. (N.S.) **62** (2005), no. 1, 55–72.
13. _____, *Sets with more sums than differences*, Integers **7** (2007), A5.
14. M. B. Nathanson, K. O'Bryant, B. Orosz, I. M. Ruzsa, and M. Silva, *Binary linear forms over finite sets of integers*, Acta Arith., to appear.
15. A. Sárközy and V. T. Sós, *On additive representation functions*, The Mathematics of Paul Erdős. I (R. L. Graham and J. Nešetřil, eds.), Algorithms Combin., vol. 13, Springer, Berlin, 1997, pp. 129–150.

DEPARTMENT OF MATHEMATICS, LEHMAN COLLEGE (CUNY), BRONX, NEW YORK 10468, USA

E-mail address: melvyn.nathanson@lehman.cuny.edu

Double and Triple Sums Modulo a Prime

Katalin Gyarmati, Sergei Konyagin, and Imre Z. Ruzsa

ABSTRACT. We study the connection between the sizes of $2A$ and $3A$ (twofold and threefold sums), where A is a set of residues modulo a prime p.

1. Introduction

Lev [3] observed that for a set A of integers the quantity
$$\frac{|kA| - 1}{k}$$
is increasing. The first cases of this result assert that

(1.1) $$|2A| \geq 2|A| - 1$$

and

(1.2) $$|3A| \geq \tfrac{3}{2}|2A| - \tfrac{1}{2}.$$

Inequality (1.1) can be extended to different summands as

(1.3) $$|A + B| \geq |A| + |B| - 1,$$

and this inequality can be extended to sets of residues modulo a prime p, the only obstruction being that a cardinality cannot exceed p:

(1.4) $$|A + B| \geq \min(|A| + |B| - 1, p);$$

this familiar result is known as the Cauchy–Davenport inequality.

In this paper we deal with the possibility of extending inequality (1.2) to residues. We also have the obstruction at p, and the third author initially hoped that this is the only one, so an inequality like
$$|3A| \geq \min(\tfrac{3}{2}|2A| - \tfrac{1}{2}, p)$$

2000 *Mathematics Subject Classification.* 11B50, 11B75, 11P70.

The first author is supported by Hungarian National Foundation for Scientific Research (OTKA), Grants No. T 43631, T 43623, T 49693.

The second author is supported by the Russian Foundation for Basic Research, Grant 05-01-00066, and by Grant NSh-5813.2006.1.

The third author is supported by Hungarian National Foundation for Scientific Research (OTKA), Grants No. T 43623, T 42750, K 61908.

This is the final form of the paper.

©2007 American Mathematical Society

may hold; in particular, this would imply $3A = \mathbb{Z}_p$ for $|2A| > 2p/3$. M. Garaev asked (personal communication) whether this holds at least under the stronger assumption $|2A| > cp$ with some absolute constant $c < 1$. It turned out that the answer even to this question is negative, and the relationship between the sizes of $2A$ and $3A$ is seemingly complicated.

Theorem 1.1. *Let p be a prime.*

(a) *For every $m < \sqrt{p}/3$ there is a set $A \subset \mathbb{Z}_p$ such that $|3A| \leq p - m^2$ and $|2A| \geq p - m(2\sqrt{p} - m + 3) - Cp^{1/4}$. Here C is a positive absolute constant.*

(b) *In particular, there is a set $A \subset \mathbb{Z}_p$ such that $3A \neq \mathbb{Z}_p$ and $|2A| \geq p - 2\sqrt{p} - Cp^{1/4}$.*

Our positive results are as follows. (Since the structure of sumsets is trivial when the set has 1 or 2 elements, we assume $|A| \geq 3$.)

Theorem 1.2. *Let $p \geq 29$ be a prime, $A \subset \mathbb{Z}_p$, $|A| \geq 3$ and write $|2A| = n$, $|3A| = s$.*

(a) *There is a positive absolute constant c such that for $n < cp$ we have*
$$s \geq \frac{3n-1}{2}.$$

(b) *For $6 \leq n < p/2$ we have $s > \sqrt{2}n$.*

(c) *If $n = (p+1)/2$, then*
$$s \geq \frac{3n-1}{2} = \frac{3p+1}{4}.$$

(d) *For $n \geq (p+3)/2$ we have*
$$s \geq \frac{n(2p-n)}{p}.$$

(e) *If $n > p - \sqrt{2p} + 2$, then $3A = \mathbb{Z}_p$.*

A drawback of this theorem is the discontinuous nature of the bounds in (a)–(b)–(c). It is possible to modify the argument in the proof of (c) to get a continuously deteriorating bound for n just below $p/2$, but it is hardly worth the trouble. It is unlikely that the actual behavior of $\min s$ changes in this interval, so it seems safe to conjecture the following.

Conjecture 1.3. *If $n \leq (p+1)/2$, then $s \geq (3n-1)/2$.*

To find the smallest value of n provided by Theorem 1.1 for which $s < (3n-1)/2$ can happen we have to solve a quadratic inequality for m. This gives $m \approx \sqrt{p}/5$ and $n \approx 16p/25$.

Theorem 1.1 and part (d) of Theorem 1.2 describe the quadratic connection between n and s for large values of n. Indeed, (d) can be reformulated as follows: if $s \leq p - m^2$, then $n \leq p - m\sqrt{p}$, thus the difference is the coefficient 1 or 2 of $m\sqrt{p}$. Similarly, the theorems locate the point after which necessarily $3A = \mathbb{Z}_p$ between $p - 2\sqrt{p}$ and $p - \sqrt{2p}$. We do not have a plausible conjecture about the correct coefficient of $\sqrt{p}$ in these results.

2. Construction

We prove Theorem 1.1.

Without loss of generality we may assume that p is large enough.

We will use the integers $0, \ldots, p-1$ to represent the residues modulo p. We will write $[a, b]$ to denote a discrete interval, that is, the set of integers $a \leq i \leq b$.

Take an integer $q \sim \sqrt{p}$, and write $p = qt + r$ with $1 \leq r \leq q - 1$. We will consider sets of the form $A = K \cup L$, where

$$K = [0, k-1], \quad |K| = k \leq q - 1$$

and

$$L = \{0, q, 2q, \ldots, (l-1)q\}, \quad |L| = l \leq t - 1.$$

Our parameters will satisfy $k > q/2 + 3$ and $l \geq 2t/3 + 2$. We assume that $t > 6$.

All the sums $x + y$, $x \in K$, $y \in L$ are distinct and hence we have

$$|2A| \geq |K + L| = kl.$$

It would not be difficult to calculate $|2A|$ more exactly, but it would only minimally affect the final result.

The set $3A$ is the union of $3K$, $2K + L$, $K + 2L$ and $3L$. We consider $2K + L$ first. We have $2K = [0, 2k-2]$. Since $2k - 2 > q$, the sets $2K$, $2K + q$, ... overlap and we have

$$2K + L = [0, q(l-1) + 2k - 2] = [0, ql + (2k - 2 - q)].$$

So $3A$ contains $[0, ql]$ and we will study in detail the structure in $[ql, p-1]$.

We have $3K \subset [0, 3q]$, so we do not get any new element (assuming $l \geq 3$).

Now we study $K + 2L$. The set $2L$ contains $0, q, \ldots, qt$ and then $q(t+1) - p = q - r, 2q - r, \ldots, q(2l-2) - p = q(2l - 2 - t) - r$. By adding the set K to the second type of elements we get numbers in

$$[0, q(2l - 1 - t)] \subset [0, ql],$$

so no new elements again. By adding K to iq we stay in $[0, ql]$ as long as $i \leq l - 1$, and for $l \leq i \leq t$ we get

$$[ql, ql + k - 1] \cup [(l+1)q, (l+1)q + k - 1] \cup \cdots \cup [(t-1)q, (t-1)q + k - 1]$$
$$\cup [qt, qt + \min(k - 1, r)].$$

If $r \leq k - 1$, the last of the above intervals covers $[qt, p]$, so we can restrict our attention to $[ql, qt - 1]$. If $r > k - 1$, then some elements near $p - 1$ may not be in $K + 2L$, but as $r \leq k - 1$ will typically hold in our choice, we will not try to exploit this possible gain. Note that the final segment of $2K + L$, that is, $[ql, ql + (2k - 2 - q)]$ is contained in the first of the above intervals.

Finally $3L$ consists of elements of the form $iq - jp$, where $0 \leq i \leq 3l - 3$ and $0 \leq j \leq 2$. Those with $j = 0$ are already listed above. Those with $j = 2$ are in $[0, ql]$, so no new element. Finally with $j = 1$ we have $iq - p = (i - t)q - r$ with $t + 1 \leq i \leq 2t$, and also with $i = 2t + 1$ if $r > q/2$. Among these elements the possible new ones are

$$(l+1)q - r, (l+2)q - r, \ldots, (t+1)q - r.$$

This gives no new element if

(2.1) $$r \geq q - k + 1.$$

So under this additional assumption the intervals $[iq+k, iq+q-1]$ are disjoint to $3A$ for $l \leq i \leq t-1$, and this gives

$$|3A| \leq p - (t-l)(q-k).$$

For a given m we will take $l = t - m$, $k = q - m$, hence the bound $|3A| \leq p - m^2$. With this choice we have

(2.2) $\qquad |2A| \geq kl = (q-m)(t-m) = p - m(q+t-m) - r.$

Now we select q, t and r. Define the integer v by

$$(v-1)^2 < p < v^2,$$

and write $p = v^2 - w$, $0 < w < 2v$. With arbitrary i we have

$$p = (v-i)(v+i) + (i^2 - w).$$

Hence $q = v - i$, $t = v + i$ and $r = i^2 - w$ may be a good choice. We have a lower bound for r given by (2.1), which now reads $r \geq m + 1$, but otherwise the smaller the value of r the better the bound on $2A$ in (2.2), so we take

$$i = 1 + [\sqrt{w + m + 1}].$$

Then $r = m + O(\sqrt{w+m+1}) = m + O(p^{1/4})$. Since $q + t = 2v < 2\sqrt{p} + 2$, (2.2) yields the bound in part (a) of Theorem 1.1.

3. Estimates

Here we prove Theorem 1.2.

We will assume that $0 \in A$ and consequently $A \subset 2A$, since this can be achieved by a translation which does not affect the studied cardinalities.

The proof will be based on certain Plünnecke-type estimates. These will be quoted from [5]; the basic ideas go back to Plünnecke [4].

PROOF OF (a).

Lemma 3.1. *Let $i < h$ be integers, U, V sets in a commutative group and write $|U| = m$, $|U + iV| = \alpha m$. We have*

$$|hV| \leq \alpha^{h/i} m.$$

This is Corollary 2.4 of [5].

Take a set $A \subset \mathbb{Z}_p$ such that $|2A| = n$, $|3A| = s$ and $s < 3n/2$. We apply the above lemma with $i = 1$, $h = 4$, $U = 2A$, $V = A$. We get

$$|4V| < (\tfrac{3}{2})^4 |U|.$$

Since $4V = 4A = 2U$, this means that the set $U = 2A$ has a small doubling property, namely $|2U| < (81/16)|U|$, and this permits us to "rectify" it. There are several ways to do this; the most comfortable is the following form, taken from [1], Theorem 1.2, with some change in the notation.

Lemma 3.2. *Let p be a prime and let $U \subseteq \mathbb{Z}/p\mathbb{Z}$ be a set with $|U| = \delta p$ and $\min(|2U|, |U - U|) = \alpha|U|$. Suppose that $\delta \leq (16\alpha)^{-12\alpha^2}$. Then the diameter of U is at most*

(3.1) $\qquad\qquad\qquad 12\delta^{1/4\alpha^2} \sqrt{\log(1/\delta)} p.$

The *diameter* in the above lemma is the length of the shortest arithmetic progression that contains the set. We apply this lemma for our set $U = 2A$. We fix $\alpha = 81/16$ and select c and then $\delta \leq c$, satisfying the hypotheses, so that the bound in (3.1) is $< p/4$. (The requirement that δ satisfies the hypotheses is the stronger constraint, and leads to the value $c = 2^{-3^9/2^4}$.) This will be the constant c in (a) of the theorem.

The lemma yields that $A \subset 2A \subset \{-kd, -(k-1)d, \ldots, -d, 0, d, 2d, \ldots, ld\}$ with a suitable d and integers k, l such that $k + l < p/4$. Let
$$A' = \{j : -k \leq j \leq l, jd \in A\}.$$
Then $4A' \subset [-4k, 4l]$, still an interval of length $< p$, hence $|4A| = |4A'|$ and the claim follows from Lev's result (1.2) on sets of integers. $\square$

PROOF OF (b).

Lemma 3.3. *Let $U, V \subset \mathbb{Z}_p$, $|U| \geq 2$, $|V| \geq 2$, $|U| + |V| \leq p - 1$. Then either $|U + V| \geq |U| + |V|$, or U, V are arithmetic progressions with a common difference.*

This is the Cauchy–Davenport inequality with Vosper's description of the extremal pairs incorporated; see, e.g., [2].

Lemma 3.4. *If $A \subset \mathbb{Z}_p$ and $2A$ is an arithmetic progression, then $s \geq \min(p, (3n-1)/2)$.*

PROOF. First, use a dilation to make the difference of the arithmetic progression 1, and then a translation to achieve $0 \in A$; these transformations do not change the size of our sets. In this case $A \subset 2A$, so we can write
$$2A = \{k, k+1, \ldots, -1, 0, 1, \ldots, l\}, \quad k \leq 0 \leq l, \ l - k = n - 1.$$
Let the first and last elements of A (in the list above) be a and b. We have $k \leq a \leq 0 \leq b \leq l$. Furthermore $2A \subset [2a, 2b]$, that is, $n = |2A| \leq 2(b-a) + 1$ and so $b - a \geq (n-1)/2$. Now $3A$ contains the residue of every integer in the set
$$[k, l] + \{a, b\} = [k+a, l+b],$$
an interval of length $l + b - k - a \geq 3(n-1)/2$ (to see that it is an interval observe that $l + a \geq k + b$), hence its cardinality is at least the cardinality of this interval or p. $\square$

Lemma 3.4 allows us to prove slightly stronger results than we would obtain by applying the Cauchy-Davenport inequality directly, the main benefit being that the statements of the results become simpler.

Lemma 3.5. *Let $i < h$ be integers, U, V sets in a commutative group and write $|U| = m$, $|U + iV| = \alpha m$. There is an $X \subset U$, $X \neq \emptyset$ such that*
$$|X + hV| \leq \alpha^{h/i}|X|.$$

This is Theorem 2.3 of [5].

Now we prove part (b). We apply the above lemma with $i = 1$, $h = 2$ for $U = 2A$, $V = A$, so that $\alpha = s/n$. We get that there is a nonempty $X \subset 2A$ such that
(3.2) $$|X + 2A| \leq \alpha^2 |X|.$$
We will now apply Lemma 3.3 to the sets X and $2A$. To check the conditions observe that $|X| + |2A| \leq 2n \leq p - 1$. The condition $|X| \geq 2$ may not hold. If

it fails, then (3.2) reduces to $n \leq \alpha^2$ and hence $\alpha \geq \sqrt{2}$. If $2A$ is an arithmetic progression, then we get (b) by Lemma 3.4. If none of these happens, then by Lemma 3.3 we know that $|X + 2A| \geq |X| + n$, and then (3.2) can be rearranged as

$$n \leq (\alpha^2 - 1)|X| \leq (\alpha^2 - 1)n,$$

that is, $\alpha \geq \sqrt{2}$ as claimed. □

PROOF OF (e). If $3A \neq \mathbb{Z}_p$, then $|2A| + |A| \leq p$ (by the Cauchy–Davenport inequality, or by an appropriate application of the pigeonhole principle). Write $|A| = m$. We have $n \leq m(m+1)/2$, hence $m \geq \sqrt{2n} - \frac{1}{2}$ and the previous inequality implies $n + \sqrt{2n} \leq p + \frac{1}{2}$. By solving this as a quadratic inequality for $\sqrt{n}$ we obtain

$$n \leq p - \sqrt{2p + 2} + \frac{3}{2} < p - \sqrt{2p} + 2.$$ □

PROOF OF (c) AND (d). We will prove that

$$s \geq \min\left(\frac{3n-1}{2}, \frac{n(2p-n)}{p}\right),$$

which implies both (c) and (d). Indeed, observe that the bound in (c), $(3n-1)/2$, is smaller than the bound $n(2p-n)/p$ in (d) for $n = (p+1)/2$ and it is larger otherwise.

If $s = p$, we are done. If $s = p - 1$, then from part (e) we get that $n < p - \sqrt{2p} + 2 < p - \sqrt{p}$ and then $n(2p - n)/p < p - 1$, and again we are done. So assume $s \leq p - 2$.

Lemma 3.6. *Let $i < h$ be positive integers, U, V, W sets in a commutative group and write $|U| = m$, $|(U + iV) \setminus (W + (i-1)V)| \leq \beta m$. There is an $X \subset U$, $X \neq \emptyset$ such that*

$$|(X + hV) \setminus (W + (h-1)V)| \leq \beta^{h/i}|X|.$$

This is Theorem 2.8 of [5].

Lemma 3.7. *Let U, V be sets in a commutative group and write $|U| = m$, $|U + V| \leq \alpha m$. There is an $X \subset U$, $X \neq \emptyset$ such that*

$$|X + 2V| \leq \alpha m + (\alpha - 1)^2|X|.$$

PROOF. We apply the previous lemma with $i = 1$, $h = 2$, $W = U + v$ with an arbitrary $v \in V$; clearly $\beta = \alpha - 1$. We obtain the existence of an $X \subset U$, $X \neq \emptyset$ such that

$$|(X + 2V) \setminus (U + V + v)| \leq (\alpha - 1)^2|X|.$$

The claim follows by observing that $|U + V + v| \leq \alpha m$. □

Consider the set $D = \mathbb{Z}_p \setminus (-3A)$. We have $m = |D| = p - s \geq 2$. The set $D + A$ is disjoint to $-2A$, hence $|D + A| \leq p - n$. We apply the previous lemma with $U = D$, $V = A$ and $\alpha = (p-n)/(p-s)$. We obtain the existence of a nonempty $X \subset D$ such that

(3.3) $$|X + 2A| \leq p - n + (\alpha - 1)^2|X|.$$

We have $|X| + |2A| \leq p - s + n \leq p - 1$. By Lemma 3.3 either we have

(3.4) $$|X + 2A| \geq |X| + |2A|,$$

or $|X| = 1$, or $2A$ is an arithmetic progression. In the last case the claim follows from Lemma 3.4, since $n(2n-p)/p < (3n-1)/2$ for $n > (p+1)/2$.

If (3.4) holds, then (3.3) implies

$$(3.5) \qquad 2n - p \leq \alpha(\alpha - 2)|X|.$$

Since the left side is positive, so is the right side, that is, necessarily $\alpha \geq 2$, and then using that $|X| \leq |D| = p - s$, (3.5) becomes

$$(3.6) \qquad 2n - p \leq \alpha(\alpha - 2)(p - s).$$

Substituting $\alpha = (p-n)/(p-s)$ and $\alpha - 2 = (2s - n - p)/(p - s)$ this becomes

$$(2n-p)(p-s) \leq (p-n)(2s-n-p)$$

which can be rearranged to give the bound in (d).

If (3.4) fails, then $|X| = 1$ and (3.3) becomes

$$(3.7) \qquad 2n - p \leq (\alpha - 1)^2.$$

If α is such that $(\alpha-1)^2 \leq 2\alpha(\alpha-2)$, then, as $p - s \geq 2$, (3.6) holds again and we complete the proof as before. If this is not the case, then $\alpha < 1 + \sqrt{2}$, and (3.7) yields $2n - p < 2$. Since p is odd, this leaves the only possibility $n = (p+1)/2$. Now (3.7) becomes $\alpha \geq 2$, that is, $p - n \geq 2(p - s)$,

$$s \geq \frac{p+n}{2} = \frac{3p+1}{4}$$

as wanted. $\square$

Acknowledgment. The authors are grateful to a referee for remarks and corrections.

References

1. B. Green and I. Z. Ruzsa, *Sets with small sumsets and rectification*, Bull. London Math. Soc. **38** (2006), 43–52.
2. H. Halberstam and K. F. Roth, *Sequences*, Clarendon, Oxford, 1966; 2nd ed., Springer, New York–Berlin, 1983.
3. V. F. Lev, *Structure theorem for multiple addition and the Frobenius problem*, J. Number Theory **58** (1996), 79–88.
4. H. Plünnecke, *Eine zahlentheoretische Anwendung der Graphtheorie*, J. Reine Angew. Math. **243** (1970), 171–183.
5. I. Z. Ruzsa, *Cardinality questions about sumsets*, in this book.

ALFRÉD RÉNYI INSTITUTE OF MATHEMATICS, HUNGARIAN ACADEMY OF SCIENCES, POB 127, 1364 BUDAPEST, HUNGARY
E-mail address: gykati@cs.elte.hu

DEPARTMENT OF MECHANICS AND MATHEMATICS, MOSCOW STATE UNIVERSITY, MOSCOW, 119992, RUSSIA
E-mail address: konyagin@ok.ru

ALFRÉD RÉNYI INSTITUTE OF MATHEMATICS, HUNGARIAN ACADEMY OF SCIENCES, POB 127, 1364 BUDAPEST, HUNGARY
E-mail address: ruzsa@renyi.hu

Additive Properties of Product Sets in Fields of Prime Order

A. A. Glibichuk and S. V. Konyagin

1. Introduction

Let $p > 2$ be a prime, $\mathbb{Z}_p$ be the field of residues modulo p, and $\mathbb{Z}_p^*$ be the multiplicative group of $\mathbb{Z}_p$, so that $\mathbb{Z}_p^* = \mathbb{Z}_p \setminus \{0\}$. For sets $X \subset \mathbb{Z}_p$, $Y \subset \mathbb{Z}_p$, and for a (possibly, partial) binary operation $*\colon \mathbb{Z}_p \times \mathbb{Z}_p \to \mathbb{Z}_p$ we let

$$X * Y = \{x * y : x \in X, y \in Y\}.$$

We will write XY instead of $X * Y$ if $*$ is multiplication mod p; and, for an element $\lambda \in \mathbb{Z}_p$, we write

$$\lambda * A = \{\lambda\}A.$$

For a set $X \subset \mathbb{Z}_p$ and $k \in \mathbb{N}$ let

$$kX = \{x_1 + \cdots + x_k : x_1, \ldots, x_k \in A\},$$
$$X^k = \{x_1 \cdots x_k : x_1, \ldots, x_k \in A\}.$$

A set A is called an (additive) basis of order k (for $\mathbb{Z}_p$) if $kA = \mathbb{Z}_p$. Observe that any basis of order k is also a basis of any order $k' > k$. The general problem that will be discussed in this paper is whether, for given integers $t < p, n, N$, the set A^n is a basis of order N for each set $A \subset \mathbb{Z}_p$ of cardinality $\geq t$?

This problem is easily answered for $n = 1$: The Cauchy–Davenport theorem [5, Theorem 5.4] implies that for any sets $X_1, \ldots, X_N \subset \mathbb{Z}_p$ we have

$$|X_1 + \cdots + X_N| \geq \min(|X_1| + \cdots + |X_N| - N + 1, p).$$

Therefore any set A with $|A| \geq 1 + (p-1)/N$ is a basis of order N for $\mathbb{Z}_p$. On the other hand, if t is an integer with $1 \leq t < 1+(p-1)/N$ then the set $A = \{0, \ldots, t-1\}$ has t elements, but $NA \neq \mathbb{Z}_p$.

For $n = 2$, some useful information can be obtained by using exponential sums: various facts related to harmonic analysis in $\mathbb{Z}_p$ can be found in [5, Chapter 4].

For fixed $k \in \mathbb{N}$ and $\varepsilon > 0$, it is known that a random subset of $\mathbb{Z}_p$ of cardinality $> p^{1/k+\varepsilon}$ is a basis of order k with high probability (tending to 1 as

---

2000 *Mathematics Subject Classification.* 11B75.

This research was carried out while the authors were visitors of the Centre de recherches mathématiques. It is our pleasure to thank the CRM for its hospitality and the Clay Institute for its generous support of our visit.

This is the final form of the paper.

$p \to \infty$). Thus, if a large subset of $\mathbb{Z}_p$ is not a basis of small order, then it is unlike a random set and so, in this sense, has special additive structure. On the other hand, we believe that any large subset of $\mathbb{Z}_p$ with multiplicative structure must be a basis of small order: often one can obtain nontrivial estimates for exponential sums over this kind of set and deduce various additive properties. Probably, the simplest result of this type is the following [6, Chapter VI, Problem 8, α].

Proposition 1.1. *If $X, Y \subset \mathbb{Z}_p$, $a \in \mathbb{Z}_p^*$, then*

$$\left| \sum_{x \in X, y \in Y} \exp(2\pi i a x y / p) \right| \le \sqrt{p|X||Y|}.$$

One can easily deduce from Proposition 1.1 that for any $\varepsilon > 0$ there is an integer $N = N(\varepsilon)$ such that for any $A \subset \mathbb{Z}_p$ with $|A| > p^{1/2+\varepsilon}$ we have

(1) $$NA^2 = \mathbb{Z}_p.$$

In particular, it is known that (1) holds for $|A| > p^{3/4} + 1$ and $N = 3$. We do not, however, see a way to prove (1) for bounded N via exponential sums since there is no upper bound for the above exponential sum, under the weaker restriction

(2) $$|A| > \sqrt{p},$$

which is significantly better than the trivial estimate $|A|^2$. (Note that the restriction (2) cannot be relaxed significantly, since A^2 is not a basis of fixed order if $A = \{1, \ldots, [f(p)]\}$ where $f(p) = o(p^{1/2})$ as $p \to \infty$. Similarly the condition $|A| \le f(p)$ is insufficient to guarantee that A^n is a basis of a fixed order, if $f(p) = o(p^{1/n})$.)

By using combinatorial arguments the first author proved in [4], that $8XY = \mathbb{Z}_p$ for any $X \subset \mathbb{Z}_p$, $Y \subset \mathbb{Z}_p$ provided that $|X||Y| > p$ and either $Y = -Y$ or $Y \cap (-Y) = \emptyset$. This implies that (1) with $N = 16$ holds whenever (2) holds (see Section 2 for a proof).

The estimates for exponential sums established in [1] (Theorem 5) imply that for any fixed $\delta > 0$ there exist integers $n = n(\delta)$ and $N = N(\delta)$ such that

(3) $$NA^n = \mathbb{Z}_p$$

for any set $A \subset \mathbb{Z}_p$ with $|A| > p^\delta$. Moreover, when $\delta < \frac{1}{2}$ we have

$$n \le \delta^{-C}, \quad N \le \exp(\delta^{-C})$$

where C is a constant that was not evaluated in [1]. It is natural to ask whether one can obtain sharper bounds for n and N. The main result of this paper is the following.

Theorem 1.2. *For any integer $n > 1$, and any δ and ε satisfying $\delta \ge 1/(n-\varepsilon)$ and $0 < \varepsilon < n$, there exists an integer*

(4) $$N \le 15 \cdot 4^n \log(2 + 1/\varepsilon),$$

such that (3) holds for any subset $A \subset \mathbb{Z}_p$ with $|A| > p^\delta$, for any prime p.

We know that the restriction for $n \ge 1/\delta + \varepsilon$ is essentially best possible, as we have seen that one cannot take $n < 1/\delta$ in general. Moreover N must grow at least as fast as an exponential function of $1/\delta$ as $\delta \to 0$: To see this let $A = \{0, 1\}$ so that we must select δ so that $p < 2^{1/\delta}$, and note that $A^n = A$ for all n so that (3) holds only if $N \ge p - 1$; and we can therefore choose δ so that $N \ge 2^{1/2\delta}$.

Corollary 1.3. *For any $\delta > 0$ there is an integer $N \leq 100 \cdot 4^{1/\delta}$ for which $NG = \mathbb{Z}_p$ for any subgroup G of $\mathbb{Z}_p^*$ of cardinality $> p^\delta$, for any prime p.*

To deduce Corollary 1.3 from Theorem 1.2 let $A = G$ and n be the smallest integer $\geq 1/\delta + 0.1$, $\varepsilon = n - 1/\delta$, and note that $G^n = G$ as G is a group and so is closed under multiplication and that $15 \cdot 4^n \log(2 + 1/\varepsilon) < 100 \cdot 4^{1/\delta}$ for $\varepsilon \in [0.1, 1.1)$.

We have been unable to prove that (3) holds for every A with $|A| > p^{1/n}$, for some $N = N(n)$, when $n > 2$.

We will obtain preliminary results in Sections 2–5, leaving the proof of Theorem 1.2 until Section 6.

The authors are grateful to Andrew Granville for a careful reading of the paper. Due to his remarks, some mistakes and misprints have been corrected.

2. On additive properties of a product of two sets

Lemma 2.1. *If $A \subset \mathbb{Z}_p$, $B \subset \mathbb{Z}_p$, and $|A| \cdot \lceil |B|/2 \rceil > p$ then $8AB = \mathbb{Z}_p$.*

Proof. Let $B_1 = \{b \in B : -b \in B\}$ and $B_2 = \{b \in B : -b \notin B\}$, which is a partition of B, so that $|B_i| \geq \lceil |B|/2 \rceil$ for $i = 1$ or $i = 2$. Therefore $8AB_i = \mathbb{Z}_p$ by [4], so that $8AB = \mathbb{Z}_p$, as required.

Lemma 2.2. *If $A \subset \mathbb{Z}_p$, $B \subset \mathbb{Z}_p$, and $|A||B| > p$ then $16AB = \mathbb{Z}_p$.*

Proof. If $2B = \mathbb{Z}_p$ then $a * (B + B) = \mathbb{Z}_p$ for any $a \in A \cap \mathbb{Z}_p^*$, so that $AB + AB = \mathbb{Z}_p$ and hence $16AB = \mathbb{Z}_p$. Otherwise if $2B$ is a proper subset of $\mathbb{Z}_p$ then the Cauchy–Davenport theorem implies that $|2B| \geq 2|B| - 1$, so that $\lceil |2B|/2 \rceil \geq |B|$. But then Lemma 2.1 implies that $8A(2B) = \mathbb{Z}_p$, and we are done. $\square$

3. Main lemmata

For sets $X, Y \subset \mathbb{Z}_p$, $|Y| > 1$, we let

$$Q[X, Y] = \frac{X - X}{(Y - Y) \setminus \{0\}}.$$

We will use the following observation.

Lemma 3.1. *Let $\xi \in \mathbb{Z}_p$. Then $\xi \in Q[X, Y]$ if and only if $|X + \xi * Y| < |X||Y|$.*

Proof. Consider the mapping $F : X \times Y$ onto $X + \xi * Y$ defined by $F(x, y) = x + \xi y$. F is not an injection if and only if $|X + \xi * Y| < |X||Y|$. On the other hand, the condition that F is not an injection means that there are $x_1, x_2 \in X$, $y_1, y_2 \in Y$ with $(x_1, y_1) \neq (x_2, y_2)$ such that $F(x_1, y_1) = F(x_2, y_2)$. Note that $y_1 \neq y_2$ else $x_1 = x_2$ contradicting the fact that $(x_1, y_1) \neq (x_2, y_2)$. Therefore $\xi = (x_1 - x_2)/(y_2 - y_1) \in (X - X)/((Y - Y) \setminus \{0\})$, which completes the proof of the lemma. $\square$

For $X = Y$, Lemma 3.1 is Lemma 2.50 from [5].

The key ingredient for the proof of Theorem 1.2 is the following lemma based on ideas of T. Tao and V. Vu [5, Section 2.8].

Lemma 3.2. *If $X, Y \subset \mathbb{Z}_p$, $a \in \mathbb{Z}_p^*$, $|X| \geq 1$, $|Y| > 1$, and $Q[X, Y] \neq \mathbb{Z}_p$ then*

$$|2XY - 2XY + a * Y^2 - a * Y^2| \geq |X||Y|.$$

PROOF. The hypotheses imply that $Q[X,Y] \neq \varnothing$ and $Q[X,Y] \neq \mathbb{Z}_p$, from which one can deduce that there exists $\xi \in Q[X,Y]$ such that $\xi + a \notin Q[X,Y]$. Now since $\xi \in Q[X,Y]$, there exists $x_1, x_2 \in X$, $y_1 \neq y_2 \in Y$, $y_1 \neq y_2$ for which $\xi = (x_1 - x_2)/(y_1 - y_2)$. Moreover since $(x_1 - x_2)/(y_1 - y_2) + a = \xi + a \notin Q[X,Y]$ one can deduce from Lemma 3.1 that

$$\left|\left\{x + \left(\frac{x_1 - x_2}{y_1 - y_2} + a\right)y : x \in X, y \in Y\right\}\right| = |X||Y|.$$

Multiplying through by $y_1 - y_2$ we get

$$|\{x(y_1 - y_2) + (x_1 - x_2)y + (y_1 - y_2)ay : x \in X, y \in Y\}| = |X||Y|.$$

But

$$x(y_1 - y_2) + (x_1 - x_2)y + (y_1 - y_2)ay \in 2XY - 2XY + a*Y^2 - a*Y^2,$$

and the result follows from combining the last two lines. $\square$

Lemma 3.3 ([2,3]). *Let $X \subset \mathbb{Z}_p$, $Y \subset \mathbb{Z}_p$, $G \subset \mathbb{Z}_p^*$, and $G \neq \varnothing$. Then there exists $\xi \in G$ such that*

$$|X + \xi * Y| \geq \frac{|X||Y||G|}{|X||Y| + |G|}.$$

Lemma 3.4. *If $\xi \in Q[X,Y]$, then*

$$|2XY - 2XY| \geq |X + \xi * Y|.$$

PROOF. By the condition on ξ, there are elements $x_1, x_2 \in X$ and $y_1, y_2 \in Y$, $y_1 \neq y_2 \in Y$ such that

(5) $$x_1 - x_2 = (y_1 - y_2)\xi.$$

Let

$$S = (y_1 - y_2) * (X + \xi * Y),$$

and note that

$$|S| = |X + \xi * Y|.$$

Any element of S can be written in the form

$$(y_1 - y_2)x + (y_1 - y_2)\xi y = x(y_1 - y_2) + (x_1 - x_2)y \in 2XY - 2XY$$

for some $x \in X$, $y \in Y$, by (5). Therefore $S \subset 2XY - 2XY$, and the result follows by combining the last two displayed equations. $\square$

Corollary 3.5. *If $X, Y \subset \mathbb{Z}_p$, $a \in \mathbb{Z}_p^*$ and $|Y| > 1$, then*

$$|2XY - 2XY + a*Y^2 - a*Y^2| \geq \frac{|X||Y|(p-1)}{|X||Y| + p - 1}.$$

PROOF. We may assume that $Q[X,Y] = \mathbb{Z}_p$ else the result follows from Lemma 3.2. By Lemma 3.3, there is a $\xi \in \mathbb{Z}_p^*$ such that

$$|X + \xi * Y| \geq \frac{|X||Y|(p-1)}{|X||Y| + p - 1}.$$

Since $\xi \in Q[X,Y]$, we have

$$|2XY - 2XY + a*Y^2 - a*Y^2| \geq |2XY - 2XY| \geq |X + \xi * Y|,$$

by Lemma 3.4, and the result follows by combining the last two displayed equations. $\square$

Corollary 3.6. *If $|Y| > 1$ then*

$$|3Y^2 - 3Y^2| \geq \frac{|Y|^2(p-1)}{|Y|^2 + p - 1}.$$

This follows from Corollary 3.5 by taking $a = 1$ and $X = Y$.

Corollary 3.7. *If $|Y| > 1$ and $X = KY^k - KY^k$, then*

$$|(4K+1)Y^{k+1} - (4K+1)Y^{k+1}| \geq \frac{|X||Y|(p-1)}{|X||Y| + p - 1}.$$

This follows from Corollary 3.5 by taking $a = y^{k-1}$ where $y \in Y$ with $y \neq 0$.

4. Ruzsa triangle inequality and its corollaries

The following nice result was first proved by I. Ruzsa (see [5, Lemma 2.6]).

Lemma 4.1. *For any subsets X, Y, Z of $\mathbb{Z}_p$ we have*

$$|X||Y - Z| \leq |X - Y||X - Z|.$$

Lemma 4.1 is often called the *Ruzsa triangle inequality* since it can be reformulated as follows: The binary function ρ, defined on nonzero subsets of $\mathbb{Z}_p$ by

$$\rho(X, Y) = \log(|X - Y|^2/(|X||Y|)),$$

satisfies the triangle inequality. Lemma 4.1 can be proved for subsets of an arbitrary abelian group, but we need it only for subsets of $\mathbb{Z}_p$.

Corollary 4.2. *If $X, Y \subset \mathbb{Z}_p$, then*

$$|X + Y| \geq |X|^{1/2}|Y - Y|^{1/2}.$$

PROOF. By Lemma 4.1 with $-Y$ substituted for both Y and Z, we have

$$|X||(-Y) - (-Y)| \leq |X + Y||X + Y|,$$

and the result follows. $\square$

Corollary 4.3. *If $X \subset \mathbb{Z}_p$, $k \in \mathbb{N}$, then*

(6) $$|kX| \geq |X|^{2^{1-k}}|X - X|^{1 - 2^{1-k}}.$$

PROOF. By induction on k. For $k = 1$ the result is trivial. Now let $k > 1$ and assume that (6) is true for $k - 1$: By Corollary 4.2 we have

$$|kX| = |(k-1)X + X| \geq |(k-1)X|^{1/2}|X - X|^{1/2},$$

and the result follows by induction. $\square$

5. Some inequalities

In this section we will establish lower bounds for $|KA^k|$ and $|KA^k - KA^k|$ where $A \subset \mathbb{Z}_p$ with

(7) $$|A| \geq 5.$$

We construct the sequence of sets: $A_1 = A$, and

$$A_k = N_k A^k - N_k A^k, \quad \text{with } N_k = \frac{5}{24} 4^k - \frac{1}{3},$$

for every $k \geq 2$. Thus
$$A_2 = 3A^2 - 3A^2, \quad A_3 = 13A^3 - 13A^3, \ldots$$
Corollaries 3.6 and 3.7 show that for $k \geq 2$
$$|A_k| \geq \frac{|A|\,|A_{k-1}|(p-1)}{|A|\,|A_{k-1}| + p - 1}. \tag{8}$$
We can deduce from (8) an explicit lower bound for $|A_k|$.

Lemma 5.1. *For any k and $0 \leq U \leq |A|^k$ we have*
$$|A_k| \geq U - \frac{5}{4}\frac{U^2}{p-1}.$$

PROOF. Observe, that for any $u > 0$ we have
$$\frac{u}{1 + u/(p-1)} \geq u\left(1 - \frac{u}{p-1}\right);$$
the inequality can be rewritten as
$$\frac{u(p-1)}{u + p - 1} \geq u - \frac{u^2}{p-1}. \tag{9}$$
We use induction on k. For $k = 1$ the assertion of Lemma 5.1 is obvious. We assume that it holds for $k - 1 \geq 1$ and then prove it for k. By the induction hypothesis,
$$|A_{k-1}| \geq V := \frac{U}{|A|} - \frac{5}{4}\frac{U^2}{|A|^2(p-1)}.$$
If $V < 0$ then
$$U - \frac{5}{4}\frac{U^2}{p-1} < 0,$$
and the assertion of the lemma is trivial. If $V \geq 0$, then, applying (8) and (9), we have
$$|A_k| \geq \frac{|A|\,|A_{k-1}|(p-1)}{|A|\,|A_{k-1}| + p - 1} \geq \frac{|A|V(p-1)}{|A|V + p - 1}$$
$$\geq U - \frac{5}{4}\frac{U^2}{|A|(p-1)} - \frac{U^2}{p-1} = U - \frac{U^2}{p-1}\left(1 + \frac{5}{4|A|}\right),$$
which completes the proof of the lemma, by (7). $\square$

Lemma 5.2. *For any k we have*
$$|A_k| \geq \tfrac{3}{8}\min(|A|^k, (p-1)/2).$$

PROOF. This follows from Lemma 5.1 by taking
$$U = \min(|A|^k, (p-1)/2),$$
since we then have
$$U - \frac{5}{4}\frac{U^2}{p-1} \geq \frac{3}{8}U. \qquad \square$$

Lemma 5.3. *If*
$$2 \leq k \leq 1 + \frac{\log\bigl((p-1)/2\bigr)}{\log|A|}$$
then
$$|N_k A^k| \geq \tfrac{3}{8}|A|^{k-8/7}.$$

PROOF. By induction on k. For $k = 2$ the assertion is trivial, so we now prove the result for $k+1$ assuming that it is true for k. The hypothesis $k+1 \leq 1 + \log\bigl((p-1)/2\bigr)/(\log|A|)$ can be rewritten as $|A|^k \leq (p-1)/2$, so that Lemma 5.2 implies
$$|N_k A^k - N_k A^k| \geq \tfrac{3}{8}|A|^k.$$
Applying Corollary 4.3 with $k = 4$, we obtain from the induction hypothesis and the last displayed equation,
$$|4N_k A^k| \geq \bigl(\tfrac{3}{8}|A|^{k-8/7}\bigr)^{1/8}\bigl(\tfrac{3}{8}|A|^k\bigr)^{7/8} = \tfrac{3}{8}|A|^{k-1/7}.$$
Therefore
$$|N_{k+1}A^{k+1}| \geq |N_{k+1}A^k| = |(4N_k+1)A^k| \geq |4N_k A^k| \geq \tfrac{3}{8}|A|^{k-1/7},$$
as required. $\square$

6. The proof of Theorem 1.2

We may assume that $|A| \geq 5$ for, if not, the inequalities $|A| > p^\delta > p^{1/n}$ imply that $4^n > p$ and $|A| \geq 2$, so that $|A^n| \geq 2$. But then, by the Cauchy–Davenport theorem, we have
$$|NA^n| \geq \min(N|A^n| - N + 1, p) \geq \min(N+1, p) = p,$$
for any $N \geq 4^n > p$, and the theorem follows.

We may also assume that $4A \neq \mathbb{Z}_p$, else the theorem is trivial. Therefore the Cauchy–Davenport theorem implies that

(10) $$p - 1 \geq |4A| \geq 4|A| - 3 > 3|A|.$$

Let
$$n_0 = \left\lceil \frac{\log\bigl((p-1)/2\bigr)}{\log|A|} \right\rceil,$$
which is ≥ 1 by (10). We also have $n_0 \leq n-1$ since
$$|A|^n > p > (p-1)/2 \geq |A|^{n_0}.$$

We now prove the theorem when $n_0 = n-1$: By Lemma 5.2 and Lemma 5.3 we have
$$|N_{n-1}A^{n-1} - N_{n-1}A^{n-1}| \geq \tfrac{3}{8}|A|^{n-1},$$
$$|N_{n-1}A^{n-1}| \geq \tfrac{3}{8}|A|^{n-15/7},$$
respectively, for all $n \geq 3$. These inequalities hold also for $n = 2$ if we define $N_1 = 1$. Let
$$k = \lceil \log(1/\varepsilon)/\log 2 \rceil + 2.$$
By Corollary 4.3 we obtain
$$|kN_{n-1}A^{n-1}| \geq \tfrac{3}{8}|A|^{n-1-2^{4-k}/7} \geq \tfrac{3}{8}|A|^{n-1-\varepsilon}.$$
This inequality and (10) imply
$$|kN_{n-1}A^{n-1}||4A| > \tfrac{9}{8}|A|^{n-\varepsilon} > p,$$
and we are in position to use Lemma 2.2:
$$16(kN_{n-1}A^{n-1})(4A) = \mathbb{Z}_p.$$
Thus,
$$64kN_{n-1}A^n = \mathbb{Z}_p.$$

We observe that $64N_{n-1} \leq (10/3)4^n$ for $n \geq 3$ and $k \leq (9/2)\log(2+1/\varepsilon)$, and the theorem follows.

We now prove the theorem when $n_0 < n-1$: The definition of n_0 implies that $|A|^{n-1} > (p-1)/2$. By Lemma 5.2 and Lemma 5.3 we have

$$|N_{n-1}A^{n-1} - N_{n-1}A^{n-1}| \geq \frac{3}{16}(p-1),$$

$$|N_{n-1}A^{n-1}| \geq |N_{n_0}A^{n_0}| \geq \frac{3}{8}|A|^{n_0-8/7} > \frac{3}{16}(p-1)|A|^{-15/7},$$

respectively. By Corollary 4.3 we obtain

$$|3N_{n-1}A^{n-1}| \geq \frac{3}{16}(p-1)|A|^{-15/28}.$$

This inequality and (10) imply

$$|3N_{n-1}A^{n-1}||4A| > \frac{9}{16}(p-1)|A|^{13/28}.$$

By (7) and (10),
$$|A|^{13/28} > 2, \quad p-1 > \tfrac{8}{9}p.$$

Therefore
$$|3N_{n-1}A^{n-1}||4A| > p,$$
and we are in position to use Lemma 2.2:
$$16(3N_{n-1}A^{n-1})(4A) = \mathbb{Z}_p.$$

Thus
$$192N_{n-1}A^n = \mathbb{Z}_p.$$

This completes the proof of the theorem. $\square$

References

1. J. Bourgain, A. A. Glibichuk, and S. V. Konyagin, *Estimates for the number of sums and products and for exponential sums in fields of prime order*, J. London Math. Soc. (2) **73** (2006), 380–398.
2. J. Bourgain, N. Katz, and T. Tao, *A sum-product estimate in finite fields and their applications*, Geom. Funct. Anal. **14** (2004), no. 1, 27–57.
3. J. Bourgain and S. Konyagin, *Estimates for the number of sums and products and for exponential sums over subgroups in fields of prime order*, C. R. Math. Acad. Sci. Paris **337** (2003), no. 2, 75–80.
4. A. A. Glibichuk, *Combinational properties of sets of residues modulo a prime and the Erdős–Graham problem*, Mat. Zametki **79** (2006), no. 3, 384–395 (Russian); English transl., Math. Notes **79** (2006), no. 3-4, 356–365.
5. T. C. Tao and V. H. Vu, *Additive combinatorics*, Cambridge Stud. Adv, Math, vol. 105, Cambridge Univ. Press, Cambridge, 2006.
6. I. M. Vinogradov, *Elements of number theory*, translated by S. Kravetz, Dover Publ. Inc., New York, 1954.

Department of Mechanics and Mathematics, Moscow State University, Moscow, 119992, Russia

E-mail address, A. A. Glibichuk: `aanatol@mail.ru`

E-mail address, S. V. Konyagin: `konyagin@ok.ru`

Many Sets Have More Sums Than Differences

Greg Martin and Kevin O'Bryant

1. Introduction

Nathanson opined [6]:

> "Even though there exist sets A that have more sums than differences, such sets should be rare, and it must be true with the right way of counting that the vast majority of sets satisfies $|A - A| > |A + A|$."

The origin of this sentiment is that addition is commutative but subtraction is not — the set of sums

$$S + S := \{s_1 + s_2 : s_i \in S\}$$

of a finite set S is predisposed to be smaller than the set of differences

$$S - S := \{s_1 - s_2 : s_i \in S\}.$$

More precisely, the sizes of these two sets satisfy the bounds

(1) $\quad 2|S| - 1 \leq |S + S| \leq \frac{1}{2}|S|^2 + \frac{1}{2}|S|, \quad 2|S| - 1 \leq |S + S| \leq |S|^2 - |S| + 1.$

Moreover, the sizes of $S + S$ and $S - S$ are correlated: both lower bounds are achieved exactly for arithmetic progressions, and if either upper bound is achieved then both are achieved (such sets are called Sidon sets). Following this reasoning, one would expect that a vanishingly small proportion of the 2^n subsets of $\{0, 1, 2, \ldots, n-1\}$ have more sums than differences.

Our purpose, however, is to show that this is not the case.

The following terminology will be used throughout this article:

Definition. A finite set S is *difference-dominant* if $|S - S| > |S + S|$, *sum-dominant* if $|S + S| > |S - S|$, and *sum-difference-balanced* if $|S + S| = |S - S|$.

2000 *Mathematics Subject Classification.* Primary 11B05, 11B13, 11B25, 11B75.

The first author was supported in part by grants from the Natural Sciences and Engineering Research Council.

The second author was supported in part by a grant from The City University of New York PSC-CUNY Research Award Program.

The second author also acknowledges helpful discussions with Natella V. O'Bryant.

This is the final form of the paper.

Nathanson [7] calls sum-dominant sets "MSTD" sets, short for "More Sums Than Differences." We refer the reader to [6, 7] for the history of this problem.

The following examples show that none of the three categories is empty for $n \geq 15$:

Example. The set $S = \{0, 1, 3\}$ has $S + S = \{0, 1, 2, 3, 4, 6\}$ and $S - S = \{-3, -2, -1, 0, 1, 2, 3\}$; therefore S is difference-dominant, since $|S - S| = 7 > 6 = |S + S|$.

Example. The set $S = \{0, 2, 3, 4, 7, 11, 12, 14\}$ has $S + S = \{0, \ldots, 28\} \setminus \{1, 20, 27\}$ and $S - S = \{-14, \ldots, 14\} \setminus \{-13, -6, 6, 13\}$; therefore S is sum-dominant, since $|S + S| = 26 > 25 = |S - S|$.

Example. A set S is symmetric if $S = a^* - S$ for some $a^* \in \mathbb{R}$. Any symmetric set has $S + S = S + (a^* - S) = a^* + (S - S)$; therefore symmetric sets are sum-difference-balanced. In particular, any interval or arithmetic progression is sum-difference-balanced.

Our main theorem shows that, perhaps contrary to intuition, all three types of set in the above definition are ubiquitous.

Theorem 1. *Let P be any arithmetic progression of length n. A positive proportion of the subsets of P are difference-dominant, a positive proportion are sum-dominant, and a positive proportion are sum-difference-balanced. More precisely, there exists $c > 0$ such that for all $n \geq 15$,*

$$\#\{S \subseteq P : S \text{ is difference-dominant}\} > c2^n,$$
$$\#\{S \subseteq P : S \text{ is sum-dominant}\} > c2^n,$$
$$\#\{S \subseteq P : S \text{ is sum-difference-balanced}\} > c2^n.$$

We observe that the sizes of $S + S$ and $S - S$ are invariant under translation and dilation of S, so that without loss of generality we can restrict our attention to $P = \{0, 1, 2, \ldots, n - 1\}$.

The idea behind the proof of Theorem 1 is the following. The typical subset S of $\{0, \ldots, n - 1\}$ has about $n/2$ elements. Thus each $k \in \{0, \ldots, 2n - 2\}$ has, on average, roughly $n/4 - |n - k|/4$ representations as a sum of two elements of S. Not only is this expression positive, it is quite large except when k is near 0 or $2n - 2$. Similarly, each nonzero $k \in \{-(n-1), \ldots, n-1\}$ has, on average, roughly $n/4 - |k|/4$ representations as a difference of two elements of S. Not only is this expression positive, it is quite large except when $|k|$ is near $n - 1$. Putting these observations together, we see that the sizes of the sumset and difference set are predominantly impacted by the elements of S that are near 0 or near n.

If we choose the "fringe" of S cleverly, say by deciding in advance which of the first ℓ and the final u elements of $\{0, \ldots, n - 1\}$ will be in S, then the middle of S will be largely irrelevant. More precisely, we define specific sets $L \subseteq \{0, \ldots, \ell - 1\}$ and $U \subseteq \{n-u, \ldots, n-1\}$ such that $L+L$ omits exactly j elements of $\{0, \ldots, 2\ell-2\}$ and $U-L$ omits exactly k elements of $\{n-u-\ell, \ldots, n-1\}$. We then use probabilistic arguments to show that a positive proportion of the sets $A \subseteq \{0, 1, 2, \ldots, n - 1\}$ that satisfy $A \cap [0, \ell) = L$ and $A \cap [n - \ell, n) = U$ are missing exactly j sums and exactly $2k$ differences. By choosing L and U deliberately, we control the sumsets and difference sets of a positive proportion of subsets of $\{0, 1, 2, \ldots, n - 1\}$.

This philosophy suggests the following conjecture; see Section 7 for a more refined conjecture.

Conjecture 2. *Let P be any arithmetic progression with length n. The limiting proportions*

$$\rho_- = \lim_{n\to\infty} 2^{-n} \#\{S \subseteq P : S \text{ is difference-dominant}\}$$

$$\rho_+ = \lim_{n\to\infty} 2^{-n} \#\{S \subseteq P : S \text{ is sum-dominant}\}$$

$$\rho_= = \lim_{n\to\infty} 2^{-n} \#\{S \subseteq P : S \text{ is sum-difference-balanced}\}$$

all exist and are positive.

The following result, on the other hand, supports Nathanson's instinct as quoted above, with one interpretation of "the right way" and a suitably humble understanding of "vast." Theorem 3 is proved in Section 4.

Theorem 3. *Let P be any arithmetic progression with length n. On average, the difference set of a subset of P has 4 more elements than its sumset. More precisely,*

$$\frac{1}{2^n} \sum_{S \subseteq P} |S - S| \sim 2n - 7,$$

$$\frac{1}{2^n} \sum_{S \subseteq P} |S + S| \sim 2n - 11.$$

Nathanson [7] also asks for the possible values of $|A + A| - |A - A|$. We show by construction in Section 5 that the range of $|A + A| - |A - A|$ is $\mathbb{Z}$; in fact our constructions are economical, in the sense of the following theorem, which is the subject of Section 5:

Theorem 4. *For every integer x, there is a set $S \subseteq \{0, 1, \ldots, 17|x|\}$ with $|S + S| - |S - S| = x$.*

2. Sums and differences in randomly chosen sets

In this section, we establish several ancillary results on the probabilities that particular sums and differences are present or absent in sets chosen randomly from certain classes of sets. We will consider in particular the following classes: Let n, ℓ, and u be integers with $n \geq \ell + u$. Fix $L \subseteq \{0, \ldots, \ell - 1\}$ and $U \subseteq \{n - u, \ldots, n - 1\}$. We will consider the set of all subsets $A \subseteq \{0, \ldots, n - 1\}$ satisfying $A \cap \{0, \ldots, \ell - 1\} = L$ and $A \cap \{n - u, \ldots, n - 1\} = U$ as a probability space endowed with the uniform probability, where each such set A occurs with the probability $2^{-(n-\ell-u)}$.

All of the calculations in this section are straightforward, but the details depend upon the size and sometimes the parity of the particular sum or difference we are investigating, and so the lemmas herein are rather ugly. The reader with limited tolerance could scan Propositions 8 and 12 and move on to the next section without significantly interrupting the flow of ideas.

We begin with three lemmas describing the probabilities of particular sums missing from $A + A$, where A is chosen randomly from a class of the type indicated above.

Lemma 5. Let n, ℓ, and u be integers with $n \geq \ell + u$. Fix $L \subseteq \{0, \ldots, \ell - 1\}$ and $U \subseteq \{n - u, \ldots, n - 1\}$. Suppose that R is a uniformly randomly chosen subset of $\{\ell, \ldots, n - u - 1\}$, and set $A := L \cup R \cup U$. Then for any integer k satisfying $2\ell - 1 \leq k \leq n - u - 1$, the probability

$$\mathbb{P}\left[k \notin A + A\right] = \begin{cases} \left(\frac{1}{2}\right)^{|L|} \left(\frac{3}{4}\right)^{(k+1)/2-\ell}, & \text{if } k \text{ is odd,} \\ \left(\frac{1}{2}\right)^{|L|+1} \left(\frac{3}{4}\right)^{k/2-\ell}, & \text{if } k \text{ is even.} \end{cases}$$

PROOF. Define random variables X_j by setting $X_j = 1$ if $j \in A$ and $X_j = 0$ otherwise. By the definition of A, the variables X_j are independent random variables for $\ell \leq j \leq n - u - 1$, each taking the values 0 and 1 with probability $1/2$ each, while the variables X_j for $0 \leq j \leq \ell - 1$ and $n - u \leq j \leq n - 1$ have values that are fixed by the choices of L and U.

We have $k \notin A + A$ if and only if $X_j X_{k-j} = 0$ for all $0 \leq j \leq k/2$; the key point is that these variables $X_j X_{k-j}$ are independent of one another. Therefore

$$\mathbb{P}\left[k \notin A + A\right] = \prod_{0 \leq j \leq k/2} \mathbb{P}\left[X_j X_{k-j} = 0\right].$$

If k is odd, this becomes

$$\mathbb{P}\left[k \notin A + A\right] = \prod_{j=0}^{\ell-1} \mathbb{P}\left[X_j X_{k-j} = 0\right] \prod_{j=\ell}^{(k-1)/2} \mathbb{P}\left[X_j X_{k-j} = 0\right]$$

$$= \prod_{j \in L} \mathbb{P}\left[X_{k-j} = 0\right] \prod_{j=\ell}^{(k-1)/2} \mathbb{P}\left[X_j = 0 \text{ or } X_{k-j} = 0\right]$$

$$= \left(\frac{1}{2}\right)^{|L|} \left(\frac{3}{4}\right)^{(k+1)/2-\ell}.$$

On the other hand, if k is even then

$$\mathbb{P}\left[k \notin A + A\right] = \prod_{j=0}^{\ell-1} \mathbb{P}\left[X_j X_{k-j} = 0\right] \left(\prod_{j=\ell}^{k/2-1} \mathbb{P}\left[X_j X_{k-j} = 0\right]\right) \mathbb{P}\left[X_{k/2} X_{k/2} = 0\right]$$

$$= \prod_{j \in L} \mathbb{P}\left[X_{k-j} = 0\right] \left(\prod_{j=\ell}^{k/2-1} \mathbb{P}\left[X_j = 0 \text{ or } X_{k-j} = 0\right]\right) \mathbb{P}\left[X_{k/2} = 0\right]$$

$$= \left(\frac{1}{2}\right)^{|L|} \left(\frac{3}{4}\right)^{k/2-\ell} \cdot \frac{1}{2}. \qquad \square$$

Lemma 6. Let n, ℓ, and u be integers with $n \geq \ell + u$. Fix $L \subseteq \{0, \ldots, \ell - 1\}$ and $U \subseteq \{n - u, \ldots, n - 1\}$. Suppose that R is a uniformly randomly chosen subset of $\{\ell, \ldots, n - u - 1\}$, and set $A := L \cup R \cup U$. Then for any integer k satisfying $n + \ell - 1 \leq k \leq 2n - 2u - 1$, the probability

$$\mathbb{P}\left[k \notin A + A\right] = \begin{cases} \left(\frac{1}{2}\right)^{|U|} \left(\frac{3}{4}\right)^{n-(k+1)/2-u}, & \text{if } k \text{ is odd,} \\ \left(\frac{1}{2}\right)^{|U|+1} \left(\frac{3}{4}\right)^{n-1-k/2-u}, & \text{if } k \text{ is even.} \end{cases}$$

PROOF. This follows from Lemma 5 applied to the parameters $\ell' = u$ and $L' = n - 1 - U$, $u' = \ell$ and $U' = n - 1 - L$, and $A' = n - 1 - A$ and $k' = 2n - 2 - k$. $\square$

Lemma 7. *Suppose that A is a uniformly randomly chosen subset of $\{0, \ldots, n-1\}$. Then for any integer $0 \leq k \leq n-1$, the probability*

$$\mathbb{P}[k \notin A + A] = \begin{cases} \left(\frac{3}{4}\right)^{(k+1)/2}, & \text{if } k \text{ is odd,} \\ \frac{1}{2}\left(\frac{3}{4}\right)^{k/2}, & \text{if } k \text{ is even;} \end{cases}$$

while for any integer $n-1 \leq k \leq 2n-2$, the probability

$$\mathbb{P}[k \notin A + A] = \begin{cases} \left(\frac{3}{4}\right)^{n-(k+1)/2}, & \text{if } k \text{ is odd,} \\ \frac{1}{2}\left(\frac{3}{4}\right)^{n-1-k/2}, & \text{if } k \text{ is even.} \end{cases}$$

PROOF. This follows immediately from Lemmas 5 and 6 upon setting $\ell = u = 0$ and $L = U = \varnothing$. □

We now use these lemmas to establish the following proposition, in which we want a positive probability that many integers k appear in the sumset $A+A$. While these events, varying over k, are not independent, we need only a lower bound on the probability; hence it suffices to combine crudely the exact probabilities given in Lemmas 5 and 6. We emphasize that we have made no effort to optimize the lower bound given in the following proposition.

Proposition 8. *Let n, ℓ, and u be integers with $n \geq \ell + u$. Fix $L \subseteq \{0, \ldots, \ell-1\}$ and $U \subseteq \{n-u, \ldots, n-1\}$. Suppose that R is a uniformly randomly chosen subset of $\{\ell, \ldots, n-u-1\}$, and set $A := L \cup R \cup U$. Then the probability that*

$$\{2\ell - 1, \ldots, n - u - 1\} \cup \{n + \ell - 1, \ldots, 2n - 2u - 1\} \subseteq A + A$$

is greater than $1 - 6(2^{-|L|} + 2^{-|U|})$.

PROOF. We employ the crude inequality

$$\mathbb{P}[\{2\ell - 1, \ldots, n - u - 1\} \cup \{n + \ell - 1, \ldots, 2n - 2u - 1\} \not\subseteq A + A]$$
$$\leq \sum_{k=2\ell-1}^{n-u-1} \mathbb{P}[k \notin A + A] + \sum_{k=n+\ell-1}^{2n-2u-1} \mathbb{P}[k \notin A + A].$$

The first sum can be bounded, using Lemma 5, by

$$\sum_{k=2\ell-1}^{n-u-1} \mathbb{P}[k \notin A + A] < \sum_{\substack{k \geq 2\ell-1 \\ k \text{ odd}}} \left(\tfrac{1}{2}\right)^{|L|} \left(\tfrac{3}{4}\right)^{(k+1)/2-\ell} + \sum_{\substack{k \geq 2\ell-1 \\ k \text{ even}}} \left(\tfrac{1}{2}\right)^{|L|+1} \left(\tfrac{3}{4}\right)^{k/2-\ell}$$
$$= \left(\tfrac{1}{2}\right)^{|L|} \sum_{m=0}^{\infty} \left(\tfrac{3}{4}\right)^m + \left(\tfrac{1}{2}\right)^{|L|+1} \sum_{m=0}^{\infty} \left(\tfrac{3}{4}\right)^m = 6\left(\tfrac{1}{2}\right)^{|L|}.$$

The second sum can be bounded in a similar way using Lemma 6, yielding

$$\sum_{k=n+\ell-1}^{2n-2u-1} \mathbb{P}[k \notin A + A] < 6\left(\tfrac{1}{2}\right)^{|U|}.$$

Therefore $\mathbb{P}[\{2\ell - 1, \ldots, n - u - 1\} \cup \{n + \ell - 1, \ldots, 2n - 2u - 1\} \not\subseteq A + A]$ is bounded above by $6(1/2)^{|L|} + 6(1/2)^{|U|}$, which is equivalent to the statement of the proposition. □

We turn now to three lemmas describing the probabilities that particular differences are missing from $A - A$, where A is chosen randomly from one of our classes. A new obstacle appears: while the random variables $X_j X_{k-j}$ controlling the presence of the sum k in $A+A$ are always mutually independent, the same is not true of the random variables $X_j X_{k+j}$ controlling the presence of the difference k in $A - A$, at least when k is small enough that j, $k+j$, and $2k+j$ can all lie between 0 and $n-1$. Fortunately, when k is this small the probabilities in question are already minuscule, so a simple argument provides a serviceable bound (Lemma 10 below).

Lemma 9. *Let n, ℓ, and u be integers with $n \geq \ell + u$. Fix $L \subseteq \{0, \ldots, \ell-1\}$ and $U \subseteq \{n-u, \ldots, n-1\}$. Suppose that R is a uniformly randomly chosen subset of $\{\ell, \ldots, n-u-1\}$, and set $A := L \cup R \cup U$. Then for any integer k satisfying $n/2 \leq k \leq n-u-\ell$, the probability*

$$\mathbb{P}[k \notin A - A] = \left(\tfrac{1}{2}\right)^{|L|+|U|} \left(\tfrac{3}{4}\right)^{n-\ell-u-k}.$$

PROOF. Define random variables X_j by setting $X_j = 1$ if $j \in A$ and $X_j = 0$ otherwise, as in the proof of Lemma 5. We have $k \notin A - A$ if and only if $X_j X_{k+j} = 0$ for all $0 \leq j \leq n-1-k$, and again these variables $X_j X_{k+j}$ are independent of one another. Therefore

$$\mathbb{P}[k \notin A - A] = \prod_{j=0}^{n-1-k} \mathbb{P}[X_j X_{k+j} = 0]$$

$$= \prod_{j=0}^{\ell-1} \mathbb{P}[X_j X_{k+j} = 0] \prod_{j=\ell}^{n-u-1-k} \mathbb{P}[X_j X_{k+j} = 0] \prod_{j=n-u-k}^{n-1-k} \mathbb{P}[X_j X_{k+j} = 0]$$

$$= \prod_{j \in L} \mathbb{P}[X_{k+j} = 0] \prod_{j=\ell}^{n-u-1-k} \mathbb{P}[X_j = 0 \text{ or } X_{k+j} = 0] \prod_{j \in U-k} \mathbb{P}[X_j = 0]$$

$$= \left(\tfrac{1}{2}\right)^{|L|} \left(\tfrac{3}{4}\right)^{n-\ell-u-k} \left(\tfrac{1}{2}\right)^{|U|}. \qquad \square$$

Lemma 10. *Let a and b be integers with $a < b$. Suppose that R is a uniformly randomly chosen subset of $\{a, \ldots, b-1\}$. Then for any integer k satisfying $1 \leq k \leq 2(b-a)/3$, the probability*

$$\mathbb{P}[k \notin R - R] \leq \left(\tfrac{3}{4}\right)^{(b-a)/3}.$$

Remark. In fact, the probability in question can be written exactly in terms of products of Fibonacci numbers: in the simplest case, $\mathbb{P}[1 \notin R - R] = F_{b-a+2}/2^{b-a}$. However, the resulting expressions would become too tedious to handle in our applications below. When $b - a$ is large and k is small, the actual value of the probability $\mathbb{P}[k \notin R - R]$ is proportional to $((1+\sqrt{5})/4)^{b-a-k} \approx 0.809^{b-a}$, whereas the bound in Lemma 9 gives $(3/4)^{(b-a)/3} \approx 0.909^{b-a}$. However, in the particular case $k = (b-a)/2$, the probability in question is exactly $(3/4)^{(b-a)/2} \approx 0.866^{b-a}$, so the bound in Lemma 9 is not too unreasonable.

PROOF. Define the set

$$J := \left\{ a \leq j < b - k : \left\lfloor \tfrac{j-a}{k} \right\rfloor \text{ is even} \right\}.$$

In other words, J contains the first k integers starting at a, then omits the following k integers, then contains the next k integers, and so on until the upper bound $a + 2(b-a)/3$ is reached. The following properties of J can be easily verified:

(i) if $j \in J$, then $j + k \notin J$;
(ii) $|J| \geq (b-a)/3$.

Now define random variables X_j by setting $X_j = 1$ if $j \in R$ and $X_j = 0$ otherwise, as in the proof of Lemma 10. We have $k \notin R - R$ if and only if $X_j X_{k+j} = 0$ for all $a \leq j < b - k$.

$$\mathbb{P}\left[k \notin R - R\right] = \mathbb{P}\left[X_j X_{k+j} = 0 \text{ for all } a \leq j < b - k\right]$$
$$\leq \mathbb{P}\left[X_j X_{k+j} = 0 \text{ for all } j \in J\right].$$

However, property (i) above ensures that the random variables $X_j X_{k+j}$ are independent of one another as j ranges over J. Therefore

$$\mathbb{P}\left[k \notin R - R\right] \leq \prod_{j \in J} \mathbb{P}\left[X_j X_{k+j} = 0\right] = \left(\tfrac{3}{4}\right)^{|J|} \leq \left(\tfrac{3}{4}\right)^{(b-a)/3}$$

by property (ii) above. $\square$

Lemma 11. *Suppose that A is a uniformly randomly chosen subset of $\{0, \ldots, n-1\}$. Then for any integer $1 \leq k \leq n/2$, the probability $\mathbb{P}\left[k \notin A - A\right] \leq (3/4)^{n/3}$, while for any integer $n/2 \leq k \leq n - 1$, the probability $\mathbb{P}\left[k \notin A - A\right] = (3/4)^{n-k}$.*

PROOF. The first assertion follows immediately from Lemma 10 upon setting $a = 0$ and $b = n$, while the second assertion follows immediately from Lemma 9 upon setting $\ell = u = 0$ and $L = U = \emptyset$. $\square$

We now use these lemmas to establish the following proposition, in which we want a positive probability that many integers k appear in the difference set $A - A$. Again it suffices to combine crudely the results of Lemmas 9 and 10, since we need only a lower bound on the probability. Once again we have emphasized ease of exposition over optimization of the lower bound itself; in particular, we could have achieved better constants at the expense of uglier technicalities.

Proposition 12. *Let n, ℓ, and u be integers with $n \geq 4(\ell + u)$. Fix $L \subseteq \{0, \ldots, \ell - 1\}$ and $U \subseteq \{n - u, \ldots, n - 1\}$. Suppose that R is a uniformly randomly chosen subset of $\{\ell, \ldots, n - u - 1\}$, and set $A := L \cup R \cup U$. Then the probability that*

$$\{-(n - \ell - u), \ldots, n - \ell - u\} \subseteq A - A$$

is greater than $1 - 4(1/2)^{|L|+|U|} - (n/2)(3/4)^{(n-\ell-u)/3}$.

PROOF. By the symmetry of $A - A$ about 0 and the fact that $0 \in A - A$ for any nonempty set A, it suffices to show that $A - A$ contains $\{1, \ldots, n - \ell - u\}$. We employ the crude inequality

$$\mathbb{P}\left[\{1, \ldots, n - \ell - u\} \not\subseteq A - A\right] \leq \sum_{k=1}^{n-\ell-u} \mathbb{P}\left[k \notin A - A\right]$$
$$\leq \sum_{1 \leq k \leq n/2} \mathbb{P}\left[k \notin R - R\right] + \sum_{n/2 < k \leq n - \ell - u} \mathbb{P}\left[k \notin A - A\right].$$

The first sum can be bounded using Lemma 10 with $a = \ell$ and $b = n - u$; it is here that we use the hypothesis $n \geq 4(\ell + u)$, to guarantee that every k in the range

$1 \le k \le n/2$ satisfies $k \le 2(n-\ell-u)/3$. We obtain
$$\sum_{1 \le k \le n/2} \mathbb{P}[k \notin R-R] \le \tfrac{n}{2}\left(\tfrac{3}{4}\right)^{(n-\ell-u)/3}.$$

The second sum can be bounded using Lemma 6, yielding
$$\sum_{n/2 < k \le n-\ell-u} \mathbb{P}[k \notin A-A] < \sum_{k=-\infty}^{n-\ell-u} \left(\tfrac{1}{2}\right)^{|L|+|U|}\left(\tfrac{3}{4}\right)^{n-\ell-u-k} = 4\left(\tfrac{1}{2}\right)^{|L|+|U|}.$$

Therefore $\mathbb{P}[\{-(n-\ell-u),\ldots,n-\ell-u\} \not\subseteq A-A]$ is bounded above by $4(1/2)^{|L|+|U|} + (n/2)(3/4)^{(n-\ell-u)/3}$, which is equivalent to the statement of the proposition. □

3. Proof of Theorem 1

In this section we show that the collections of sum-dominant sets, difference-dominant sets, and sum-difference-balanced sets all have positive lower density. Our strategy is to fix the "fringes" of a subset of $\{0,1,2,\ldots,n-1\}$ (that is, stipulate which integers close to 0 and $n-1$ are and are not in the set) in a way that forces the set to have missing differences (or sums). We then use the probabilistic lemmas of the previous section to show that for many sets with the prescribed fringes, all other sums (or differences) will be present. We have not attempted to optimize the constants appearing in the following three theorems, in part because the previous section would have become even more technical and ugly, and in part because we were unlikely to have come close to the true constants (see Conjecture 18 below) in any event.

We begin by showing that a positive proportion of sets are sum-dominant. Here, choosing appropriate fringes is most non-trivial, compared to the two theorems that follow.

Theorem 13. *For $n \ge 15$, the number of sum-dominant subsets of $\{0,1,2,\ldots, n-1\}$ is at least $(2 \times 10^{-7})2^n$.*

PROOF. First, note that the bound $(2 \times 10^{-7})2^n$ is less than 1 for $15 \le n \le 22$; the existence of the single sum-dominant set $\{0,2,3,4,7,11,12,14\}$ is enough to verify the theorem in that range. Henceforth we can assume that $n \ge 23$.

Define $L := \{0,2,3,7,8,9,10\}$ and $U := \{n-11, n-10, n-9, n-8, n-6, n-3, n-2, n-1\}$. We show that the number of sum-dominant subsets $A \subseteq \{0,1,2,\ldots, n-1\}$ satisfying $A \cap \{0,\ldots,10\} = L$ and $A \cap \{n-11,\ldots,n-1\} = U$ is at least $(2 \times 10^{-7})2^n$. For any such A, the fact that $U-L$ does not contain $n-7$ implies that $A-A$ contains neither $n-7$ nor $-(n-7)$; since $A-A \subseteq \{-(n-1),\ldots,n-1\}$, we see that
$$|A-A| \le 2n-3.$$

Therefore it suffices to show that there are at least $(2 \times 10^{-7})2^n$ sets A, satisfying $A \cap \{0,\ldots,10\} = L$ and $A \cap \{n-11,\ldots,n-1\} = U$, for which $|A+A| \ge 2n-2$.

For any such A, we see by direct calculation that $A+A$ contains
$$L+L = \{0,\ldots,20\} \setminus \{1\},$$
$$L+U = \{n-11,\ldots,n+9\},$$
$$U+U = \{2n-22,\ldots,2n-2\}.$$

In particular, if $23 \leq n \leq 32$ then $A+A$ automatically equals $\{0,\ldots,2n-2\}\setminus\{1\}$, giving $|A+A|=2n-2$; the number of such A is exactly $2^{n-22} > (2\times 10^{-7})2^n$, since there are $n-22$ numbers between 11 and $n-12$ inclusive.

For $n \geq 33$, Proposition 8 (applied with $\ell = r = 11$) tells us that when A is chosen uniformly randomly from all such sets, the probability that $A+A$ contains $\{21,\ldots,n-12\} \cup \{n+10,\ldots,2n-23\}$ is at least

$$1 - 6(2^{-|L|} + 2^{-|U|}) = 1 - 6(2^{-7} + 2^{-8}) = \frac{119}{128}.$$

In other words, there are at least $2^{n-22} \cdot 119/128 > (2 \times 10^{-7})2^n$ such sets A. For all these sets, we see that $A+A$ again equals $\{0,\ldots,2n-2\}\setminus\{1\}$, and hence all such sets are sum-dominant. □

The next two theorems carry out a similar approach to showing that a positive proportion of sets are difference-dominant or sum-difference-balanced. These two results appeal to the serviceable but crude Lemma 10, and consequently the constants that appear, as well as the computation needed to take care of smaller values of n, are likewise far from optimal.

Theorem 14. *For $n \geq 4$, the number of difference-dominant subsets of $\{0,1,2,\ldots,n-1\}$ is at least $0.0015 \cdot 2^n$.*

PROOF. The bound can be verified computationally for small n: we have computed by exhaustive search for $n \leq 27$ the number of difference-dominant subsets $\{0,1,2,\ldots,n-1\}$ that contain both 0 and $n-1$. Counting just these sets and their translates is enough to prove this theorem for $n \leq 39$. Henceforth, we assume that $n \geq 40$.

Define $L := \{0,2,3\}$ and $U := \{n-2,n-1\}$. We show that the number of difference-dominant subsets $A \subseteq \{0,1,2,\ldots,n-1\}$ satisfying $A \cap \{0,1,2,3\} = L$ and $A \cap \{n-2,n-1\} = U$ is at least $0.0015 \cdot 2^n$. For any such A, the fact that $L+L$ does not contain 1 implies that $A+A$ does not contain 1, and so $|A+A| \leq 2n-2$. Therefore it suffices to show that there are at least $0.0015 \cdot 2^n$ sets A, satisfying $A \cap \{0,1,2,3\} = L$ and $A \cap \{n-2,n-1\} = U$, for which $|A-A| = 2n-1$.

For any such A, we see by direct calculation that $A - A$ contains

$$(L-U) \cup (U-L) = \{-(n-5),\ldots,-(n-1)\} \cup \{n-5,\ldots,n-1\}.$$

Furthermore, Proposition 12 (applied with $\ell = 4$, $u = 2$, and $n \geq 24$) tells us that when A is chosen uniformly randomly from all such sets, the probability that $A - A$ contains $\{-(n-6),\ldots,n-6\}$ is at least

$$1 - 4\left(\frac{1}{2}\right)^{|L|+|U|} - \left(\frac{n}{2}\right)\left(\frac{3}{4}\right)^{(n-\ell-u)/3} = 1 - 4\left(\frac{1}{2}\right)^5 - \left(\frac{n}{2}\right)\left(\frac{3}{4}\right)^{(n-6)/3}$$

$$= \frac{7}{8} - \frac{8n}{9}\left(\frac{3}{4}\right)^{n/3}.$$

As a function of n, this expression is increasing for $n \geq 11$, and at $n = 40$ its value is larger than 0.107536. In other words, there are at least $2^{n-6} \cdot 0.107536 > 0.0015 \cdot 2^n$ such sets A. For all these sets, we see that $A - A$ equals $\{-(n-1),\ldots,n-1\}$, and hence all such sets are difference-dominant. □

Theorem 15. *For $n \geq 1$, the number of sum-difference-balanced subsets of $\{0,1,2,\ldots,n-1\}$ is at least $(8.7 \times 10^{-7})2^n$.*

PROOF. The bound can be verified computationally for small n: for $n \leq 27$ we have computed the exact number of sum-difference-balanced subsets of $\{0, 1, 2, \ldots, n-1\}$ that contain both 0 and $n-1$. Counting only these sets and their translates proves the theorem for $n \leq 47$. Henceforth, we assume that $n \geq 48$.

Define $L := \{0, \ldots, 5\}$ and $U := \{n-6, \ldots, n-1\}$. We give a lower bound for the number of sum-difference-balanced subsets $A \subseteq \{0, 1, 2, \ldots, n-1\}$ satisfying $L \cup U \subseteq A$; in fact we show that the number of such subsets with $|A+A| = |A-A| = 2n-1$, the maximum possible size, is at least $(8.7 \times 10^{-7}) 2^n$. Combining Propositions 8 and 12 (applied with $\ell = u = 6$), we find that when A is chosen uniformly randomly from all such sets, the probability that both $A+A$ and $A-A$ are as large as possible is at least

$$(2) \qquad 1 - 6(2^{-|L|} + 2^{-|U|}) - 4\left(\frac{1}{2}\right)^{|L|+|U|} - \left(\frac{n}{2}\right)\left(\frac{3}{4}\right)^{(n-\ell-u)/3} = \frac{831}{1024} - \frac{128n}{81}\left(\frac{3}{4}\right)^{n/3}.$$

This function is increasing for $n \geq 11$ and takes a value larger than 0.05129 when $n = 48$. In other words, there are at least $2^{n-12} \cdot 0.05129 > (8.7 \times 10^{-7}) 2^n$ such sets A. For all these sets, we see that $A+A$ equals $\{0, \ldots, 2n-2\}$ and $A-A$ equals $\{-(n-1), \ldots, n-1\}$, and hence all such sets are sum-difference-balanced. □

4. Average values

In this section, we prove Theorem 3 by calculating the average values of $|S-S|$ and $|S+S|$ as S ranges over an arithmetic progression P of length n. Since the problem is invariant under dilations and translations, it suffices to prove the theorem in the case $P = \{0, 1, 2, \ldots, n-1\}$.

We begin by addressing the average cardinality of the sumset $S+S$. In fact, we can give an exact formula for the average size of the sumset, or equivalently for the sum of the sizes of all sumsets as S ranges over subsets of $\{0, 1, 2, \ldots, n-1\}$. The reason we can do so is essentially because of the linearity of expectations of random variables.

Theorem 16. *For any positive integer n, we have*

$$(2) \qquad \sum_{S \subseteq \{0,\ldots,n-1\}} |S+S| = 2^n(2n-11) + \begin{cases} 19 \cdot 3^{(n-1)/2}, & \text{if } n \text{ is odd,} \\ 11 \cdot 3^{n/2}, & \text{if } n \text{ is even.} \end{cases}$$

PROOF. We begin with the manipulation

$$(3) \qquad \sum_{S \subseteq \{0,\ldots,n-1\}} |S+S| = \sum_{S \subseteq \{0,\ldots,n-1\}} \sum_{\substack{0 \leq k \leq 2n-2 \\ k \in S+S}} 1 = \sum_{k=0}^{2n-2} \sum_{\substack{S \subseteq \{0,\ldots,n-1\} \\ k \in S+S}} 1$$

$$= \sum_{k=0}^{2n-2} 2^n \mathbb{P}[k \in S+S] = 2^n(2n-1) - 2^n \sum_{k=0}^{2n-2} \mathbb{P}[k \notin S+S].$$

We suppose that $n = 2m+1$ is odd, the case where n is even being similar. We begin by considering only the lower half of possible values for k. By Lemma 7, we have

$$\sum_{k=0}^{n-2} \mathbb{P}[k \notin S+S] = \sum_{j=0}^{m-1} \mathbb{P}[2j \notin S+S] + \sum_{j=0}^{m-1} \mathbb{P}[2j+1 \notin S+S]$$

$$= \sum_{j=0}^{m-1} \tfrac{1}{2}\bigl(\tfrac{3}{4}\bigr)^j + \sum_{j=0}^{m-1} \bigl(\tfrac{3}{4}\bigr)^{j+1} = 5\bigl(1 - \bigl(\tfrac{3}{4}\bigr)^m\bigr).$$

By the symmetry of $S+S$ about $n-1$, the same calculation holds for $\sum_{k=n}^{2n-2} \mathbb{P}[k \notin S+S]$. Therefore, appealing to Lemma 7 again for $k = n-1 = 2m$,

$$\sum_{k=0}^{2n-2} \mathbb{P}[k \notin S+S] = 5\bigl(1-\bigl(\tfrac{3}{4}\bigr)^m\bigr) + \tfrac{1}{2}\bigl(\tfrac{3}{4}\bigr)^m + 5\bigl(1-\bigl(\tfrac{3}{4}\bigr)^m\bigr) = 10 - \tfrac{19}{2}\bigl(\tfrac{3}{4}\bigr)^{(n-1)/2}.$$

Inserting this value into the right-hand side of equation (3) establishes the lemma for odd n. A similar calculation gives the result for even n. $\square$

While it is possible to write down an exact formula for the average size of the difference set $S - S$ as S ranges over all subsets of $\{0, 1, 2, \ldots, n-1\}$, the formula would be far too ugly to be of use. We prefer in this case to present a simple asymptotic formula with a reasonable error term.

Theorem 17. *For any positive integer n, we have*

(4) $$\sum_{S \subseteq \{0,\ldots,n-1\}} |S - S| = 2^n(2n - 7) + O\bigl(n 6^{n/3}\bigr).$$

PROOF. As in the proof of the previous theorem, we have

(5) $$\sum_{S \subseteq \{0,\ldots,n-1\}} |S - S| = \sum_{S \subseteq \{0,\ldots,n-1\}} \sum_{\substack{-(n-1) \le k \le n-1 \\ k \in S - S}} 1 = \sum_{k=-(n-1)}^{n-1} \sum_{\substack{S \subseteq \{0,\ldots,n-1\} \\ k \in S - S}} 1$$

$$= \sum_{k=-(n-1)}^{n-1} 2^n \mathbb{P}[k \in S - S]$$

$$= 2^n(2n-1) - 2^n \sum_{k=-(n-1)}^{n-1} \mathbb{P}[k \notin S - S]$$

$$= 2^n(2n-1) - 1 - 2^{n+1} \sum_{k=1}^{n-1} \mathbb{P}[k \notin S - S],$$

the last equality following from the symmetry of $S - S$ around 0 and the fact that 0 is in $S - S$ for nonempty S. From Lemma 11 we have

$$\sum_{k=1}^{\lceil n/2 \rceil - 1} \mathbb{P}[k \notin S - S] \le \sum_{k=1}^{\lceil n/2 \rceil - 1} \bigl(\tfrac{3}{4}\bigr)^{n/3} \le n\bigl(\tfrac{3}{4}\bigr)^{n/3}$$

and

$$\sum_{k=\lceil n/2 \rceil}^{n-1} \mathbb{P}[k \notin S - S] = \sum_{k=\lceil n/2 \rceil}^{n-1} \bigl(\tfrac{3}{4}\bigr)^{n-k} = 3\bigl(1 - \bigl(\tfrac{3}{4}\bigr)^{n - \lceil n/2 \rceil}\bigr),$$

which combine to give

$$\sum_{k=1}^{n-1} \mathbb{P}[k \notin S-S] = 3 + O\left(n\left(\tfrac{3}{4}\right)^{n/3}\right).$$

Inserting this expression into the right-hand side of equation (5) establishes the theorem. □

Examining the derivations of these two theorems reveals that it really is the commutativity $s_1 + s_2 = s_2 + s_1$ that causes the difference in the average sizes of $S+S$ and $S-S$: a typical potential element of $S+S$ has only about half as many chances to be realized as a sum as the corresponding potential element of $S-S$ has at being realized as a difference. To further emphasize this observation, we note that if the single set S is replaced by two sets S and T, the disparity disappears: for an arithmetic progression P of length n, we have

$$\frac{1}{2^{2n}} \sum_{S \subseteq P} \sum_{T \subseteq P} |S-T| \sim \frac{1}{2^{2n}} \sum_{S \subseteq P} \sum_{T \subseteq P} |S+T| \sim 2n - 7.$$

5. Sets with prescribed imbalance between sums and differences

In this section we prove that the range of possible values for $|S+S|-|S-S|$ is all of $\mathbb{Z}$. Furthermore, as asserted in Theorem 4, our constructions show that for every integer x, we can find a subset S of $\{0,\ldots,17|x|\}$ such that $|S+S|-|S-S|=x$. As one might expect from the foregoing discussion, the case where x is negative is easiest.

Negative values of x. For any integer $x < 0$, set $S_x = \{0,\ldots,|x|+1\} \cup \{2|x|+2\}$. Then $S_x + S_x = \{0,1,\ldots,3|x|+3\} \cup \{4|x|+4\}$ while $S_x - S_x = \{-(2|x|+2),\ldots,2|x|+2\}$, whereupon

$$|S_x + S_x| - |S_x - S_x| = (3|x|+5) - (4|x|+5) = -|x| = x.$$

Even more generally, take any integer $n \geq |x|+2$ and set $S = \{0,\ldots,n-1\} \cup \{n+|x|\}$. Then $S+S = \{0,\ldots,2n+|x|-1\} \cup \{2n+2|x|\}$ and $S-S = \{-(n+|x|),\ldots,n+|x|\}$, which again yields $|S+S|-|S-S| = x$.

We turn now to nonpositive values of x. Our general construction works for larger values of x, but we need to handle a few small values of x individually.

A few special cases. For a few small values of x, we find suitable sets S_x simply by computation: if we set

(6)
$$\begin{aligned} S_0 &:= \varnothing \\ S_1 &:= \{0, 2, 3, 4, 7, 11, 12, 14\} \\ S_2 &:= \{0, 1, 2, 4, 5, 9, 12, 13, 14, 16, 17\} \\ S_4 &:= \{0, 1, 2, 4, 5, 9, 12, 13, 17, 20, 21, 22, 24, 25\}, \end{aligned}$$

then in each case it can be checked that $|S_x + S_x| - |S_x - S_x| = x$. In fact, these examples are all minimal in the sense that the diameter $\max S - \min S$

is as small as possible. (Vishaal Kapoor and Erick Wong confirmed computationally the fact that S_4 is the unique, up to reflection, set of integers of diameter at most 25 for which the sumset has four more elements than the difference set. We note that Pigarev and Freĭman [9] gave the slightly larger example $S'_4 = \{0, 1, 2, 4, 5, 9, 12, 13, 14, 16, 17, 21, 24, 25, 26, 28, 29\}$, which also satisfies $|S'_4 + S'_4| - |S'_4 - S'_4| = 4$.)

In fact, these diameter-minimal examples are unique, up to reflection, except for S_1: there are two other subsets of $\{0, \ldots, 14\}$, namely

$$S'_1 = \{0, 1, 2, 4, 5, 9, 12, 13, 14\}$$

and its reflection, for which the sumset has one element more than the difference set. The first set S_1 has only eight elements, as compared with the nine elements of S'_1. In fact, Hegarty [2] has shown that S_1 is also *the* sum-dominant set with the smallest cardinality, unique up to dilation, translation, and reflection. On the other hand, there are tantalizing similarities among the sets S'_1, S_2, S_4, and S'_4 that might admit a clever generalization.

We note that Ruzsa [11] claimed that $U = \{0, 1, 3, 4, 5, 6, 7, 10\}$ is sum-dominant, but this is incorrect: both $U + U$ and $U - U$ have 19 elements. We also mention the following observation of Hegarty: if one sets

$$A = S_4 \cup (S_4 + 20)$$
$$= \{0, 1, 2, 4, 5, 9, 12, 13, 17, 20, 21, 22, 24, 25, 29, 32, 33, 37, 40, 41, 42, 44, 45\},$$

then one has $|A + A| = 91$ and $|A - A| = 83$, providing the statistic $\log 91 / \log 83 = 1.0208\ldots$ which is important when using the elements of A as "digits." More precisely, considering sets of the form $A_n = A + bA + b^2 A + \cdots + b^{n-1} A$ for suitably large fixed b, we have $|A_n + A_n| = |A_n - A_n|^{1.0208\cdots}$, which is currently the best exponent known.

For other positive values of x, the basic general construction is an adaptation of the subset $S_1 \times \{0, \ldots, k\}$ of $\mathbb{Z} \times \mathbb{Z}$, embedded in $\mathbb{Z}$ itself by a common technique of regarding the coordinates as digits in a base-b representation for suitably large b. In our simple case, we can be completely explicit from the start.

Odd values of x exceeding 1. Let $x = 2k + 1$ with $k \geq 1$. With S_1 defined as in equation (6), set

(7) $\quad S_{2k+1} = S_1 + \{0, 29, 58, \ldots, 29k\}$
$$= \{0 \leq s \leq 29k + 14 \colon s \equiv 0, 2, 3, 4, 5, 11, 12, \text{or } 14 \pmod{29}\}.$$

Then we find that

$$S_{2k+1} + S_{2k+1} = (S_1 + S_1) + \{0, 29, 58, \ldots, 58k\}$$
$$= \{0 \leq s < 29(2k+1) \colon s \not\equiv 1, 20, \text{or } 27 \pmod{29}\},$$

which reveals that $|S_{2k+1} + S_{2k+1}| = 26(2k+1)$. On the other hand,

$$S_{2k+1} - S_{2k+1} = \{-29\bigl(k + \tfrac{1}{2}\bigr) < s < 29\bigl(k + \tfrac{1}{2}\bigr) \colon s \not\equiv -13, -6, 6, \text{or } 13 \pmod{29}\},$$

showing that $|S_{2k+1} - S_{2k+1}| = 25(2k+1)$, and so $|S_{2k+1} + S_{2k+1}| - |S_{2k+1} - S_{2k+1}| = 2k + 1$ as desired.

Even values of x exceeding 4. Let $x = 2k$ with $k \geq 3$. With S_{2k+1} defined as in equation (7), set $S_{2k} = S_{2k+1} \setminus \{29\}$. One can check that $S_{2k} - S_{2k}$ still equals all of $S_{2k+1} - S_{2k+1}$ but that $S_{2k} + S_{2k} = (S_{2k+1} + S_{2k+1}) \setminus \{29\}$. Therefore

$$|S_{2k} + S_{2k}| - |S_{2k} - S_{2k}| = |S_{2k+1} + S_{2k+1}| - |S_{2k+1} - S_{2k+1}| - 1 = 2k$$

as desired. Notice that S_{2k} is indeed contained in $\{0, \ldots, 17(2k)\}$ as asserted by Theorem 4, the closest call being the comparison between $\max S_6 = 101$ and $17 \cdot 6 = 102$.

We note that as this manuscript was in preparation, Hegarty [2, Theorem 9] independently proved that $|S + S| - |S - S|$ can take all integer values t. In fact he proved, extending ideas originating in our proof of Theorem 1, somewhat more: for each fixed integer t, if n is sufficiently large then a positive proportion of subsets S of $\{0, 1, 2, \ldots, n-1\}$ satisfy $|S + S| - |S - S| = t$.

6. Analysis of data

Theorem 3 gave the expected values of $|S + S|$ and $|S - S|$, which seems most naturally phrased as saying that the expected number of missing sums is asymptotically 10, while the expected number of missing differences is asymptotically 6. One is naturally led to enquire as to the details of the joint distribution of these two quantities. Let $c_n(x, y)$ be the number of subsets of $\{0, 1, 2, \ldots, n-1\}$ with $|S + S| = x$ and $|S - S| = y$. Figure 1 shows a square centered at $(x, y) \in \mathbb{Z}^2$ whose area is proportional to $\log(1 + c_{25}(x, y))$. Also shown are the lines $x = 2^{-25} \sum_S |S + S|$ (the average size of a sumset), $y = 2^{-25} \sum_S |S - S|$ (the average size of a difference set), and $y = x$.

Figure 2 shows the observed distribution of $X := 2n - 1 - |S + S|$ (that is, the number of missing sums) for three million randomly generated subsets of $\{0, 1, 2, \ldots, 999\}$. For example, the histogram shows that approximately 1.4% of these subsets S have the largest possible sumset $S + S = \{0, \ldots, 1998\}$, approximately 2.1% of them have exactly one element of $\{0, \ldots, 1998\}$ missing from their sumsets, and so on. The histogram is essentially identical to one generated from the complete data set for subsets of $\{0, \ldots, 26\}$.

Notice that there is a "divot" at the top of the histogram: the observed frequencies of sets missing exactly 6 or 8 sums are both larger than the observed frequency of sets missing exactly 7 sums. In fact, the frequency for every even value seems to be larger than the average of its two neighbors, while the opposite is true for the frequencies of the odd values; in other words, the piecewise linear graph that connected the points at the tops of the histogram's bars would alternate between being convex and concave.

Recall that the missing sums are typically very near the edges of the interval of possible sums. In particular, the missing sums for a subset S of $\{0, \ldots, 999\}$ tend to be near either 0 or 1998, and are therefore so far apart that their numbers are independent. Therefore the distribution shown in Figure 2 is the sum of two independent, identically distributed (by symmetry) random variables that count the number of missing sums near one end. This is also essentially the same distribution as the number Y of missing sums in randomly chosen (infinite) subsets A of the nonnegative integers $\{0, 1, \ldots\}$. That is, if Y_1, Y_2 are independent with the same distribution as Y, then X and $Y_1 + Y_2$ have approximately the same distribution (for large n).

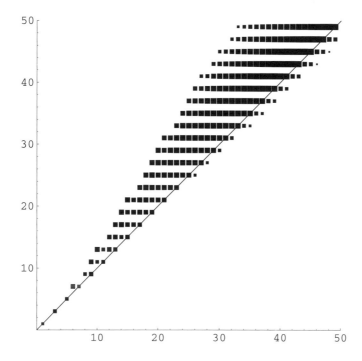

FIGURE 1. The size of the square centered at (x,y) indicates the number of subsets of $\{0,\ldots,24\}$ with $(|S+S|, |S-S|) = (x,y)$.

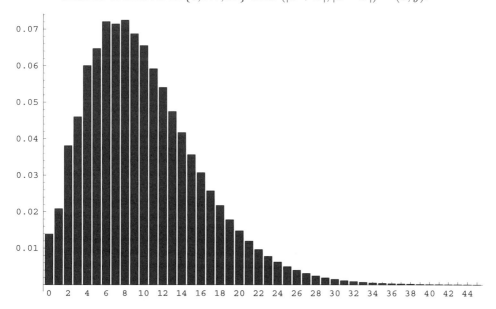

FIGURE 2. The observed frequencies of the number of missing sums

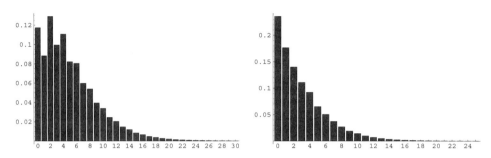

FIGURE 3. The observed frequencies of the number of missing sums for randomly chosen subsets of $\{0, 1, \dots\}$, with no restriction (left) and with the restriction that 0 belongs to the set (right)

At first one might think, then, that the parity phenomenon in Figure 2 is caused by that distribution being the sum of two independent copies of a simpler distribution. However, in this latter distribution (the first histogram in Figure 3), the disparity between odd and even values is even more apparent.

Fortunately, the phenomenon here is easy to analyze: if 0 is not in our randomly chosen subset of $\{0, 1, \dots\}$, then there are automatically 2 missing sums, namely 0 and 1, and the rest of the random subset can be shifted downwards by 1 to find the distribution of other missing sums:

$$\mathbb{P}[Y = k] = \sum_{i=0}^{\lfloor k/2 \rfloor} \mathbb{P}[Y = k - 2i \mid \min A = 0] \, 2^{-i}.$$

In other words, there is a yet more fundamental distribution (the second histogram in Figure 3), given by the number of missing subsums in a randomly chosen subset of $\{0, 1, \dots\}$ containing 0. For example, that histogram shows that if a subset S of $\{0, 1, \dots\}$ containing 0 is chosen at random, there is about a 23.6% chance that $S + S = \{0, 1, \dots\}$.

The parity discrepancy seems to be absent in this last distribution, suggesting that it should be the focus of further analysis; the two more complicated preceding distributions can be reconstructed from suitable manipulations of this most fundamental one. The histogram suggests the existence of a function $f(x)$, smooth and decaying faster than exponentially, such that the probability of a randomly chosen subset of $\{0, 1, \dots\}$ that contains 0 missing exactly n subsums is $f(n)$.

It would of course be interesting to do a similar empirical analysis for the distribution of the number of missing differences; perhaps their joint distribution could even be reduced to a simpler one using similar observations.

7. Conjectures and open problems

We have already conjectured, in Conjecture 2, that the limiting proportions of difference-dominant, sum-difference-balanced, and sum-dominant subsets of $\{0, 1, 2, \dots, n-1\}$ approach nonzero limits as n tends to infinity. (As long as the limits do in fact exist, Theorem 1 shows that they are necessarily nonzero.) Figure 4 shows the observed proportions, for $n \leq 27$, of the subsets of $\{0, 1, 2, \dots, n-1\}$ that are difference-dominant, sum-difference-balanced, and sum-dominant, respectively. Note particularly that each graph is monotonic in n, supporting our conjecture that

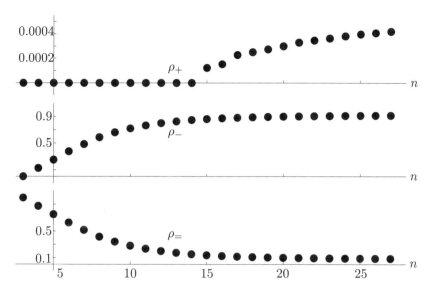

FIGURE 4. The probability of a random subset of $\{0,\ldots,n-1\}$ being sum-dominant (top graph), difference-dominant (middle graph), or sum-difference-balanced (bottom graph)

the limits exist. Using ten million randomly chosen subsets of $\{0,1,\ldots,999\}$, we estimate:

Conjecture 18. *Using the notation of Conjecture 2,*

$$\rho_- \approx 0.93, \quad \rho_+ \approx 0.00045, \quad \text{and} \quad \rho_= \approx 0.07.$$

In fact the philosophy behind Theorem 1 suggests somewhat more: a typical subset of $\{0,1,2,\ldots,n-1\}$ will achieve virtually all possible sums and differences, and the ones that aren't achieved are due to the edges of the subset. Since a positive proportion of sets have any prescribed edges, we make the following conjecture. Define

(8) $\quad \rho_{j,k} := \lim_{n\to\infty} \big(2^{-n} \#\{S \subset \{0,1,2,\ldots,n-1\} :$

$$|S+S| = 2n-1-j, |S-S| = 2n-1-k\}\big),$$

assuming the limit exists. Since the different set $S - S$ is symmetric about 0 and thus always has odd cardinality, we never have $|S - S| = 2n - 1 - k$ with k odd. Therefore we conjecture:

Conjecture 19. *For any nonnegative integers j and k with k even, the limiting proportion $\rho_{j,k}$ defined above in (8) exists and is positive; furthermore,*

$$\sum_{j=0}^{\infty} \sum_{\substack{k=0 \\ k \text{ even}}}^{\infty} \rho_{j,k} = 1.$$

Remark. Given Theorem 3, it seems reasonable to conjecture also that

$$\sum_{j=0}^{\infty} \sum_{\substack{k=0 \\ k \text{ even}}}^{\infty} k\rho_{j,k} = 6 \quad \text{and} \quad \sum_{j=0}^{\infty} \sum_{\substack{k=0 \\ k \text{ even}}}^{\infty} j\rho_{j,k} = 10.$$

For any particular pair j, k, if a single finite configuration of edges could be found that omitted exactly j possible sums and k possible differences, the methods of this paper would then show that $\rho_{j,k} > 0$ (technically, that the analogous expression with $\liminf_{n\to\infty}$ in place of $\lim_{n\to\infty}$ is positive).

The last remark suggests as well the following open problem, for which a simple proof might exist, though we have not been able to find one.

Conjecture 20. *For any nonnegative integers j and k with k even, there exists a positive integer n, and a set $S \subset \{0, 1, 2, \ldots, n-1\}$ with $0 \in S$ and $n-1 \in S$, such that $|S+S| = 2n - 1 - j$ and $|S-S| = 2n - 1 - k$.*

Hegarty points out that his methods from [2] can establish both Conjecture 19 and Conjecture 20.

We know [1, 5] that essentially all subsets of $\{0, 1, 2, \ldots, n-1\}$ of cardinality $O(n^{1/4})$ are Sidon sets and hence difference-dominant sets. More generally, we can show (perhaps in a sequel paper) that if $m = o(n^{1/2})$, then almost all subsets of $\{0, 1, 2, \ldots, n-1\}$ of cardinality m are difference-dominant sets.

This result may indicate the presence of a threshhold. Set p_n to vary with n, and define n independent random variables X_i, with $X_i = 1$ with probability p_n. This defines a random set $A := \{i \in \{0, 1, 2, \ldots, n-1\}\colon X_i = 1\}$. The observations above can then be rephrased in the following way: if $p_n = o(n^{-1/2})$, then A is a difference-dominant set with probability 1 (as $n \to \infty$). We showed in this article that if $p_n = 1/2$, then A is a sum-dominant set with positive probability (as $n \to \infty$), and our result is easily extended to $p_n = c > 0$. An important unanswered question is "Which sequences p_n generate a sum-dominant set with positive probability?" Perhaps our last conjecture captures the correct notion:

Conjecture 21. *For each $n \geq 1$, let $X_{n,0}, X_{n,1}, \ldots, X_{n,n-1}$ be independent identically distributed random variables, and set $A_n := \{i\colon 0 \leq i < n, X_{n,i} = 1\}$. If both $|A_n| \to \infty$ and $|A_n|/n \to 0$ with probability 1, then the probability that A_n is difference-dominant also goes to 1.*

References

1. A. P. Godbole, S. Janson, N. W. Locantore Jr., and R. Rapoport, *Random Sidon sequences*, J. Number Theory **75** (1999), no. 1, 7–22.
2. P. V. Hegarty, *Some explicit constructions of sets with more sums than differences*, available at arXiv:math.NT/0611582.
3. F. Hennecart, G. Robert, and A. Yudin, *On the number of sums and differences*, Structure Theory of Set Addition, Astérisque, vol. 258, Soc. Math. France, Paris, 1999, pp. 173–178.
4. J. Marica, *On a conjecture of Conway*, Canad. Math. Bull. **12** (1969), 233–234.
5. M. B. Nathanson, *On the ubiquity of Sidon sets*, Number Theory (New York, 2003), Springer, New York, 2004, pp. 263–272.
6. ———, *Problems in additive number theory. 1*, available at arXiv:math.NT/0604340.
7. ———, *Sets with more sums than differences*, Integers **7** (2007), A5.
8. K. O'Bryant, *A complete annotated bibliography of work related to Sidon sequences*, Electron. J. Combin. (2004), DS11.
9. V. P. Pigarev and G. A. Freĭman, *The relation between the invariants R and T*, Number-Theoretic Studies in the Markov Spectrum and in the Structural Theory of Set Addition, Kalinin. Gos. Univ., Moscow, 1973, pp. 172–174 (Russian).
10. F. Roesler, *A mean value density theorem of additive number theory*, Acta Arith. **96** (2000), no. 2, 121–138.
11. I. Z. Ruzsa, *On the number of sums and differences*, Acta Math. Hungar. **59** (1992), no. 3-4, 439–447.

12. _____, *Sets of sums and differences*, Seminar on Number Theory (Paris, 1982/1983), Progr. Math., vol. 51, Birkhäuser Boston, Boston, MA, 1984, pp. 267–273.

DEPARTMENT OF MATHEMATICS, UNIVERSITY OF BRITISH COLUMBIA, ROOM 121, 1984 MATHEMATICS ROAD, VANCOUVER, BC V6T 1Z2, CANADA
E-mail address: `gerg@math.ubc.ca`

DEPARTMENT OF MATHEMATICS, COLLEGE OF STATEN ISLAND (CUNY), 2800 VICTORY BLVD., STATEN ISLAND, NY 10314, USA
E-mail address: `kevin@member.ams.org`

Davenport's Constant for Groups of the Form $\mathbb{Z}_3 \oplus \mathbb{Z}_3 \oplus \mathbb{Z}_{3d}$

Gautami Bhowmik and Jan-Christoph Schlage-Puchta

ABSTRACT. We determine Davenport's constant for all groups of the form $\mathbb{Z}_3 \oplus \mathbb{Z}_3 \oplus \mathbb{Z}_{3d}$.

1. Introduction and notation

For a finite abelian group G let $D(G)$ be the Davenport's constant, that is, the least integer n such that among each sequence g_i in G there exists a non-empty subsequence g_{i_k} with sum 0. Writing G as

$$G \cong \mathbb{Z}_{d_1} \oplus \mathbb{Z}_{d_2} \oplus \cdots \oplus \mathbb{Z}_{d_r}, \quad d_1 \mid d_2 \mid \cdots \mid d_r,$$

we obtain a sequence of $\sum_i d_i - r$ elements without a zerosum subsequence, thus we have the trivial bound $D(G) \geq M(G) = \sum_i d_i - r + 1$. It has been conjectured that $D(G) = M(G)$ holds true for all finite groups, and this conjecture was proved for various special cases, including finite p-groups, and groups of rank $r \leq 2$. However, there are infinitely many counterexamples known for every rank $r \geq 4$. It is unknown whether $D(G) = M(G)$ holds true for all groups of rank 3, the authors are inclined to believe that this is always the case. The simplest undecided case up to now is $G = \mathbb{Z}_3 \oplus \mathbb{Z}_3 \oplus \mathbb{Z}_{15}$, which was already mentioned by van Emde Boas and Kruyswijk [5]. In the present note this case is solved, more generally, we show the following.

Theorem 1. *Let d be an integer, $A \subseteq \mathbb{Z}_3 \oplus \mathbb{Z}_3 \oplus \mathbb{Z}_{3d}$ be a multiset consisting of $3d + 4$ elements. Then there exists a multiset $B \subseteq A$, such that $\sum_{b \in B} b = 0$.*

Our approach is inspired by an idea of Delorme, Ordaz and Quiroz [1]. Suppose we are given a sequence A of $3d + 4$ points in $\mathbb{Z}_3 \oplus \mathbb{Z}_3 \oplus \mathbb{Z}_{3d}$. Consider the image $\tilde{A}$ of this sequence under the canonical projection $\mathbb{Z}_3 \oplus \mathbb{Z}_3 \oplus \mathbb{Z}_{3d} \to \mathbb{Z}_3^3$. If this sequence contains a family of d pairwise disjoint subsequences adding up to zero, we obtain a sequence of d elements in $\mathbb{Z}_d$, each of which is represented as a sum of certain elements in A. Among these elements we choose a subsequence adding up to 0, and find that A contains a subsequence adding up to 0. Using this method Delorme, Ordaz and Quiroz showed that for groups of the form $G = \mathbb{Z}_3 \oplus \mathbb{Z}_3 \oplus \mathbb{Z}_{3d}$ we have $D(G) \leq M(G) + 2$. Unfortunately, this inequality is the best possible,

---

2000 *Mathematics Subject Classification.* Primary 11B75; Secondary 20K01, 20D60.

This is the final form of the paper.

©2007 American Mathematical Society

since for every $d \geq 3$ there exists a sequence $A \subseteq G$ with $3d + 5$ elements, which does not contain d pairwise disjoint zerosum subsets. To remedy this, we note that for $(d, 3) = 1$ we have

$$\mathbb{Z}_3 \oplus \mathbb{Z}_3 \oplus \mathbb{Z}_{3d} \cong \mathbb{Z}_3^3 \oplus \mathbb{Z}_d,$$

thus we can represent a sequence $A \subseteq \mathbb{Z}_3 \oplus \mathbb{Z}_3 \oplus \mathbb{Z}_{3d}$ as a pair $(\tilde{A}, f)$, where $\tilde{A} \subseteq \mathbb{Z}_3^3$ is the image of A under the canonical projection, and $f : \tilde{A} \to \mathbb{Z}_d$ is the function such that the element $a_i \in A$ is represented as $(\tilde{a}_i, f(\tilde{a}_i))$ in $\mathbb{Z}_3^3 \oplus \mathbb{Z}_d$. This idea allows us to concentrate only on the small group $\mathbb{Z}_3^3$; in fact, the remainder of this article deals only with combinatorial properties of $\mathbb{Z}_3^3$.

Since the order of elements plays no rôle, we will henceforth speak of multisets instead of sequences. To visualize the combinatorial considerations, we view $\mathbb{Z}_3^3$ as the elements of a $3 \oplus 3 \oplus 3$-cube, and this cube again as three $3 \oplus 3$-squares placed side by side. The origin is placed in the lower left corner of the leftmost rectangle and coordinates are associated to cells in the order of board, row and column. Cells marked with a black circle are elements that are known to be contained in the set under consideration, cells marked with a white circle denote elements known not to be contained in the set. Cells marked with a black circle and a white number n denote elements which are known to be contained in the multiset under consideration at least n times. For example, in the following visualization of some information on a multiset $A \subseteq \mathbb{Z}_3 + \mathbb{Z}_3 + \mathbb{Z}_3$, the cell marked o denotes the neutral element of the group, whereas the cell marked a is the element $(1, 2, 1)$, and the multiset A contains the element $(1, 1, 1)$, and the element $(0, 1, 2)$ at least twice.

One advantage of this notation is the fact that one can often read off the existence of zero-sums from the picture. For example, 3 distinct elements add up to 0 if and only if they lie on an affine line, thus, in the following picture the three cells marked a as well as the three cells marked b are zerosum subsets.

We will use this argument repeatedly without further explanation.

The technique of Delorme, Ordaz, and Quiroz requires the study of certain auxiliary functions. Denote by $D_k(G)$ [2] the least integer n, such that every multiset of n elements contains k disjoint zerosum subsets, and by $D^k(G)$ the least integer n, such that every multiset of n elements contains a zerosum subset consisting of at most k elements. Note that $D^k(G)$ is finite only if k is at least the exponent of G; the case that k equals the exponent of G has received particular interest. Adding a star always means that we ask for the least n, such that each subset of n distinct elements has the required property, for example, $D_2^*(\mathbb{Z}_4) = 4$, since 4 distinct points in $\mathbb{Z}_4$ contain the element 0 as well as the set $\{1, 3\}$, and therefore 2 disjoint zerosum subsets. In particular, $D^*(G)$ is known as Olson's constant, which was determined for cyclic groups by Olson [4].

This article is organized as follows. In the next section we give several rather special results for the variations of $D(\mathbb{Z}_3^3)$ just mentioned. These results are of limited interest, but will save us a lot of work later on. In the last section we describe the splitting of the set A into the pair $(\tilde{A}, f)$, and prove Theorem 1.

Our approach is not restricted to groups of the form $\mathbb{Z}_3 \oplus \mathbb{Z}_3 \oplus \mathbb{Z}_{3d}$, but can be applied to all sequences of the form
$$\mathbb{Z}a_1 \oplus \mathbb{Z}a_2 \oplus \cdots \oplus \mathbb{Z}_{a_{r-1}} \oplus \mathbb{Z}_{a_r d}, \quad a_1 \mid a_2 \mid \cdots \mid a_r,$$
where $a_1, \ldots, a_r$ are fixed, and d runs over all integers coprime to a_r. However, soon the computational effort becomes too large for a treatment as explicit as given here. In work in progress, we hope to automate parts of the proof to deal with larger groups as well.

2. Some special values of D_k and related functions

The results of this section are summarized in the following.

Proposition 1. *We have*

$$D^3(\mathbb{Z}_3^3) = 17 \qquad D^4(\mathbb{Z}_3^3) = 10 \qquad D^5(\mathbb{Z}_3^3) = 9$$
$$D^{3*}(\mathbb{Z}_3^3) = 9 \qquad D^{4*}(\mathbb{Z}_3^3) = 8 \qquad D^{5*}(\mathbb{Z}_3^3) = 8$$
$$D_2(\mathbb{Z}_3^3) = 11 \qquad D_k(\mathbb{Z}_3^3) = 3k + 6 \ (k \geq 3)$$
$$D^*(\mathbb{Z}_3^3) = 7 \qquad D_2^*(\mathbb{Z}_3^3) = 10$$

Lemma 1. *Set $G = \mathbb{Z}_3 + \mathbb{Z}_3 + \mathbb{Z}_3$.*

(1) *Let $A = \{a_1, \ldots, a_6\}$ be a set of distinct elements of G such that there does not exist a zerosum subset Z of A with at most 3 elements. Then there are distinct indices i, j, k, such that $a_i + a_j = a_k$.*

(2) *Let $A = \{a_1, \ldots, a_8\}$ be a set of distinct elements of G such that there does not exist a zerosum subset Z of A with at most 3 elements. Then, up to linear equivalence, A is the set*

(3) *[3] Let $A = \{a_1, \ldots, a_9\}$ be a set of distinct elements of G. Then there exists a zerosum subset Z of A with at most 3 elements.*

PROOF. Let $A = \{a_1, \ldots, a_6\}$ be a set of 6 elements, and suppose that none of the equations $x + y = z$ and $x + y + z = 0$ is solvable within A. Then A cannot be contained in a plane, thus, we may choose a basis in A. We therefore obtain the following description of A.

Any two of the three points $(1, 1, 2)$, $(1, 2, 1)$ and $(2, 1, 1)$ form together with one of the points $(0, 0, 1)$, $(0, 1, 0)$ and $(1, 0, 0)$ a zerosum subset, thus, by symmetry we may assume that $(1, 1, 2)$ and $(1, 2, 1)$ are not contained in A. Moreover, if $(2, 1, 1)$ were in A, the only remaining position would be $(1, 2, 2)$, and we would have $|A| \leq 5$, that is, $(2, 1, 1)$ is not in A as well. From the remaining 4 positions, 3 have to be taken by elements in A, but $(1,1,1)$ and $(1,2,2)$ cannot be taken at the same time, thus, both $(2,1,2)$ and $(2,2,1)$ have to be in A. Then $(1,1,1)$ cannot be contained in A, and we obtain the following situation.

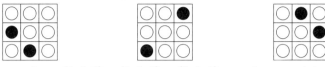

But now we have $(2,1,2) + (1,2,2) = (0,0,1)$, proving our claim.

Now let A be a set of size 8 without a zerosum of length 3. By part 1 we may assume that $a_1 + a_2 = a_3$, moreover, not all elements of A are contained in the plane generated by a_1 and a_2, and we obtain the following situation.

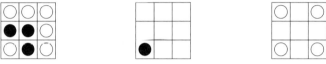

Moreover, we may suppose that there are more elements in the middle layer than in the uppermost one. Suppose first that $(1,1,1)$ is in A.

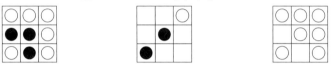

If both the remaining cells in the uppermost layer were contained in A, then no further cell in the middle layer could be contained in A; if on the other hand both cells were not contained in A, then there are three more cells in the middle layer, and it is easily seen that this implies the existence of a zerosum sequence of length 3 in the middle layer. Hence, precisely one of $(2,0,1)$ and $(2,1,0)$ is in A, and we may assume that this element is $(2,0,1)$. Then we reach the following situation.

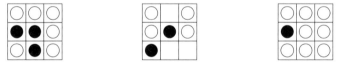

From the remaining three cells two have to be taken, but $(1,1,0)$ would yield with one of the other two cells and $(1,0,0)$ resp. $(1,1,1)$ a zerosum, thus, we obtain the constellation given in the Lemma.

Now suppose that $(1,1,1)$ is not in A. Since any element in the two upper layers can be interchanged with $(1,0,0)$ by a linear transformation leaving the lower layer fixed, we can avoid this case unless for all elements $x, y, z \in A$ with $x + y \in A$ and $z \neq x + y$ we have $x + y + z \notin A$. In particular, in our situation this implies that we may suppose that for each $z \in A$ which is not in the lower layer both elements $z \pm (0,1,1)$ are not in A. Assume that $(1,1,0)$ is in A. The

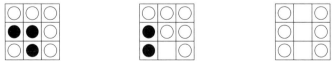

Sine there are at least 3 elements in the middle layer, we deduce that $(1,1,0) \in A$, which contradicts the fact that there are two more elements in the uppermost layer. Next suppose that $(1,0,2)$ was in A. Then we obtain the following.

In the middle layer there has to be another element of A, however, for both possible places we see that there could be at most one other cell in the uppermost layer. Hence, $(1,0,2)$ and $(1,2,0)$ are not in A. Then we obtain the following constellation.

Here, not both cells in the middle layer can be taken, thus there have to be three cells in the uppermost layer, contrary to our assumption that there are more cells in the middle layer then in the uppermost one. Hence, the second statement of the lemma is proved. $\square$

Lemma 2. *Let A be a sequence of 14 points which does not contain a zerosum subset of length[] ≤ 3 or of length ≥ 12. Then A contains 7 distinct points, each of which is taken twice. Moreover, there exists a multiset A with these properties, and it is unique up to linear equivalence.*

PROOF. The existence of A is given by the following example.

It is easy to check that A does not contain a zerosum of length ≤ 3. Next, the sum of all elements in A equals $(2,2,2)$; the inverse of this element being the element marked a in the picture above. This element is neither 0 nor contained in A, hence, there is no zerosum of length ≥ 13. Suppose that a was the sum of two elements in A. Then a is either the sum of an element in the lowest layer with an element of the middle layer, or the sum of two elements in the upper layer, but both possibilities are easily dismissed.

We now show that all sets of such 14 elements have indeed the form described above. Thus, let A be a set consisting of 14 elements, 8 of which are distinct. Denote by B the configuration described in Lemma 1, part (ii). Then up to some linear transformation, A consists of 6 points of B taken twice, and the 2 remaining points of B taken once. Note that the sum of all elements in B is 0, and the sum of any two elements is non-zero, hence, the sum of all elements in A is non-zero as well. But B is maximal among all sets of distinct elements without zerosum subsets of length ≤ 3, thus, every non-zero element, and in particular the sum of all elements in A, can be represented as the sum of 1 or 2 elements in B. Hence, by deleting 1 or 2 elements of A we obtain a zerosum subset consisting of 12 or 13 elements. $\square$

Proposition 2. *We have $D_2^*(\mathbb{Z}_3^3) = 10$.*

PROOF. Consider the example

Clearly there is no zerosum subset in the lowermost layer. Hence, if there are two disjoint zero sum subsets, each of them must contain precisely 3 elements of the second layer. Now consider the sum of all elements not contained in the two zerosum subsets. This sum is equal to the sum of all elements in A, and therefore $(2, 0, 0)$, on the other hand, it equals a subset sum of the elements in the lowermost layer. However, $(2, 0, 0)$ cannot be represented by elements in the lowermost layer. Hence, $D_2^*(\mathbb{Z}_3^3) \geq 10$. On the other hand, $D^{3*}(\mathbb{Z}_3^3) = 9$, thus among 10 points there is a zerosum subset of length ≤ 3, and among the remaining 7 points, there is always another zerosum subset. $\square$

Proposition 3. *We have $D^{*4}(\mathbb{Z}_3^3) = 8$.*

PROOF. Let A be a set of 8 distinct elements which does not contain a zerosum subset of length 4. Then A cannot be contained in one plane, hence, we may choose a basis of $\mathbb{Z}_3^3$ in A, which without loss is the standard basis. Moreover, there has to be a sum in A, which we may assume, without any loss, to be $(0, 1, 1)$. Finally, we may change the third element of the basis in such a way that the middle layer contains at least as many elements as the upper layer, thus, we obtain the following picture, where at least two elements of A in the middle layer are not yet drawn.

Suppose that $a = (1, 0, 1)$ is contained in A. Then several other elements of G can be excluded, since they would immediately give zerosum subsets of size ≤ 4, and we obtain the following situation.

Hence, there is at most one element in the uppermost layer, that is, there are at least two more in the middle layer. But any two elements in the middle row of the middle layer would give a contradiction, hence, $(1, 2, 0)$ is in A. But then no other element of A could be in the middle layer, giving a contradiction. Hence, $(1, 0, 1)$, and by symmetry $(1, 1, 0)$ are not contained in A. Now assume that $(1, 0, 2)$ is in A. Then we obtain the following situation.

By direct inspection we see that if both possible elements in the uppermost layer are contained in A, then no additional element in the second layer could be chosen, and A would have at most 7 elements. Hence, two more elements of A in the middle layer are not yet shown. If $(1, 2, 2)$ was in A, this is impossible, thus, we find that $(1, 1, 1)$ and $(1, 1, 2)$ are both in A. Then we reach the following situation, which immediately implies $|A| = 7$, thus showing that $(1, 0, 2)$ and therefore by symmetry $(1, 2, 0)$, are not in A and hence, we obtain the following situation.

By assumption, there are two more elements in the middle layer, but on the cells marked a and b, respectively, there can be at most one element of A. Hence, without loss, we may assume that $(1, 1, 2)$ is in A, and, since $(1, 0, 0) + (1, 1, 1) + (1, 1, 2) + (0, 1, 0) = (0, 0, 0)$, that $(1, 2, 2)$ is in A as well, that is, we reach the following situation.

Clearly, $|A| = 6$, contrary to our assumption, and we see that the initial assumption on the existence of A is wrong. □

Proposition 4. *We have* $D^4(\mathbb{Z}_3^3) = 10$.

PROOF. The fact that $D^4(\mathbb{Z}_3^3) \geq 10$ is proved by the following example.

Hence, it remains to show that every set of 10 elements contains a zerosum of length at most 4. By means of contraction, let A be a set consisting of 10 elements of G without a zerosum subset of size at most 4. Since $D^{*4}(\mathbb{Z}_3^3) = 8$, at least 3 elements of A are repeated, and these 3 elements form a basis of G. Hence, we have the following situation.

Suppose that 2 of the three elements $(0, 1, 1)$, $(1, 0, 1)$ and $(1, 1, 0)$ are in A. Then we may suppose without loss that these elements are $(1, 1, 0)$ and $(1, 0, 1)$. Then we obtain the following.

Since $2 \cdot (1, 1, 0) + (1, 0, 0) + (0, 1, 0) = (0, 0, 0)$, both $(1, 1, 0)$ and $(1, 0, 1)$ are taken at most once. Hence, there are two elements in A of the form $(a, 1, 1)$. If they are distinct, we have $(a, 1, 1) + (b, 1, 1) + (x, 1, 0) + (y, 0, 1) = (0, 0, 0)$ for appropriate values $x, y \in \{0, 1\}$, hence, there is one value which is taken twice. However, all three remaining choices lead to contradictions, and we conclude that at most one of $(0, 1, 1)$, $(1, 0, 1)$ and $(1, 1, 0)$ is contained in A; without loss we may assume that $(1, 0, 1)$ and $(1, 1, 0)$ are not in A. Next we note that any two of $(1, 1, 2)$, $(1, 2, 1)$ and $(2, 1, 1)$ together with the elements already placed give a zerosum subset of size

3, hence, at most one of these elements can be contained in A. By symmetry we may suppose that $(1,2,1)$ is not in A; moreover, if $(0,1,1)$ was not in A, then we may also assume that $(1,1,2)$ is not in A.

We shall now assume that $(0,1,1)$ is not in A. Then we have the following situation.

There are at most 8 elements in the lower two layers, hence, there are at least two elements in the uppermost layer; in particular, $(1,1,1)$ is not in A. Suppose that $(1,2,2)$ does not occur twice in A. Then there are at least 3 elements of A in the uppermost layer, and since $(2,1,2)$ and $(2,2,1)$ cannot both occur in A, we deduce that $(2,1,1)$ and one of $(2,1,2)$ and $(2,2,1)$ is in A; without loss we may assume the former, and obtain the following situation.

Since $(2,1,2) + (1,2,2) + 2 \cdot (0,0,1) = (0,0,0)$, we see that $(1,2,2)$ is not contained in A, and we find that the elements in the uppermost layer are contained twice in A. But then we obtain the contradiction $2 \cdot (2,1,2) + (2,1,1) + (0,0,1) = (0,0,0)$.

Hence, our initial assumption that $(0,1,1)$ was not in A is false, and we obtain the following situation.

Note that $(0,1,1)$ cannot be repeated in A, because $D^4(\mathbb{Z}_3^2) = 6$, thus, there are 3 elements of A on the remaining three cells. But $(1,1,1)$ and $(2,1,1)$ cannot be simultaneously in A, thus, $(1,1,2)$ is in A. Because of $(1,0,0) + (0,1,0) + (1,1,1) + (1,1,2) = (0,0,0)$, $(1,1,1)$ cannot be contained in A, and therefore $(2,1,1)$ has to be contained in A. But then we obtain the zerosum $(2,1,1) + (1,1,2) + (0,1,0)$, and obtain a contradiction proving our claim. $\square$

Proposition 5. *We have $D_2(\mathbb{Z}_3^3) = 11$.*

PROOF. Let A be a subset of $\mathbb{Z}_3^3$ containing 11 elements. Then A contains a zerosum subset of size ≤ 4, and the complement of this zerosum is a set with ≥ 7 elements, which therefore contains another zerosum subset. Hence, $D_2(\mathbb{Z}_3^3) \leq 11$. On the other hand, the inequality $D_2(\mathbb{Z}_3^3) \geq 11$ is proved by the following example.

In fact, every zerosum subset contains either no elements of the middle layer, or precisely 3, thus, if there were to distinct zerosum subsets, one of them has to be contained in the lowermost layer. Obviously, it has to contain all points of this layer. But there is no zerosum subset of size three in the middle layer, and we find that this set does not contain two disjoint zerosum subsets, which proves $D_2(\mathbb{Z}_3^3) \geq 11$. □

Proposition 6. *We have* $D_3(\mathbb{Z}_3^3) = 15$.

PROOF. The upper bound $D_3(\mathbb{Z}_3^3) \leq 15$ follows from Proposition 4 and 5, whereas the lower bound follows from the configuration given in Lemma 2. □

Proposition 7. *We have* $D^5(\mathbb{Z}_3^3) = 9$.

PROOF. The lower bound $D^5(\mathbb{Z}_3^3) \geq 8$ is given by the following example.

Let A be a set consisting of 9 elements without a zerosum subset of length 5. Suppose first that there are at most 5 distinct elements in A. Then there are at least 4 elements twice in A, and these elements cannot be in one plane, hence, there is a basis of elements taken twice in A. We therefore obtain the following.

If $(2,1,1) \in A$, then there are no further elements in the middle layer, which would imply $|A| \leq 8$; thus, by symmetry, we have $(1,1,2), (1,2,1), (2,1,1) \notin A$. But then A could only have $(1,1,1)$ as a possible element left, and therefore $|A| \leq 8$.

Hence, we may assume that A contains 6 distinct elements, and therefore there is some basis $\{a,b,c\} \in A$, such that $a+b \in A$, that is, we have the following situation.

Moreover, we can choose the third base element in such a way that it is either twice in A, or that no element outside the lowest plane is twice in A. Consider first the case that A occurs twice.

Note that $(0,1,1)$ cannot be contained twice in A, and that $(0,0,1)$ and $(0,1,0)$ cannot be both twice in A, that is, there are at most 4 elements in the lowest layer. Moreover, all elements in the middle layer not yet depicted can occur at most once, and not all three empty places can be taken, thus, there is an element in the uppermost layer; without loss we may suppose that this element is $(2,0,1)$. We then obtain the following.

Here, $(2,0,1)$ can only be once in A, since otherwise we had the zerosum $2 \cdot (2,0,1) + 2 \cdot (1,0,0) + (0,0,1)$, thus, there are at most 2 points in the uppermost layer. Hence, $(1,1,0) \in A$, which implies that $(2,1,0) \notin A$, and we find that $|A| \leq 8$. Thus, from now on, we shall assume that all elements outside the lowest layer occur only once in A.

Suppose that $(1,0,1) \in A$. Then we have the following.

Obviously, there can only be 4 elements outside the first layer, and at most 4 inside the first layer, which gives a contradiction to $|A| = 9$.

Next, suppose that $(1,1,2) \in A$. Then we obtain

Again, there are only 5 places for elements outside the first layer, and $(1,0,2)$ and $(1,2,2)$ cannot occur at the same time, thus, $(1,1,2) \notin A$ as well.

Hence, we are led to the following constellation.

There are only 7 places left for elements outside the first layer. Moreover, among these $(1,1,1)$ and $(1,2,2)$ as well as $(1,2,0)$ and $(1,0,2)$ are mutually exclusive, and we conclude that $(2,1,0), (2,0,1) \in A$, and, without loss, $(1,2,0) \in A$. But then we obtain the zerosum $(2,0,1) + (1,2,0) + (0,1,1) + (0,0,1)$, and this contradiction proves our claim. □

Proposition 8. *For $k \geq 3$ we have $D_k(\mathbb{Z}_3^3) = 3k + 6$. Moreover, the set of all multisets A of size $3k+5$ which do not have k disjoint zerosum subsets can be constructed as follows: Take all sets $B = \{b_1, \ldots, b_7\}$ of 7 distinct points without a subsum of length ≤ 3, such that the multiset C obtained from B by taking each point twice does not contain a zerosum subset of length ≤ 12. Choose a partition $k - 3 = \kappa_1 + \cdots + \kappa_7$, and set $A = C \cup \{b_1^{3\kappa_1}, \ldots, b_7^{3\kappa_7}\}$.*

PROOF. We first show that none of the sets described here contain k disjoint zerosum subsets. In fact, since there is no zerosum of length ≤ 3 in B, every zerosum of length 3 in A must contain the same element three times, thus, any collection of disjoint zerosum subsets can contain at most $k - 2$ zerosums of length 3. next, note that the sum of all elements of A equals twice the sum of all elements of B, and that the set of elements representable by 0, 1 or 2 elements of A is equal to the set of elements representable by 0, 1, or 2 elements of $B \cup B$; thus, as in the proof of Lemma 2 we see that there is no zerosum of length $\geq 3k + 3$. Hence, every collection of disjoint zerosum subsets can contain $3k+2$ points at most. Thus, such a collection contains no zerosum subset of length ≤ 2, at most $k-3$ of length 3, and all together consists of at most $3k+2$ points, which implies that the total number of zerosum subsets is $\leq k-1$.

Now we show that there are no examples different from the one described here. Let A be a multiset consisting of $3k+5$ elements of $\mathbb{Z}_3^3$ which does not have k disjoint zerosum subsets. If there is a zerosum of length ≤ 2 in A, then removing this zerosum yields $k-1$ zerosums in the remaining $3(k-1)+6$ points, which gives a contradiction. Next suppose there is an element repeated 4 times in A. Then we can remove this element 3 times, and may assume by induction that the new set has the form described. Hence, it suffices to consider the case that every element in A occurs at most 3 times.

Suppose there is one element a occurring once. Then we remove as many zerosum subsets of length 3 from A as possible without removing this point. We end up with a set B of 14 or 17 points, which contains one element precisely once, and does not contain 3 resp. 4 disjoint zerosum subsets. However, we already saw that this is impossible for a set of 14 elements. If $|B| = 17$, then either B contains one element three times, contradicting the assumption that we removed all zerosums of length three avoiding a, or there are at least 9 distinct points in B. In the latter case let $\ell \leq 8$ be the number of points occurring twice in B. Collect each point which is twice in B and as many points that occur once in B necessary to reach 9 points in a set C. Then C contains a zerosum of length 3, we claim that removing this zerosum of B yields a set which contains one element only once. This is clear if $b < 8$, for then some element occurring once in B is not contained in C, and can therefore also not be a part of the zerosum removed. If on the other hand $b = 8$, then at least two points occurring twice in B are removed once, thus there are at least 2 points in the new set with multiplicity 1. In any case, we have removed $k-3$ zerosum subsets of size 3 from the beginning set A, and ended up in a set of 14 elements, one of which occurring only once. Hence, there are 3 more zerosums, contradicting the assumption that A does not contain k zerosum subsets.

Next, suppose there is a zerosum of length 3 which consists of distinct elements. Then we can remove this zerosum once or twice to reach a situation with one element occurring precisely once, a situation we just dealt with. In particular, there are at most 8 distinct elements in A. If there are only 7 distinct elements, we reached the position described in Lemma 2 and are done. Otherwise we have a set with 17 elements, 8 distinct ones among them forming the set described in Lemma 1 (ii), and one point occurs three times, whereas the other points occur exactly twice. Removing the three times repeated point once, we obtain a set of 16 elements with sum 0, thus, among the 15 remaining points there are 3 disjoint zerosums, whereas the complement of these three sums constitute a fourth one, yielding k disjoint zerosums for the original set A.

Hence, our claim follows. □

3. Proof of Theorem 1

Lemma 3. *In every set of* 5 *distinct elements of* $\mathbb{Z}_3^3$ *there is either a zerosum of length* ≤ 3, *or there are* 3 *elements* x, y, z *satisfying the equation* $x + y = z$.

PROOF. It is easy to check that the analogous statement in two dimensions holds true for all sets of 3 elements, hence, we may assume that A does not contain

3 points in any plane passing through the origin, and we obtain the following situation.

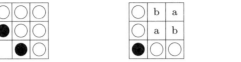

Without loss we may suppose that the middle plane contains at least one other element of A. If one of the cells marked a is in A, then none of the cells marked b is in A, and vice versa. On the other hand, if both cells marked a are in A, we would obtain a zerosum of length 3, and similar for b, thus, precisely one cell in the middle plane is contained in A. Moreover, if a cell marked a is in A, the cells marked d cannot be in A, and similarly with b, and we are left with the possibilities that there is precisely one element a and one c, or one b and one d.

If $(1,1,1) \in A$, both elements marked c yield zerosums of length 3, similarly, if $(2,1,1) \in A$. If $(1,1,2) \in A$, then the only remaining possibility for the last element is $(2,2,2)$, but $(1,1,2) + (2,2,2) = (0,0,1) \in A$, and the same argument applies to the case that $(2,1,1)$ is in A. Hence, no such set A can exist. □

Theorem 2. *Let n be an integer coprime to 6. Then there does not exist a multiset $A \subseteq \mathbb{Z}_3^3$ with 10 elements together with a function $f \colon A \to \mathbb{Z}_n$, such that A does not contain 2 disjoint zerosum subsets, that for every zerosum subset B of A we have $\sum_{b \in B} f(b) = 1$, and that there is some element $a \in A$ with $3f(a) = 1$.*

PROOF OF THE THEOREM. Suppose that A and f are as in the statement. Observe that A does not have a zerosum of length ≤ 3 or ≥ 8. In particular, there are at most 8 distinct elements in A, that is, there are at least 3 elements occurring twice. We distinguish cases according to the constellation of these elements.

(i) Suppose that there exist 3 elements a, b, c occurring twice within one plane passing through the origin. Without loss we may suppose that $c = a + b$. Then we have the zerosums $2a + 2b + c = a + b + 2c = 0$, thus $f(a) + f(b) = f(c)$ and $3f(c) = 1$, in particular, $f(c) \neq 0$. Suppose that in A there are elements x_i outside this plane adding up to $2a$. Then we have the zerosums $a + \sum x_i = 2a + b + 2c + \sum x_i = 0$, which implies the equation

$$f(a) + \sum f(x_i) = 4f(a) + 3f(b) + \sum f(x_i),$$

which in turn implies $f(c) = 0$, a contradiction. Next suppose that $2a + 2b$ can be represented as a sum of elements x_i of A outside the given plane. Then we have the zerosums $2a + 2b + 2c + \sum x_i$ and $c + \sum x_i$, which in the same way implies $f(c) = 0$. If $2a + b$ can be represented, we have the zerosums $a + 2b + \sum x_i$ and $2a + 2c + \sum x_i$, which implies $3f(a) = 0$, that is, $f(a) = 0$. If a can be represented, we obtain the zerosums $2a + \sum x_i$ and $b + 2c + \sum x_i$, which implies $f(b) = 0$. Noting that not both $f(a)$ and $f(b)$ can vanish, since otherwise $f(c)$ would be zero, we can summarize these considerations as follows: None of $2a, 2b, 2c$ can be represented as a sum of elements of A outside the plane spanned by a and b, and if one of $a, a + 2b$ is represented in such a way, none of $2a + b, b$ is, and vice versa.

(ii) As an immediate consequence we obtain that A cannot consist of 5 elements, each taken twice. In fact, without loss, we have the following situation.

If there is an element x in the middle layer occurring twice in A, $2 \cdot (1,0,0) + x$ and $(1,0,0) + 2x$ are elements in the lowest layer; if there is an element x in the uppermost layer occurring twice, both $(1,0,0) + x$ and $2 \cdot (1,0,0) + 2x$ are elements in the lowest layer. In any case, there is some y in the lowest layer, such that both y and $2y$ can be represented as a sum of elements outside the lowermost layer. If $y = 0$, we have a zerosum of length 2, if $y = (0,1,2)$, both $(0,1,2)$ and $(0,2,1)$ are representable, which gives a contradiction, and in all other cases one of $(0,0,2)$, $(0,2,0)$ and $(0,2,2)$ is representable, which gives also a contradiction.

(iii) Suppose that $(0,0,1), (0,1,0)$ and $(0,1,1)$ occur twice, and that none of the equations $x + (0,1,0) = y$, $x + (0,0,1) = y$ is solvable with $x, y \in A$. Without loss we may assume that $(1,0,0)$ is in A, and that, if some other element occurs twice in A, then so does $(1,0,0)$. We obtain the following situation.

If there are two elements in the uppermost layer, we may assume without loss that $(2,1,0) \in A$. Then $(2,0,1)$ and $(2,2,1)$ cannot be in A since otherwise $(0,1,0)$ and one of $(0,0,1), (0,2,1)$ is representable by elements outside the lowest layer; and $(2,1,1)$ and $(2,1,2)$ cannot be contained in A, since we would obtain a solution of the equation $x + (0,1,0 = y$ in A. Hence, there is at most one element of A in the uppermost layer. On the other hand, there can be at most 3 elements in the middle layer, thus, there is precisely one element in the uppermost layer, and 3 elements in the middle layer, two of which are $(1,0,0)$. Moreover, $(2,1,1) \notin A$, since otherwise there would be no place left for the last element in the middle layer, and $(1,2,2) \notin A$, since otherwise $(0,2,2)$ was representable. We therefore get the following.

We now check that $(2,1,2) \in A$ is impossible, since there is no place for the last element in the middle layer, and we may assume without loss that $(2,1,0)$ is the unique element in the uppermost layer.

The sum of all elements in A is therefore $(2,0,1)$ or $(2,0,2)$, and both these elements can be represented as a sum of 1 or 2 elements of A; thus, A contains a zerosum of length 8 or 9, which gives a contradiction.

(iv) Now suppose that $(0,0,1), (0,1,0)$ and $(0,1,1)$ occur twice. By the preceding argument we know that for some elements $x, y \in A$ outside the lowermost plane we have $x + (0,1,0) = y$ or $x + (0,0,1) = y$, and we may therefore assume without loss that both $(1,0,0)$ and $(1,0,1)$ are contained in A. We obtain the following situation.

Suppose first that $(1,0,0)$ is not twice in A. Then $(1,2,0)$ is in A, since $(1,0,1)$ cannot be twice in A, and therefore all remaining elements are in the plane $(s,1,t)$, which would imply that the sum of all elements outside the lowermost plane is of the form $(0,2,t)$, which for $t \neq 1$ yields a contradiction, whereas for $t = 1$ we deduce that the sum of all elements in A is $(0,0,2)$, which is a sum of two elements in A, thus, A contains a zerosum of length 8, which also yields a contradiction. Hence, we conclude that $(1,2,0)$ is in A, and obtain the following.

Now $(0,2,1)$ is a sum of elements outside the lowermost plane, hence, $(0,1,0)$ and $(0,1,2)$ are not, which implies that $(1,1,1), (2,1,1), (2,1,2), \notin A$, and that $(1,2,0)$ is only once in A, thus, A consists only of 9 elements. This contradiction shows that $(1,0,0)$ occurs twice in A, and we find the following.

We now can represent $(0,0,1)$ as a sum of elements outside the lowermost layer, hence, $(0,1,0)$ and $(0,1,2)$ cannot be represented this way. The remaining element of A has the form $(s,1,t)$. If $t \neq 1$, we add $(1,0,0)$ once or twice, to obtain a forbidden sum in the lowermost layer, whereas if $t = 1$, we add $(1,0,1)$, and, if necessary, $(1,0,0)$ to obtain a forbidden sum as well. Hence, in any case, we obtain a contradiction.

Thus, we see that the equation $x+y=z$ is not solvable among distinct elements occurring twice in A, in particular, we may assume that $(1,0,0), (0,1,0)$ and $(0,0,1)$ all occur twice in A.

(v) Suppose that there are 4 elements occurring twice in A. We already saw that three of them may be taken to be $(0,0,1), (0,1,0)$ and $(1,0,0)$, and that the fourth cannot be contained in one of the planes generated by this three. Hence, there are 8 possibilities left. These 8 points fall into 4 equivalence classes under rotation around the spatial diagonal, and taking the fourth point to be $(2,2,2)$ would yield a zerosum of length 8. Moreover, a linear transformation fixing the plane $(0,s,t)$ and interchanging $(1,0,0)$ and $(1,1,1)$ shows that the case of the fourth point being $(1,1,1)$ is equivalent to the case $(1,2,2)$. Hence, up to symmetry we may suppose that the fourth point is $(1,1,1)$ or $(1,1,2)$.

(vi) Suppose that $(0,0,1), (0,1,0), (1,0,0)$ and $(1,1,1)$ occur twice in A. Then we have the following.

Up to symmetry all elements of $\mathbb{Z}_3^3$ except $(0,1,2)$, $(0,2,2)$ and $(1,2,2)$ can be represented as the sum of at most 2 elements already depicted, thus we deduce that the sum of all elements of A is equal to one of these, for otherwise we would obtain a zerosum of length ≥ 8. Subtracting the elements depicted we find that the sum of the two remaining elements equals $(2,0,1)$, $(2,1,1)$ or $(0,1,1)$.

Suppose first the sum is $(2,0,1)$. Deleting all elements x such that $(2,0,1) - x$ is impossible, only the following two possibilities remain.

The same argument applied to the second case yields only one case.

Finally, the last sum gives two cases, which are symmetric to each other, thus, we only have to consider the following.

Here, the sum of all elements different from $(1,1,1)$ is 0, thus, there is a zerosum of length 8, and it suffices to consider the 3 previous cases.

We set $x = (0,0,1)$, $y = (0,1,0)$, $z = (1,0,0)$, $w = (1,1,1)$. Then we have the zerosums $2x + 2y + 2z + w$, $x + y + z + 2w$, which imply that $f(w) = f(x) + f(y) + f(z)$ and $3f(w) = 1$. Moreover, if $a = (r,s,t)$ is one of the other elements of A, we have the zerosum $a + [-r]z + [-s]y + [-t]x$, similarly, we have the zerosums $a + w + [-1-r]z + [-1-s]y + [-1-t]x$ and $a + 2w + [-2-r]z + [-2-s]y + [-2-t]x$, and we obtain the equations

$$(1 - [-r] + [-1-r])f(z) + (1 - [-s] + [-1-s])f(y) \\ + (1 - [-t] + [-1-t])f(x) = 0$$

$$(2 - [-r] + [-2-r])f(z) + (2 - [-s] + [-2-s])f(y) \\ + (2 - [-t] + [-2-t])f(x) = 0$$

Every coefficient is divisible by 3, and in the interval $[0,5]$, hence, dividing by 3 we obtain equations with all coefficients 0 and 1. We can apply this argument to both points in A occurring only once as well as to their sum, thus, in each of the four cases we obtain a $3 \oplus 6$-matrix with the property that $(f(z), f(y), f(x))$ is in the kernel of this matrix. To compute these matrices, note that

$$\frac{1 - [-a] + [-1 - a]}{3} = \begin{cases} 0, & a = 1, 2 \\ 1, & a = 0 \end{cases}, \quad \frac{2 - [-a] + [-2 - a]}{3} = \begin{cases} 0, & a = 1 \\ 1, & a = 0, 2 \end{cases}$$

for all $a \in \mathbb{Z}_3$. We therefore obtain the matrices

$$\begin{pmatrix} 1 & 0 & 0 \\ 1 & 1 & 0 \\ 0 & 0 & 1 \\ 1 & 0 & 1 \\ 0 & 1 & 0 \\ 1 & 1 & 0 \end{pmatrix}, \begin{pmatrix} 0 & 0 & 1 \\ 0 & 0 & 1 \\ 0 & 0 & 0 \\ 0 & 1 & 0 \\ 0 & 1 & 0 \\ 1 & 1 & 0 \end{pmatrix}, \begin{pmatrix} 0 & 0 & 1 \\ 0 & 0 & 1 \\ 0 & 1 & 0 \\ 0 & 1 & 0 \\ 0 & 0 & 0 \\ 1 & 0 & 0 \end{pmatrix}$$

Obviously, all these matrices have rank 3, which implies that $f(x) = f(y) = f(z) = 0$, and we obtain the contradiction

$$1 = 3f(w) = 3(f(x) + f(y) + f(z)) = 0.$$

(vii) Suppose that $(0, 0, 1)$, $(0, 1, 0)$, $(1, 0, 0)$ and $(1, 1, 2)$ occur twice in A. Then we have the following situation.

As in the previous argument we find that $(r, s, t) \in A$ implies the equations

$$(1 - [-r] + [-1 - r])f(z) + (1 - [-s] + [-1 - s])f(y)$$
$$+ (2[-t] + [-2 - t])f(x) = 0$$
$$(2 - [-r] + [-2 - r])f(z) + (2 - [-s] + [-2 - s])f(y)$$
$$+ (1 - [-t] + [-1 - t])f(x) = 0$$

If $r = 1$ and $s, t \in \{0, 2\}$, these equations imply that $f(x) = f(y) = 0$, thus we obtain a contradiction unless all the remaining equations do not involve $f(z)$. However, this would imply that both points occurring once in A as well as their sum have z-coordinate 1, which is absurd. The same argument applies if $s = 1$ and $r, t \in \{0, 2\}$, and we deduce the following.

Moreover, applying this argument to the sum of the two remaining elements, we see that this sum cannot have x-coordinate $\neq 1$ and precisely one of y and z-coordinate equal to 1.

Suppose that $(0, 1, 1) \in A$. Then this argument gives the following.

The sum of all elements equals $(1, 2, 1)$, thus if the remaining element is in the uppermost layer, there is a zerosum of length 9. Hence, there is only one possibility left, which leads to the matrix

$$\begin{pmatrix} 1 & 0 & 0 \\ 1 & 0 & 0 \\ 0 & 1 & 0 \\ 0 & 1 & 0 \\ 0 & 1 & 0 \\ 0 & 1 & 1 \end{pmatrix}$$

which has rank 3. Hence, $(0, 1, 1) \notin A$.

Next suppose that $(1, 1, 1) \in A$. Then we have only the following possibility left.

From this we obtain the matrix (deleting the zero-rows coming from $(1, 1, 1)$)

$$\begin{pmatrix} 0 & 1 & 0 \\ 1 & 1 & 0 \\ 1 & 0 & 1 \\ 1 & 0 & 0 \end{pmatrix},$$

which has rank 3.

Now suppose that $(1, 0, 1) \in A$. Then the only possibility is

which gives the matrix

$$\begin{pmatrix} 0 & 1 & 0 \\ 0 & 1 & 0 \\ 0 & 0 & 1 \\ 0 & 0 & 1 \\ 0 & 0 & 0 \\ 1 & 0 & 0 \end{pmatrix}.$$

It is easy to check that none of the remaining possibilities fits with $(2, 2, 2)$, thus $(2, 2, 2) \notin A$. By now we have reached

where two of the three cells marked a, b, c are in A. These elements give the following $2 \oplus 3$-matrices:

$$a \rightsquigarrow \begin{pmatrix} 1 & 0 & 0 \\ 1 & 1 & 0 \end{pmatrix}, \quad b \rightsquigarrow \begin{pmatrix} 0 & 0 & 1 \\ 0 & 0 & 1 \end{pmatrix}, \quad c \rightsquigarrow \begin{pmatrix} 0 & 1 & 0 \\ 1 & 1 & 0 \end{pmatrix},$$

obviously, b cannot be in A, and the fact that the x-coordinate of $a + c$ is not 1 yields a matrix of rank 3 again. Hence, in each case we obtain a contradiction.

(viii) We have seen so far that a counterexample to our statement would have precisely 3 elements occurring twice, which generate $\mathbb{Z}_3^3$. Hence, we may suppose that $(0,0,1)$, $(0,1,0)$ and $(1,0,0)$ appear twice in A, whereas all other elements of A occur with multiplicity 1. On one hand, having many distinct elements makes the non-existence of short zerosums a more stringent restriction, on the other hand, the number of cases increases dramatically. To deal with this number we use a simple computer program to list all possible sets of 10 elements without zerosums of length ≤ 3 or ≥ 8 having precisely 3 elements taken twice. With a rather un-sophisticated approach we obtain a list of 84 configurations, many of which are symmetric. One easily sees from this list that always one of $(0,1,1)$, $(1,0,1)$ and $(1,1,0)$ is contained in A, prescribing this element by a rotation around the spatial diagonal to be $(0,1,1)$ reduces the number of cases to 41, taking care of the remaining symmetries by hand gives the following list of 16 cases.

$(0,1,1)$	$(1,0,1)$	$(1,1,0)$	$(1,1,1)$
$(0,1,1)$	$(1,0,1)$	$(1,1,0)$	$(1,1,2)$
$(0,1,1)$	$(1,0,1)$	$(1,1,1)$	$(1,1,2)$
$(0,1,1)$	$(1,0,1)$	$(1,1,1)$	$(1,2,0)$
$(0,1,1)$	$(1,0,1)$	$(1,1,2)$	$(1,2,2)$
$(0,1,1)$	$(1,0,1)$	$(1,2,0)$	$(1,2,1)$
$(0,1,1)$	$(1,0,1)$	$(1,2,0)$	$(1,2,2)$
$(0,1,1)$	$(1,0,1)$	$(1,2,1)$	$(1,2,2)$
$(0,1,1)$	$(1,0,2)$	$(1,1,1)$	$(1,1,2)$
$(0,1,1)$	$(1,0,2)$	$(1,1,1)$	$(1,2,1)$
$(0,1,1)$	$(1,0,2)$	$(1,2,0)$	$(1,2,1)$
$(0,1,1)$	$(1,0,2)$	$(1,2,0)$	$(1,2,2)$
$(0,1,1)$	$(1,0,2)$	$(1,2,1)$	$(1,2,2)$
$(0,1,1)$	$(1,0,2)$	$(2,1,0)$	$(2,1,1)$
$(0,1,1)$	$(1,0,2)$	$(2,1,0)$	$(2,1,2)$
$(0,1,1)$	$(1,0,2)$	$(2,1,1)$	$(2,1,2)$

Let (r,s,t) be an element of A occurring once, distinct from $(0,1,1)$. Then we have the zerosums $(0,1,1) + 2x + 2y$, $(r,s,t) + [-t]x + [-s]y + [-t]z$ and $(0,1,1) + (r,s,t) + [-1-t]x + [-1-s]y + [-1-t]z$, which together imply the equations

$$f\bigl((0,1,1)\bigr) + 2f(y) + 2f(x) = 1$$
$$f\bigl((r,s,t)\bigr) + [-r]f(z) + [-s]f(y) + [-t]f(x) = 1$$
$$f\bigl((0,1,1)\bigr) + f\bigl((r,s,t)\bigr) + [-1-r]f(z) + [-1-s]f(y) + [-1-t]f(x) = 1$$

Subtracting the third equation from the sum of the other two equations we obtain

$$(2 + [-s] - [-1-s])f(y) + (2 + [-t] - [-1-t])f(x) = 1.$$

Note that for $a \in \mathbb{Z}_3^3$ we have

$$\frac{2 + [-a] - [-1-a]}{3} = \begin{cases} 0, & a = 0 \\ 1, & a = 1, 2 \end{cases}.$$

Now consider the first entry in the table above. The second element is $(1,0,1)$, which gives the equation $f(x) = 1$, the third element yields $f(y) = 1$, whereas the third one implies $f(x) + f(y) = 1$, and we obtain a contradiction. In the same way we can deal with all cases in which the third an fourth entry contains an entry 0, which is the case for all but the following.

$$\begin{array}{cccc}
(0,1,1) & (1,0,1) & (1,1,1) & (1,1,2) \\
(0,1,1) & (1,0,1) & (1,1,2) & (1,2,2) \\
(0,1,1) & (1,0,1) & (1,2,1) & (1,2,2) \\
(0,1,1) & (1,0,2) & (1,1,1) & (1,1,2) \\
(0,1,1) & (1,0,2) & (1,1,1) & (1,2,1) \\
(0,1,1) & (1,0,2) & (1,2,1) & (1,2,2) \\
(0,1,1) & (1,0,2) & (2,1,1) & (2,1,2)
\end{array}$$

Finally, there is no need to take (r,s,t) to be one of the three elements occurring once distinct from $(0,1,1)$, we could also take the sum of two distinct ones among them. Since in the 7 remaining cases we already have found equations coming from $s=0, t \neq 0$ and $s, t \neq 0$, it suffices to obtain (r,s,t) with $s \neq 0, t = 0$. For the first three cases this is achieved by adding the second element to the fourth one, whereas for the other 4 cases we add the second element to the third one. Hence, in all cases we reach a contradiction, which finishes our proof. $\square$

From Theorem 2 and Proposition 1 we can now deduce Theorem 1.

PROOF OF THEOREM 1. We may suppose that $(6, d) = 1$, since for all other cases this is already known. Let A be a set of $3d + 4 \geq 13$ points such that there is a function as described in Theorem 2. We claim that A contains a zerosum sequence of length ≤ 3.

We remove zero-sums of length 3 as long as possible, ending in a set with $3k+1$ points, k at most 5. We obtain $n - k + 1$ zero-sums this way. If among $3k + 1$ points, we can find $k - 1$ disjoint zero-sums, we obtain n zero-sums, which suffices. Unfortunately, this is only the case for $k = 0, 1, 2$. For $k = 3, 4, 5$, we only find $k - 2$ disjoint zero-sums, hence, altogether we obtain a system of $n - 1$ zero-sums, each inducing an element in Z_n, such that the induced set is zero-sum free. Hence, the induced elements have to be equal.

In each case we obtain $3k + 1$ points in Z_3^3 containing no system of $k - 1$ disjoint zero-sums, no zero-sum of length 3 together with a function such that for every system of $k - 2$ disjoint zero-sums the sum over each subset is equal to some generator for Z_n which does not depend on the choice of the system of zero-sums but is determined by any of the three-sums removed in the beginning. We may call this generator 1, thus, we always obtain a function f as described above.

Now as k decreases we have to make more and more use of f, which leads to more and more complicated arguments.

For $k = 5$ things are easy, since up to linear equivalence there is only one set of 16 points containing no zero-sum of length 3 (Lemma 1(2)), and the sum over all elements in this set is 0. Since a set of size 15 contains 3 disjoint zero-sums, a set of 16 points with sum 0 contains 4 disjoint zero-sums and we are done.

For $k = 4$ we have 13 points containing no zero-sum of length at most 3 and since 9 distinct points contain a zero-sum of length at most 3 (Proposition 1), 5 points have to occur with multiplicity 2 but this is ruled out at the beginning of the proof of Theorem 2 steps (i) and (ii).

For $k = 3$, as above, we reach a set of 10 points admitting a function f, which does not have 2 disjoint zerosum subsets. But Theorem 2 asserts that such a function does not exist. Hence, there does not exist a zerosum free set of size $3d + 4$, and we deduce $D(G) \leq 3d + 4$. The other inequality is trivial. □

References

1. C. Delorme, O. Ordaz, and D. Quiroz, *Some remarks onDavenport constant*, Discrete Math. **237** (2001), 119–128.
2. F. Halter-Koch, *A generalization of Davenport's constant and its arithetical applications*, Colloq. Math. **63** (1992), 203–210.
3. A. Kemnitz, *On a lattice point problem*, Ars Combin. **16** (1983), B, 151–160.
4. J. E. Olson, *An addition theorem modulo p*, J. Combinatorial Theory **5** (1968), 45–52.
5. P. van Emde Boas and D. Kruswjik, *A combinatorial problem on finite Abelian groups. III*, Math. Centrum Amsterdam Afd. Zuivere Wisk. **1969**, ZW-008.

LABORATOIRE PAUL PAINLEVÉ, U.M.R. CNRS 8524, UNIVERSITÉ DE LILLE 1, 59655 VILLENEUVE D'ASCQ CEDEX, FRANCE
E-mail address: bhowmik@math.univ-lille1.fr

MATHEMATISCHES INSTITUT, ALBERT-LUDWIGS-UNIVERSITÄT, ECKERSTR. 1, 79104 FREIBURG, GERMANY
E-mail address: jcp@math.uni-freiburg.de

Some Combinatorial Group Invariants and Their Generalizations with Weights

S. D. Adhikari, R. Balasubramanian, and P. Rath

1. Introduction

In the present article, we consider the weighted versions of some combinatorial invariants related to finite abelian groups. The study of these invariants is at present a flourishing area in the domain of combinatorial group theory. We first give a brief state-of-the-art of these invariants generally referred to as zero-sum constants. Unless otherwise stated, all our groups are assumed to be finite.

For a finite abelian group G, the Davenport constant $D(G)$ is defined to be the smallest natural number k such that any sequence of k elements in G has a non-empty subsequence whose sum is zero (the identity element). Clearly, by the pigeonhole principle, the value of this constant doesn't exceed the cardinality of the group G.

This constant, though attributed to Davenport, seems to have been first studied by K. Rogers [38] in 1962. This particular reference was somehow missed out by most of the authors in this area.

Apart from its interest in zero-sum problems of additive number theory and non-unique factorization in algebraic number theory (see [25]), this constant also plays an important role in graph theory (see, for instance, [16] or [24]). One of the most important applications of this can be seen in the proof of the infinitude of Carmichael numbers by Alford, Granville and Pomerance [7], where some knowledge of zero-sum sequences in the group of units of $\mathbb{Z}_n$ was required. Zero-sum results have been also seen to be useful in an interesting paper of Brüdern and Godinho [14].

Determining the Davenport constant for a general finite abelian group, however, does not seem to be possible in the near future. One is not even sure about the exact invariants which should appear in a general formula for the Davenport constant of an abelian group. It is perceived, from whatever little is known, that the following invariants ought to have a bearing in such a general formula: (i) the rank of the group, (ii) the number of prime factors of the order of the group; (iii) the distribution of orders of the various group elements, for instance the ratio of the largest and the smallest (greater than 1) possible orders.

2000 *Mathematics Subject Classification.* Primary 11B50; Secondary 11B75.
This is the final form of the paper.

©2007 American Mathematical Society

Given a finite abelian group $G = \mathbb{Z}_{n_1} \times \mathbb{Z}_{n_2} \times \cdots \times \mathbb{Z}_{n_d}$, with $n_1|n_2|\cdots|n_d$, writing $M(G) = 1 + \sum_{i=1}^{d}(n_i - 1)$, it is trivial to see that $M(G) \leq D(G) \leq |G|$. The equality $D(G) = |G|$ holds if and only if $G = \mathbb{Z}_n$, the cyclic group of order n. Olson [34, 35] proved that $D(G) = M(G)$ for all finite abelian groups of rank 2 and for all p-groups. It is also known that $D(G) > M(G)$ for infinitely many finite abelian groups of rank $d > 3$ (see [26], for instance). For some interesting results on the upper bound of $D(G)$ for a non-cyclic abelian group G, one may look into the papers of Alon, Bialostocki and Caro [8], Caro [15], Ordaz and Quiroz [36] and Balasubramanian and Bhowmik [12].

Chronologically, Emde Boas and Kruyswijk [43], Baker and Schmidt [11] and Meshulam [31] gave upper bounds for Davenport constant which involves the exponent of the group and the cardinality of the group G. In this direction, the best known bound is due to Emde Boas and Kruyswijk [43] who proved that

$$(1) \qquad D(G) \leq m\left(1 + \log \frac{|G|}{m}\right),$$

where m is the exponent of G. This was again proved by Alford, Granville and Pomerance [7].

Obtaining a good upper bound for the Davenport constant is very important and the current state of knowledge is expected to be far from the truth. However, we do have the following conjectures (with notations as above).

Conjectures. (1) $D(G) = M(G)$ for all G with rank $d = 3$ or $G = \mathbb{Z}_n^d$ [22, 23].
(2) $D(G) \leq \sum_{i=1}^{d} n_i$ [32].

Another combinatorial invariant $E(G)$ for a finite abelian group G of n elements is defined to be the smallest integer t such that for any sequence of t elements of G, there exists a subsequence of exactly n elements whose sum is the identity element 0 of G.

The constants $D(G)$ and $E(G)$ were being studied independently until Gao [21] (see also [25, Proposition 5.7.9]) established the following beautiful result connecting these two invariants.

Theorem 1. *If G is a finite abelian group of order n, then $E(G) = D(G) + n - 1$.*

Thus knowing one of the above constants amounts to knowing the other. In certain cases, it is easier to find $D(G)$ while in some other cases $E(G)$ is easier to determine.

The final invariant we want to introduce here is denoted by $f(n, d)$ which is defined for abelian groups of the form $(\mathbb{Z}/n\mathbb{Z})^d$. For any positive integer d, we define $f(n, d)$ to be the smallest positive integer N such that every sequence of N elements in $(\mathbb{Z}/n\mathbb{Z})^d$ has a subsequence of n elements which add up to the identity element $\underbrace{(0, 0, \ldots, 0)}_{d \text{ times}}$.

This constant was initially considered by Harborth [28] and Kemnitz [29]. Harborth observed the following general bounds for $f(n, d)$.

$$(2) \qquad 1 + 2^d(n-1) \leq f(n, d) \leq 1 + n^d(n-1).$$

For $d = 1$, noting that $f(n,1)$ and $E(G)$ are identical for the group $(\mathbb{Z}/n\mathbb{Z})$, Gao's theorem gives the exact value

$$f(n,1) = 2n - 1.$$

This result, first established by Erdős, Ginzburg and Ziv [20], can be regarded as the starting point of this rather active area in combinatorial group theory, generally referred to as zero-sum combinatorics. Apart from the original paper of Erdős, Ginzburg and Ziv [20], there are many proofs of the above theorem available in the literature (see [1, 6, 9, 13, 27, 33, 44] for instance).

For the case $d = 2$ also, the lower bound in (2) was expected to give the right magnitude of $f(n,2)$ and this expectation, which had been known as *Kemnitz conjecture* in the literature, has been recently established by Reiher [37]. It should be mentioned that Alon, Dubiner and Rónyai had obtained very good upper bounds for $f(n,2)$. More precisely, Alon and Dubiner [9] had established that $f(n,2) \leq 6n - 5$ (may also look into [1]) and Rónyai [39] had proved that for a prime p, one has $f(p,2) \leq 4p - 2$.

The lower bound given in (2) is known not to be tight in general. Harborth [28] proved that $f(3,3) = 19$; this is strictly greater than the lower bound 17 which one obtains from (2). However, Harborth's result on $f(3,3)$ did not rule out the possibility that for a fixed dimension d and for a sufficiently large prime p the lower bound in (2) might determine the exact value for $f(p,d)$. But the following recent result of Elsholtz [19] in this direction rules out such possibilities.

Theorem 2. *For an odd integer $n \geq 3$, the following inequality holds:*

$$f(n,d) \geq (\tfrac{9}{8})^{[d/3]}(n-1)2^d + 1.$$

Now, one observes that the gap is quite large between the lower and the upper bounds given in (2). A very important result of Alon and Dubiner [10] says that the growth of $f(n,d)$ is linear in n; when d is fixed and n is increasing, this is much better as compared to the upper bound given by (2). More precisely, Alon and Dubiner [10] prove the following.

Theorem 3. *There is an absolute constant $c > 0$ such that for all n we have,*

$$f(n,d) \leq (cd \log_2 d)^d n.$$

The proof of Theorem 3 due to Alon and Dubiner combines techniques from additive number theory with results about the expansion properties of Cayley graphs with given eigenvalues. In the same paper [10] the authors conjecture that the estimate in Theorem 3 can possibly be improved. More precisely, the existence of an absolute constant c is predicted such that

$$f(n,d) \leq c^d n, \quad \text{for all } n \text{ and } d.$$

It should be mentioned that Balasubramanian and Bhowmik [12] furnished a new proof of a lemma in [10] which avoids the graph theoretic techniques.

In the next section, we define the weighted versions of the constants $D(G)$ and $E(G)$, the starting point of which was the line of investigation taken up in [4].

In Section 3, we consider the question of establishing a weighted version of Theorem 1, linking two of the weighted versions of the constants and enlist some of the recent results. In that section we also discuss some of the progress made towards finding the exact values of the weighted versions of $E(G)$ and $D(G)$ and their upper bounds for certain classes of weights.

Finally in Section 4, we consider a weighted version of the constant $f(n,d)$ and discuss a recent result [2] regarding its exact value in the case $d = 2$ when n is an odd integer.

2. Weighted versions of the constants $D(G)$ and $E(G)$

For the particular group $\mathbb{Z}/n\mathbb{Z}$, certain generalizations of $D(G)$ and $E(G)$ were considered in [4, 5].

Following these generalizations, for a finite abelian group G and any non-empty $A \subseteq \mathbb{Z}$, the following definition was given in [3]. The *Davenport constant of G with weight A*, denoted by $D_A(G)$, is defined to be the least natural number k such that for any sequence $(x_1, \cdots, x_k)$ with $x_i \in G$, there exists a non-empty subsequence $(x_{j_1}, \ldots, x_{j_l})$ and $a_1, \ldots, a_l \in A$ such that $\sum_{i=1}^{l} a_i x_{j_i} = 0$. Clearly, if G is of order n, it is equivalent to consider A to be a nonempty subset of $\{0, 1, \ldots, n-1\}$ and cases with $0 \in A$ are trivial.

Similarly, for any such set A, for a finite abelian group G of order n, the constant $E_A(G)$ is defined to be the least $t \in \mathbb{N}$ such that for any sequence $(x_1, \ldots, x_t)$ of t elements with $x_i \in G$, there exist indices $j_1, \ldots, j_n \in \mathbb{N}$, $1 \leq j_1 < \cdots < j_n \leq t$, and $\vartheta_1, \ldots, \vartheta_n \in A$ with $\sum_{i=1}^{n} \vartheta_i x_{j_i} = 0$.

For the group $G = \mathbb{Z}/n\mathbb{Z}$, we write $E_A(n)$ and $D_A(n)$ respectively for $E_A(G)$ and $D_A(G)$.

When $A = \{1\}$, $D_A(G)$ and $E_A(G)$ are nothing but $D(G)$ and $E(G)$ respectively.

If $A = \{1, -1\}$, by the pigeonhole principle (see [4]), $D_A(n) \leq [\log_2 n] + 1$; and considering the sequence $(1, 2, \ldots, 2^r)$, where r is defined by $2^{r+1} \leq n < 2^{r+2}$, it follows that $D_A(n) \geq [\log_2 n] + 1$.

In this case, it was shown in [4] that $E_A(n) = n + [\log_2 n]$.

The result follows rather easily in some particular cases. For instance, if $n = p$ is an odd prime, then for any integer $N \geq p - 1$ and any sequence of integers $(x_1, \ldots, x_N) \in \mathbb{Z}^N$ (no one divisible by p), by choosing $A_i = \{x_i, -x_i\}$, by Cauchy–Davenport inequality [17, 18] (see also [33]) it follows that

$$|\{x_1, -x_1\} + \{x_2, -x_2\} + \cdots + \{x_N, -x_N\}| \geq p.$$

A conditional result which yields to comparatively easier techniques [4] is the following. Let $n \in \mathbb{N}$ and assume that $N \geq n + [\log_2 n]$ is an integer. Given any sequence $(x_1, \ldots, x_N) \in \mathbb{Z}^N$ with at least one multiple of n, there exist $m = N - [\log_2 n]$ indices $\{j_1, \ldots, j_m\} \subseteq \{1, 2, \ldots, N\}$ and signs $\varepsilon_1, \ldots, \varepsilon_m \in \{1, -1\}$ such that

$$\varepsilon_1 x_{j_1} + \cdots + \varepsilon_m x_{j_m} \equiv 0 \pmod{n}.$$

By the method employed for the above result, when n is an even integer it follows unconditionally that $E_A(n) = n + [\log_2 n]$.

To tackle the situation when n is any odd integer, the following concept was introduced in [4]. Given a sequence $\underline{x} = (x_1, \ldots, x_N) \in \mathbb{Z}^N$, we say that the sequence $\underline{x}$ is *complete with respect to a positive integer m* if for every positive $d \mid m$ we have

$$|\{j \in \{1, 2, \ldots, N\} \mid x_j \not\equiv 0 \pmod{d}\}| \geq d - 1.$$

The following theorem is then proved by different arguments depending on whether the given sequence is complete with respect to m or not:

Theorem 4. *Assume that $m \in \mathbb{N}$ is odd. If $N \geq m + [\log_2 m]$ and $\underline{x} = (x_1, \ldots, x_N) \in \mathbb{Z}^N$, then there exists $I_0 = \{j_1, \ldots, j_t\} \subseteq \{1, 2, \ldots, N\}$ with $|I_0| = t = N - [\log_2 m]$ and some choice of coefficients $\varepsilon_1, \ldots, \varepsilon_t \in \{\pm 1\}$, so that*

$$\sum_{i=1}^{t} \varepsilon_i x_{j_i} \equiv 0 \pmod{m}.$$

We remark that with $A = \{1, -1\}$, for the group $G = (\mathbb{Z}/n\mathbb{Z})^d$, it is not difficult to observe that

$$2 - d + [\log_2 |G|] \leq D_A(G) \leq 1 + [\log_2 |G|].$$

For the case $A = \{1, 2, \ldots, n-1\}$ it was also observed in [4] (see Theorem 6.1) that $E_A(n) = n + 1$. In this case it is very easy to see that $D_A(n) = 2$.

The case $A = \{a \in \{1, 2, \ldots, n-1\} | (a, n) = 1\}$ was considered in [30] where, settling a conjecture from [4], it was proved that

(3) $$E_A(n) = n + \Omega(n),$$

where $\Omega(n)$ denotes the number of prime factors of n, multiplicity included.

In this case, writing $n = p_1 \cdots p_s$ as product of $s = \Omega(n)$ not necessarily distinct primes, the sequence $(1, p_1, p_1 p_2, \ldots, p_1 p_2 \cdots p_{s-1})$ gives the lower bound

$$D_A(n) \geq 1 + \Omega(n).$$

On the other hand, it is not difficult to observe that

(4) $$E_A(n) \geq D_A(n) + n - 1 \quad \text{for any } A \subset \mathbb{Z}/n\mathbb{Z} \setminus \{0\}.$$

Therefore, for the case $A = \{a \in \{1, 2, \ldots, n-1\} | (a, n) = 1\}$ it follows from (3) and (4) that

$$D_A(n) \leq 1 + \Omega(n).$$

3. Relation between the constants $D_A(G)$ and $E_A(G)$

All the known results described in the previous section, lead [5] to the expectation that for any set $A \subset \mathbb{Z} \setminus n\mathbb{Z}$ of weights, the equality $E_A(n) = D_A(n) + n - 1$ holds. In [5], for the special case where $n = p$ is a prime, the values of $D_A(p)$ and $E_A(p)$ were determined for the cases where A is $\{1, 2, \ldots, r\}$ and the set of quadratic residues (mod p). It was observed that in these cases too, the equality $E_A(p) = D_A(p) + p - 1$ holds.

Using some methods of Troi and Zannier [42], Thangadurai [41] has recently obtained good upper bounds for Davenport's constant with weights for finite abelian p-groups.

Later in [3], the following theorem which exhibits the expected relation between the constants $D_A(G)$ and $E_A(G)$ has been established for any finite abelian group G under certain conditions.

Theorem 5. *Let G be a finite abelian group of order n, and $A = \{a_1, a_2, \ldots, a_r\}$ be a finite subset of $\mathbb{Z}$ with $|A| = r \geq 2$ and*

(5) $$\gcd(a_2 - a_1, a_3 - a_1, \ldots, a_r - a_1, n) = 1.$$

Then we have $E_A(G) = D_A(G) + n - 1$.

Remark. The notion of a complete sequence which had been introduced in [4] (see the previous section) was suitably generalized in order to prove the above result. For any finite abelian group G of order n, the above theorem covers a large class of subsets A of $\mathbb{Z}$. For instance, when $A \subset \mathbb{Z}$ is such that A contains two integers a, b with $a - b \equiv 1 \pmod{n}$ (this surely happens if A contains elements belonging to more than $n/2$ residue classes modulo n), Theorem 5 is applicable. We also note that by Theorem 1, the conclusion of Theorem 5 holds unconditionally for the case $|A| = 1$.

If $G = \mathbb{Z}/p\mathbb{Z}$, where p is a prime, then the condition (5) is trivially satisfied if there are $a_i \not\equiv a_j \pmod{p}$ in A and hence the result $E_A(p) = D_A(p) + p - 1$ follows from Theorem 5. If $a_1 \equiv a_2 \cdots \equiv a_r \not\equiv 0 \pmod{p}$, then it follows from the EGZ Theorem (or Theorem 1); and the case $a_1 \equiv a_2 \equiv \cdots \equiv a_r \equiv 0 \pmod{p}$ is trivial. Hence the result $E_A(p) = D_A(p) + p - 1$ follows for any nonempty A in this case. As can be seen from [5], if A is $\{1, 2, \ldots, r\}$ or the set of quadratic residues (mod p), deriving the exact value of $D_A(p)$ is much easier; we can therefore use this relation to obtain the exact value of $E_A(p)$ from that of $D_A(p)$ in these cases. We should remark that the method employed in the proof of Theorem 2 of the above mentioned paper [5] can also be used to derive the result $E_A(p) = D_A(p) + p - 1$ in this particular case for any nonempty $A \subseteq \{1, 2, \ldots, p - 1\}$.

4. Weighted version of Harborth's constant

Let $n \in \mathbb{N}$ and $A \subseteq \{1, \ldots, n-1\}$. Then we define $f_A(n, d)$ to be the smallest positive integer k such that given a sequence $(\bar{x}_1, \ldots, \bar{x}_k)$ of k not necessarily distinct elements of $(\mathbb{Z}/n\mathbb{Z})^d$, there exists a subsequence $(\bar{x}_{j_1}, \ldots, \bar{x}_{j_n})$ of length n and $a_1, \ldots, a_n \in A$ such that

$$\sum_{i=1}^{n} a_i \bar{x}_{j_i} = \bar{0},$$

where $\bar{0}$ is the zero element of the group $(\mathbb{Z}/n\mathbb{Z})^d$ and the multiplication of a vector $\bar{x} = (x_1, \ldots, x_d)$ in $(\mathbb{Z}/n\mathbb{Z})^d$ by an element $a \in A$ is the standard scalar multiplication defined by $a(x_1, \ldots, x_d) = (ax_1, \ldots, ax_d)$. We note that if the weight set A is closed under multiplication, then recovering sequences from quotients becomes possible and one obtains the following upper bound:

(6) $\quad f_A(mn, d) \leq \min(f_A(n, d) + n(f_A(m, d) - 1), f_A(m, d) + m(f_A(n, d) - 1))$.

It is worthwhile to mention that while size of the weight set is crucial in each of the three combinatorial constants defined before, the structure of the weight set is also rather relevant. Just to illustrate this, we note that while for the weight $A = \{1, 2\}$ the value of $D_A(Z/nZ)$ is $1 + [n/2]$, the value of $D_A(Z/nZ)$ is $[\log_2 n] + 1$ for the weight $A = \{1, -1\}$.

We here consider the particular case $A = \{1, -1\}$; for this case, in [4] it had already been established that

$$f_{\{1,-1\}}(n, 1) = n + [\log_2 n].$$

Recently, the following theorem has been established [2]:

Theorem 6. *For an odd integer n, we have*

$$f_{\{1,-1\}}(n, 2) = 2n - 1.$$

SKETCH OF THE PROOF. Since the weight set $\{1,-1\}$ is closed under multiplication, the result follows by induction on the number of prime factors (counted with multiplicity) of n.

Therefore, it is enough to establish the result for any odd prime p.

The lower bound
$$f_{\{1,-1\}}(p,2) \geq 2p-1$$
is obtained by considering the sequence of $2p-2$ elements where the elements $(1,0)$, $(0,1)$ are repeated $(p-1)$ times each.

For the upper bound, given any sequence of vectors $v_1 = (c_1,d_1), v_2 = (c_2,d_2)$, $\ldots, v_m = (c_m,d_m)$ in $(\mathbb{Z}/p\mathbb{Z})^2$ with $m = 2p-1$, one considers the system of equations

$$\sum_{i=1}^{2p-1} c_i x_i^{(p-1)/2} = 0,$$

$$\sum_{i=1}^{2p-1} d_i x_i^{(p-1)/2} = 0,$$

$$\sum_{i=1}^{2p-1} x_i^{p-1} = 0.$$

Since $2(p-1) < 2p-1$ and $x_1 = x_2 = \cdots = x_{2p-1} = 0$ is a solution, by Chevalley's theorem (see [40], for instance), there is another solution. If $J \subset \{1,2,\ldots,2p-1\}$ is the set of all indices of the non-zero entries of such a solution, from the first two equations it follows that

$$\sum_{i \in J} a_i v_i = (0,0), \quad \text{in } (\mathbb{Z}/p\mathbb{Z})^2,$$

where $a_i \in \{1,-1\}$. From the third equation we have $|J| = p$.

This proves that $f_{\{1,-1\}}(p,2) \leq 2p-1$. □

For $d > 2$, it is rather difficult to obtain a good upper bound by using the above technique. The Chevalley–Warning theorem is no longer useful since the number of variables involved exceeds $2p$ and hence the possibility of zero-sum subsequences of length $2p$ can not be ruled out. The upper bound obtained by using (6) is rather large while we expect the order of the function to be nearer to the trivial lower bound. In fact, in all the three constants considered here, even when the weight set A is $\{1\}$, beyond rank 2, our present knowledge is rather limited.

References

1. S. D. Adhikari, *Aspects of combinatorics and combinatorial number theory*, Narosa, New Delhi, 2002.
2. S. D. Adhikari, R. Balasubramanian, F. Pappalardi, and P. Rath, *Some zero-sum constants with weights*, Proc. Indian Acad. Sci. Math. Sci., to appear.
3. S. D. Adhikari and Y. G. Chen, *Davenport constant with weights and some related questions. II*, J. Combin. Theory Ser. A, in press.
4. S. D. Adhikari, Y. G. Chen, J. B. Friedlander, S. V. Konyagin, and F. Pappalardi, *Contributions to zero-sum problems*, Discrete Math. **306** (2006), no. 1, 1–10.
5. S. D. Adhikari and P. Rath, *Davenport constant with weights and some related questions*, Integers **6** (2006), A30.

6. _____ , *Zero-sum problems in combinatorial number theory*, The Riemann Zeta Function and Related Themes (Bangalore, 2003) (R. Balasubramanian and K. Srinivas, eds.), Ramanujan Math. Soc. Lect. Notes Ser., vol. 2, Ramanujan Math. Soc., Mysore, 2006, pp. 1–14.
7. W. R. Alford and A. Granville, *There are infinitely many Carmichael numbers*, Ann. of Math. (2) **139** (1994), no. 3, 703–722.
8. N. Alon, A. Bialostocki, and Y. Caro, *Extremal zero-sum problems*, manuscript.
9. N. Alon and M. Dubiner, *Zero-sum sets of prescribed size*, Combinatorics, Paul Erdős Is Eighty, Vol. 1 (D. Miklós, V. T. Sós, and T. Szönyi, eds.), Bolyai Soc. Math. Stud., János Bolyai Math. Soc., Budapest, 1993, pp. 33–50.
10. _____ , *A lattice point problem and additive number theory*, Combinatorica **15** (1995), no. 3, 301–309.
11. R. C. Baker and W. M. Schmidt, *Diophantine problems in variables restricted to the values of 0 and 1*, J. Number Theory **12** (1980), no. 4, 460–486.
12. Balasubramanian R. and G. Bhowmik, *Upper bounds for the Davenport constant*, Integers, to appear.
13. B. Bollobás and I. Leader, *The number of k-sums modulo k*, J. Number Theory **78** (1999), no. 1, 27–35.
14. J. Brüdern and H. Godinho, *On Artin's conjecture. II: Pairs of additive forms*, Proc. London Math. Soc. (3) **84** (2002), no. 3, 513–538.
15. Y. Caro, *Zero-sum subsequences in abelian non-cyclic groups*, Israel J. Math. **92** (1995), no. 1-3, 221–233.
16. _____ , *Zero-sum problems — a survey*, Discrete Math. **152** (1996), no. 1-3, 93–113.
17. A. L. Cauchy, *Recherches sur les nombres*, J. École Polytech. **9** (1813), 99–123.
18. H. Davenport, *On the addition of residue classes*, J. London Math. Soc. **22** (1947), 100–101.
19. C. Elsholtz, *Lower bounds for multidimensional zero sums*, Combinatorica **24** (2004), no. 3, 351–358.
20. P. Erdős, A. Ginzburg, and A. Ziv, *Theorem in the additive number theory*, Bull. Res. Council Israel Sect. F, Math. Phys. **10** (1961), 41–43.
21. W. D. Gao, *A combinatorial problem on finite abelian groups*, J. Number Theory **58** (1996), no. 1, 100–103.
22. _____ , *On Davenport's constant of finite abelian groups with rank three*, Discrete Math. **222** (2000), no. 1-3, 111–124.
23. W. D. Gao and A. Geroldinger, *Zero-sum problems and coverings by proper cosets*, European J. Combin. **24** (2003), no. 5, 531–549.
24. _____ , *Zero-sum problems in finite abelian groups; a survey*, Expo. Math. **24** (2006), no. 4, 337–369.
25. A. Geroldinger and F. Halter-Koch, *Non-unique factorizations*, Pure Appl. Math. (Boca Raton), vol. 278, Chapman & Hall/CRC, Boca Raton, FL, 2006.
26. A. Geroldinger and R. Schneider, *On Davenport's constant*, J. Combin. Theory Ser. A **61** (1992), no. 1, 147–152.
27. Y. O. Hamidoune, *On weighted sums in abelian groups*, Discrete Math. **162** (1996), no. 1-3, 127–132.
28. H. Harborth, *Ein Extremalproblem für Gitterpunkte*, J. Reine Angew. Math. **262/263** (1973), 356–360.
29. A. Kemnitz, *On a lattice point problem*, Ars Combin. **16** (1983), B, 151–160.
30. F. Luca, *A generalization of a classical zero-sum problem*, Discrete Math. **307** (2007), no. 13, 1672–1678.
31. R. Meshulam, *An uncertainty inequality and zero subsums*, Discrete Math. **84** (1990), no. 2, 197–200.
32. W. Narkiewicz and J. Śliwa, *Finite abelian groups and factorization problems. II*, Colloq. Math. **46** (1982), no. 1, 115–122.
33. M. B. Nathanson, *Additive number theory: Inverse problems and the geometry of sumsets*, Grad.Texts in Math., vol. 165, Springer, New York, 1996.
34. J. E. Olson, *A combinatorial problem in finite Abelian groups. I*, J. Number Theory **1** (1969), 8–10.
35. _____ , *A combinatorial problem in finite Abelian groups. II*, J. Number Theory **1** (1969), 195–199.

36. O. Ordaz and D. Quiroz, *The Erdős–Ginzburg–Ziv theorem in abelian non-cyclic groups*, Divulg. Mat. **8** (2000), no. 2, 113–119.
37. C. Reiher, *On Kemnitz' conjecture concerning lattice-points in the plane*, Ramanujan J. **13** (2007), no. 1-3, 333–337.
38. K. Rogers, *A combinatorial problem in abelian groups*, Proc. Cambridge Philos. Soc. **59** (1963), 559–562.
39. L. Rónyai, *On a conjecture of Kemnitz*, Combinatorica **20** (2000), no. 4, 569–573.
40. J.-P. Serre, *A course in arithmetic*, Grad. Texts in Math., vol. 7, Springer, New York, 1973.
41. R. Thangadurai, *A variant of Davenport constant*, Proc. Indian Acad. Sci. Math. Sci., to appear.
42. G. Troi and U. Zannier, *On a theorem of J. E. Olson and an application (vanishing sums in finite abelian p-groups)*, Finite Fields Appl. **3** (1997), no. 4, 378–384.
43. P. van Emde Boas and D. Kruswjik, *A combinatorial problem on finite Abelian groups. III*, Math. Centrum Amsterdam Afd. Zuivere Wisk. **1969**, ZW-008.
44. H. B. Yu, *A simple proof of a theorem of Bollobás and Leader*, Proc. Amer. Math. Soc. **131** (2003), no. 9, 2639–2640.

HARISH-CHANDRA RESEARCH INSTITUTE, CHHATNAG ROAD, JHUSI, ALLAHABAD 211 019, INDIA
E-mail address: adhikari@mri.ernet.in

INSTITUTE OF MATHEMATICAL SCIENCES, C.I.T CAMPUS, TARAMANI, CHENNAI 600 113, INDIA
E-mail address, R. Balasubramanian: balu@imsc.res.in
E-mail address, P. Rath: rath@imsc.res.in